COMPUTATIONAL THEORY OF ITERATIVE METHODS

STUDIES IN COMPUTATIONAL MATHEMATICS 15

Editors:

C.K. CHUI
Stanford University
Stanford, CA, USA

L. WUYTACK
University of Antwerp
Antwerp, Belgium

ELSEVIER

Amsterdam • Boston • Heidelberg • London • New York • Oxford • Paris
San Diego • San Francisco • Singapore • Sydney • Tokyo

COMPUTATIONAL THEORY OF ITERATIVE METHODS

IOANNIS K. ARGYROS
Department of Mathematical Sciences
Cameron University
Lawton, OK 73505, USA

ELSEVIER
Amsterdam • Boston • Heidelberg • London • New York • Oxford • Paris
San Diego • San Francisco • Singapore • Sydney • Tokyo

Elsevier
Radarweg 29, PO Box 211, 1000 AE Amsterdam, The Netherlands
Linacre House, Jordan Hill, Oxford OX2 8DP, UK

First edition 2007

Library of Congress Cataloging-in-Publication Data
A catalog record for this book is available from the Library of Congress

British Library Cataloguing in Publication Data
A catalogue record for this book is available from the British Library

ISBN: 978-0-444-53162-9
ISSN: 1570-579X

Transferred to Digital Printing in 2010

DEDICATED TO MY MOTHER
ANASTASIA

Introduction

Researchers in computational sciences are faced with the problem of solving a variety of equations. A large number of problems are solved by finding the solutions of certain equations. For example, dynamic systems are mathematically modelled by difference or differential equations, and their solutions represent usually the states of the systems. For the sake of simplicity, assume that a time-invariant systems is driven by the equation $x' = f(x)$, where x is the state, then the equilibrium states are determined by solving the equations $f(x) = 0$. Similar equations are used in the case of discrete systems. The unknowns of engineering equations can be functions (difference, differential, integral equations), vectors (systems of linear or nonlinear algebraic equations), or real or complex numbers (single algebraic equations with single unknowns). Except special cases, the most commonly used solutions methods are iterative, when starting from one or several initial approximations a sequences is constructed, which converges to a solution of the equation. Iteration methods are applied also for solving optimization problems. In such cases the iteration sequences converge to an optimal solution of the problem in hand. since all of these methods have the same recursive structure, they can be introduced and discussed in a general framework.

To complicate the matter further, many of these equations are nonlinear. However, all may be formulated in terms of operators mapping a linear space into another, the solutions being sought as points in the corresponding space. Consequently, computational methods that work in this general setting for the solution of equations apply to a large number of problems, and lead directly to the development of suitable computer programs to obtain accurate approximate solutions to equations in the appropriate space.

This monograph is intended for researchers in computational sciences, and as a reference book for an advanced numerical-functional analysis or computer science course. The goal is to introduce these powerful concepts and techniques at the earliest possible stage. The reader is assumed to have had basic courses in numerical analysis, computer programming, computational linear algebra, and an introduction to real, complex, and functional analysis.

We have divided the material into twelve chapters. Although the monograph is of a theoretical nature, with optimization and weakening of existing hypotheses considerations each chapter contains several new theoretical results and important applications in engineering, in dynamic economic systems, in input-output systems, in the solution of nonlinear and linear differential equations, and optimization problems. The applications appear in the form of Examples or Applications or Exercises or study cases or they are implied since our results improve earlier one that have already been applied in concrete problems. Sections have been written as independent of each other as possible. Hence the interested reader can go directly to a certain section and understand the material without having to go back and forth in the whole textbook to find related material.

There are four basic problems connected with iterative methods.

Problem 1 *Show that the iterates are well defined. For Example, if the algorithm requires the evaluation of F at each x_n, it has to be guaranteed that the iterates remain in the domain of F. It is, in general, impossible to find the exact set of all initial data for which a given process is well defined, and we restrict ourselves to giving conditions which guarantee that an iteration sequence is well defined for certain specific initial guesses.*

Problem 2 *Concerns the convergence of the sequences generated by a process and the question of whether their limit points are, in fact, solutions of the equation. There are several types of such convergence results. The first, which we call a local convergence theorem, begins with the assumption that a particular solution x^* exists, and then asserts that there is a neighborhood U of x^* such that for all initial vectors in U the iterates generated by the process are well defined and converge to x^*. The second type of convergence theorem, which we call semilocal, does not require knowledge of the existence of a solution, but states that, starting from initial vectors for which certain-usually stringent-conditions are satisfied, convergence to some (generally nearby) solutions x^* is guaranteed. Moreover, theorems of this type usually include computable (at least in principle) estimates for the error $x_n - x^*$, a possibility not afforded by the local convergence theorems. Finally, the third and most elegant type of convergence result, the global theorem, asserts that starting anywhere in a linear space, or at least in a large part of it, convergence to a solution is assured.*

Problem 3 *Concerns the economy of the entire operations, and, in particular, the question of how fast a given sequence will converge. Here, there are two approaches, which correspond to the local and semilocal convergence theorems. As mentioned above, the analysis which leads to the semilocal type of theorem frequently produces error estimates, and these, in turn, may sometimes be reinterpreted as estimates of the rate of convergence of the sequence. Unfortunately, however, these are usually overly pessimistic. The second approach deals with the behavior of the sequence $\{x_n\}$ when n is large, and hence when x_n is near the solutions x^*. This behavior may then be determined, to a first approximation, by the properties of the iteration function near x^* and leads to so-called asymptotic rates of convergence.*

Problem 4 *Concerns with how to best choose a method, algorithm, or software program to solve a specific type of problem and its descriptions of when a given algorithm or method succeeds or fails.*

We have included a variety of new results dealing with Problems 1–4. Results by other authors (or us) are either explicitely stated or implicitely when compared with the corresponding ones of ours or they are left as exercises

This monograph is an outgrowth of research work undertaken by us and complements/updates earlier works of ours focusing on in depth treatment of convergence theory for iterative methods [7]–[100]. Such a comprehensive study of optimal iterative procedures appears to be needed and should benefit not only those working in the field but also those interest in, or in need of, information about specific results or techniques. We have endeavored to make the main text as self contained as possible, to prove all results in full detail and to include a number of exercises throughout the monograph. In order to make the study useful as a reference source, we have complemented each section with a set of "Remarks" in which literature citations are given, other related results are discussed, and various possible extensions of the results of the text are indicated. For completion, the monograph ends with a comprehensive list of references and an appendix containing useful contemporary numerical algorithms. Because we believe our readers come from diverse backgrounds and have varied interests, we provide "recommended reading" throughout the textbook. Often a long monograph summarizes knowledge in a field. This monograph, however, may be viewed as a report on work in progress. We provide a foundation for a scientific field that is rapidly changing. Therefore we list numerous conjectures and open problems as well as alternative models which need to be explored.

In this monograph we do not attempt to cover all the aspects of what is commonly understood to be computational methods. Although we make reference, and briefly study (e.g. homotopy methods), our main concern is restricted to certain nongeometrical computational methods. For a geometrical approach to computational methods we refer the reader to the excellent book by Krasnosel'skii and Zabrejko [214].

The monograph is organized as follows. Chapter 1: The essentials on linear spaces are provided. Chapter 2: Some results on monotone convergence are reported. Chapter 3: Stationary and nonstationary Mann and Ishikawa-type iterations are studied. Recent results of ours are also compared favourably with earlier ones. Chapters 4–9: Newton-type methods and their implications/applications are covered here. The Newton-Kantorovich Theorem 4.2.4 is one of the most important tools in nonlinear analysis and in classical numerical analysis. This theorem has been successfully used by numerical analysts for obtaining optimal bounds for many iterative procedures. The original paper of Kantorovich [206] contains optimal a-priori bounds for the Newton's method, albeit not in explicit form. Explicit forms of those a-priori bounds were obtained independently by Ostrowski [264] and Gragg and Tapia [170] (see [284] for the equivalence of the bounds).

The paper of Gragg and Tapia [170] also contains sharp a-posteriori bounds for Newton's method. By using different techniques and/or different a-posteriori information, the a-posteriori error bounds were refined by other authors [6], [121], [141], [143], [209]–[212], [274], [284] and us [11]–[99]. Various extensions of the Kantorovich theorem have been used to obtain error bounds for Newton-like methods [16], [29], [34], [35], [48], [121], [274], [351], inexact Newton methods [42], [58], [359], the secant method [13], [21], [275], Halley's method [41], [160], etc. A survey of such results can be found in [68], [99], [350].

The Newton-Kantorovich theorem has also been used for proving existence and uniqueness of solutions of nonlinear equations arising in various fields. The spectrum of applications of this theorem is immense. One can sit in front of a computer and make a search for the "Newton-Kantorovich theorem" and will find hundreds if not thousands of works related/based on this theorem.

Our list below is therefore very incomplete.

However we mention such diverse problems as the chaotic behavior of maps [314], optimal shape design [220], nonlinear Riemann-Hilbert problems [104], optimal control [172], nonlinear finite element [324], integral equations [4], [141], [160], flow problems [226], nonlinear elliptic problems [249], semidefinite programming [251], LP methods [294], interior point methods [282], multiparameter singularly perturbed systems [245]. We also mention that this theorem has been extended for analyzing Newton's method on manifolds [162], [265] (see also our section 8.7). The relation between the Kantorovich Theorem and other existence theorems for solutions of nonlinear equations like Moore and Miranda theorems have been investigated in [1], [369] (see also the improvements in our chapter 8, sections 8.1 and 8.2).

The foundation of the Newton-Kantorovich theorem is the famous for its simplicity and clarity Newton-Kantorovich hypothesis (4.2.51). This condition is the sufficient hypothesis for the convergence of Newton's method. However, it is common knowledge as numerical experiments indicate that the convergence of the Newton method can be obtained even if that the initial guess is such that this condition is violated. Therefore weakening this condition is of extreme importance in computational mathematics, since the trust region will be enlarged. Recently [92] we showed that the Newton-Kantorovich hypothesis can always be replaced by weaker hypothesis (5.1.56) which doubles (at most, if $l_0 = 0$) the application of this theorem. Note that the verification of condition (5.1.56) requires the same information and computational cost as (4.2.51), since in practice the computation of Lipschitz constant l requires that of center Lipschitz constant l_0.

Moreover the following advantages hold:

semilocal case: finer error bounds on the distances involved and an at least as precise information on the location of the solution;

local case: finer error bounds and a larger trust region.

The following advantages carry over if our approach is extended to Newton-like methods (see Chapter 5).

For all the above reasons it is worth it for somebody to revisit all existing results using the Newton-Kantorovich hypothesis and replace it with our weaker condition (5.1.56). Such a task is obviously impossible to be undertaken by us and we leave it to the motivated readers to pick on specific problems mentioned (or not) above. Here we simply have attempted (Chapters 5–9) to provide usefull but by far a complete list of such applications/implications.

We report on some results on the study of variational inequalities on Chapter 10.

Chapters 11 and 12 deal with the convergence of point to set mappings and their applications.

Finally in Chapter 13 we improve the mesh independence principle [145] and also provide a more efficient conjugate gradient solver for the normal equations [194].

Contents

Chapter 1

Linear Spaces

The basic background for solving equations is introduced here.

1.1 Linear Operators

Some mathematical operations have certain properties in common. These properties are given in the following definition.

Definition 1.1.1 *An operator T which maps a linear space X into a linear space Y over the same scalar field S is said to be additive if*

$$T(x+y) = T(x) + T(y), \qquad \textit{for all } x, y \in X,$$

and homogeneous if

$$T(sx) = sT(x), \qquad \textit{for all } x \in X, s \in S.$$

An operator that is additive and homogeneous is called a linear operator.

Many examples of linear operators exist.

Example 1.1.1 *Define an operator T from a linear space X into it self by $T(x) = sx$, $s \in S$. Then T is a linear operator.*

Example 1.1.2 *The operator $D = \frac{d}{dt}$ mapping $X = C^1[0,1]$ into $Y = C[0,1]$ given by*

$$D(x) = \frac{dx}{dt} = y(t), \ 0 \le t \le 1,$$

is linear.

If X and Y are linear spaces over the same scalar field S, then the set $L(X,Y)$ containing all linear operators from X into Y is a linear space over S if addition is defined by

$$(T_1 + T_2)(x) = T_1(x) + T_2(x), \qquad \text{for all } x \in X,$$

and scalar multiplication by

$$(sT)(x) = s(T(x)), \qquad \text{for all } x \in X, \; s \in S.$$

We may also consider linear operators B mapping X into $L(X,Y)$. For an $x \in X$ we have

$$B(x) = T,$$

a linear operator from X into Y. Hence, we have

$$B(x_1, x_2) = (B(x_1))(x_2) = y \in Y.$$

B is called a bilinear operator from X into Y. The linear operators B from X into $L(X,Y)$ form a linear space $L(X, L(X,Y))$. This process can be repeated to generate j-linear operators ($j > 1$ an integer).

Definition 1.1.2 *A linear operator mapping a linear space X into its scalar S is called a linear functional in X.*

Definition 1.1.3 *An operator Q mapping a linear space X into a linear space Y is said to be nonlinear if it is not a linear operator from X into Y.*

1.2 Continuous Linear Operators

Some metric concepts of importance are introduced here.

Definition 1.2.1 *An operator F from a Banach space X into a Banach space Y is continuous at $x = x^*$ if*

$$\lim_{n\to\infty} \|x_n - x^*\|_X = 0 \implies \lim_{n\to\infty} \|F(x_n) - F(x^*)\|_Y = 0$$

Theorem 1.2.2 *If a linear operator T from a Banach space X into a Banach space Y is continuous at $x^* = 0$, then it is continuous at every point x of space X.*

Proof. We have $T(0) = 0$, and from $\lim_{n\to\infty} \|x_n\| = 0$ we get $\lim_{n\to\infty} \|T(x_n)\| = 0$. If sequence $\{x_n\}$ $(n \geq 0)$ converges to x^* in X, by setting $y_n = x_n - x^*$ we obtain $\lim_{n\to\infty} \|y_n\| = 0$. By hypothesis this implies that

$$\lim_{n\to\infty} \|T(x_n)\| = \lim_{n\to\infty} \|T(x_n - x^*)\| = \lim_{n\to\infty} \|T(x_n) - T(x^*)\| = 0.$$

■

Definition 1.2.3 *An operator F from a Banach space X into a Banach space Y is bounded on the set A in X if there exists a constant $c < \infty$ such that*

$$\|F(x)\| \leq c\|x\|, \qquad \text{for all } x \in A.$$

The greatest lower bound (infimum) of numbers c satisfying the above inequality is called the bound of F on A. An operator which is bounded on a ball (open) $U(z,r) = \{x \in X \mid \|x - z\| < r\}$ is continuous at z. It turns out that for linear operators the converse is also true.

Theorem 1.2.4 *A continuous linear operator T from a Banach space X into a Banach space Y is bounded on X.*

Proof. By the continuity of T there exists $\varepsilon > 0$ such that $\|T(z)\| < 1$, if $\|z\| < \varepsilon$. For $0 \neq z \in X$

$$\|T(z)\| \leq \tfrac{1}{\varepsilon}\|z\|, \tag{1.2.1}$$

since $\|cz\| < \varepsilon$ for $|c| < \frac{\varepsilon}{\|z\|}$, and $\|T(cz)\| = |c| \cdot \|T(z)\| < 1$. Letting $c = \varepsilon^{-1}$ in (1.2.1), we conclude that operator T is bounded on X. ∎

The bound on X of a linear operator T denoted by $\|T\|_X$ or simply $\|T\|$ is called the norm of T. As in Theorem 1.2.4 we get

$$\|T\| = \sup_{\|x\|=1} \|T(x)\|. \tag{1.2.2}$$

Hence, for any bounded linear operator T

$$\|T(x)\| \leq \|T\| \cdot \|x\|, \qquad \text{for all } x \in X. \tag{1.2.3}$$

From now on, $L(X,Y)$ denotes the set of all bounded linear operators from a Banach space X into another Banach space Y. It also follows immediately that $L(X,Y)$ is a linear space if equipped with the rules of addition and scalar multiplication introduced in Section 1.1.

The proof of the following result is left as an exercise (see also [129]).

Theorem 1.2.5 *The set $L(X,Y)$ is a Banach space for the norm* (1.2.2).

1.3 Equations

In a Banach space X solving a linear equation can be stated as follows: given a bounded linear operator T mapping X into itself and some $y \in X$, find an $x \in X$ such that

$$T(x) = y. \tag{1.3.1}$$

The point x (if it exists) is called a solution of Equation (1.3.1).

Definition 1.3.1 *If T is a bounded linear operator in X and a bounded linear operator T_1 exists such that*

$$T_1 T = T T_1 = I, \tag{1.3.2}$$

where I is the identity operator in X (i.e., $I(x) = x$ for all $x \in X$), then T_1 is called the inverse of T and we write $T_1 = T^{-1}$. That is,

$$T^{-1} T = T T^{-1} = I. \tag{1.3.3}$$

If T^{-1} exists, then Equation (1.3.1) *has the unique solution*

$$x = T^{-1}(y). \tag{1.3.4}$$

The proof of the following result is left as an exercise (see also [147], [211], [214]).

Theorem 1.3.2 *(Banach Lemma on Invertible Operators). If T is a bounded linear operator in X, T^{-1} exists if and only if there is a bounded linear operator P in X such that P^{-1} exists and*

$$\|I - PT\| < 1. \tag{1.3.5}$$

If T^{-1} exists, then

$$T^{-1} = \sum_{n=0}^{\infty} (I - PT)^n P \qquad \textit{(Neumann Series)} \tag{1.3.6}$$

and

$$\|T^{-1}\| \leq \frac{\|P\|}{1 - \|I - PT\|}. \tag{1.3.7}$$

Based on Theorem 1.3.2 we can immediately introduce a computational theory for Equation (1.3.1) composed by three factors:

(A) *Existence and Uniqueness.* Under the hypotheses of Theorem 1.3.2 Equation (1.3.1) has a unique solution x^*.

(B) *Approximation.* The iteration

$$x_{n+1} = P(y) + (I - PT)(x_n) \quad (n \geq 0) \tag{1.3.8}$$

gives a sequence $\{x_n\}$ $(n \geq 0)$ of successive approximations, which converges to x^* for any initial guess $x_0 \in X$.

(C) *Error Bounds.* Clearly the speed of convergence of iteration $\{x_n\}$ $(n \geq 0)$ to x^* is governed by the estimate:

$$\|x_n - x^*\| \leq \frac{\|I - PT\|^n}{1 - \|I - PT\|} \|P(y)\| + \|I - PT\|^n \|x_0\|. \tag{1.3.9}$$

1.4 The Inverse of a Linear Operator

Let T be a bounded linear operator in X. One way to obtain an approximate inverse is to make use of an operator sufficiently close to T.

Theorem 1.4.1 *If T is a bounded linear operator in X, T^{-1} exists if and only if there is a bounded linear operator P_1 in X such that P_1^{-1} exists, and*

$$\|P_1 - T\| \le \left\|P_1^{-1}\right\|^{-1}. \tag{1.4.1}$$

If T^{-1} exists, then

$$T^{-1} = \sum_{n=0}^{\infty} \left(I - P_1^{-1}T\right)^n P_1^{-1} \tag{1.4.2}$$

and

$$\left\|T^{-1}\right\| \le \frac{\left\|P^{-1}\right\|}{1 - \left\|I - P_1^{-1}T\right\|} \le \frac{\left\|P_1^{-1}\right\|}{1 - \left\|P_1^{-1}\right\| \|P_1 - T\|}. \tag{1.4.3}$$

Proof. Let $P = P_1^{-1}$ in Theorem 1.3.2 and note that by (1.4.1)

$$\left\|I - P_1^{-1}T\right\| = \left\|P_1^{-1}(P_1 - T)\right\| \le \left\|P_1^{-1}\right\| \cdot \|P_1 - T\| < 1. \tag{1.4.4}$$

That is, (1.3.5) is satisfied. The bounds (1.4.3) follow from (1.3.7) and (1.4.4). That proves the sufficiency. The necessity is proved by setting $P_1 = T$, if T^{-1} exists. ∎

The following result is equivalent to Theorem 1.3.2.

Theorem 1.4.2 *A bounded linear operator T in a Banach space X has an inverse T^{-1} if and only if linear operators P, P^{-1} exist such that the series*

$$\sum_{n=0}^{\infty} (I - PT)^n P \tag{1.4.5}$$

converges. In this case we have

$$T^{-1} = \sum_{n=0}^{\infty} (I - PT)^n P.$$

Proof. If series (1.4.5) converges, then it converges to T^{-1} (see Theorem 1.3.2). The existence of P, P^{-1} and the convergence of series (1.4.5) is again established as in Theorem 1.3.2, by taking $P = T^{-1}$, when it exists. ∎

Definition 1.4.3 *A linear operator N in a Banach space X is said to be nilpotent if*

$$N^m = 0, \tag{1.4.6}$$

for some positive integer m.

Theorem 1.4.4 *A bounded linear operator T in a Banach space X has an inverse T^{-1} and only if there exist linear operators P, P^{-1} such that $I - PT$ is nilpotent.*

Proof. If P, P^{-1} exists and $I - PT$ is nilpotent, then series

$$\sum_{n=0}^{\infty} (I - PT)^n P = \sum_{n=0}^{m-1} (I - PT)^n P$$

converges to T^{-1} by Theorem 1.4.2. Moreover, if T^{-1} exists, then $P = T^{-1}$, $P^{-1} = T$ exists, and $I - PT = I - T^{-1}T = 0$ is nilpotent. ■

1.5 Fréchet Derivatives

The computational techniques to be considered later make use of the derivative in the sense of Fréchet [211], [212], [264].

Definition 1.5.1 *Let F be an operator mapping a Banach space X into a Banach space Y. If there exists a bounded linear operator L from X into Y such that*

$$\lim_{\|\Delta x\| \to 0} \frac{\|F(x_0 + \Delta x) - F(x_0) - L(\Delta x)\|}{\|\Delta x\|} = 0, \tag{1.5.1}$$

then P is said to be Fréchet differentiable at x_0, and the bounded linear operator

$$P'(x_0) = L \tag{1.5.2}$$

is called the first Fréchet-derivative of F at x_0. The limit in (1.5.1) is supposed to hold independently of the way that Δx approaches 0. Moreover, the Fréchet differential

$$\delta F(x_0, \Delta x) = F'(x_0)\Delta x \tag{1.5.3}$$

is an arbitrary close approximation to the difference $F(x_0 + \Delta x) - F(x_0)$ relative to $\|\Delta x\|$, for $\|\Delta x\|$ small.

If F_1 and F_2 are differentiable at x_0, then

$$(F_1 + F_2)'(x_0) = F_1'(x_0) + F_2'(x_0). \tag{1.5.4}$$

Moreover, if F_2 is an operator from a Banach space X into a Banach space Z, and F_1 is an operator from Z into a Banach space Y, their composition $F_1 \circ F_2$ is defined by

$$(F_1 \circ F_2)(x) = F_1(F_2(x)), \qquad \text{for all } x \in X. \tag{1.5.5}$$

It follows from Definition 1.5.1 that $F_1 \circ F_2$ is differentiable at x_0 if F_2 is differentiable at x_0 and F_1 is differentiable at $F_2(x_0)$ of Z, with (chain rule):

$$(F_1 \circ F_2)'(x_0) = F_1'(F_2(x_0))F_2'(x_0). \tag{1.5.6}$$

In order to differentiate an operator F we write:

$$F(x_0 + \Delta x) - F(x_0) = L(x_0, \Delta x)\Delta x + \eta(x_0, \Delta x), \tag{1.5.7}$$

where $L(x_0, \Delta x)$ is a bounded linear operator for given $x_0, \Delta x$ with

$$\lim_{\|\Delta x\| \to 0} L(x_0, \Delta x) = L, \tag{1.5.8}$$

and

$$\lim_{\|\Delta x\| \to 0} \frac{\|\eta(x_0, \Delta x)\|}{\|\Delta x\|} = 0. \tag{1.5.9}$$

Estimates (1.5.8) and (1.5.9) give

$$\lim_{\|\Delta x\| \to 0} L(x_0, \Delta x) = F'(x_0). \tag{1.5.10}$$

If $L(x_0, \Delta x)$ is a continuous function of Δx in some ball $U(0, R)$ $(R > 0)$, then

$$L(x_0, 0) = F'(x_0). \tag{1.5.11}$$

We need the definition of a mosaic:
Higher-order derivatives can be defined by induction:

Definition 1.5.2 *If F is $(m-1)$-times Fréchet-differentiable ($m \geq 2$ an integer), and an m-linear operator A from X into Y exists such that*

$$\lim_{\|\Delta x\| \to 0} \frac{\left\|F^{(m-1)}(x_0 + \Delta x) - F^{(m-1)}(x_0) - A(\Delta x)\right\|}{\|\Delta x\|} = 0, \tag{1.5.12}$$

then A is called the m-Fréchet-derivative of F at x_0, and

$$A = F^{(m)}(x_0) \tag{1.5.13}$$

Higher partial derivatives in product spaces can be defined as follows: Define

$$X_{ij} = L(X_j, X_i), \tag{1.5.14}$$

where X_1, $X_2, \ldots$ are Banach spaces and $L(X_j, X_i)$ is the space of bounded linear operators from X_j into X_i. The elements of X_{ij} are denoted by L_{ij}, etc. Similarly,

$$X_{ijm} = L(X_m, X_{ij}) = L(X_m, L(X_j, X_i)) \tag{1.5.15}$$

denotes the space of bounded bilinear operators from X_k into X_{ij}. Finally, we write

$$X_{ij_1j_2\cdots j_m} = L\left(X_{jk}, X_{ij_1j_2\cdots j_{m-1}}\right), \tag{1.5.16}$$

which denotes the space of bounded linear operators from X_{jm} into $X_{ij_1j_2\cdots j_{m-1}}$. The elements $A = A_{ij_1j_2\cdots j_m}$ of $X_{ij_1j_2\cdots j_m}$ are a generalization of m-linear operators [10], [54].

Consider an operator F_i from space

$$X = \prod_{p=1}^{n} X_{j_p} \tag{1.5.17}$$

into X_i, and that F_i has partial derivatives of orders $1, 2, \ldots, m-1$ in some ball $U(x_0, R)$, where $R > 0$ and

$$x_0 = \left(x_{j_1}^{(0)}, x_{j_2}^{(0)}, \ldots, x_{j_n}^{(0)}\right) \in X. \tag{1.5.18}$$

For simplicity and without loss of generality we renumber the original spaces so that

$$j_1 = 1, j_2 = 2, \ldots, j_n = n. \tag{1.5.19}$$

Hence, we write

$$x_0 = (x_1^{(0)}, x_2^{(0)}, \ldots, x_n^{(0)}). \tag{1.5.20}$$

A partial derivative of order $(m-1)$ of F_i at x_0 is an operator

$$A_{iq_1q_2\cdots q_{m-1}} = \frac{\partial^{(m-1)} F_i(x_0)}{\partial x_{q_1} \partial x_{q_2} \cdots \partial x_{q_{m-1}}} \tag{1.5.21}$$

(in $X_{iq_1q_2\cdots q_{m-1}}$) where

$$1 \le q_1, q_2, \ldots, q_{m-1} \le n. \tag{1.5.22}$$

Let $P(X_{q_m})$ denote the operator from X_{q_m} into $X_{iq_1q_2\cdots q_{m-1}}$ obtained from (1.5.21) by letting

$$x_j = x_j^{(0)}, \quad j \ne q_m, \tag{1.5.23}$$

for some q_m, $1 \le q_m \le n$. Moreover, if

$$P'(x_{q_m}^{(0)}) = \frac{\partial}{\partial x_{q_m}} \cdot \frac{\partial^{m-1} F_i(x_0)}{\partial x_{q_1} \partial x_{q_2} \cdots \partial x_{q_{m-1}}} = \frac{\partial^m F_i(x_0)}{\partial x_{q_1} \cdots \partial x_{q_m}}, \tag{1.5.24}$$

exists, it will be called the partial Fréchet-derivative of order m of F_i with respect to $x_{q_1}, \ldots, x_{q_m}$ at x_0.

Furthermore, if F_i is Fréchet-differentiable m times at x_0, then

$$\frac{\partial^m F_i(x_0)}{\partial x_{q_1}\cdots\partial x_{q_m}} x_{q_1}\cdots x_{q_m} = \frac{\partial^m F_i(x_0)}{\partial x_{s_1}\partial x_{s_2}\cdots\partial x_{s_m}} x_{s_1}\cdots x_{s_m} \tag{1.5.25}$$

for any permutation $s_1, s_2, \dots, s_m$ of integers $q_1, q_2, \dots, q_m$ and any choice of points $x_{q_1}, \dots, x_{q_m}$, from $X_{q_1}, \dots, X_{q_m}$ respectively. Hence, if $F = (F_1, \dots, F_t)$ is an operator from $X = X_1 \times X_2 \times \cdots \times X_n$ into $Y = Y_1 \times Y_2 \times \cdots \times Y_t$, then

$$F^{(m)}(x_0) = \left(\frac{\partial^m F_i}{\partial x_{j_1}\cdots\partial x_{j_m}}\right)_{x=x_0} \tag{1.5.26}$$

$i = 1, 2, \dots, t$, $j_1, j_2, \dots, j_m = 1, 2, \dots, n$, is called the m-Fréchet derivative of F at $x_0 = (x_1^{(0)}, x_2^{(0)}, \dots, x_n^{(0)})$.

1.6 Integration

In this section we state results concerning the mean value theorem, Taylor's theorem, and Riemannian integration. The proofs are left out as exercises.

The mean value theorem for differentiable real functions f:

$$f(b) - f(a) = f'(c)(b - a), \tag{1.6.1}$$

where $c \in (a, b)$, does not hold in a Banach space setting. However, if F is a differentiable operator between two Banach spaces X and Y, then

$$\|F(x) - F(y)\| \leq \sup_{\bar{x}\in L(x,y)} \|F'(\bar{x})\| \cdot \|x - y\|, \tag{1.6.2}$$

where

$$L(x, y) = \{z : z = \lambda y + (1 - \lambda)x,\ 0 \leq \lambda \leq 1\}. \tag{1.6.3}$$

Set

$$z(\lambda) = \lambda y + (1 - \lambda)x, \quad 0 \leq \lambda \leq 1, \tag{1.6.4}$$

and

$$F(\lambda) = F(z(\lambda)) = F(\lambda y + (1 - \lambda)x). \tag{1.6.5}$$

Divide the interval $0 \leq \lambda \leq 1$ into n subintervals of lengths $\Delta\lambda_i$, $i = 1, 2, \dots, n$, choose points λ_i inside corresponding subintervals and as in the real Riemann integral consider sums

$$\sum_{\sigma} F(\lambda_i)\Delta\lambda_i = \sum_{i=1}^{n} F(\lambda_i)\Delta\lambda_i, \tag{1.6.6}$$

where σ is the partition of the interval, and set

$$|\sigma| = \max_{(i)} \Delta\lambda_i. \tag{1.6.7}$$

Definition 1.6.1 *If*

$$S = \lim_{|\sigma| \to 0} \sum_{\sigma} F(\lambda_i) \Delta \lambda_i \tag{1.6.8}$$

exists, then it is called the Riemann integral from $F(\lambda)$ from 0 and 1, denoted by

$$S = \int_0^1 F(\lambda)\, d\lambda = \int_x^y F(\lambda)\, d\lambda. \tag{1.6.9}$$

Definition 1.6.2 *A bounded operator $P(\lambda)$ on $[0,1]$ such that the set of points of discontinuity is of measure zero is said to be integrable on $[0,1]$.*

We now state the famous Taylor theorem [171].

Theorem 1.6.3 *If F is m-times Fréchet-differentiable in $U(x_0, R)$, $R > 0$, and $F^{(m)}(x)$ is integrable from x to any $y \in U(x_0, R)$, then*

$$F(y) = F(x) + \sum_{n=1}^{m-1} \tfrac{1}{n!} F^{(n)}(x)(y-x)^n + R_m(x,y), \tag{1.6.10}$$

$$\left\| F(y) - \sum_{n=0}^{m-1} \tfrac{1}{n!} F^{(n)}(x)(y-x)^n \right\| \le \sup_{\bar{x} \in L(x,y)} \left\| F^{(m)}(\bar{x}) \right\| \frac{\|y-x\|^m}{m!}, \tag{1.6.11}$$

where

$$R_m(x,y) = \int_0^1 F^{(m)}\left(\lambda y + (1-\lambda) x\right) (y-x)^m \tfrac{(1-\lambda)^{m-1}}{(m-1)!} d\lambda. \tag{1.6.12}$$

1.7 Exercises

1.1 Show that the operators introduced in Examples 1.1.1 and 1.1.2 are indeed linear.

1.2. Show that the Laplace transform

$$\Delta = \frac{\partial^2}{\partial x_1^2} + \frac{\partial^2}{\partial x_2^2} + \frac{\partial^2}{\partial x_3^2}$$

is a linear operator mapping the space of real functions $x = x(x_1, x_2, x_3)$ with continuous second derivatives on some subset D of R^3 into the space of continuous real functions on D.

1.3. Define $T : C''[0,1] \times C'[0,1] \to C[0,1]$ by

$$T(x,y) = \left(\alpha \frac{d^2}{dt^2} \; \beta \frac{d}{dt} \right) \begin{pmatrix} x \\ y \end{pmatrix} = \alpha \frac{d^2 x}{dt^2} + \beta \frac{dy}{dt}, \quad 0 \le t \le 1.$$

Show that T is a linear operator.

1.4. In an inner product $\langle \cdot, \cdot \rangle$ space show that for any fixed z in the space

$$T(x) = \langle x, z \rangle$$

is a linear functional.

1.5. Show that an additive operator T from a real Banach space X into a real Banach space Y is homogeneous if it is continuous.

1.6. Show that matrix $A = \{a_{ij}\}$, $i, j = 1, 2, \ldots, n$ has an inverse if

$$|a_{ii}| > \tfrac{1}{2} \sum_{j=1}^{n} |a_{ij}| > 0, \quad i = 1, 2, \ldots, n.$$

1.7. Show that the linear integral equation of second Fredholm kind in $C[0,1]$

$$x(s) - \lambda \int_0^1 K(s,t)x(t)dt = y(s), \quad 0 \le \lambda \le 1,$$

where $K(s,t)$ is continuous on $0 \le s, t \le 1$, has a unique solution $x(s)$ for $y(s) \in C[0,1]$ if

$$|\lambda| < \left[\max_{[0,1]} \int_0^1 |K(s,t)|dt \right]^{-1}.$$

1.8. Prove Theorem 1.2.5.

1.9. Prove Theorem 1.3.2.

1.10. Show that the operators defined below are all linear:

(a) Identity operator. The identity operator $I_X : X \to X$ given by $I_X(x) = x$, for all $x \in X$.

(b) Zero operator. The zero operator $O : X \to Y$ given by $O(x) = 0$, for all $x \in X$.

(c) Integration. $T : C[a,b] \to C[a,b]$ given by $T(x(t)) = \int_0^1 x(s)ds$, $t \in [a,b]$.

(d) Differentiation. Let X be the vector space of all polynomials on $[a,b]$. Define T on X by $T(x(t)) = x'(t)$.

(e) Vector algebra. The cross product with one factor kept fixed. Define $T_1 : \mathbb{R}^3 \to \mathbb{R}^5$. Similarly, the dot product with one fixed factor. Define $T_2 : \mathbb{R}^3 \to \mathbb{R}$.

(f) Matrices. A real matrix $A = \{a_{ij}\}$ with m rows and n columns. Define $T : \mathbb{R}^n \to \mathbb{R}^m$ given by $y = Ax$.

1.11. Let T be a linear operator. Show:

(a) the $R(T)$ (range of T) is a vector space;

(b) if $\dim(T) = n < \infty$, then $\dim R(T) \leq n$;

(c) the null/space $N(T)$ is a vector space.

1.12. Let X, Y be vector spaces, both real or both complex. Let $T : D(T) \to Y$ (domain of T) be a linear operator with $D(T) \subseteq X$ and $R(T) \subseteq Y$. Then, show:

(a) the inverse $T^{-1} : R(T) \to D(T)$ exists if and only if

$$T(x) = 0 \Rightarrow x = 0;$$

(b) if T^{-1} exists, it is a linear operator;

(c) if $\dim D(T) = n < \infty$ and T^{-1} exists, then $\dim R(T) = \dim D(T)$.

1.13. Let $T : X \to Y$, $P : Y \to Z$ be bijective linear operators, where X, Y, Z are vector spaces. Then, show: the inverse $(ST)^{-1} : Z \to X$ of the product ST exists, and

$$(ST)^{-1} = T^{-1}S^{-1}.$$

1.14. If the product (composite) of two linear operators exists, show that it is linear.

1.15. Let X be the vector space of all complex 2×2 matrices and define $T : X \to X$ by $T(x) = cx$, where $c \in X$ is fixed and cx denotes the usual product of matrices. Show that T is linear. Under what conditions does T^{-1} exist?

1.16. Let $T : X \to Y$ be a linear operator and $\dim X = \dim Y = n < \infty$. Show that $R(T) = Y$ if and only if T^{-1} exists.

1.17. Define the integral operator $T : C[0,1] \to C[0,1]$ by $y = T(x)$, where $y(t) = \int_0^1 k(x,s)x(s)ds$ and k is continuous on $[0,1] \times [0,1]$. Show that T is linear and bounded.

1.18. Show that the operator T defined in Exercise 1.10(f) is bounded.

1.19. If a normed space X is finite dimensional then show that every linear functional on X is bounded.

1.20. Let $T : D(T) \to Y$ be a linear operator, where $D(T) \subseteq X$ and X, Y are normed spaces. Show:

(a) T is continuous if and only if it is bounded;

(b) if T is continuous at a single point, it is continuous.

1.21. Let T be a bounded linear operator. Show:

(a) $x_n \to x$ (where $x_n, x \in D(T)$) $\Rightarrow T(x_n) \to T(x)$;

(b) the null space $N(T)$ is closed.

1.22. If $T \neq 0$ is a bounded linear operator, show that for any $x \in D(T)$ such that $\|x\| < 1$, we have $\|T(x)\| < \|T\|$.

1.23. Show that the operator $T : \ell^\infty \to \ell^\infty$ defined by $y = (y_i) = T(x)$, $y_i = \frac{x_i}{i}$, $x = (x_i)$, is linear and bounded.

1.24. Let $T : C[0,1] \to C[0,1]$ be defined by

$$y(t) = \int_0^t x(s)ds.$$

Find $R(T)$ and $T^{-1} : R(T) \to C[0,1]$. Is T^{-1} linear and bounded?

1.25. Show that the functionals defined on $C[a,b]$ by

$$f_1(x) = \int_a^b x(t)y_0(t)dt \quad (y_0 \in C[a,b])$$
$$f_2(x) = c_1 x(a) + c_2 x(b) \quad (c_1, c_2 \text{ fixed})$$

are linear and bounded.

1.26. Find the norm of the linear functional f defined on $C[-1,1]$ by

$$f(x) = \int_{-1}^0 x(t)dt - \int_0^1 x(t)dt.$$

1.27. Show that

$$f_1(x) = \max_{t \in J} x(t), \quad f_2(x) = \min_{t \in J} x(t), \quad J = [a,b]$$

define functionals on $C[a,b]$. Are they linear? Bounded?

1.28. Show that a function can be additive and not homogeneous. For example, let $z = x + iy$ denote a complex number, and let $T : \mathbb{C} \to \mathbb{C}$ be given by

$$T(z) = \bar{z} = x - iy.$$

1.29. Show that a function can be homogeneous and not additive. For example, consider the operator $T : \mathbb{R}^2 \to \mathbb{R}$ given by

$$T((x_1, x_2)) = \frac{x_1^2}{x_2}.$$

1.30. Let F be an operator in $C[0,1]$ defined by

$$F(x)(s) = x(s) \int_0^1 \frac{s}{s+t} x(t)dt, \quad 0 \le \lambda \le 1.$$

Show that for $x_0, z \in C[0,1]$

$$F'(x_0)z = x_0(s) \int_0^1 \frac{s}{s+t} z(t)dt + z(s) \int_0^1 \frac{s}{s+t} x_0(t)dt.$$

1.31. Find the Fréchet-derivative of the operator F in $\mathbb{R}^2_\infty$ given by

$$F\binom{x}{y} = \begin{pmatrix} x^2 + 7x + 2xy - 3 \\ x + y^3 \end{pmatrix}.$$

1.32. Find the first and second Fréchet-derivatives of the Uryson operator

$$U(x) = \int_0^1 k(s, t, x(t))dt$$

in $C[0,1]$ at $x_0 = x_0(s)$.

1.33. Find the Fréchet-derivative of the Riccati differential operator

$$R(z) = \frac{dz}{dt} + p(t)z^2 + q(t)z + r(t),$$

from $C'[0,s]$ into $C[0,s]$ at $z_0 = z_0(t)$ in $C'[0,s]$.

1.34. Find the first two Fréchet-derivatives of the operator

$$F\binom{x}{y} = \begin{pmatrix} x^2 + y^2 - 3 \\ x \sin y \end{pmatrix} \quad \text{in } \mathbb{R}^2.$$

1.35. Consider the partial differential operator

$$F(x) = \Delta x - x^2$$

from $C^2(I)$ into $C(I)$, the space of all continuous function on the square $0 \le \alpha, \beta \le 1$. Show that

$$F'(x_0)z = \Delta z(\alpha, \beta) - 2x_0(\alpha, \beta)z(\alpha, \beta),$$

where Δ is the usual Laplace operator.

1.36. Let $F(L) = L^3$, in $L(x)$. Show:

$$F'(L_0) = L_0[\ ]L_0 + L_0^2[\ ] + [\ ]L_0.$$

1.37. Let $F(L) = L^{-1}$, in $L(x)$. Show:

$$F'(L_0) = -L_0^{-1}[\ ]L_0^{-1},$$

provided that L_0^{-1} exists.

1.38. Show estimates (1.6.1) and (1.6.2).

1.39. Show Taylor's Theorem 1.6.3.

1.40. Integrate the operator

$$F(L) = L^{-1} \quad \text{in } L(X)$$

from $L_0 = I$ to $L_1 = A$, where $\|I - A\| < 1$.

Chapter 2

Monotone convergence

This chapter introduces the fundamentals of the theory of divided differences of a nonlinear operator. Several results are also provided involving differences as well as Fréchet derivatives satisfying Lipschitz or monotone-type conditions that will be used later. Results on monotone convergence are discussed here and also in chapters 11 and 12.

2.1 Partially Ordered Topological Spaces

Let X be a linear space. We introduce the following definition:

Definition 2.1.1 *a partially ordered topological linear space (POTL-space) is a locally convex topological linear space X which has a closed proper convex cone.*

A proper convex cone is a subset K such that $K + K \subset K$, $\alpha K \subset K$ for $\alpha > 0$, and $K \cap (-K) = \{0\}$. Thus the order relation $\leq$, defined by $x \leq y$ if and only if $y - x \in K$, gives a partial ordering which is compatible with the linear structure of the space. The cone K which defines the ordering is called the positive cone since $K = \{x \in X \mid x \geq 0\}$. The fact that K is closed implies also that intervals, $[a, b] = \{z \in X \mid a \leq z \leq b\}$, are closed sets.

Example 2.1.1 *Some simple examples of POTL-spaces are:*

(1) $X = E^n$, *n-dimensional Euclidean space, with*

$$K = \{(x_1, x_2, ..., x_n) \in E^n \mid x_i \geq 0,\ i = 1, 2, ..., n\};$$

(2) $X = E^n$ *with*

$$K = \{(x_1, x_2, ..., x_n) \in E^n \mid x_i \geq 0,\ i = 1, 2, ..., n-1, x_n = 0\};$$

(3) $X = C^n[0,1]$, *continuous functions, maximum norm topology, pointwise ordering;*

(4) $X = C^n[0,1]$, *$n-$times continuously differentiable functions with*

$$\|f\| = \sum_{k=0}^{n} \max \left| f^{(K)}(t) \right|, \text{ and pointwise ordering;}$$

(5) $C = L^p[0,1]$, $0 \le p \le \infty$ *usual topology,* $K = \{f \in L^p[0,1] \mid f(t) \le 0 \text{ a.e.}\}$.

Remark 2.1.1 *Using the above examples, it is easy to see that the closedness of the positive cone is not, in general, a strong enough connection between the ordering and the topology. Consider, for example, the following properties of sequences of real numbers:*

(1) $x_1 \le x_2 \le \cdots \le x^*$, *and* $\sup\{x_n\} x^*$ *implies* $\lim_{n\to\infty} x_n = x^*$;

(2) $\lim_{n\to\infty} x_n = 0$ *implies that there exists a sequence* $\{y_n\}$ *with* $y_1 \ge y_2 \ge \cdots \ge 0$, $\inf\{y_n\} = 0$ *and* $-y_n \le x_n \le y_n$;

(3) $0 \le x_n \le y_n$, *and* $\lim_{n\to\infty} y_n = 0$ *imply* $\lim_{n\to\infty} x_n = 0$.

Unfortunately, these statements are not trues for all POTL-spaces:

(a) In $X = C[0,1]$ let $x_n(t) = -t^n$. Then $x_1 \le x_2 \le \cdots \le 0$, and $\sup\{x_n\} = 0$, but $\|x_n\| = 1$ for all n, so $\lim_{n\to\infty} x_n$ does not exist. Hence (1) does not hold.

(b) In $X = L^1[0,1]$ let $x_n(t) = n$ for $\frac{1}{n+1} \le t \le \frac{1}{n}$ and zero elsewhere. Then $\lim_{n\to\infty} \|x_n\| = 0$ but clearly property (2) does not hold.

(c) In $X = C^1[0,1]$ let $x_n(t) = \frac{t^n}{n}$, $y_n(t) = \frac{1}{n}$. then $0\check{S}x_n \le y_n$, and $\lim_{n\to\infty} y_n = 0$, but $\|x_n\| = \max\left|\frac{t^n}{n}\right| + \max\left|t^{n-1}\right| = \frac{1}{n} + 1 > 1$; hence x_n does not converge to zero.

We will now devote a brief discussion of certain types of POTL spaces in which some of the above statements are true.

Definition 2.1.2 *A POTL-space is called regular if every order-bounded increasing sequence has a limit.*

Remark 2.1.2 *Examples of regular POTL-spaces are* E^n *and* L^p, $0 \le p \le \infty$, *where as* $C[0,1]$, $C^n[0,1]$ *and* $L^\infty[0,1]$ *are not regular, as was shown in (a) of the above remark. If* $\{x_n\}$ $n \ge 0$ *is a monotone increasing sequence and* $\lim_{n\to\infty} x_n = x^*$ *exists, then for any* k_0, $n \ge k_0$ *implies* $x_n \ge x_{k_0}$. *Hence* $x^* = \lim_{n\to\infty} x_n \ge x_{k_0}$, *i.e.,*

x^* is an upper bound on $\{x_n\}$ $n = 0$. Moreover, if y is any other upper bound, then $x_n \leq y$, and hence $x^* = \lim_{n\to\infty} x_n \leq y$, i.e., $x^* = \sup\{x_n\}$. This shows that in any POTL-space, the closedness of the positive cone guarantees that, if a monotone increasing sequence has a limit, then it is also a supremum. In a regular space, the converse of this is true; i.e., if a monotone increasing sequence has a supremum, then it also has a limit. It is important to note that the definition of regularity involves both an order concept (monotone boundedness) and a topological concept (limit).

Definition 2.1.3 *A POTL-space is called normal if, given a local base U for the topology, there exists a positive number η so that if $0 \leq x \in V \in U$ then $[0, x] \subseteq \eta^U$.*

Remark 2.1.3 *If the topology of a POTL-space is given by a norm then this space is called a partially ordered normed space (PON)-space. If a PON-space is complete with respect to its topology then it is called a partially ordered Banach space (POB)-space. According to Definition 2.1.3. A PON-space is normal if and only if there exists a positive number α such that*

$$\|x\| \leq \alpha \|y\| \quad \text{for all} \quad x, y \in X \quad \text{with} \quad 0 \leq x \leq y.$$

Let us note that any regular POB-space is normal. The converse is not true. For example, the space $C[0,1]$, ordered by the cone of nonnegative functions, is normal but is not regular. All finite dimensional POTL-spaces are both normal and regular.

Remark 2.1.4 *Let us now define some special types of operators acting between two POTL-spaces. First we introduce some notation if X and Y are two linear spaces then we denote by (X, Y) the set of all operators from X into Y and by $L(X, Y)$ the set of all linear operators from X into Y. If X and Y are topological linear spaces then we denote by $LB(X, Y)$ the set of all continuous linear operators from X into Y. For simplicity the spaces $L(X, X)$ and $LB(X, X)$ will be denoted by $L(X)$ and $LB(X)$. Now let X and Y be two POTL-spaces and consider an operator $G \in (X, Y)$. G is called isotone (resp. antitone) if $x \geq y$ implies $G(x) \leq G(y)$ (resp. $G(x) \leq G(y)$). G is called nonnegative if $x \geq 0$ implies $G(x) \geq 0$. For linear operators the nonnegativity is clearly equivalent with the isotony. Also, a linear operator is inverse nonnegative if and only if it is invertible and its inverse is nonnegative. If G is a nonnegative operator then we write $G \geq 0$. If G and H are two operators from X into Y such that $H - G$ is nonnegative then we write $G \leq H$. If Z is a linear space then we denote by $I = I_z$ the identity operator in Z (i.e., $I(x) = x$ for all $x \in Z$). If Z is a POTL-space then we have obviously $I \geq 0$. Suppose that X and Y are two POTL-spaces and consider the operators $T \in L(X, Y)$ and $S \in L(Y, X)$. If $ST \leq I_x$ (resp. $ST \geq I_x$) then S is called a left subinverse (resp. superinverse) of T and T is called a right sub-inverse (resp. superinverse) of S. We say that S is a subinverse of T is S is a left-as a right subinverse of T.*

We finally end this section by noting that for the theory of partially ordered linear spaces the reader may consult M.A. Krasnosel'skii [210]–[213], Vandergraft [332] or Argyros and Szidarovszky [99].

2.2 Divided Differences in a Linear Space

The concept of a divided difference of a nonlinear operator generalizes the usual notion of a divided difference of a scalar function in the same way in which the Fréchet-derivative generalizes the notion of a derivative of a function.

Definition 2.2.1 *Let F be a nonlinear operator defined on a subset D of a linear space X with values in a linear space Y, i.e., $F \in (D, Y)$ and let x, y be two points of D. A linear operator from X into Y, denoted $[x, y]$, which satisfies the condition*

$$[x, y](x - y) = F(x) - F(y) \tag{2.2.1}$$

is called a divided difference of F at the points x and y.

Remark 2.2.1 *If X and Y are topological linear spaces then we shall always assume the continuity of the linear operator $[x, y]$. (Generally, $[x, y] \in L(X, Y)$ if X, Y are POTL-spaces then $[x, y] \in LB(X, Y)$).*

Obviously, condition (2.2.1) does not uniquely determine the divided difference, with the exception of the case when X is one-dimensional. An operator $[\cdot, \cdot] : D \times D \to L(X, Y)$ satisfying (2.2.1) is called a divided difference of F on D. If we fix the first variable, we get an operator

$$[x^0, \cdot] : D \to L(X, Y). \tag{2.2.2}$$

Let x^1, x^2 be two points of D. A divided difference of the operator (2.2.2) at the points x^1, x^2 will be called a divided difference of the second order of F at the points x^0, x^1, x^2 and will be denoted by $[x^0, x^1, x^2]$. We have by definition

$$[x^0, x^1, x^2](x^1 - x^2) = [x^0, x^1] - [x^0, x^2]. \tag{2.2.3}$$

Obviously, $[x^0, x^1, x^2] \in L(X, L(X, Y))$.

Let us now state a well known result due to Kantorovich concerning the location of fixed points which will be used extensively later.

Theorem 2.2.2 *Let X be a regular POTL-space and let x, y be two points of X such that $x \leq y$. If $H : [x, y] \to X$ is a continuous isotone operator having the property that $x \leq H(x)$ and $y \geq H(y)$, then there exists a point $z \in [x, y]$ such that $H(z) = z$.*

2.3 Divided Difference in a Banach Space

In this section we will assume that X and Y are Banach spaces. Accordingly we shall have $[x,y] \in LB(X,Y)$, $[x,y,z] \in LB(X, LB(X,Y))$. As we will see in later chapters, most convergence theorems in a Banach space require that the divided differences of F satisfy Lipschitz conditions of the form:

$$\|[x,y]-[x,z]\| \le c_0 \|y-z\| \tag{2.3.1}$$

$$\|[y,x]-[z,x]\| \le c_1 \|y-z\| \tag{2.3.2}$$

$$\|[x,y,z]-[u,y,z]\| \le c_2 \|x-y\| \quad \text{for all} \quad x,y,z,u \in D. \tag{2.3.3}$$

It is a simple exercise to show that if $[\cdot,\cdot]$ is a divided difference of F satisfying (2.3.1) or (2.3.2) then F is Fréchet differentiable on D and we have

$$F'(x) = [x,x] \quad \text{for all} \quad x \in D. \tag{2.3.4}$$

Moreover, if (2.3.1) and (2.3.2) are both satisfied then the Fréchet derivative F' is Lipschitz continuous on D with Lipschitz constant $I = c_0 + c_1$.

At the end of this section we shall give an example of divided differences of the first and of the second order in the finite dimensional case. We shall consider the space $\mathbf{R}^q$ equipped with the Chebysheff norm which is given by

$$\|x\| = \max\{|x_i| \in \mathbf{R} | 1 \le I \le q\} \quad \text{for} \quad x = (x_1, x_2, ..., x_q) \in \mathbf{R}^q. \tag{2.3.5}$$

It follows that the norm of a linear operator $L \in LB(\mathbf{R}^q)$ represented by the matrix with entries I_{ij} is given by

$$\|L\| = \max\left\{\sum_{j=1}^{q} |I_{ij}| \mid |1 \le i \le q\right\}. \tag{2.3.6}$$

We cannot give a formula for the norm of a bilinear operator. However, if B is a bilinear operator with entries b_{ijk} then we have the estimate

$$\|B\| \le \max\left\{\sum_{j=1}^{q}\sum_{k=1}^{q} |b_{ijk}| \; |1 \le i \le q\right\}. \tag{2.3.7}$$

Let U be an open ball of $\mathbf{R}^q$ and let F be an operator defined on U with values in $\mathbf{R}^q$. We denote by f_1, ..., f_q the components of F. For each $x \in U$ we have

$$F(x) = (f_1(x), ..., f_q(x))^T. \tag{2.3.8}$$

Moreover, we introduce the notation

$$D_j f_i(x) = \frac{\partial f(x)}{\partial x_j}, \quad D_{kj} f_i(x) = \frac{\partial^2 f_i(x)}{\partial x_j \partial x_k}. \tag{2.3.9}$$

Let x, y be two points of U and let us denote by $[x, y]$ the matrix with entries

$$[x,y]_{ij} = \tfrac{1}{x_j - y_j}\left(f_i\left(x_1,...,x_j,y_{j+1},...,y_q\right) - f_i\left(x_1,...,x_{j-1},y_j,...,y_q\right)\right). \quad (2.3.10)$$

The linear operator $[x, y] \in LB(\mathbf{R}^q)$ defined in this way obviously satisfies condition (2.2.1). If the partial derivatives $D_j f_i$ satisfy some Lipschitz conditions of the form

$$|D_j f_i(x_1,...,x_k+t,...,x_q) - D_j f_i(x_1,...,x_k,...,x_q)| \le p^i_{jk}|t| \quad (2.3.11)$$

then condition (2.3.1) and (2.3.2) will be satisfied with

$$c_0 = \max\left\{\frac{1}{2}\textstyle\sum_{j=1}^q\left(p^i_{jj} + \sum_{k=j+1}^q p^i_{jk}\right)|1 \le i \le q\right\} \quad (2.3.12)$$

and

$$c_1 = \max\left\{\frac{1}{2}\textstyle\sum_{j=1}^q\left(p^i_{jj} + \sum_{k=1}^{j-1} p^i_{jk}\right)|1 \le i \le q\right\}. \quad (2.3.13)$$

We shall prove (2.3.1) only since (2.3.2) can be proved similarly.

Let x, y, z be three points of U. We shall have in turn

$$\begin{aligned}[x,y]_{ij} - [x,z] = \textstyle\sum_{k=1}^q &\left\{[x,(y_1,...,y_k,z_{k+1},...,z_q)]_{ij}\right.\\ &\left.-[x(y_1,...,y_{k-1},z_k,...,z_q)]_{ij}\right\}.\end{aligned} \quad (2.3.14)$$

by (2.3.10).

If $k \le j$ then we have

$$\begin{aligned}&[x,(y_1,...,y_k,z_{k+1},...,z_q)]_{ij} - [x,(y_1,...,y_{k-1},z_k,...,z_q)]_{ij} =\\ &= \tfrac{1}{x_j-z_j}\{f_i(x_1,...,x_j,z_{j+1},...,z_q) - f_i(x_1,...,x_{j-1},z_j,...,z_q)\}\\ &-\tfrac{1}{x_j-z_j}\{f_i(x_1,...,x_j,z_{j+1},...,z_q) - f_i(x_1,...,x_{j-1},z_j,...,z_q)\} = 0.\end{aligned}$$

For $k = j$ we have

$$\begin{aligned}&\left|[x,(y_1,...,y_j,z_{j+1},...,z_q)]_{ij} - \left[x,(y_1,...,y_{j-1},z_j,...,z_q)_{ij}\right]\right| =\\ &= \left|\tfrac{1}{x_j-y_j}\{f_i(x_1,...,x_j,z_{j+1},...,z_q) - f_i(x_1,...,x_{j-1},y_i,z_{j+1},...,z_q)\}\right.\\ &\left.-\tfrac{1}{x_j-y_j}\{f_i(x_1,...,x_j,z_{j+1},...,z_q) - f_i(x_1,...,x_{j-1},z_j,...,z_q)\}\right|\\ &= \left|\int_0^1\{D_j f_i(x_1,...,x_j,y_j+t(x_j-y_j),z_{j+1},...,z_q)\right.\\ &-D_j f_i(x_1,...,x_j,z_j+t(x_j-z_j),z_{j+1},...,z_q)\}\,dt|\\ &\le |y_j - z_j|\,p^i_{jj}\int_0^1 t\,dt = \tfrac{1}{2}|x_j - z_j|\,p^i_{jj}\end{aligned}$$

(by (2.3.11)).

Finally for $k > j$ we have using (2.3.10) and (2.3.14) again

$$\begin{aligned}
&\left|[x,(y_1,...,y_k,z_{k+1},...,z_q)]_{ij} - [x,(y_1,...,y_{k-1},z_k,...,z_q)]_{ij}\right| = \\
&= \Big|\tfrac{1}{x_j - y_j}\{f_i(x_1,...,x_j,y_{j+1},...,y_k,z_{k+1},...,z_q) \\
&\quad - f_i(x_1,...,x_{j-1},y_j,...,y_k,z_{k+1},...,z_q) \\
&\quad - f_i(x_1,...,x_j,y_{j+1},...,y_{k-1},z_k,...,z_q) \\
&\quad + f_i(x_1,...,x_{j-1},y_j,...,y_{k-1},z_k,...,z_q)\}\Big| \\
&= \left|\int_0^1 \{f_i(x_1,...,x_{j-1},y_j + t(x_j - y_j), y_{j+1},...,y_k,z_{k+1},...,z_q)\right. \\
&\quad \left. - f_i(x_1,...,x_{j-1},y_j + t(x_j - y_j), y_{j+1},...,y_{k-1},z_k,...,z_q)\}\, dt\right| \\
&\le |y_k - z_k|\, p^i_{jk}.
\end{aligned}$$

By adding all the above we get

$$\begin{aligned}
\left|[x,y]_{ij} - [x,z]_{ij}\right| &\le \frac{1}{2}|y_j - z_j|\, p^i_{jj} + \sum_{k=j+1}^{q} |y_k - z_k|\, p^i_{jk} \\
&\le \|y - z\| \left\{\frac{1}{2}\sum_{j=1}^{q}\left(p^i_{jj} + \sum_{k=j+1}^{q} p^i_{jk}\right)\right\}.
\end{aligned}$$

Consequently condition (2.3.1) is satisfied with c_0 given by (2.3.12). If each f_j has continuous second order partial derivatives which are bounded on U we have

$$p^i_{jk} = \sup\{|D_{jk} f_i(x)| \ |x \in U\}.$$

In this case $p^i_{jk} = p^i_{kj}$ so that $c_0 = c_1$.

Moreover, consider again three points x, y, z of U. Similarly with (2.3.14) the second divided difference of F at x, y, z is the bilinear operators defined by

$$[x,y,z]_{ijk} = \tfrac{1}{y_k - z_k}\left\{[x,(y_1,...,y_k,z_{k+1},...,z_q)]_{ij} - [x,(y_1,...,y_{k-1},z_k,...,z_q)]_{ij}\right\}. \tag{2.3.15}$$

It is easy to see as before that $[x,y,z]_{ijk} = 0$ for $k < j$. For $k = j$ we have

$$[x,y,z]_{ijj} = [x_j,y_j,z_j]_t\, f_i(x_1,...,x_{i-1},t,z_{j+1},...,z_q) \tag{2.3.16}$$

where the right hand side of (2.3.16) represents the divided difference of $f_i(x_1,...,x_{j-1},t,z_{j+1},...,z_q)$ as a function of t, at the points x_j, y_j, z_j. Using Genocchi's integral representation of divided differences of scalar functions we get

$$[x,y,z]_{ijj} = \\ = \int_0^1\int_0^1 tD_{jj}f_i\left(x_1,...,x_{j-1},x_j+t\left(y_j-x_j\right)+ts\left(z_j-y_j\right),z_{j+1},...,z_q\right)dsdt. \tag{2.3.17}$$

Hence, for $k>j$ we obtain

$$\begin{aligned}
&[x,y,z]_{ijk} = \\
&= \tfrac{1}{(y_k-z_k)(x_j-y_j)}\left\{f_i\left(x_1,...,x_j,y_{j+1},...,y_k,z_{k+1},...,z_q\right)\right. \\
&\quad - f_i\left(x_1,...,x_j,x_{j+1},...,y_{k-1},z_k,...,z_q\right) \\
&\quad - f_i\left(x_1,...,x_{j-1},y_j,...,y_k,z_{k+1},...,z_q\right) \\
&\quad \left.+f_i\left(x_1,...,x_{j-1},y_j,...,y_{k-1},z_k,...,z_q\right)\right\} \\
&\quad \tfrac{1}{x_j-y_j}\int_0^1\left\{D_kf_i\left(x_1,...,x_j,y_{j+1},...,y_{k-1},z_k+t\left(y_k-z_k\right),z_{k+1},...,z_q\right)\right. \\
&\quad \left.-D_kf_i\left(x_1,...,x_{j-1},y_j,...,y_{k-1},z_k+t\left(y_k-z_k\right),z_{k+1},...,z_q\right)\right\}dt \\
&= \int_0^1\int_0^1 D_{kj}f_i(x_1,...,x_{j-1},y_j+s\left(x_j-y_i\right),y_{j+1},...,y_{k-1},z_k \\
&\quad +t\left(y_k-z_k\right),z_{k+1},...,z_q)dsdt.
\end{aligned} \tag{2.3.18}$$

We now want to show that if

$$\begin{aligned}
&\left|D_{kj}f_i\left(v_1,...,v_m+t,...,v_q\right)-D_{kj}f_i\left(v_1,...,v_m,...v_q\right)\right| \le q_{km}^{ij}\left|t\right| \\
&\text{for all}\quad v=\left(v_1,...,v_q\right)\in U,\quad 1\le i,j,k,m\le q,
\end{aligned} \tag{2.3.19}$$

then the divided difference of F of the second order defined by (2.3.15) satisfies condition (2.3.3) with the constant

$$c_2 = \max_{1\le i\le q}\sum\nolimits_{j=1}^q\left\{\frac{1}{6}q_{jj}^{ij}+\frac{1}{2}\sum\nolimits_{m=1}^{j-1}q_{jm}^{ij}+\frac{1}{2}\sum\nolimits_{k=j+1}^q q_{kj}^{ij}+\sum\nolimits_{k=j+1}^q\sum\nolimits_{m=1}^{j-1}q_{km}^{ij}\right\}. \tag{2.3.20}$$

Let u,x,y,z be four points of U. Then using (2.3.15) we can easily have

$$[x,y,z]_{ijk}-[u,y,z]_{ijk} = \sum\nolimits_{m=1}^q\left\{\left[\left(x_1,...,x_m,u_{m+1},...,u_q\right),y,z\right]_{ijk}\right. \\ \left.\left[\left(x_1,...,x_{m-1},u_m,...,u_q\right),y,z\right]_{ijk}\right\}. \tag{2.3.21}$$

If $m = j$ the terms in (2.3.21) vanish so that using (2.3.18) and (2.3.19) we deduce that for $k > j$

$$
\begin{aligned}
&\left|[x, y, z]_{ijk} - [u, y, z]_{ijk}\right| = \\
&= \Big|\textstyle\sum_{m=1}^{j-1} \displaystyle\int_0^1 \int_0^1 \{D_{kj} f_i (x_1, ..., x_m, u_{m+1}, ..., u_{j-1}, y_j + s (x_j - y_j), \\
&y_{j+1}, ..., y_{k-1}, z_k + t (y_k - z_k), z_{k+1}, ..., z_q) \\
&- D_{kj} f_i(x_1, ..., x_{m-1}, u_m, ..., u_{j-1}, y_j + s (x_j - y_j), y_{j+1}, ..., \\
&\quad y_{k-1} z_k + t (y_k - z_k), z_{k+1}, ..., z_q)\} \, dsdt \\
&+ \int_0^1 \int_0^1 \{D_{kj} f_i (x_1, ..., x_{j-1}, y_j + s (x_j - y_j), y_{j+1}, ...,) \\
&y_{k-1}, z_k + t (y_k - z_k), z_{k+1}, ..., z_q\} \\
&- D_{kj} f_i(x_1, ..., x_{j-1}, y_j + s (u_j - y_j), y_{j+1}, ..., \\
&y_{k-1}, z_k + t (y_k - z_k), z_{k+1}, ..., z_q)\} \, dsdt| \\
&\leq \frac{1}{2} |x_j - u_j| \, q_{kj}^{ij} + \textstyle\sum_{m=1}^{j-1} |x_m - u_m| \, q_{km}^{ij}.
\end{aligned}
$$

Similarly for $k = j$ we obtain in turn

$$
\begin{aligned}
&\left|[x, y, z]_{ijj} - [u, y, z]_{ijj}\right| = \\
&= \Big|\int_0^1 \int_0^1 t \{D_{jj} f_i (x_1, ..., x_{j-1}, x_j + t (y_j - x_j) + ts (z_j - y_j), z_{j+1}, ..., z_q) \\
&-D_{jj} f_i (x_1, ..., x_{j-1}, u_j + t (y_j - u_j) + ts (z_j - y_j), z_{j+1}, ..., z_q)\} \, dsdt \\
&+ \textstyle\sum_{m=1}^{j-1} \displaystyle\int_0^1 \int_0^1 t\{D_{jj} f_i(x_1, ..., x_m, u_{m+1}, ..., \\
&u_{j-1}, x_j + t (x_j - y_j) + ts (z_j - y) j, z_{j+1}, ..., z_q) \\
&- D_{jj} f_i(x_1, ..., x_{m-1}, u_m, ..., u_{j-1}, \\
&x_j + t (y_j - x_j) + ts (z_j - y_j), z_{j+1}, ..., z_q)\} \, dsdt| \\
&\leq \tfrac{1}{6} |x_j - u_j| \, q_{jj}^{ij} + \tfrac{1}{2} \textstyle\sum_{m=1}^{j-1} |x_m - u_m| \, q_{jm}^{ij}.
\end{aligned}
$$

Finally using the estimate (2.3.7) of the norm of a bilinear operator, we deduce that condition (2.3.3) holds with c_2 given by (2.3.20).

2.4 Divided Differences and Monotone Convergence

In this section we make an introduction to the problem of approximating a locally unique solution x^* of the nonlinear operator equation $F(x) = 0$, in a POTL-space X. In particular, consider an operator $F : D \subseteq X \to Y$ where X is a POTL-space with values in a POTL-space Y. Let x_0, y_0, y_{-1} be three points of D such that

$$x_0 \leq y_0 \leq y_{-1}, \quad [x_0, y_{-1}],$$

and denote

$$\begin{aligned} D_1 &= \left\{(x,y) \in X^2 \mid x_0 \leq x \leq y \leq y_0\right\}, \\ D_2 &= \left\{(y,y_{-1}) \in X^2 \mid x_0 \leq u \leq y_0\right\}, \\ D_3 &= D_1 \cup D_2. \end{aligned} \tag{2.4.1}$$

Assume there exist operators $A_0 : D_3 \to LB\left(X,Y\right)$, $A : D_1 \to L\left(X,Y\right)$ such that:

(a)

$$F\left(y\right) - F\left(x\right) \leq A_0\left(w,z\right)\left(y-x\right) forall \left(x,y\right), \left(y,w\right) \in D_1, \left(w,z\right) \in D_3; \tag{2.4.2}$$

(b) the linear operator $A_0\left(u,v\right)$ has a continuous nonsingular nonnegative left subinverse;

(c)

$$F\left(y\right) - F\left(x\right) \geq A\left(x,y\right)\left(y-x\right) \quad \text{for all} \quad \left(x,y\right) \in D_1; \tag{2.4.3}$$

(d) the linear operator $A\left(x,y\right)$ has a nonegative left superinverse for each $\left(x,y\right) \in D_1$

$$F\left(y\right) - F\left(x\right) \leq A_0\left(y,z\right)\left(y-x\right) \quad \text{for all} \quad x,y \in D_1, \ \left(y,z\right) \in D_3. \tag{2.4.4}$$

Moreover, let us define approximations

$$F\left(y_n\right) + A_0\left(y_n, y_{n-1}\right)\left(y_{n+1} - y_n\right) = 0 \tag{2.4.5}$$

$$F\left(x_n\right) + A_0\left(y_n, y_{n-1}\right)\left(x_{n+1} - x_n\right) = 0 \tag{2.4.6}$$

$$y_{n+1} = y_n - B_n F\left(y_n\right) n \geq 0 \tag{2.4.7}$$

$$x_{n+1} = x_n - B_n^1 F\left(x_n\right) \quad n \geq 0, \tag{2.4.8}$$

where B_n and B_n^1 are nonnegative subinverses of $A_0\left(y_n, y_{n-1}\right)$ $n \geq 0$.

Under very natural conditions, hypotheses the form (2.4.2) or (2.4.3) or (2.4.4) have been used extensively to show that the approximations (2.4.2) and (2.4.6) or (2.4.7) and (2.4.8) generate two sequences $\{x_n\}$ $n \geq 1$, $\{y_n\}$ $n \geq 1$ such that

$$x_0 \leq x_1 \leq \cdots \leq x_n \leq x_{n+1} \leq y_{n+1} \leq y_n \leq \cdots \leq y_1 \leq y_0 \tag{2.4.9}$$

$$\lim_{n\to\infty} x_n = x^* = y^* = \lim_{n\to\infty} y_n \quad \text{and} \quad F\left(x^*\right) = 0. \tag{2.4.10}$$

For a complete survey on these results we refer to the works of [279], Argyros and Szidarovszky [97]–[99].

Here we will use similar conditions (i.e. like (2.4.2), (2.4.3), (2.4.4)) for two-point approximations of the form (2.4.5) and (2.4.6) or (2.4.7) and (2.4.8).

Consequently a discussion must follow on the possible choices of the linear operators A_0 and A.

Remark 2.4.1 *Let us now consider an operator $F : D \subseteq X \to Y$, where X, Y are both POTL-spaces. The operator F is called order-complex on an interval $[x_0, y_0] \subseteq D$ if*

$$F(\lambda x + (1-\lambda) y) \leq \lambda F(x) + (1-\lambda) F(y) \tag{2.4.11}$$

for all comparable $x, y \in [x_0, y_0]$ and $\lambda \in [0,1]$. If F has a linear G-derivative $F'(x)$ at each point $[x_0, y_0]$ then (2.4.11) *holds if and only if*

$$F'(x)(y-x) \leq F(y) - F(x) \leq F'(y)(y-x) \quad \text{for} \quad x_0 \leq x \leq y \leq y_0. \tag{2.4.12}$$

(See e.g. Ortega and Rheinboldt [263] or our Exercise 2.25 for the properties of the Gateaux-derivative).

Hence, for order-convex G-differentiable operators conditions (2.4.3) and (2.4.4) are satisfied with $A_0(y,v) = A(y,v) = F'(u)$. In the unidimensional case (2.4.12) is equivalent with the isotony of the operator $x \to F'(x)$ but in general the latter property is stronger. Assuming the isotony of the operator $x \to F'(x)$ it follows that

$$F(y) - F(x) \leq F'(w)(y-x) \quad \text{for} \quad x_0 \leq x \leq y \leq w \leq y_0$$

Hence, in this case condition (2.4.2) is satisfied for $A_0(w,z) = F'(w)$.

The above observations show to choose A and A_0 for single or two-step Newton-methods. We note that the iterative algorithm (2.4.5)-(2.4.8) with

$$A_0(u,v) = F'(u)$$

is the algorithm proposed by Fourier in 1818 in the unidimensional case and extended by Baluev in 1952 in the general case. The idea of using an algorithm of the form (2.4.7)-(2.4.8) goes back to Slugin [311]. In Ortega and Rheinboldt [263] it is shown that with B_n properly chosen (2.4.7) reduces to a general Newton-SOR algorithm. In particular,suppose (in the finite dimensional case) that $F'(y_n)$ is an M -matrix and let $F'(y) = D_n - L_n - U_n$ be the partition of $F'(y_n)$ into diagonal, strictly lower - and strictly upper-triangular parts respectively for all $n \geq 0$. Consider an integer $m_n \geq 1$, a real parameter $w_n \in (0,1]$ and denote

$$P_n = w_n^{-1}(D_n - w_n L_n), \quad Q_n = w_n^{-1}[(1-w_n) D_n + w_n U_n], \tag{2.4.13}$$

$$H_n = P_n^{-1} Q_n, \quad \text{and} \quad B_n = \left(I + H_n + \cdots + H_n^{m_n - 1}\right) P_n^{-1}. \tag{2.4.14}$$

It can easily be seen that B_n $n \geq 0$ is a nonnegative subinverse of $F'(y_n)$ (see, also [99]). If $f : [a,b] \to \mid R$ is a real function of a real variable, then f is (order) convex if and only if

$$\frac{f(x) - f(y)}{x - y} \leq \frac{f(u) - f(v)}{u - v}, \quad \forall x, y, u, v \in [a,b] \text{ such that } x \leq u \text{ and } y \leq v.$$

This fact motivates the notion of convexity with respect to a divided difference considered by J.W. Schmidt and H. Leonhardt [307]. Let $F : D \subseteq X \to Y$ be a nonlinear operator between the POTL-spaces X and Y. Assume that the nonlinear operator F has a divided difference $[\cdot,\cdot]$ on D. F is called convex with respect to the divided difference $[\cdot,\cdot]$ on D if

$$[x,y] \leq [u,v] \quad \text{for all} \quad x, y, u, v \in D \quad \text{with} \quad x \leq u \quad \text{and} \quad y \leq v. \tag{2.4.15}$$

In the above quoted study, Schmidt and Leonhardt studied (2.4.5)-(2.4.6) with $A_0(u,v) = [u,v]$ in case the nonlinear operator F is convex with respect to $[\cdot,\cdot]$. Their result was extended by N. Schneider [308] who assumed instead of (1.2.1) the milder condition

$$[u,v](u - v) \geq F(u) - F(v) \text{ for all comparable } u, v \in D. \tag{2.4.16}$$

An operator $[\cdot,\cdot] : D \times D \to L(X,Y)$ satisfying (2.4.16) is called a generalized divided difference of F on D. If both (2.4.15) and (2.4.16) are satisfied, then we say that F is convex with respect to the generalizedivided difference of $[\cdot,\cdot]$. It is easily seen that if (2.4.15) and (2.4.16) are satisfied on $D = [x_0, y_{-1}]$ then conditions (2.4.2) and (2.4.3) are satisfied with $A = A_0 = [\cdot,\cdot]$. Indeed for $x_0 \leq x \leq y \leq w \leq z \leq y_{-1}$ we have

$$[x,y](y - x) \leq F(y) - F(x) \leq [y,x](y - x) \leq [w,z](y - x).$$

Moreover, concerning general secant-SOR methods, in case the generalized difference $[y_n, y_{n-1}]$ is an M-matrix and if B_n $n \geq 0$ is computed according to (2.4.13) and (2.4.14) where $[y_n, y_{n-1}] = D_n - L_n - U_n$ $n \geq 0$ is the partition of $[y_n, y_{n-1}]$ into its diagonal, strictly lower- and strictly upper-triangular parts.

We remark that an operator which is convex with respect to a generalized divided difference is also order-convex. To see that, consider $x, y \in D$, $x \leq y$, $\lambda \in [0,1]$ and set $z = \lambda x + (I - \lambda) y$. Observing that $y - x = (1 - \lambda)^{-1}(z - x) = \lambda^{-1}(y - z)$ and applying (2.4.16) we have in turn:

$$\begin{aligned} (1 - \lambda)^{-1}(F(z) - F(x)) &\leq (1 - \lambda)^{-1}[z,x](z - x) \\ &= [z,x](y - x) \leq [z,y](y - x) \\ &= \lambda^{-1}[z,y](y - z) \leq^{1} (F(y) - F(z)). \end{aligned}$$

By the first and last term we deduce that $F(z) \leq \lambda F(x) + (1 - \lambda) F(y)$. Thus, Schneider's result can be applied only to order-convex operators and its importance resides in the fact that the use of a generalized divided difference instead of the G-derivative may be more advantageous from a numerical point of view. We note however, that conditions (2.4.2) and (2.4.3) do not necessarily imply convexity. For example, if f is a real function of a real variable such that

$$\inf_{x,y\in[x_0,y_0]} \frac{f(x)-f(y)}{x-y} = m > 0, \qquad \sup_{x,y\in[x_0,y_0]} \frac{f(x)-f(y)}{x-y} = M < \infty$$

then (2.4.2) and (2.4.3) are satisfied for $A_0(u,v) = M$ and $A(u,v) = m$. It is not difficult to find examples of nonconvex operators in the finite (or even in the infinite) dimensional case satisfying a condition of the form

$$A(y-x) \le F(y) - F(x) \le A_0(y,x), \quad x_0 \le x \le y \le y_0$$

where A_0 and A are fixed linear operators. If A_0 has a continuous nonsingular nonnegative left subinverse and A has a nonnegative right superinverse, then convergence of the algorithm (2.4.5)-(2.4.6) can be discussed. This algorithm becomes extremely simple in this case. The monotone convergence of such an iterative procedure seems to have been first investigated by S. Slugin [256].

In the end of this section we shall consider a class of nonconvex operators which satisfy condition (2.4.2) but do not necessarily satisfy condition (2.4.3). Consequently from convergence theorems involving (2.4.5) and (2.4.6) it will follow that Jacobi-Newton and the Jacobi-secant methods have monotonous convergence for operators belonging to this class (see, also the elegenat papers by F. Potra[274], [277], [279].

Let $F = (f_1, ..., f_q)^{T0}$ *be* an operator acting in the finite dimensional space $\mathbf{R}^q$, endowed with the natural (componentwise) partial ordering. Let us denote by e_i the i–th coordinate vector of $\mathbf{R}^q$. We say that F is off-diagonally antitone if the functions

$$g_{ij} : \mathbf{R} \to \mathbf{R}, \quad g_{ij}(t) = f_i(x + te_j), \quad i \ne j, \quad i,j = 1, ..., q$$

are antitone. Suppose that at each point x belonging to an interval $U = [x_0, y_{-1}]$ the partial derivatives $\partial_i F_i(x)$, $i = 1,2,...,q$, exist and are positive. For any two points $x, y \in U$ we consider the quotients

$$[x,y]_i = \begin{cases} \frac{f_i(y) - f_i\left(y - e_i^T(y-x)e_i\right)}{e_i^T(y-x)}, & \text{if} \quad e_i^T(y-x) \ne 0 \\ \partial_i F_i(x) & \text{if} \quad e_i^T(y-x) = 0. \end{cases} \tag{2.4.17}$$

Let us denote $\Delta[x,y]$ the diagonal matrix having as elements the number $[x,y]_i$, $i = 1,2,...,q$. For the diagonal matrix $\Delta[x,y]$ formed by the partial derivatives $\partial_i f_i(x)$, $i = 1,2,...,q$, we shall also use the notation $DF(x)$.

Suppose now that F is off-diagonally antitone and that the operator $DF : U \to LB(\mathbf{R}^q)$ is isotone (i.e. all functions $\partial_i F_i : \mathbf{R} \to \mathbf{R}$) are isotone. In this case for all

$x_0 \leq x \leq y \leq w \leq z \leq y_{-1}$ and all $i \in \{1, 2, ..., q\}$ there exist $\lambda, \mu \in [0, 1]$ such that

$$\begin{aligned} f_i(y) - f_i(x) &\leq f_i(y) - f_i\left(y - e_i^T(y-x)\,e_i\right) = \partial_i f_i\left(y - \lambda e_i^T(y-x)\,e_i\right), \\ e_i^T(y-x) &\leq \partial_i f_i(y)\, e_i^T(y-x) \leq \partial_i f_i(w)\, e_i^T(y-x) \\ &\leq \partial_i f_i\left(z - \mu e_i^T(z-w)\right) e_i^T(y-x) = [w, z]_i\, e_i^T(y-x). \end{aligned}$$

It follows that condition (2.4.2) is satisfied for $A_0(w, z) = DF(w)$ as well as for $A_0(w, z) = [\Delta w, z]$. With the choice $A_0(w, z) = \Delta[w, z]$ the iterative procedure (2.4.5) is a Jacobi-secant method while with the choice $A_0(w, z) = DF(w)$ it reduces to the Jacobi-Newton method. For some applications of the latter method, see W. Torning [320].

2.5 Divided Differences and Fréchet-derivatives

Let F be a nonlinear operator defined on an open subset D of a Banach space X with values in a Banach space Y. Choose also $x_0 \in D$ to be fixed.

Definition 2.5.1 *The operator $F : D \subseteq X \to Y$ is Fréchet - differentiable at the point $x_0 \in D$ if there exists a linear operator $L \in LB(X, Y)$ such that*

$$\lim_{\|h\| \to 0} \frac{1}{\|h\|} \|F(x_0 + h) - F(x_0) - A(h)\| = 0.$$

It is easy to see that such an operator is unique if it exists. This operator L is called the first Fréchet-derivative of F at the point x_0 and is denoted by $F'(x_0)$.

The following facts are left as exercises:

1. If F is Fréchet differentiable at the point x_0 then F is continuous at this point.

2. If $A \in LB(X, Y)$ and $F(x)$ for all $x \in X$, then F is Fréchet-differentiable at every point $x \in X$ and $F'(x) = A$.

3. Let F and G be two operators defined on D with values in Y. If F and G are Fréchet-differentiable at a point $x_0 \in D$, then the operator $F + G$ is also Fréchet-differentiable at x_0 and $(F + G)'(x_0) = F'(x_0) + G'(x_0)$.

4. Let X, y and Z be three Banach spaces and consider the operators $F : D_1 \subseteq X \to Y$, $G : D_2 \subseteq Y \to Z$. Let $x_0 \in D_1^0$ such that $y_0 = F(x_0)$ is a point in D_2^0. If F is Fréchet-differentiable at x_0 and G is Fréchet differentiable at y_0 then the mapping $H = G_0 F$ is Fréchet-differentiable at x_0 and $H'(x_0) = G'(y_0)\, F'(x_0)$.

If the operator F is Fréchet-differentiable at all $x \in D$, then we shall say that F is Fréchet-differentiable on D. In this case we may consider an operator $F' : D \to LB(X, Y)$ which associates to each point $x \in D$ the Fréchet-derivative of F at x. This operator will be called the first Fréchet-derivative of F.

Let x, y be two points in D and suppose that the segment

$$S = \{x + t(y - x) \,|t \in [0,1]\} \subseteq D.$$

Let y' be a continuous linear functional, set $h = y - x$ and define

$$\varphi(t) = (F(x + th), y').$$

If F is fréchet-differentiable at each point of the segment S, then φ is differentiable on $[0,1]$ and

$$\varphi(t) = (F'(x + th), y').$$

Let us now suppose that

$$\alpha = \sup_{t \in [0,1]} \|F'(x + t(y - x))\| < \infty;$$

then we have

$$\begin{aligned} \|(F(y) - F(x), y')\| &= \|\varphi(1) - \varphi(0)\| \\ &\leq \sup_{t\in[0,1]} \|\varphi'(t)\| \leq \alpha \|y'\| \, \|y - x\|. \end{aligned}$$

But, we also have

$$\|F(y) - F(x)\| = \sup_{\|y'\| \leq 1} \|(F(y) - F(x), y')\|,$$

we deduce that

$$\|F(y) - F(x)\| \leq \sup_{t\in[0,1]} \|F'(x + t(y - x))\| \cdot \|y - x\| \tag{2.5.1}$$

so, we proved:

Theorem 2.5.2 *Let D be a convex subset of a Banach space X and $F : D \subseteq X \to Y$. If F is Fréchet-differentiable on D and if there exists a constant c such that*

$$\|F'(x)\| \leq M \quad \textit{for all} \quad x \in D \Rightarrow \|F(x) - F(y)\| \leq c\|x - y\| \quad \textit{for all} \quad x \in D. \tag{2.5.2}$$

The estimate (2.5.1) is the analogue of the famous mean value formula from real analysis. If the operator F' is Riemann integrable on the segment S we can give the following integral representation of the mean value formula

$$F(x) - F(y) = \int_0^1 F'(x + t(y - x))\, dt\, (x - y). \tag{2.5.3}$$

Let now D be a convex open subset of X and let us suppose that we have associated to each pair (x, y) of distinct points from D a divided difference $[x, y]$ of F at these points. In applications one often has to require that the operator $(x, y) \to [x, y]$ satisfy a Lipschitz condition (see also Section 2.3). We suppose that there exists a nonnegative $c > 0$ such that

$$\|[x, y] - [x_1, y_1]\| \leq c\,(\|x - x_1\| + \|y - y_1\|) \tag{2.5.4}$$

for all $x, y, x_1, y_1 \in D$ with $x \neq y$ and $x_1 = y_1$.

We say in this case that F has a Lipschitz continuous difference on D. This condition allows us to extend by continuity the operator $(x, y) \to [x, y]$ to the whole Cartesian product $D \times D$. From (1.2.1) and (2.5.4) it follows that F is Fréchet-differentiable on D and that $[x, x] = F'(x)$. It also follows that

$$\|F'(x) - F'(y)\| \leq c_1\,\|x - y\| \quad \text{with} \quad c_1 = 2c \tag{2.5.5}$$

and

$$\|[x, y] - F'(z)\| \leq c\,(\|x - z\| + \|y - z\|) \tag{2.5.6}$$

for all $x, y \in D$. conversely if we assume that F is Fréchet-differentiable on D and that its Frechet derivative satisfies (2.5.5) then it follows that F has a Lipschitz continuous divided difference on D. We can certainly take

$$[x, y] = \int_0^1 F'(x + t\,(y - x))\,dt. \tag{2.5.7}$$

We now want to give the definition of the second Frechet derivative of F. We must first introduce the definition of bounded multilinear operators (which will also be used later).

Definition 2.5.3 *Let X and Y be two Banach spaces. An operator $A : X^n \to y$ $(n \in N)$ will be called n-linear operator from X to Y if the following conditions are satisfied:*

(a) *The operator $(x_1, ..., x_n) \to A(x_1, ..., x_n)$ is linear in each variable x_k $k = 1, 2, ..., n$.*

(b) *There exists a constant c such that*

$$\|A(x_1, x_2, ..., x_n)\| \leq c\,\|x_1\| \dots \|x_n\| \tag{2.5.8}$$

The norm of a bounded $n-$linear operator can be defined by the formula

$$\|A\| = \sup\{\|A(x_1, ..., x_n)\| \mid \|x_n\| = 1\}. \tag{2.5.9}$$

Set $LB^{(1)}(X, Y) = LB(X, Y)$ and define recursively

$$LB^{(k+1)}(X, Y) = LB\left(X, LB^{(k)}(X, Y)\right), \quad k \geq 0. \tag{2.5.10}$$

In this way we obtain a sequence of Banach spaces $LB^{(n)}(X,Y)$ $(n \geq 0)$. *Every* $A \in LB^{(n)}(X,Y)$ *can be viewed as a bounded* n*-linear operator if one takes*

$$A(x_1,...,x_n) = (...(A(x_1)(x_2)(x_3))...)(x_n). \tag{2.5.11}$$

In the right hand side of (2.5.11) *we have*

$$A(x_1) \in LB^{(n-1)}(X,Y), \quad (A(x_1))(x_2) \in LB^{(n-2)}(X,Y), \quad \textit{etc.}$$

Conversely, any bounded n*-linear operator* A *drom* X *to* Y *can be interpreted as an element of* $B^{(n)}(X,Y)$. *Moreover the norm of* A *as a bounded* n*-linear operator coincides with the norm as an element of the space* $LB^{(n)}(X,Y)$. *Thus we may identify this space with the space of all bounded* $n-$*linear operators from* X *to* Y. *In the sequel we will identify* $A(x,x,...,x) = Ax^n$, *and*

$$A(x_1)(x_2)...(x_n) = A(x_1,x_2,...,x_n) = Ax_1x_2...x_n.$$

Let us now consider a nonlinear operator $F : D \subseteq X \to Y$ *where* D *is open. Suppose that* F *is Fréchet differentiable on* D. *Then we may consider the operator* $F' : D \to LB(X,Y)$ *which associated to each point* x *the Fréchet derivative of* F *at* x. *If the operator* F' *is Fréchet differentiable at a point* $x_0 \in D$ *then we say that* F *is twice Fréchet differentiable at* x_0. *The Fréchet derivative of* F' *at* x_0 *will be denoted by* $F''(x_0)$ *and will be called the second Fréchet derivative of* F *at* x_0. *Note that* $F''(x_0) \in LB^{(2)}(X,Y)$. *Similarly we can define Fréchet derivatives of higher order. Finally by analogy with* (2.5.7)

$$\begin{aligned}&[x_0,...,x_k] = \\ &= \int_0^1 \cdots \int_0^1 t_1^{k-1} t_2^{k-2} \cdots t_{k-1} F(x_0 + t_1(x_1 - x_0) + t_1t_2(x_2 - x_1) \\ &+ \cdots + t_1t_2,...,t_k(x_k - x_{k-1}))\, dt_1dt_2...dt_k.\end{aligned} \tag{2.5.12}$$

It is easy to see that the multilinear operators defined above verify

$$[x_0,...,x_{k-1},x_k,x_{k+1}](x_k - x_{k+1}) = [x_0,...,x_{k-1},x_{k+1}]. \tag{2.5.13}$$

We note that throughout this sequel a 2−linear operator will also be called bilinear.

Finally, we will also need the definition of a $n-$linear symmetric operator. Given a n-linear operator $A : X^n \to Y$ and a permutation $i = (i_1,i_2,...,i_n)$ of the integers $1,2,...,n$, the notation $A(i)$ (or $A_n(i)$ if we want to emphasize the $n-$linearity of A) can be used for the $n-$linear operator $A(i) = A_n(i)$ such that

$$A(i)(x_1,x_2,...,x_n) = A_n(i)(x_1,x_2,...,x_n) = A_n(x_{i_1},x_{i_2},...,x_{i_n})\, A_n x_{i_1} x_{i_2} ... x_{i_n} \tag{2.5.14}$$

for all $x_1,x_2,...,x_n \in X$. Thus, there are $n!$ $n-$linear operators $A(i) = A_n(i)$ associated with a given $n-$linear operator $A = A_n$.

Definition 2.5.4 *A n-linear operator $A = A_n : X^n \to Y$ is said to be symmetric if*

$$A = A_n = A_n(i) \tag{2.5.15}$$

for all i belonging in R_n which denotes the set of all permutations of the integers $1, 2, ..., n$. The symmetric $n-$linear operator

$$\overline{A} = \overline{A}_n = \frac{1}{n!} \textstyle\sum_{i \in R_n} A_n(i) \tag{2.5.16}$$

is called the mean of $A = A_n$.

Definition 2.5.5 *A $n-$linear operator $A = A_n : X^n \to Y$ is said to be symmetric if*

$$A = A_n = A_n(i)$$

for all i belonging in R_n which denotes the set of all permutations of the integers $1, 2, ..., n$. The symmetric $n-$linear operator

$$\overline{A} = \overline{A}_n = \frac{1}{n!} \sum_{i \in R_n} A_n(i) \tag{2.5.17}$$

is called the mean of $A = A_n$.

Notation 2.5.6 *The notation*

$$A_n X^P = A_n xx...x \text{ (p-times)} \tag{2.5.18}$$

$p \leq n$, $A = A_n : X^n \to Y$, *for the result of applying A_n to $x \in X$ $p-$times will be used. If $p < n$, then* (2.5.17) *will represent a $(n-p)$-linear operator. For $p = n$, note that*

$$A_k x^k = \overline{A}_k x^k = A_k(i)\, x^k \tag{2.5.19}$$

for all $i \in R_k$, $x \in X$. It follows from (2.5.18) *that whenever we are dealing with an equation involving n-linear operators A_n, we may assume that they are symmetric without loss of generality, since each A_n may be replaced by $\overline{A}_n$ without changing the value of the expression at hand.*

2.6 Enclosing the Root of an Equation

We consider the equation

$$L(x) = T(x) \tag{2.6.1}$$

where L is a linear operator and T is a nonlinear operator defined on some convex subset D of a linear space E with values in a linear space Y.

We study the convergence of the iterations

$$L(y_{n+1}) = T(y_n) + A_n(y,x)(y_{n+1} - y_n) \tag{2.6.2}$$

and

$$L(x_{n+1}) = T(x_n) + A_n(y,x)(x_{n+1} - x_n) \tag{2.6.3}$$

to a solution x^* of equation (2.6.1), where $A_n(y,x)$, $n \geq 0$ is a linear operator.

If P is a linear projection operator $(P^2 = P)$, that projects the space Y into $Y_P \subseteq Y$, then the operator PT will be assumed to be Fréchet differentiable on D and its derivative $PT'_x(x)$ corresponds to the operator $PB(y,x), y, x \in D \times D$, $PT'_X(x) = PB(x,x)$ for all $x \in D$.

We will assume that

$$A_n(y,x) = APB(y_n, x_n), \quad n \geq 0.$$

Iterations (2.6.2) and (2.6.3) have been studied extensively under several assumptions [68], [99], [100], [263], [277], [279], when $P = L = I$, the identity operator on D. However, the iterates $\{x_n\}$ and $\{y_n\}$ can rarely be computed in infinite dimensional spaces in this case. But if the space Y_P is finite dimensional with $\dim(Y_P) = N$, then iterations (2.6.2) and (2.6.3) reduce to systems of linear algebraic equations of order at most N. This case has been studied in [100], [205] and in particular in [225]. The assumptions in [230] involve the positivity of the operators $PB(y,x) - APB(y,x)$, $QT'_x(x)$ with $Q = I - P$ and $L(y) - A_n(y,x)y$ on some interval $[y_0, x_0]$, which is difficult to verify.

In this section we simplify the above assumptions and provide some further conditions for the convergence of iterations (2.6.2) and (2.6.3) to a solution x^* of equation (2.6.1).

We finally illustrate our results with an example.

We can now formulate our main result.

Theorem 2.6.1 *Let $F = D \subset X \to Y$, where X is a regular POTL-space and Y is a POTL-space. Assume*

(a) *there exist points x_0, y_0, $y_{-1} \in D$ with*

$$x_0 \leq y_0 \leq y_{-1}, \quad [x_0, y_{-1}] \subset D, \quad L(x_0) - T(x_0) \leq 0 \leq L(y_0) - T(y_0).$$

Set

$$S_1 = \left\{(x,y) \in X^2; x_0 \leq x \leq y \leq y_0\right\},$$
$$S_2 = \left\{(u, y_{-1}) \in X^2; x_0 \leq u \leq y_0\right\}$$

and

$$S_3 = S_1 \cup S_2.$$

(b) *Assume that there exists an operator* $A : S_3 \to B(X,Y)$ *such that*

$$(L(y) - T(y)) - (L(x) - T(x)) \le A(w,z)(y-x) \tag{2.6.4}$$

for all (x,y), $(y,w) \in S_2$, $(w,z) \in S_3$.

(c) *Suppose that for any* $(u,v) \in S_3$ *the linear operator* $A(u,v)$ *has a continuous nonsingular nonnegative left subinverse.*

Then there exist two sequences $\{x_n\}$, $\{y_n\}$, $n \ge 1$ *satisfying* (2.6.2), (2.6.3),

$$L(x_n) - T(x_n) \le 0 \le L(y_n) - T(y_n), \tag{2.6.5}$$

$$x_0 \le x_1 \le \cdots \le x_n \le x_{n+1} \le y_{n+1} \le y_n \le \cdots \le y_1 \le y_0 \tag{2.6.6}$$

and

$$\lim_{n\to\infty} x_n = x^*, \qquad \lim_{n\to\infty} y_n = y^*. \tag{2.6.7}$$

Moreover, if the operators $A_n = A(y_n, y_{n-1})$ *are inverse nonnegative then any solution of the equation* (2.6.1) *from the interval* $[x_0, y_0]$ *belongs to the interval* $[x^*, y^*]$.

Proof. Let us define the operator $M : [0, y_0 - x_0] \to X$ by

$$M(x) = x - L_0(L(x_0) - T(x_0) + A_0(x))$$

where L_0 is a continuous nonsingular nonnegative left subinverse of A_0. It can easily be seen that M is isotone, continuous with

$$M(0) = -L_0(L(x_0) - T(x_0)) \ge 0$$

and

$$\begin{aligned} M(y_0 - x_0) &= y_0 - x_0 - L_0(L(y_0) - T(y_0)) \\ &\quad + L_0[(L(y_0) - T(y_0)) - (L(x_0) - T(x_0)) - A_0(y_0 - x_0)] \\ &\le y_0 - x_0 - L_0(L(y_0) - T(y_0)) \\ &\le y_0 - x_0. \end{aligned}$$

It now follows from the theorem of L.V. Kantorovich 2.2.2 on fixed points that the operator M has a fixed point $w \in [0, y_0 - x_0]$. Set $x_1 = x_0 + w$ to get

$$L(x_0) - T(x_0) + A_0(x_1 - x_0) = 0, \quad x_0 \le x_1 \le y_0.$$

By (2.6.4), we get

$$L(x_1) - T(x_1) = (L(x_1) - T(x_1)) - (L(x_0) - T(x_0)) + A_0(x_0 - x_1) \le 0.$$

Let us now define the operator $M_1 : [0, y_0 - x_1] \to X$ by

$$M_1(x) = x + L_0(L(y_0) - T(y_0) - A_0(x)).$$

It can easily be seen that M_1 is continuous, isotone with

$$M_1(0) = L_0(L(y_0) - T(y_0)) \geq 0$$

and

$$\begin{aligned} M_1(y_0 - x_1) = &\, y_0 - x_1 + L_0(L(x_1) - T(x_1)) \\ &+ L_0[(L(y_0) - T(y_0)) - (L(x_1) - T(x_1)) - A_0(y_0 - x_1)] \\ &\leq y_0 - x_1 + L_0(L(x_1) - T(x_1)) \\ &\leq y_0 - x_1. \end{aligned}$$

As before, there exists $z \in [0, y_0 - x_1]$ such that $M_1(z) = z$. Set $y_1 = y_0 - z$ to get

$$L(y_0) - T(y_0) + A_0(y_1 - y_0) = 0, \quad x_1 \leq y_1 \leq y_0.$$

But from (2.6.4) and the above

$$L(y_1) - T(y_1) = (L(y_1) - T(y_1)) - (L(y_0) - T(y_0)) - A_0(y_1 - y_0) \geq 0.$$

Using induction on n we can now show, following the above technique, that there exist sequences $\{x_n\}, \{y_n\}, n \geq 1$ satisfying (2.6.1), (2.6.2), (2.6.5) and (2.6.6). since the space X is regular (2.6.6) we get that there exist $x^*, y^* \in E$ satisfying (2.6.7), with $x^* \leq y^*$. Let $x_0 \leq z \leq y_0$ and $L(z) - T(z) = 0$ then we get

$$\begin{aligned} A_0(y_1 - z) &= A_0(y_0) - (L(y_0) - T(y_0)) - A_0(z) \\ &= A_0(y_0 - z) - [(L(y_0) - T(y_0)) - (L(z) - T(z))] \geq 0 \end{aligned}$$

and

$$\begin{aligned} A_0(x_1 - z) &= A_0(x_0) - (L(x_0) - T(x_0)) - A_0(z) \\ &= A_0(x_0 - z) - [(L(x_0) - T(x_0)) - (L(z) - T(z))] \leq 0. \end{aligned}$$

If A_0 is inverse isotone, then $x_1 \leq z \leq y_1$ and by induction $x_n \leq z \leq y_n$. Hence $x^* \leq z \leq y^*$. That completes the proof of the theorem. ■

Using (2.6.1), (2.6.2), (2.6.5), (2.6.6) and (2.6.7) we can easily prove the following theorem which gives us natural conditions under which the points x^* and y^* are solutions of the equation (2.6.1).

Theorem 2.6.2 *Let $L - T$ be continuous at x^* and y^* and the hypotheses of Theorem 2.6.1 be true. Assume that one of the conditions is satisfied:*

(a) $x^* = y^*$;

(b) *X is normal and there exists an operator $H : X \to Y$ $(H(0) = 0)$ which has an isotone inverse continuous at the origin and $A_n \leq H$ for sufficiently large n;*

(c) *Y is normal and there exists an operator $G : X \to Y$ $(G(0) = 0)$ continuous at the origin and such that $A_n \leq G$ for sufficiently large n.*

(d) *The operators L_n, $n \geq 0$ are equicontinuous.*

Then $L(x^*)-T(x^*) = L(y^*)-T(y^*) = 0$. Moreover, assume that there exists an operator $G_1 : S_1 \to L(X,Y)$ such that $G_1(x,y)$ has a nonnegative left superinverse for each $(x,y) \in S_1$ and

$$L(y) - T(y) - (L(x) - T(x)) \geq G_1(x,y)(y-x) \quad \text{for all} \quad (x,y) \in S_1.$$

Then if $(x^*, y^*) \in S_1$ and $L(x^*) - T(x^*) = L(y^*) - T(y^*) = 0$ then $x^* = y^*$.

We now complete this paper with an application.

2.7 Applications

Let $X = Y = \mathbb{R}^k$ with $k = 2N$. We define a projection operator P_N by

$$P_N(v) = \begin{cases} v_i, & i = 1,2,\ldots,N \\ 0, & i = N+1,\ldots,k, \end{cases} \quad v = (v_1, v_2, \ldots, v_k) \in X.$$

We consider the system of equations

$$v_i = f_i(v_1, ..., v_k), \quad i = 1,2,...,k. \tag{2.7.1}$$

Set $T(v) = \{f_i(v_1, ..., v_k)\}$, $i = 1,2,\ldots,k$, then

$$P_N T(v) = \begin{cases} f_i(v_1,\ldots,v_k) & i = 1,\ldots,N, \\ 0 & i = N+1,\ldots,k, \end{cases}$$

$$P_N T'(v) u = \begin{cases} \sum\limits_{j=1}^{k} f'_{ij}(v_1,\ldots,v_k) u_j & i = 1,2,\ldots,N, \\ 0 & i = N+1,\ldots,k, \quad f'_{ij} = \frac{\partial f_i}{\partial v_j}, \end{cases}$$

$$P_N B(w,z) u = \left\{ \begin{array}{l} \sum\limits_{j=1}^{k} F_{ij}(w_1,\ldots,w_k,z_1,\ldots,z_k) u_j, \quad i = 1,\ldots,N \\ 0, \quad i = N+1,\ldots,k \end{array} \right\}$$
$$= C_N^i(w,z) u,$$

where $F_{ij}(v_1, ..., v_k, v_1, ..., v_k) = \partial f_i(v_1, ..., v_k) / \partial v_j$. Choose

$$A_n(y,x) = C_N^i(y,x),$$

then iterations (2.6.2) and (2.6.3) become

$$y_{i,n+1} = f_i(y_{1,n},\dots,y_{k,n}) + C_N^i(y_{i,n},x_{i,n})(y_{i,n+1}-y_{i,n}) \tag{2.7.2}$$

$$x_{i,n+1} = f_i(x_{1,n},\dots,x_{k,n}) + C_N^i(y_{i,n},x_{i,n})(x_{i,n+1}-x_{i,n}). \tag{2.7.3}$$

Let us assume that the determinant D_n of the above N–th order linear systems is nonzero, then (2.7.2) and (2.7.3) become

$$y_{i,n+1} = \frac{\sum\limits_{m=1}^{N} D_{im}F_m^1(y_n,x_n)}{D_n}, \quad i=1,\dots,N \tag{2.7.4}$$

$$y_{i,n+1} = f_i(y_{1n},\dots,y_{kn}), \quad i=N+1,\dots,k \tag{2.7.5}$$

and

$$x_{i,n+1} = \frac{\sum\limits_{m=1}^{N} D_{im}F_m^2(y_n,x_n)}{D_n}, \quad i=1,\dots,N, \tag{2.7.6}$$

$$x_{i,n+1} = f_i(x_{1,n},\dots,x_{k,n}), \quad i=N+1,\dots,k \tag{2.7.7}$$

respectively. Here D_{im} is the cofactor of the element in the i-th column and m-th row of D_n and $F_m^1(y_n,x_n)$, $i=1,2$ are given by

$$F_m^1(y_n,x_n) = f_m(y_{1,n},\dots,y_{k,n}) + \sum_{i=N+1}^{k} \alpha_{mj}^n f_j(y_{1,n},\dots,y_{k,n}) - \sum_{j=1}^{k}\alpha_{mj}^n y_{j,n},$$

and

$$F_m^2(y_n,x_n) = f_m(x_{1,n},\dots,x_{k,n}) + \sum_{i=N+1}^{k} \alpha_{mj}^n f_j(x_{1,n},\dots,x_{k,n}) - \sum_{j=1}^{k}\alpha_{mj}^n x_{j,n},$$

where $\alpha_{mj}^n = F_{mj}(y_n,x_n)$.

If the hypotheses of Theorem 2.6.1 and 2.6.2 are now satisfied for the equation (2.7.1) then the results apply to obtain a solution x^* of equation (2.7.1) $[y_0,x_0]$.

In particular consider the system of differential equations

$$q_i'' = f_i(t,q_1,q_2), \quad i=1,2, \quad 0\le t\le 1 \tag{2.7.8}$$

subject to the boundary conditions

$$q_i(0)=d_i, \quad q_i(1)=e_i, \quad i=1,2,. \tag{2.7.9}$$

The functions f_1 and f_2 are assumed to be sufficiently smooth. For the discretization a uniform mesh

$$t_j = jh, \quad j=0,1,...,N+1, \quad h=\frac{1}{N+1}$$

and the corresponding central-difference approximation of the second derivatives are used. Then the discretized given by

$$x = T(x) \tag{2.7.10}$$

with

$$T(x) = (B + I)(x) = h^2 \varphi(x) - b, \quad x \in \mathbb{R}^{2N}$$

where

$$B = \begin{bmatrix} A+I & 0 \\ 0 & A+I \end{bmatrix}, \quad A + \begin{bmatrix} 2 & -1 & 0 & & \\ -1 & 2 & \ddots & & \\ & & \ddots & -1 \\ 0 & -1 & & 2 \end{bmatrix},$$

$$\varphi(x) = \begin{bmatrix} \varphi_1(x) \\ \varphi_2(x) \end{bmatrix}, \quad \varphi_i(x) = (f_i(t, x_j, x_{n+j})), \quad j = 1, 2, ..., N, \quad i = 1, 2,$$

$x \in \mathbb{R}^{2N}$, and $b \in \mathbb{R}^{2N}$ is the vector of boundary values that has zero components except for $b_1 = d_1$, $b_n = e_1$, $b_{n+1} = d_2$, $b_{2n} = e_2$. That is (2.7.10) plays the role of (2.7.1) (in vector form).

As a numerical example, consider the problem (2.7.8)-(2.7.9) with

$$\begin{aligned} f_1(t, q_1, q_2) &= q_1^2 + q_1 + .1q_2^2 - 1.2 \\ f_2(t, q_1, q_2) &= .2q_1^2 + q_2^2 + 2q_2 - .6 \\ d_1 = d_2 &= e_1 = e_2 = 0. \end{aligned}$$

Choose $N = 49$ and starting points

$$x_{i,0} = 0, \quad y_{i,0} = (t_j(1 - t_j), \quad j = 1, ..., N), \text{ with } t = .1\,(.1)\,.5.$$

It is trivial to check that the hypotheses of Theorem 2.6.1 are satisfied with the above values. Furthermore the components of the first two iterates corresponding to the above values using the procedure described above and (2.7.4)-(2.7.5), (2.7.6)-(2.7.7) we get the following values.

The computations were performed in double precision on a PRIME-850.

2.8 Exercises

2.1. Show that the spaces defined in Examples 2.1.1 are POTL.

2.2. Show that any regular POB-space is normal but the converse is not necessarily true.

2.3. Prove Theorem 2.2.2.

	t	$p=1$	$p=2$
$x_{1,p}$	.1	.0478993317265	.0490944353538
	.2	.0843667040291	.0866974354188
	.3	.1099518629493	.1132355483832
	.4	.1251063839064	.1290273001442
	.5	.1301240123325	.1342691068706
$x_{2,p}$	.1	.0219768501208	.0227479238400
	.2	.0384462112803	.0399528292723
	.3	.0498537074028	.0519796383151
	.4	.0565496306877	.0590905187490
	.5	.0587562590344	.0614432572165
$y_{1,p}$	.1	.0494803602542	.0490951403091
	.2	.0874507511044	.0866988216544
	.3	.1142981809478	.1132375255317
	.4	.1302974325097	.1290296859551
	.5	.1356123753407	.1342716394060
$y_{2,p}$	.1	.0235492475283	.0227486289905
	.2	.0415200498433	.0399542200344
	.3	.0541939935471	.0519816281202
	.4	.0617399319012	.0590929252230
	.5	.0642461600398	.0614458137439

2.4. Show that if (2.3.1) or (2.3.2) are satisfied then $F'(x) = [x, x]$ for all $x \in D$. Moreover show that if both (2.3.1) and (2.3.2) are satisfied, then F' is Lipschitz continuous with $I = c_0 + c_1$.

2.5. Find sufficient conditions so that estimates (2.4.9) and (2.4.10) are both satisfied.

2.6. Show that B_n $(n \geq 0)$ in (2.4.14) is a nonnegative subinverse of $F'(y_n)$ $(n \geq 0)$.

2.7. Let $x_0, x_1, ..., x_n$ be distinct real numbers, and let f be a given real-valued function. Show that:

$$[x_0, x_1, ..., x_n] = \sum_{j=0}^{n} \frac{f(x_j)}{g_n'(x_j)}$$

and

$$[x_0, x_1, ..., x_n](x_n - x_0) = [x_1, ..., x_n] - [x_0, ..., x_{n-1}]$$

where

$$g_n(x) = (x - x_0) \ldots (x - x_n).$$

2.8. Let $x_0, x_1, ..., x_n$ be distinct real numbers, and let f be n times continuously differentiable function on the interval $I\{x_0, x_1, ..., x_n\}$. Then show that

$$[x_0, x_1, ..., x_n] = \int_{\tau_n} \cdots \int f^{(n)}(t_0 x_0 + \cdots + t_n x_n)\, dt_1 \cdots dt_n$$

in which

$$\tau_n = \{(t_1, ..., t_n) \,|t_1 \geq 0, ..., t_n \geq 0, \; \sum_{i=1}^{n} ti \leq 1\}$$
$$t_0 = 1 - \sum_{i=1}^{n} t_i.$$

2.9. If f is a real polynomial of degree m, then show:

$$[x_0, x_1, ..., x_n, x] = \begin{cases} \text{polynomial of degree } m-n-1, & n \leq m-1 \\ a_m & n = m-1 \\ 0 & n > m-1 \end{cases}$$

where $f(x) = a_m x^n +$ lower-degree terms.

2.10. The tensor product of two matrices $M, N \in L(\mathbf{R}^n)$ is defined as the $n^2 \times n^2$ matrix $M \times N = (m_{ij}N \mid i, j = 1, ..., n)$, where $M = (m_{ij})$. Consider two F-differentiable operators $H, K : L(\mathbf{R}^n) \to L(\mathbf{R}^n)$ and set $F(X) = H(X)K(X)$ for all $X \in L(\mathbf{R}^n)$. Show that $F'(X) = [H(X) \times I] K'(X) + \left[I \times K(X)^T\right] H'(X)$ for all $X \in L(\mathbf{R}^n)$.

2.11. Let $F : \mathbf{R}^2 \to \mathbf{R}^2$ be defined by $f_1(x) = x_1^3$, $f_2(x) = x_2^2$. Set $x = 0$ and $y = (1,1)^T$. Show that there is no $z \in [x, y]$ such that $F(y) - F(x) = F'(z)(y - x)$.

2.12. Let $F : D \subset \mathbf{R}^n \to \mathbf{R}^m$ and assume that F is continuously differentiable on a convex set $D_0 \subset D$. For and $x, y \in D_0$, show that

$$\|F(y) - F(x) - F'(x)(y - x)\| \leq \|y - x\| \, w(\|y - x\|),$$

where w is the modulus of continuity of F' on $[x, y]$. That is

$$w(t) = \sup\{\|F'(x) - F'(y)\| \mid x, y \in D_0, \|x - y\| \leq t\}.$$

2.13. Let $F : D \subset \mathbf{R}^n \to \mathbf{R}^m$. Show that F'' is continuous at $z \in D$ if and only if all second partial derivatives of the components $f_1, ..., f_m$ of F are continuous at z.

2.14. Let $F : D \subset \mathbf{R}^n \to \mathbf{R}^m$. Show that $F''(z)$ is symmetric if and only if each Hessian matrix $H_1(z), ..., H_m(z)$ is symmetric.

2.15. Let $M \in L(\mathbf{R}^n)$ be symmetric, and define $f : \mathbf{R}^n \to \mathbf{R}$ by $f(x) = x^T M x$. Show, directly from the definition that f is convex if and only if M is positive semidefinite.

2.16. Show that $f : D \subset \mathbf{R}^n \to \mathbf{R}$ is convex on the set D if and only if, for any $x, y \in D$, the function $g : [0,1] \to \mathbf{R}$, $g(t) = g(tx + (1-t)y)$, is convex on $[0,1]$.

2.17. Show that if $g_i : \mathbf{R}^n \to \mathbf{R}$ is convex and $c_i \geq 0$, $i = 1, 2, ..., m$, then $g = \sum_{t=1}^{m} c_i g_i$ is convex.

2.18. Suppose that $g : D \subset \mathbf{R}^n \to \mathbf{R}$ is continuous on a convex set $D_0 \subset D$ and satisfies

$$\frac{1}{2} g(x) + \frac{1}{2} g(y) - g\left(\frac{1}{2}(x+y)\right) \geq \gamma \|x - y\|^2$$

for all $x, y \in D_0$. Show that g is convex on D_0 if $\gamma = 0$.

2.19. Let $M \in L(\mathbf{R}^n)$. Show that M is a nonnegative matrix if and only if it is an isotone operator.

2.20. Let $M \in L(\mathbf{R}^n)$ be diagonal, nonsingular, and nonnegative. Show that $\|x\| = \|D(x)\|$ is a monotonic norm on $\mathbf{R}^n$.

2.21. Let $M \in L(\mathbf{R}^n)$. Show that M is invertible and $M^{-1} \geq 0$ if and only if there exist nonsingular, nonnegative matrices $M_1, M_2 \in L(\mathbf{R}^n)$ such that $M_1 M M_2 = 1$.

2.22. Let $[\cdot,\cdot] : D \times D$ be an operator satisfying conditions (2.2.1) and (2.5.3). The following two assertions are equivalent:

(a) Equality (2.5.7) holds for all $x, y \in D$.

(b) For all points $u, v \in D$ such that $2v - u \in D$ we have

$$[u, v] = 2[u, 2v - u] - [v, 2v - u].$$

2.23. If $\delta\ F$ is a consistent approximation F' on D show that each of the following four expressions in an estimate for

$$\begin{aligned}
&\|F(x) - F(y) - \delta F(u,v)(x - y)\| \\
&c_1 = h(\|x - u\| + \|y - u\| + \|u - v\|)\|x - y\|, \\
&c_2 = h(\|x - v\| + \|y - v\| + \|u - v\|)\|x - y\|, \\
&c_3 = h(\|x - y\| + \|y - u\| + \|y - v\|)\|x - y\|,
\end{aligned}$$

and

$$c_4 = h(\|x - y\| + \|x - u\| + \|x - v\|)\|x - y\|.$$

2.24. Show that the integral representation of $[x_0, ..., x_k]$ is indeed a divided difference of k-th order of F. Let us assume that all divided differences have such an integral representation. In this case for $x_0 = x_1, = ... = x_k = x$ we shall have

$$\underbrace{[x, x, ..., x]}_{k+1 \text{ times}} = \frac{1}{k} f^{(k)}(x).$$

Suppose now that the n-th Fréchet-derivative of F is Lipschitz continuous on D, i.e. there exists a constant c_{n+1} such that

$$\left\| F^{(n)}(u) - F^{(n)}(v) \right\| \leq c_{n+1} \|u - v\|$$

for all $u, v \in D$. In this case, set

$$R_n(y) = ([x_0, ..., x_{n-1}, y] - [x_0, ..., x_{n-1}, x_n])(y - x_{n-1}), ..., (y - x_0)$$

and show that

$$\|R_n(y)\| \leq \frac{c_{n+1}}{(n+1)!} \|y - x_n\| \cdot \|y - x_{n-1}\| \cdots \|y - x_0\|$$

and

$$\left\| F(x+h) - \left(F(x) + F'(x)h + \frac{1}{2}F''(x)h^2 + \cdots + \frac{1}{n!}F^{(n)}(x)h^n \right) \right\|$$
$$\leq \frac{c_{n+1}}{(n+1)!} \|h\|^{n+1}.$$

2.25. We recall the definitions:

(a) An operator $F : D \subset \mathbf{R}^n \to \mathbf{R}^m$ is Gateaux - (or G -) differentiable at an interior point x of D if there exists a linear operator $L \in L(\mathbf{R}^n, \mathbf{R}^m)$ such that, for any $h \in \mathbf{R}^n$

$$\lim_{t \to 0} \frac{1}{t} \|F(x+th) - F(x) - tL(h)\| = 0.$$

L is denoted by $F'(x)$ and called the G-derivative of F at x.

(b) An operator $F : D \subset \mathbf{R}^n \to \mathbf{R}^m$ is hemicontinuous at $x \in D$ if, for any $h \in \mathbf{R}^n$ and $\varepsilon > 0$, there is a $\delta = \delta(\varepsilon, h)$ so that whenever $|t| < \delta$ and $x + th \in D$, then $\|F(x+th) - F(x)\| < \varepsilon$.

(c) If $F : D \subset \mathbf{R}^n \to \mathbf{R}^m$ and if for some interior point x of D, and $h \in \mathbf{R}^n$, the limit

$$\lim_{t \to 0} \frac{1}{t} [F(x+th) - F(x)] = A(x, h)$$

exists, then F is said to have a Gateaux-differential at x in the direction h.

(d) If the G-differential exists at x for all h and if, in addition

$$\lim_{h\to 0}\frac{1}{\|h\|}\|f(x+H)-F(x)-A(x,h)\|=0,$$

then F has a Fréchet differential at x.
Show:

(i) The linear operator L is unique;

(ii) If $F : D \subset \mathbf{R}^n \to \mathbf{R}^m$ is G-differentiable at $x \in D$, then F is hemicontinuous at x.

(iii) G-differential and "uniform in h" implies F-differential;

(iv) F-differential and "linear in h" implies F-derivative;

(v) G-differential and "linear in h" implies G-derivative;

(vi) G-derivative and "uniform in h" implies F-derivative;
Here "uniform in h" indicated the validity of (d).
Linear in h means that $A(x,h)$ exists for all $h \in \mathbf{R}^n$ and

$$A(x,h)=M(x)h, \quad \text{where} \quad M(x)\in L(\mathbf{R}^n,\mathbf{R}^m).$$

Define $F : \mathbf{R}^2 \to \mathbf{R}$ by $F(x) = sgn(x_2)\min(|x_1|,|x_2|)$.
Show that, for any $h \in \mathbf{R}^2$ $A(0,h) = F(h)$, bu F does not have a G-derivative at 0.

(viii) Define $F : \mathbf{R}^2 \to \mathbf{R}$ by $F(0) = 0$ if $x = 0$ and

$$F(x)=x_2\left(x_1^2+x_2^2\right)^{\frac{3}{2}}\Big/\left[\left(x_1^2+x_2^2\right)^2+x_2^2\right], \quad \text{if} \quad x\neq 0.$$

Show that F has a G-derivative at 0, but not an F-derivative. Show, moreover, that the G-derivative is hemicontinuous at 0.

(ix) If the g-differential $A(x,h)$ exists for all x in an open neighborhood of an interior point x_0 of D and for all $h \in \mathbf{R}^n$, then F has an F-derivative at x_0 provided that for each fixed h, $A(x,h)$ is continuous in x at x_0.

(e) Assume that $F : D \subset \mathbf{R}^n \to \mathbf{R}^m$ has a G-derivative at each point of an open set $D_0 \subset D$. If the operator $F' : D_0 \subset \mathbf{R}^n \to L(\mathbf{R}^n,\mathbf{R}^m)$ has a G-derivative at $x \in D_0$, then $(F')'(x)$ is denoted by $F''(x)$ and called the second G-derivative of F at x.
Show:

(i) If $F : \mathbf{R}^n \to \mathbf{R}^m$ has a G-derivative at each point of an open neighborhood of x, then F' is continuous at x if and only if all partial derivatives $\partial_i F_i$ are continuous at x.

(ii) F'' is continuous at $x_0 \in D$ if and only if all second partial derivatives of the components $f_1, ..., f_m$ of F are continuous at x_0. $F''(x_0)$ is symmetric if and only if each Hessian matrix $H_1(x_0), ..., H_m(x_0)$ is symmetric.

Chapter 3

Contractive Fixed Point Theory

The ideas of a contraction operator and its fixed points are fundamental to many questions in applied mathematics. In Section 1 we outline the essential ideas, whereas in Section 2 some examples are provided. An extension of the contraction mapping theorem with an application to a boundary value problem is provided in Section 3. Stationary and nonstationary projection methods are studied in Section 5. A special iteration method is analysed in Section 5 whose convergence results improves earlier ones. Finally, in Section 6 the stability of the Ishikawa iteration is discussed. Further results on fixed points are given in Chapter 12.

3.1 Fixed Points

Definition 3.1.1 *Let F be an operator mapping a set X into itself. A point $x \in X$ is called a fixed point of F if*

$$x = F(x). \tag{3.1.1}$$

Equation (3.1.1) leads naturally to the construction of the method of successive approximations or substitutions

$$x_{n+1} = F\left(x_n\right) \quad (n \geq 0) \quad x_0 \in X. \tag{3.1.2}$$

If sequence $\{x^n\}\,(n \geq 0)$ converges to some point $x^* \in X$ for some initial guess $x_0 \in X$, and F is a continuous operator in a Banach space X, we can have

$$x^* = \lim_{n\to\infty} x_{n+1} = \lim_{n\to\infty} F\left(x_n\right) = F\left(\lim_{n\to\infty} x_n\right) = F\left(x^*\right).$$

That is, x^* is a fixed point of operator F. Hence, we showed:

Theorem 3.1.2 *If F is a continuous operator in a Banach space X, and the sequence $\{x_n\}\,(n \geq 0)$ generated by (3.1.2) converges to some point $x^* \in X$ for some initial guess $x_0 \in X$, then x^* is a fixed point of operator F.*

We need information on the uniqueness of x^* and the distances $\|x_n - x^*\|$, $\|x_{n+1} - x_n\|$ $(n \geq 0)$. That is why we introduce the concept:

Definition 3.1.3 *Let* $(X, \| \, \|)$ *be a metric space and* F *a mapping of* X *into itself. The operator* F *is said to be a contraction or a contraction mapping if there exists a real number* c, $0 \leq c < 1$, *such that*

$$\|F(x) - F(y)\| \leq c\|x - y\|, \quad \text{for all } x, y \in X. \tag{3.1.3}$$

It follows immediately from (3.1.3) that every contraction mapping F is uniformly continuous. Indeed, F is Lipschitz continuous with a Lipschitz constant c. The point c is called the contraction constant for F.

We now arrive at the Banach contraction mapping theorem [207].

Theorem 3.1.4 *Let* $(X, \|\cdot\|)$ *be a Banach space, and* $F : X \to X$ *be a contraction mapping. Then* F *has a unique fixed point.*

Proof. *Uniqueness*: Suppose there are two fixed points x, y of F. Since F is a contraction mapping,

$$\|x - y\| = \|F(x) - F(y)\| \leq c\|x - y\| < \|x - y\|,$$

which is impossible. That shows uniqueness of the fixed point of F. *Existence*: Using (3.1.2) we obtain

$$\begin{aligned} &\|x_2 - x_1\| \leq c\|x_1 - x_0\| \\ &\|x_3 - x_2\| \leq x\|x_2 - x_1\| \leq c^2\|x_1 - x_0\| \\ &\qquad\vdots \\ &\|x_{n+1} - x_n\| \leq c^n\|x_1 - x_0\| \end{aligned} \tag{3.1.4}$$

and, so

$$\begin{aligned} \|x_{n+m} - x_n\| &\leq \|x_{n+m} - x_{n+m-1}\| + \cdots + \|x_{n+1} - x_n\| \\ &\leq \left(c^{m+1} + \cdots + c + 1\right) c^n \|x_1 - x_0\| \\ &\leq \frac{c^n}{1-c}\|x_1 - x_0\|. \end{aligned}$$

Hence, sequence $\{x_n\}(n \geq 0)$ is Cauchy in a Banach space X and such it converges to some x^*. The rest of the theorem follows from Theorem 3.1.2. ■

Remark 3.1.1 *It follows from* (3.1.1), (3.1.3) *and* $x^* = F(x^*)$ *that*

$$\|x_n - x^*\| = \|F(x_{n-1}) - F(x^*)\| \leq c\|x_{n-1} - x^*\| \leq c^n\|x_0 - x^*\| \quad (n \geq 1). \tag{3.1.5}$$

Inequality (3.1.5) *describes the convergence rate. This is convenient only when an a priori estimate for* $x_0 - x^*$ *is available. Such an estimate can be derived from the inequality*

$$\|x_0 - x^*\| \leq \|x_0 - F(x_0)\| + \|F(x_0) - F(x^*)\| \leq \|x_0 - F(x_0)\| + c\|x_0 - x^*\|,$$

which leads to

$$\|x_0 - x^*\| \leq \frac{1}{1-c}\|x_0 - F(x_0)\|. \tag{3.1.6}$$

By (3.1.5) *and* (3.1.6), *we obtain*

$$\|x_n - x^*\| \leq \frac{c^n}{1-c}\|x_0 - F(x_0)\| \quad (n \geq 1). \tag{3.1.7}$$

Estimates (3.1.5) *and* (3.1.7) *can be used to determine the number of steps needed to solve Equation* (3.1.1). *For example if the error tolerance is* $\varepsilon > 0$, *that is, we use* $\|x_n - x^*\| < \varepsilon$, *then this will certainly hold if*

$$n > \frac{1}{\ln c} \ln \frac{\varepsilon(1-c)}{\|x_0 - F(x_0)\|}. \tag{3.1.8}$$

3.2 Applications

These are examples of operators with fixed points that are not contraction mappings:

Example 3.2.1 *Let* $F : \mathbb{R} \to \mathbb{R}$, $q > 1$, $F(x) = qx + 1$. *Operator* F *is not a contraction, but it has a unique fixed point* $x = (1-q)^{-1}$.

Example 3.2.2 *Let* $F : X \to X$, $x = (0, \frac{1}{\sqrt{6}}]$, $F(x) = x^3$. *We have*

$$|F(x) - F(y)| = |x^3 - y^3| \leq \left(|x|^2 + |x| \cdot |y| + |y|^2\right)|x - y| \leq \tfrac{1}{2}|x - y|$$

That is, F *is a contraction with* $c = \frac{1}{2}$, *with no fixed points in* X. *This is not violating Theorem 3.1.4 since* X *is not a Banach space.*

Example 3.2.3 *Let* $F : [a, b] \to [a, b]$, F *differentiable at every* $x \in (a, b)$ *and* $|F'(x)| \leq c < 1$. *By the mean value theorem, if* $x, y \in [a, b]$ *there exists a point* z *between* x *and* y *such that*

$$F(x) - F(y) = F'(z)(z - y),$$

from which it follows that F *is a contraction with constant c.*

Example 3.2.4 *Let $F : [a,b] \to \mathbb{R}$. Assume there exist constants p_1, p_2 such that $p_1 p_2 < 0$ and $0 < p_1 < F'(x) \le p_2^{-1}$ and assume that $F(a) < 0 < F(b)$. How can we find the zero of $F(x)$ guaranteed to exist by the intermediate value theorem? Define a function P by*

$$P(x) = x - p_2 F(x).$$

Using the hypotheses, we obtain $P(a) = a - p_2 F(a) > a$, $P(b) = b - F(b) < b$, $P'(x) = 1 - p_2 F'(x) \ge 0$, and $P'(x) \le 1 - p_1 p_2 < 1$. Hence P maps $[a,b]$ into itself and $|p'(x)| \le 1$ for all $x \in [a,b]$. By Example 3.2.3, $P(x)$ is a contraction mapping. Hence, P has a fixed point which is clearly a zero of F.

Example 3.2.5 *(Fredholm integral equation). Let $K(x,y)$ be a continuous function on $[a,b] \times [a,b]$, $f_0(x)$ be continuous on $[a,b]$ and consider the equation*

$$f(x) = f_0(x) + \lambda \int_a^b K(x,y) f(y)\, dy.$$

Define the operator $P : C[a,b] \to C[a,b]$ by $p(f) = g$ given by

$$g(x) = f_0(x) + \lambda \int_a^b K(x,y) f(y)\, dy.$$

Note that a fixed point of P is a solution of the integral equation. We get

$$\begin{aligned}
\|P(q_1) - P(q_2)\| &= \sup_{x\in[a,b]} |p(q_1(x)) - P(q_2(x))| \\
&= |\lambda| \sup_{x\in[a,b]} \left| \int_a^b k(x,y)(q_1(y) - q_2(y))\, dy \right| \\
&\le |\lambda|\, \delta \left| \int_a^b (q_1(y) - q_2(y))\, dy \right| \quad \text{(by Weierstrass theorem)} \\
&\le |\lambda|\, \delta\, (b-a) \sup_{x\in[a,b]} |q_1(y) - q_2(y)| \\
&\le |\lambda|\, \delta\, (b-a) \|q_1 - q_2\|.
\end{aligned}$$

Hence, P is a contraction mapping if there exists a $c > 1$ such that

$$|\lambda|\, \delta\, (b-a) \le c.$$

3.3 Extending the contraction mapping theorem

An extension of the contraction mapping theorem is provided in a Banach space setting. Our approach is justified by numerical examples where our results apply whereas the contraction mapping theorem 3.1.4 cannot.

We provide the following fixed point theorem:

Theorem 3.3.1 *Let $Q\colon D \subseteq X \to X$ be an operator, p, q scalars with $p < 1$, $q \geq 0$ and x_0 a point in D such that:*

$$\|Q(x) - Q(y)\| \leq q\|x - y\|^p \quad \text{for all } x, y \in D. \tag{3.3.1}$$

Set $a = q^{\frac{1}{1-p}}$ and $b = q^{\frac{1}{p}}\|Q(x_0) - x_0\|$. Moreover assume:

$$0 \leq b < 1 \tag{3.3.2}$$

and

$$U(x_0, r) \subseteq D,$$

where,

$$r \geq \|Q(x_0) - x_0\| + a\big[b^p + \cdots + b^{p^n}\big]. \tag{3.3.3}$$

Then sequence $\{x_n\}$ $(n \geq 0)$ generated by successive approximations

$$x_{n+1} = Q(x_n) \quad (n \geq 0) \tag{3.3.4}$$

remains in $U(x_0, r)$ for all $n \geq 0$ and converges to a fixed point $x^* \in U(x_0, r)$, so that for all $n \geq 1$:

$$\|x_{n+1} - x_n\| \leq ab^{p^n} \leq ab^{np} \tag{3.3.5}$$

and

$$\begin{aligned} \|x_n - x^*\| &\leq a \lim_{m\to\infty} \big[b^{p^n} + \cdots + b^{p^{n+m-1}}\big] = c_n \\ &\leq a \lim_{m\to\infty} \big[b^{np} + \cdots + b^{(n+m-1)p}\big] \\ &\leq \frac{ab^{np}}{1 - b^p}. \end{aligned} \tag{3.3.6}$$

Moreover if

$$r < q^{-1/p} \tag{3.3.7}$$

then x^* is the unique fixed point of Q on $U(x_0, r)$.

Proof. By hypothesis (3.3.3) $Q(x_0) = x_1 \in U(x_0, r)$. Let us assume $x_k \in U(x_0, r)$ for $k = 0, 1, \ldots, n$. Then by (3.3.1), (3.3.2) and (3.3.4) we obtain in turn

$$\begin{aligned} \|x_{n+1} - x_n\| &= \|Q(x_n) - Q(x_{n-1})\| \leq q\|x_n - x_{n-1}\|^p \\ &\leq q(q\|x_{n-1} - x_{n-2}\|^p)^p \\ &\leq \cdots \leq q^{1+p+p^2+\cdots+p^{n-1}}\|x_1 - x_0\|^{p^n} \\ &\leq q^{\frac{p^n-1}{p-1}}\|x_1 - x_0\|^p \leq ab^{p^n} \leq ab^{np}, \end{aligned}$$

which shows (3.3.5). Using (3.3.3) and (3.3.5) we get

$$\begin{aligned}\|x_{n+1}-x_0\| &\le \|x_{n+1}-x_n\|+\|x_n-x_{n-1}\|+\cdots+\|x_2-x_1\|+\|x_1-x_0\|\\ &\le a\left[b^{p^n}+\cdots+b^p\right]+\|x_1-x_0\|\le r,\end{aligned} \tag{3.3.8}$$

which shows $x_{n+1}\in U(x_0,r)$. Moreover for all $m\ge 0$ we can have

$$\begin{aligned}&\|x_{n+m}-x_n\|\le\\ &\le \|x_{n+m}-x_{n+m-1}\|+\|x_{n+m-1}-x_{n+m-2}\|+\cdots+\|x_{n+1}-x_n\|\\ &\le a\left[b^{p^{n+m-1}}+\cdots+b^{p^n}\right]\\ &\le a\left[b^{(n+m-1)p}+\cdots+b^{np}\right]\\ &= ab^{np}\frac{1-b^{mp}}{1-b^p},\end{aligned} \tag{3.3.9}$$

which shows that $\{x_n\}$ is a Cauchy sequence in a Banach space X and as such it converges to some $x^*\in U(x_0,r)$ (since $U(x_0,r)$ is a closed set). By letting $m\to\infty$ in (3.3.9) we obtain (3.3.6). Furthermore by letting $n\to\infty$ in (4) we obtain $x^*=Q(x^*)$ (since Q is continuous by (3.3.1)). Finally to show uniqueness, let $y^*\in U(x_0,r)$ be a fixed point of Q. Then we can get as above

$$\begin{aligned}\|x_{n+1}-y^*\| &= \|Q(x_n)-Q(y^*)\|\le q\|x_n-y^*\|^p\\ &\le q(q\|x_{n-1}-y^*\|^p)^p\\ &\le\cdots\le ab_1^{p^n},\end{aligned} \tag{3.3.10}$$

where,

$$b_1=q^{\frac{1}{p}r}.$$

It follows by (3.3.7) and (3.3.10) that $\lim_{n\to\infty}x_n=y^*$. Hence we deduce

$$x^*=y^*.$$

■

Remark 3.3.1 *(a) Case $p>1$. Under hypotheses (3.3.1), (3.3.2) and (3.3.3) the proof of the theorem goes through since $\left[b^{p^n}+\cdots+b^{p^{n+m-1}}\right]$ is still a Cauchy sequence but estimate (3.3.6) only holds as*

$$\|x_n-x^*\|\le c_n. \tag{3.3.11}$$

(b) Case $p=1$. We simply have the contraction mapping theorem 3.1.4.

Next we provide a numerical example to show that our Theorem can apply in cases where $p\ne 1$. That is in cases where the contraction mapping principle cannot apply.

Example 3.3.1 *Let $X = \mathbf{R}^{m-1}$ with $m \geq 2$ an integer. Define operator Q on a subset D of X by*

$$Q(w) = H + h^2(p+1)M \tag{3.3.12}$$

where, h is a real parameter, $p_1 \geq 0$,

$$H = \begin{bmatrix} 2 & -1 & 0 & \cdots & 0 \\ -1 & 2 & -1 & & 0 \\ 0 & -1 & 2 & & -1 \\ 0 & \cdots & -1 & & 2 \end{bmatrix},$$

$$M = \begin{bmatrix} w_1^p & & \cdots & 0 \\ 0 & w_2^p & \cdots & 0 \\ 0 & \cdots & \cdots & w_{m-1}^p \end{bmatrix},$$

and $w = (w_1, w_2, \ldots, w_{m-1})$. Finding fixed points of operator Q is very important in many discretization studies since e.g. many two point boundary value problems can be reduced to finding such points [68], [99]. Let $w \in \mathbf{R}^{m-1}$, $M \in \mathbf{R}^{m-1} \times \mathbf{R}^{m-1}$, and define the norms of w and M by

$$\|w\| = \max_{1 \leq j \leq m-1} |w_j|$$

and

$$\|M\| = \max_{1 \leq j \leq m-1} \sum_{k=1}^{m-1} |m_{jk}|.$$

For all $w, z \in \mathbf{R}^{m-1}$ for which $|w_j| > 0$, $|z_j| > 0$, $j = 1, 2, \ldots, m-1$ we obtain, for say $p = \frac{1}{2}$

$$\begin{aligned} \|Q(w) - Q(z)\| &= \left\| diag\left\{ \frac{3}{2} h^2 \left(w_j^{1/2} - z_j^{1/2} \right) \right\} \right\| \\ &= \frac{3}{2} h^2 \max_{1 \leq j \leq m-1} \left| w_j^{1/2} - z_j^{1/2} \right| \\ &\leq \frac{3}{2} h^2 [\max |w_j - z_j|]^{1/2} \\ &= \frac{3}{2} h^2 \|w - z\|^{1/2}. \end{aligned} \tag{3.3.13}$$

Set $q = \frac{3}{2}h^2$. Then $a = q^2$ and $b = q^2\|Q(x_0) - x_0\|$. It follows by (17) that the contraction mapping principle cannot be used to obtain fixed points of operator Q since $p = \frac{1}{2} \neq 1$. However our theorem can apply provided that $x_0 \in D$ is such that

$$\|Q(x_0) - x_0\| \leq q^{-2}. \tag{3.3.14}$$

That is (2) is satisfied. Furthermore r should be chosen to satisfy (3) and we can certainly set $U(x_0, r) = D$.

3.4 Stationary and nonstationary projection-iteration methods

The aim of this work is to study the existence of the solutions of the equation

$$x = A(x, x) \tag{3.4.1}$$

and their approximations for nonlinear operator A.

It is clear that equation (3.4.1) contains the equation $x = Ax$ as a special case and the approximations

$$x_n = A_n(x_n, x_{n-1}), \quad n = 1, 2, ... \tag{3.4.2}$$

generalize a lot of approximate methods such as the Picard, Seidel and Newton-Kantorovich methods and so on.

In the sequel we study the strong and weak convergence of this type of projection-iteration method, stationary as well as non stationary for equation (3.4.1). A new generalized fixed point Edelstein's theorem [149], [150] will be stated and some examples will be shown.

Theorem 3.4.1 (Stationary) Let X *be a topological vector space,* D *a compact set in* X, A *a continuous mapping from* $D \times D$ *into* D, *and* $F = \{x \in D : A(x, x) = x\}$ *Let* $g : D \times F \to [0, \infty)$. *Assume 1.* $F \neq \emptyset$ *and* $g = 0$, $x \in D$, $p \in F \Leftrightarrow x = p \in F$; *2.* $g(A(x, y), p) < \max\{g(x, p), g(y, p)\}$, $\forall x, y \in D$, $p \in F$ *such that there are no such relations* $x = y = p$; *3. the function* $g(x, p)$ *is continuous in* x; *4. the system*

$$\begin{cases} u_n = A(u_n, v_n), \\ v_n = A_k(u_n, u_{n-1}) \end{cases} \tag{3.4.3}$$

is solvable for arbitrary element $u_0 \in D$ *anf for all* $n \geq 1$, *where*

$$A_1(x, y) = A(x, y), \quad A_i(x, y) = A(x, A_{i-1}(x, y)), \quad i = 2, 3, ..., k.$$

Then the sequences $\{u_n\}$, $\{v_n\}$ *constructed by (3.4.3) converge (in the topology of* X*) to the unique solution of equation (3.4.1).*

Proof. By assumption 2/ we have

$$g(u_n, p) < \max(g(u_n, p), g(v_n, p)), \quad \text{i.e.} \quad g(u_n, p) < g(v_n, p).$$

However

$$g(v_n, p) = g(A_k(u_n, u_{n-1}), p) < \max(g(u_n, p), g(A_{k-1}(u_n, u_{n-1}), p)) \leq ... \leq \\ \leq \max(g(u_n, p), g(u_{n-1}, p)),$$

hence

$$g(u_n, p) < g(v_n, p) < g(u_{n-1}, p).$$

Therefore $\lim g(u_n, p) = \lim g(v_n, p)$. By the compactness of D we get the subsequences $\{u_{n_j}\} \subset \{u_n\}$, $\{v_{n_j}\} \subset \{v_n\}$ such that $u_{n_j} \to u$, $v_{n_j} \to v$. Since g is continuous $g(u, p) = g(v, p)$. On the other hand

$$A(u_{n_j}, v_{n_j}) \to A(u, v)$$

and

$$g(u, p) = g(A(u, v), p) < \max\{g(u, p), g(v, p)\}.$$

This is a contradiction. So $u = v = p$. Note that F contains only one element. Indeed if $p \neq q$, $p, q \in F$, then

$$g(p, q) = g(A(p, q), p) < \max g(p, q).$$

This is impossible. Consequently the sequences $\{u_n, v_n\}$ converge to the unique solution p of equation (3.4.1). ■

Remark 3.4.1 *Instead of condition 3/ we can use the condition R: from $\lim g(u_n, p) = \lim g(v_n, p)$, $p \in F$, and v is a limit of the sequence $\{v_n\}$ it follows that either $g(u, p) \leq g(v, p)$ or $g(u, p) \geq g(v, p)$ for all limits u of the sequence $\{u_n\}$.*

To verify this Remark it suffices to use conditions 2/, 4/.

Indeed it is clear that from the sequences $\{u_{n_j}\}$, $\{v_{n_j}\}$, we can choose the subseqeunces $\{u_{n_{j_k}-1}\}$ such that $u_{n_{j_k}-1} \to \bar{u}$. Then we have

$$\begin{aligned} u_{n_{j_k}} &= A(u_{n_{j_k}}, v_{n_{j_k}}), \quad u_{n_{j_k}} \to u = A(u, v) \\ v_{n_{j_k}} &= A(u_{n_{j_k}}, u_{n_{j_k}-1}), \quad v_{n_{j_k}} \to v = A(u, \bar{u}). \end{aligned}$$

Hence

$$\begin{aligned} &g(u, p) < \max\{g(u, p), g(v, p)\} \text{ implies } g(u, p) < g(v, p) \\ &g(v, p) < \max\{g(u, p), g(\bar{u}, p)\} \text{ implies } g(v, p) < g(\bar{u}, p), \end{aligned}$$

if either $u = v = p$ or $u = \bar{u} = p$ do not hold. However the last inequalities contradict condition R). Therefore $u = v = p$ or $u = \bar{u} = p$, i.e. $v = A_k(u, u) = A_k(p, p) = p$.

Lemma 3.4.2 *[332] Let M be a bounded weakly closet subset in a reflexive Banach space, and mf a lower weakly semicontinuous functional on M. Then f is lower bounded and attains its infinimum on M.*

Theorem 3.4.3 *Let X, D, A, F, g be as in Theorem 3.4.1. Assume that conditions 1/, 3/, 4/ of Theorem 3.4.1 hold and instead of condition 2/ of this theorem the following conditions are satisfied: 5/*

$$g(A(x, y), p) \begin{cases} < \max(g(x, p), g(y, p)), & \textit{if } g(x, p) \neq g(y, p), \\ \leq \max(g(x, p), g(y, p)), & \textit{if } g(x, p) = g(y, p), \end{cases}$$

$\forall x, y \in D$, $p \in F$; 6/ $\forall y \notin F$, $x \in D$ satisfying $x = A(x,y)$, $\exists q = q(x,y) \in F$ such that $g(A(x,y),p) < \max(g(x,q), g(y,q))$. Then subsequence $\{u_{n_j}, v_{n_j}\}$ of the sequence $\{u_n, v_n\}$ constructed by (3.4.3) must converge simultaneously to the solution p of equation (3.4.1).

Proof. As in the proof of Theorem 3.4.1 we have

$$\lim g(u_n, p) = \lim g(v_n, p). \tag{3.4.4}$$

Now from $\{u_n\}, \{v_n\}$ we take $\{u_{n_j}\}, \{v_{n_j}\}$ such that $u_{n_{j_k}} \to u$, $v_{n_{j_k}} \to v$. It will be shown that $v \in F$. If it is not true, i.e. $v \notin F$, then $\exists q = q(u,v) \in F$, such that

$$g(u,p) = g(A(u,v),q) < \max(g(u,p), g(v,p))$$

i.e.

$$g(u,p) < g(v,p). \tag{3.4.5}$$

Since the function g is continuous $g(u,p) = g(v,p)$ by (3.4.4). But this contradicts (3.4.5). Hence $v \in F$. Moreover, if $g(u,v) \neq g(v,v) = 0$ then

$$g(u,p) = g(A(u,v),v) < \max(g(u,v), g(v,v)),$$

this is impossible. So $g(u,v) = 0$, i.e. $u = v \in F$. ■

Lemma 3.4.4 *Let X be a strictly convex Banach space, D a convex subset in X, $A : D \times D \to D$ a continuous mapping satisfying the condition*

$$\|A(x,y) - p\| \leq \max(\|x - p\|, \|y - p\|),$$

$\forall x, y \in D$, $p \in F = \{x \in D : A(x,x) = x\}$. If $F \neq \emptyset$, then F is convex.

The proof of this Lemma is similar to the one in a work by W.V. Petryshyn and J.R. Williamson (see [273]).

Lemma 3.4.5 *[332]. In a Banach space if the functional f is quasiconvex, i.e. for any $t \in (0,1)$*

$$f(tx - (1-t)y) \leq \max(f(x), f(y))$$

and lower semicontinuous, then f is lower weakly semicontinuous.

From now on a mapping V is said to be quasiconvex if

$$\left\| V(\frac{x_1 + x_2}{2}) \right\| \leq \max(\|V(x_1)\|, \|V(x_2)\|).$$

Theorem 3.4.6 *Let X be a strictly convex and reflexive Banach space, D a convex closed subset in X, A a continuous mapping from $D \times D$ into D and 7/ $F \neq \emptyset$; 8/*

$$\|A(x,y)-p\| \begin{cases} \leq \max(\|x-p\|, \|y-p\|) & \text{if } \|x-p\| \neq \|y-p\|, \\ \leq \max(\|x-p\|, \|y-p\|) & \text{if } \|x-p\| = \|y-p\|; \end{cases}$$

9/ the system (3.4.3) is solvable for all n and $\|u_n - v_n\| \to 0$ as $n \to \infty$; 10/ either the mapping $I_x - A$, where $(I_x - A)(x,y) = x - A(x,y)$, is quasiconvex ot the mapping $I_y - A$, where $(I_y - A)(x,y) = y - A(x,y)$, is quasiconvex; 11/ thespace mX has the property: if $\{y_n\} \subset X$, $y_n \stackrel{W}{\to} y_0 \in X$, then the $\underline{\lim}\|y_n - y\| > \underline{\lim}\|y_n - y_0\|$, $\forall y \neq y_0$, $y \in X$. Then the sequences $\{u_n\}$, $\{v_n\}$ defined by (3.4.3) weakly converge simoultaneously to the solution $p \in F$.

Proof. Similarly to the proof of Theorem 3.4.1 we obtain:

$$\|u_n - p\| \leq \|v_n - p\| \leq \|u_{n-1} - p\|.$$

Hence

$$\lim \|u_n - p\| = \lim \|v_n - p\|.$$

Consequently $\{u_n\}$, $\{v_n\} \subset B(p, \|u_0 - p\|)$, where mB is a ball with center at $p \in F$. Therefore from $\{u_n\}$, $\{v_n\}$ we can choose weakly convergent subsequences $\{u_{n_j}\}$, $\{v_{n_j}\}$ such that

$$u_{n_j} \stackrel{W}{\to} u, \quad v_{n_j} \stackrel{W}{\to} v.$$

However,

$$\|u - v\| \leq \underline{\lim} \left\|u_{n_j} - v_{n_j}\right\| = \lim \|u_n - v_n\| = 0.$$

Hence $u = v$. By condition 10/, if the mapping $I_y - a$ is quasiconvex (and continuous, of course) then it is weakly lower semicontinuous, we thus have

$$\|v - A(u,v)\| \leq \underline{\lim} \left\|u_{n_j} - A(u_{n_j}, v_{n_j})\right\| = 0.$$

Hence $u = A(u,v)$, i.e. $u = v = A(u,u) = A(v,v)$. Therefore for both cases the subsequences $\{u_{n_j}\}$, $\{v_{n_j}\}$ weakly converge to the solution $p \in F$. Now let $f(p) = \lim \|u_n - p\|$. This function is defined on the set $F_0 = F \cap B(p \, \|u_0 - p\|)$ which is convex closed bounded by the convexity and closedness of F. and hence it is weakly compact. On the other hand it is easy to see that $f(p)$ is convex and continuous and thus $f(p)$ is lower weakly semicontinuous. Therefore $f(p)$ attains on F_0 its infinimum, say, at a point p^{**}. It will be shown that p is a unique limit point of both sequences $\{u_n\}$, $\{v_n\}$. Indeed if there exists a subsequence $\{u_{n_k}\}$ weakly converging to $u \neq p$ then by condition 11/ we have

$$\underline{\lim} \|u_{n_k} - p\| > \underline{\lim} \|u_{n_k} - u\|.$$

However

$$\underline{\lim} \|u_{n_k} - p\| = \underline{\lim} \|u_n - p\| = f(p),$$

and

$$\underline{\lim} \|u_{n_k} - u\| = \underline{\lim} \|u_n - u\| = f(u).$$

That is $f(p) > f(u)$, which is a contradiction. ■

Theorem 3.4.7 *Let X be a uniformly convex Banach space, D a convex subset in X and $A : D \times D \to D$ a continuous mapping satisfying conditions 7/, 8/ and a/ assume that for a point $c \in [0,1)$ the mapping $J_c - A$, where $J_c(x,y) = cx + (1-c)y$ is quasiconvex on $D \times D$; then for any $t \in (0,1)$ the sequences*

$$\begin{aligned} u_n &= ((tJ_c - (I-t)A)(u_n, v_n), \\ v_n &= ((tJ_c - (I-t)A)_k(u_n, u_{n-1}), \end{aligned}$$

weakly converge simultaneously to the solution $p \in F$; b/ when the mapping $I_x - A$ is quasiconvex, we construct an algorithm as follows: $\bar{u}_0 = u_0$, then by means of the system

$$\begin{aligned} u_n &= A(u_n, v_n), \quad n \geq 1 \\ v_n &= A_k(u_n, \tilde{u}_{n-1}), \end{aligned} \tag{3.4.6}$$

we obtain (u_1, v_1),then we take $\tilde{u}_1 = tu_1 + (1-t)v_1$, and again using system (3.4.6) we get (u_2, v_2) etc; we claim that such obtained sequences $\{u_n\}, \{v_n\}$ will weakly converge to the solution $p \in F$.

Proof. It suffices to show that the constructed sequences $\{u_n\}, \{v_n\}$ satisfy condition 9/ i.e. $\|u_n - v_n\| \to 0$ as $n \to \infty$. i/ First we study the case $c \neq 1$. We have

$$\begin{aligned} \|u_n - p\| &= \|(tJ_c - (I-t)A)(u_n, v_n) - p\| \\ &\leq t\|J_c(u_n, v_n) - p\| + (1-t)\|A(u_n, v_n) - p\| \\ &\leq t(c\|u_n - p\| + (1-c)\|v_n - p\|) + (1-t)\|A(u_n, v_n) - p\|. \end{aligned}$$

But

$$(1-t)\|A(u_n, v_n) - p\|$$
$$\begin{cases} < (1-t)\max(\|u_n - p\|, \|v_n - p\|), & \text{if } \|u_n - p\| \neq \|v_n - p\| \\ \leq (1-t)\max(\|u_n - p\|, \|v_n - p\|), & \text{if } \|u_n - p\| = \|v_n - p\|. \end{cases}$$

It follows that

$$\|u_n - p\| \leq \|v_n - p\|.$$

Analogously we obtain

$$\|v_n - p\| = \|(tJ_c + (1-t)A)_k(u_n, u_{n-1}) - p\| \leq \|u_{n-1} - p\| .$$

Therefore there exists d such that

$$d = \lim \|u_n - p\| = \lim \|v_n - p\| .$$

If $d = 0$, then $u_n \to p$, $v_n \to p$ and, of course,

$$u_n \overset{W}{\to} p, \quad v_n \overset{W}{\to} p.$$

If $d \neq 0$, then

$$\frac{\|u_n - p\|}{\|v_n - p\|} \to 1.$$

However

$$\left\| \frac{u_n - p}{\|v_n - p\|} \right\| = \left\| t \left[\frac{c(u_n - p) + (1-c)(v_n - p)}{\|v_n - p\|} \right] + (1-t) \frac{A(u_n, v_n) - p}{\|v_n - p\|} \right\| .$$

Let

$$W_n = \left\| t \left[\frac{c(u_n - p) + (1-c)(v_n - p)}{\|v_n - p\|} \right] + (1-t) \frac{A(u_n, v_n) - p}{\|v_n - p\|} \right\| .$$

Obviously

$$W_n \leq 1, \quad \|Z_n\| \leq 1.$$

Hence from $\|tW_n + (1-t)Z_n\| \to 0$ as $n \to \infty$ and by the uniform convexity of X it follows that $\|W_n - Z_n\| \to 0$ as $n \to \infty$, i.e.

$$\|c(u_n - p) + (1-c)(v_n - p) - (A(u_n, v_n) - p)\| \to 0 \quad \text{as } n \to \infty.$$

Taking into account that

$$\begin{aligned}
&\|cu_n + (1-c)v_n - A(u_n, v_n)\| = \\
&= \frac{1}{1-t} \|(1-t)(cu_n + (1-c)v_n) - (1-t)A(u_n, v_n)\| \\
&= \frac{1}{1-t} \|cu_n + (1-c)v_n - [t(cu_n + (1-c)v_n) + (1-t)A(u_n, v_n)]\| \\
&= \frac{1}{1-t} \|cu_n + (1-c)v_n - u_n\| \\
&= \frac{1}{1-t} \|(1-c)(v_n - u_n)\| ,
\end{aligned}$$

we have $\|v_n - u_n\| \to \infty$ as $n \to \infty$. At the same time it is easily seen

$$(J_c - A)(u_n - v_n) = cu_n + (1-c)v_n - A(u_n, v_n) = \frac{1-c}{1-t}(v_n - u_n).$$

Now as in the proof of Theorem 3.4.1 we obtain

$$u_{n_j} \overset{W}{\to} u, \quad v_{n_j} \overset{W}{\to} v, \quad u = v.$$

By the quasiconvexity of the mapping $J_c - A$ we get

$$\begin{aligned}\|cu + (1-c)v - A(u_n, v_n)\| = \|u - A(u,u)\| &\leq \lim \left\|(J_c - A)(u_{n_j} - v_{n_j})\right\| \\ &= \lim \|v_n - u_n\| \frac{1-c}{1-t} = 0,\end{aligned}$$

i.e. $u = v = A(u,u) = A(v,v)$. It remains to apply the same arguments as in the proof of Theorem 3.4.1. ii/ The case $c = 1$. By condition 7/ we have

$$\|u_{n+1} - p\| \leq \|v_{n+1} - p\| \leq \|\bar{u}_n - p\|$$

and

$$\|u_n - p\| \leq \|v_n - p\|.$$

But $\bar{u}_n = tu_n + (1-t)v_n$ so

$$\|\bar{u}_n - p\| \leq \|v_n - p\|.$$

Thus there exists d such that

$$d = \lim \|\bar{u}_n - p\| = \lim \|v_n - p\|.$$

If $d = 0$, then $\|v_n - p\| \to 0$ as $n \to \infty$. If $d \neq 0$, then

$$\left\| \frac{\bar{u}_n - p}{\|v_n - p\|} \right\| \to 1.$$

Nevertheless

$$\left\| \frac{\bar{u}_n - p}{\|v_n - p\|} \right\| = \left\| \frac{t(u_n - p)}{\|v_n - p\|} + \frac{(1-t)(v_n - p)}{\|v_n - p\|} \right\|.$$

Setting

$$W_n = \frac{\bar{u}_n - p}{\|v_n - p\|}, \quad Z_n = \frac{v_n - p}{\|v_n - p\|},$$

we have

$$\|W_n\| \leq 1, \quad \|Z_n\| = 1.$$

Hence $\|W_n - Z_n\| \to 0$ as $n \to \infty$, i.e.

$$\left\| \frac{\bar{u}_n - p}{\|v_n - p\|} - \frac{v_n - p}{\|v_n - p\|} \right\| = \left\| \frac{\bar{u}_n - p}{v_n - p} \right\| \to 0 \text{ as } n \to \infty.$$

So $\|u_n - v_n\| \to 0$, as $n \to \infty$. The proof of the theorem is complete. ■

Let us deal with the algorithm

$$\begin{cases} u_n = & A_n(u_n, v_n), \quad n \geq 1, \\ v_n = & A_{n,k}(u_n, u_{n-1}), \end{cases} \tag{3.4.7}$$

where $A_{n,1}(x, y) = A_n(x, y)$, $A_{n,i}(x, y) = A_n(x, A_{n,i-1}(x, y))$, $i = 1, 2, ..., k$.

Lemma 3.4.8 *Let (X, d) be a metric space, M_n, $n = 0, 1, 2, ...$ compact subsets in X, and $M_n \xrightarrow{B} M_0$ as $n \to \infty$ i.e. $\sup\limits_{x \in M_n} \inf\limits_{y \in M_0} d(x, y) \to 0$ as $n \to \infty$. Then the set $\cup_{n=0}^{\infty} M_n$ is compact too.*

Remark 3.4.2 *Similarly to the proofs of the last Theorem and of Theorems 3.4.6, 3.4.7 it is able to show the weak convergence of the approximants defined by (3.4.7) or by*

$$\begin{cases} u_n = & (tJ_c - (I - t)A_n)(u_n, v_n), \quad n \geq 1, \\ v_n = & (tJ_c - (I - t)A_n)_k(u_n, u_{n-1}), \end{cases}$$

for any $t \in (0, 1)$, the the cases when X is a reflexive Banach space or an uniformly convex one and A_n are continuous mappings from $D \times D$ into D_n, where D_n, mn≥ 0 are convex closed subsets in $D \subset X$.

Theorem 3.4.9 *Let X be a Banach space, D a closed subset of X. Assume that A_n, $n > 0$ are continuous and weakly compact mappings from $D \times D$ into D. Moreover the following conditions are satisfied: 12/ $F = \{x : A(x, x) = x\} \neq \emptyset$; 13/*

$$\|A_n(x, y) - p\| \begin{cases} < \max(\|x - p\|, \|y - p\|) & \text{if } \|x - p\| \neq \|y - p\|, \\ \leq \max(\|x - p\|, \|y - p\|) & \text{if } \|x - p\| = \|y - p\| \end{cases}$$

$\forall x, y \in D$, $p \in F$, $n \geq 0$; 14/ for $c \in [0, 1]$ the mapping $J_c - A_0$ is quasiconvex on $D \times D$. 15/ the space X has the property: if $\{y_n\} \subset X$, $y_n \xrightarrow{W} y_0 \in X$, then $\underline{\lim}\|y_n - y\| > \underline{\lim}\|y_n - y_0\|$, $\forall y \neq y_0$, $y \in X$; 16/ The system (3.4.7) is solvable for $n \geq 1$ and $\|u_n - v_n\| \to 0$ as $n \to \infty$; 17/

$$\|A_n(x, y) - A_0(x, y)\| \leq \varepsilon, \ \varepsilon \to 0 \text{ as } n \to \infty, \forall x, y \in D.$$

Then the sequences $\{u_n\}$, $\{v_n\}$ defined by formula (3.4.7) weakly converge to the solution of equation (3.4.1) .

Proof. As in the proof of Theorem 3.4.1 we have

$$\lim \|v_n - p\| = \lim \|u_n - p\| \leq \|u_0 - p\|.$$

Hence

$$\{u_n\}, \{v_n\} \subset D_0 = B(p, \|u_0 - p\|) \cap D$$

(convex and closed in D). Let M_n be the image of the restriction of A_n on $D_0 \times D_0$, $n \geq 0$. By the weak compactness of A_n, M_n are weakly compact. By means of constructed method (3.4.7) it is clear that $u_n, v_n \in M_n$, $n \geq 1$. By condition 17/ and Lemma 3.4.8, $M = \cup_{n\geq 0} M_n$ is weakly compact. Therefore from the sequences $\{u_n\}, \{v_n\}$ we can choose weakly convergent subsequences $\{u_{n_j}\}, \{v_{n_j}\}$ such that $u_{n_j} \overset{W}{\rightarrow} u$, $v_{n_j} \overset{W}{\rightarrow} v$. But

$$\|u - v\| \leq \underline{\lim} \|u_{n_j} - v_{n_j}\| = \lim \|u_n - v_n\| = 0,$$

that is, $u = v$. On the other hand,

$$\begin{aligned}
\|(J_c - A_0)(u, v)\| &= \|cu + (1 - c) v - A_0(u, v)\| \\
&\leq \underline{\lim} \|cu_{n_j} + (1 - c) v_{n_j} - A_0 (u_{n_j}, v_{n_j})\| + \varepsilon_{n_j} \\
&\leq \underline{\lim} (\|cu_{n_j} + (1 - c) v_{n_j} - A_{n_j} (u_{n_j}, v_{n_j})\|) + \varepsilon_{n_j} \\
&\leq \underline{\lim} \|(1 - c) (u_{n_j}, v_{n_j})\| \quad (\text{since } u_{n_j} = A_{n_j} (u_{n_j}, v_{n_j})) \\
&= (1 - c) \underline{\lim} \|u_n - v_n\| = 0.
\end{aligned}$$

Hence

$$cu + (1 - c) v = A_0(u, v),$$

and $u = v$, so $u = v \in F$. It will be shown that $\{u_n\}$, $\{v_n\}$ have just one limit point (in the weak sense), say, p. In order to do it we consider the function $f(p) = \lim \|v_n - p\|$ defined on $F_0 = F \cap M_0$. Since the function f is lower weakly semicontinuous and the set F is weakly compact (as the image of a bounded closed set under a weakly compact mapping), we claim that the function f attains on F_0 its infinimum at a point, say, $\bar{p}$. Suppose that a subsequence $\{u_{n_k}\}$ weakly converges to $u \neq \bar{p}$. Then by condition 15/ we have

$$\underline{\lim} \|u_{n_k} - \bar{p}\| > \underline{\lim} \|u_{n_k} - u\|.$$

This is a contradiction. So $u_n \overset{W}{\rightarrow} p$, $v_n \overset{W}{\rightarrow} p$. The theorem is thus proved. ■

By the use of Edelstein's Theorem from [149], [150] for the operator $A_n(x) = A(x, v)$ it is easy to prove

Theorem 3.4.10 *Let D be a closed bounded subset of a normed space X, and A a mapping from $D \times D$ to D satisfying the condition*

$$\|A(x,y) - A(z,t)\| \begin{cases} < \max\{\|x-z\|, \|y-t\|\}, & \text{if } (x,y) \neq (z,t) \text{ and } x \neq y \\ & \text{or } z \neq t, \\ \leq \|x-z\| = \|y-t\|, & \text{if } x = y \text{ and } z = t, \end{cases} \tag{3.4.8}$$

for all $x, y, z, t \in D$. Assume that either D is compact or A maps $D \times D$ into a compact subset of D. Then the equation

$$A(x,x) = x \tag{3.4.9}$$

has a unique solution in D. Moreover every equation

$$x_n = A(x_n, x_{n-1}), \quad n \geq 0, \tag{3.4.10}$$

has also a unique solution x_n and the sequence $\{x_n\}$ defined by (3.4.10) converges to the solution of equation (3.4.9) for any $x_0 \in D$.

Remark 3.4.3 *If instead of (3.4.9) under the assumptions as in Theorem 3.4.10 we use the Picard iteration for $A_*(x) = A(x,x)$ as in [105] by L.P. Belluce and W.A. Kirk, in [111] by F.E. Browder, in [149] by M. Edelstein, in [273] by W.V. Petryshyn and T.E. Williamson, etc., then obviously it is impossible to assert either the existence or the convergence of the approximate solution for equation (3.4.10).*

In what follows some simple examples will be shown.

Example 3.4.1 *Let us consider the set D of $C(-\infty, +\infty)$ with elements defined as follows $x(t) \in D$ if $|x(t)| \leq M$, M is a constant, $x(k) = 0$ for all integers $k \in \mathbb{Z}$,*

$$x(t) = \begin{cases} 2(t-k)\,x(k+1/2), & \text{for } t \in [k, k+1/2] \\ 2(k+1-tx(k+1/2)), & \text{for } t \in [k+1/2, k+1] \end{cases}$$

$|x(k+1/2)| \leq |x(k+3/2)|$ for all $k \in \mathbb{Z}$. Let

$$A(x,y) = \frac{x(t+1) + y(t+1)}{2 + \|x-y\| / (\|x-y\| + M)}, \quad \forall x(t), y(t) \in D,$$

where

$$\|x-y\| = \sup_{t \in (-\infty,\infty)} \|x(t) - y(t)\|.$$

Mapping A satisfies all conditions of Theorem 3.4.10. Hence for any $x_0(t) \in D$, the sequence $x_n(t) = A(x_n(t), x_{n-1}(t))$ must converge to the solution of (3.4.10) $x(k+1/2) = x(k+3/2)$, $\forall k \in \mathbb{Z}$ (including $x(t) \equiv 0$). On the other hand the iteration $x_n = A(x_{n-1}, x_{n-1})$ may have got any limit point when, for instance, $x_0(t)$ is such that: $x_0(k_0+1/2) < x_0(k_0+3/2)$ for some $k_0 \in \mathbb{Z}$.

Example 3.4.2 *Let*

$$D = l_p,\ 1 < p < \infty,\ A(x,y) = \frac{\tau y}{\exp(\|x-y\|)}$$

where

$$\tau : x = (x_1, x_2, \ldots) \longmapsto \tau x = (0, x_1, x_2, \ldots),\ x \in l_p.$$

Finally for arbitrary $x_0 = (x_{01}, x_{02}, \ldots)$, $x_{01} \neq 0$, *the sequence defined by (3.4.10) converges to* 0 *as* $n \to \infty$.

3.5 A Special iteration method

The convergence of the iteration algorithm based on the algorithmic function $x \to \alpha x + (1-\alpha) A(x)$ is analyzed, where $A : \mathbf{R}^n \to \mathbf{R}^n$ is a point to point mapping, and the coefficient α may change during the process.

We consider iteration methods of the form

$$x_{k+1} = \alpha_k x_k + (1-\alpha_k) A(x_k) \tag{3.5.1}$$

where $A : D \to D$ with $D \subseteq \mathbf{R}^n$ being convex, and for all k, $0 \leq \alpha_k < 1$. The convexity of D implies that the iteration sequence exists for arbitrary $x_0 \in D$.

In Section 3.1 the convergence of this procedure was studied for the special case, when A is a contraction and $\alpha_k \equiv \alpha$ for all k. This result was further generalized for the nonstationary case [99]. These earlier results guarantee the convergence of process (3.5.1) in cases when the original algorithm

$$z_{k+1} = A(z_k) \tag{3.5.2}$$

also converges to the unique fixed point of mapping A. In this section we will find convergence condition even in the cases when the original process (3.5.2) diverges. In such cases the optimal selection of the weights α_k will also be determined.

The following conditions will be assumed:

(A)

$$\|A(x) - z^*\| \leq K(x)\|x - z^*\|,\ \text{for all } x \in D, \tag{3.5.3}$$

where $z^* \in D$ is a fixed point of A, $\|\cdot\|$ is the Euclidean norm, and $K : D \to \mathbf{R}$ is a real valued function;

(B)

$$(x - z^*)^T (A(x) - z^*) \leq \beta(x)\|x - z^*\|^2,\ \text{for all } x \in D, \tag{3.5.4}$$

where $\beta : D \to \mathbf{R}$ is a real valued function.

Before analyzing the convergence of process (3.5.1), some comments are in order. If mapping A is bounded on D, that is, if

$$\|A(x) - A(y)\| \leq K \|x - y\|,$$

the constant function $K(x) \equiv K$. If mapping $-A$ is monotonic, that is, if

$$(x - z)^T (A(x) - A(y)) \leq 0$$

for all $x, y \in D$, then we may select $\beta(x) \equiv 0$. The Cauchy-Schwartz inequality implies that for all $x \in D$

$$(x - z^*)^T (A(x) - z^*) \leq \|x - z^*\| \cdot \|A(x) - z^*\| \leq K(x) \|x - z^*\|^2.$$

Therefore we may assume that $\beta(x) \leq K(x)$, otherwise we may replace $\beta(x)$ by $K(x)$.

Under assumptions (A) and (B)

$$\begin{aligned}\|x_{k+1} - z^*\|^2 &= \|\alpha_k x_k + (1 - \alpha_k) A(x_k) - \alpha_k z^* - (1 - \alpha_k) z^*\|^2 \\ &= \|\alpha_k (x_k - z^*) + (1 - \alpha_k)(A(x_k) - z^*)\|^2 \\ &= \alpha_k^2 \|x_k - z^*\| + (1 - \alpha_k)^2 \|A(x_k) - z^*\|^2 \\ &+ 2\alpha_k (1 - \alpha_k)(x_k - z^*)(A(x_k) - z^*) \\ &\leq \left[\alpha_k^2 + (1 - \alpha_k)^2 K_k^2 + 2\alpha_k (1 - \alpha_k) \beta_k\right] \|x_k - z^*\|^2,\end{aligned}$$

where $K_k = K(x_k)$ and $\beta_k = \beta(x_k)$. Let $p(\alpha_k)$ denote the multiplier of $\|x_k - z^*\|^2$, then

$$p(\alpha_k) = \alpha_k^2 \left(1 + K_k^2 - 2\beta_k\right) - 2\alpha_k \left(K_k^2 - \beta_k\right) + K_k^2.$$

Notice that

$$p(0) = K_k^2 \text{ and } p(1) = 1,$$

further more p is a convex quadratic polynomial, unless $1 + K_k^2 - 2\beta_k = 0$.

Since

$$1 + K_k^2 - 2\beta_k \geq 1 + K_k^2 - 2K_k = (1 - K_k)^2 \geq 0,$$

this is the case if and only if $K_k = \beta_k = 1$. In this special case $p(\alpha_k) \equiv 1$. Otherwise the stationary point of function p is

$$\alpha^* = \frac{K_k^2 - \beta_k}{1 + K_k^2 - 2\beta_k}, \tag{3.5.5}$$

which is in $[0,1)$ if and only if $\beta_k \leq K_k^2$ and $\beta_k < 1$. Easy calculation shows that

$$p(\alpha^*) = \frac{K_k^2 - \beta_k^2}{1 + K_k^2 - 2\beta_k},$$

which is below $1-\varepsilon$ (where ε is small) when

$$\frac{(\beta_k - 1)}{1 + K_k^2 - 2\beta_k} \geq \varepsilon.$$

If $\alpha^* \notin [0,1)$, then

$$\min\{p(\alpha) : 0 \leq \alpha \leq 1\} = \begin{cases} K_k^2, & \text{if } K_k \leq 1, \\ 1, & \text{if } K_k > 1. \end{cases}$$

The above iteration can be summarized as follows.

Theorem 3.5.1 *Assume that for each $k \geq 0$, either*

$$K_k \leq 1 - \delta, \ \beta_k > K_k^2;$$

or

$$K_k \leq 1, \ \beta_k \leq K_k^2, \ \beta_k \leq 1 - \delta;$$

or

$$1 < K_k \leq Q, \ \beta_k \leq 1 - \delta,$$

where δ and Q are fixed constants. Then with appropriate selection of α_k,

$$\|x_{k+1} - z^*\| \leq (1-\varepsilon)\|x_k - z^*\|$$

for all k with some $\varepsilon > 0$, therefore the process converges to z^*.

Assume next that $K(x) \equiv K$ and $\beta(x) \equiv \beta$ for all $x \in D$. Then the theorem reduces to the following result.

Theorem 3.5.2 *Assume that either $K < 1$, or $K \geq 1$ and $\beta < 1$. Then the process converges to z^* with the appropriate selection of the coefficients α_k.*

Corollary 3.5.3 *If mapping A is bounded and $-A$ is monotonic, then all conditions of Theorem 3.5.2 are satisfied. In this special case the best selection for α_k is the following*

$$\alpha_k = \frac{K^2}{1+K^2}, \ \textit{for all } k.$$

This relation follows from (3.5.5) and the fact that in this case $K(x) \equiv K$ and $\beta(x) \equiv 0$.

Example 3.5.1 *Consider equation $x = -2x$, that is, select $D = \mathbf{R}$, $A(x) = -2x$, and $z^* = 0$. Mapping A is bounded and $K = 2$. The original iteration process $x_{k+1} = A(z_k)$ does not converge for $x_0 \neq 0$. However, by selecting $\alpha_k = \frac{4}{5}$, the modified process has the form*

$$x_{k+1} = \frac{4}{5}x_k + \frac{1}{5}(-2x_k) = \frac{2}{5}x_k,$$

which is obviously convergent, since mapping $x \to \frac{2}{5}x$ is a contraction.

Consided now the generalized algorithm

$$x_{k+1} = \sum_{i=1}^{I} \alpha_{ik} A_i(x_k), \tag{3.5.6}$$

where $A_i : D \to D$, and α_{ik} is a given nonnegative constant such that $\sum_{i=1}^{I} \alpha_{ik} = 1$ for all $k \geq 0$. It is assumed that

(C) for all i and j,

$$(A_i(x) - z^*)^T (A_j(x) - z^*) \leq \beta_{ij}(x) \|x - z^*\|^2$$

for all $x \in D$, where $\beta_{ij} : D \to \mathbf{R}$ is a real valued function for all i and j. Notice that these conditions are the straightforward generalizations of assumptions (A) and (B).

Easy calculation shows that

$$\begin{aligned}
\|x_k - z^*\|^2 &= \left\| \sum_{i=1}^{I} \alpha_{ik} (A_i(x_k) - z^*) \right\|^2 \\
&= \left(\sum_{i=1}^{I} \sum_{j=1}^{I} \alpha_{ik}\alpha_{jk}\beta_{ijk} \right) \|x_k - z^*\|^2,
\end{aligned}$$

where $\beta_{ijk} = \beta_{ij}(x_k)$. The best choice of the α_{ik} values can be obtained by solving the quadratic programming problem

$$\text{minimize} \sum_{i=1}^{I} \sum_{j=1}^{I} \alpha_{ik}\alpha_{jk}\beta_{ijk}, \tag{3.5.7}$$

subject to $\alpha_{ik} \geq 0$ $(i = 1, 2, \ldots, I)$

$$\sum_{i=1}^{I} \alpha_{ik} = 1.$$

The iteration procedure is convergent, if for all k, the optimal objective function value is below $1-\varepsilon$ with some fixed small $\varepsilon > 0$. Since for all k, the constraints are linear, the solution of problem (3.5.7) is a simple task.

Process (3.5.6) can be further generalized as

$$x_{k+1} = F_k\left(A_1\left(x_k\right), \ldots, A_I\left(x_k\right)\right), \tag{3.5.8}$$

where $A_i : D \to D$, $F_k : D^I \to D$. Let z^* be a common fixed point of mappings A_i, and assume that $z^* = F_k\left(z^*, \ldots, z^*\right)$, furthermore

(D) for all $x \in D$ and all i,

$$\left\|A_i\left(x\right) - z^*\right\| \leq K_i\left(x\right)\left\|x - z^*\right\|,$$

where $K_i : d \to \mathbf{R}$ is a real valued function;

(E) for all $x^{(i)} \in D$ $(i = 1, 2, \ldots, I)$ and $k \geq 0$,

$$\left\|F_k\left(x^{(1)}, \ldots, x^{(I)}\right)\right\| \leq \sum_{i=1}^{I} \alpha_{ik}\left\|x^{(i)} - z^*\right\|.$$

Under the above conditions,

$$\begin{aligned}
\left\|x_{k+1} - z^*\right\| &= \left\|F_k\left(A_1(x_k), \ldots, A_I(x_k)\right) - z^*\right\| \\
&\leq \sum_{i=1}^{I} \alpha_{ik}\left\|A_i\left(x_k\right) - z^*\right\| \\
&\leq \left(\sum_{i=1}^{I} \alpha_{ik} K_i\right)\left\|x_k - z^*\right\|.
\end{aligned}$$

It is easy to see that there exist coefficients α_{ik} such that $\sum_{i=1}^{I} \alpha_{ik} K_i \leq 1 - \varepsilon$ with some $\varepsilon > 0$ if and only if at least one $K_i < 1$. This observation generalizes the earlier results of Section 3.1.

Finally we note that D may be a subset of an infinite dimensional Euclidean space, and in the last result $\|\cdot\|$ is not necessarily the Euclidean norm.

3.6 Stability of the Ishikawa iteration

Let X be an uniformly real smooth Banach space and $T : X \to X$ be Φ-strongly pseudocontractive mapping with the bounded range. It is shown that Ishikawa iteration procedures are almost T-stable and half T-stable. our results improve and extend corresponding results announced by others.

Let X be a real Banach space and J the normalized duality mapping from X into 2^{X^*} given by

$$J(x) = \{f \in X^* : \langle x, f\rangle = \|x\|^2 = \|f\|^2\},$$

where X^* denotes the dual space of X and $\langle \cdot, \cdot \rangle$ denotes the generalized duality pairing. It is well known that if X^* is strictly convex then J is single-valued and such that $J(tx) = tJ(x)$ for all $t \geq 0$ and $x \in X$, and if X^* is uniformly convex, then J is uniformly continuous on the bounded subsets of X. In the sequel we shall denote the single-valued normalized duality mapping by j.

Definition 3.6.1 *A mapping T with domain $D(T)$ and range $R(T)$ in X is called Φ-strongly pseudocontraction if for all $x, y \in D(T)$ there exist $j(x-y) \in J(x-y)$ and a strictly increasing function $\Phi : [0, \infty) \to [0, \infty)$ with $\Phi(0) = 0$ such that*

$$\langle Tx - Ty, j(x-y) \rangle \leq \|x-y\|^2 - \Phi(\|x-y\|)\,\|x-y\| \tag{3.6.1}$$

T is called strongly pseudocontraction if there exists $k > 0$ such that (3.6.1) is satisfied with $\Phi(t) = kt$, and it is called Φ-hemicontractive if the nullspace of T, $N(T) = \{x \in D(T); Tx = 0\}$ is nonempty and (3.6.1) is satisfied for all $x \in D(T)$ but $y \in N(T)$. Furthermore, the class of strongly pseudocontraction mapping is a proper subclass of the class of Φ-strongly pseudocontraction mappings.

Recently, some stability results for certain classes of nonlinear maps have been researched by several authors, the study topic on stability is both of theoretical and of numerical interest. In order to prove the main results of the paper, we need the following some concepts.

Let mK be a nonempty convex subset of a real banach space X and T be a self-mapping of K. Let $\{\alpha_n\}$ be a real sequence in $[0, \infty)$. Assume that $x_0 \in K$ and $x_{n+1} = f(T, x_n, \alpha_n)$ defines an iteration procedure which yields a sequence of points $\{x_n\}$ in K. Suppose, furthermore, that $F(T) = \{x \in X : Tx = x\} \neq \emptyset$ and that $\{x_n\}$ converges strongly to $x^* \in F(T)$. Let $\{y_n\}_{n=0}^{\infty}$ be any sequence in K and $\{\varepsilon_n\}_{n=0}^{\infty}$ a sequence in $R^+ = [0, \infty)$ given by $\varepsilon_n = \|y_{n+1} - f(T, y_n, \alpha_n)\|$.

Definition 3.6.2 *1) If $\lim\limits_{n\to\infty} \varepsilon_n = 0$ implies that $\lim\limits_{n\to\infty} y_n = x^*$, then the iteration procedure $\{x_n\}$ defined by $x_{n+1} = f(T, x_n, \alpha_n)$ is said to be T-stable.*

2) $\sum\limits_{n=1}^{\infty} < \infty$ implies that $\lim\limits_{n\to\infty} y_n = x^*$, then the iteration procedure $\{x_n\}$ defined by $x_{n+1} = f(T, x_n, \alpha_n)$ is said to be almost T-stable.

3) If $\varepsilon_n = o(\alpha_n)$ implies that $\lim\limits_{n\to\infty} y_n = x^*$, then the iteration procedure $\{x_n\}$ defined by $x_{n+1} = f(T, x_n, \alpha_n)$ is said to be half T-stable.

4) If $\varepsilon_n = \varepsilon_n' + \varepsilon_n''$ with $\sum\limits_{n=1}^{\infty} \varepsilon_n' < \infty$ and $\varepsilon_n'' = o(\alpha_n)$ implies that $\lim\limits_{n\to\infty} y_n = x^*$, then the iteration procedure $\{x_n\}$ defined by $x_{n+1} = f(T, x_n, \alpha_n)$ is said to be weakly T-stable.

We remark immediatly that, if an iterative sequence $\{x_n\}$ is T-stable, then it is weakly T-stable and, if the iterative sequence $\{x_n\}$ is weakly T-stable, then it is both almost and half T-stable. Conversly, it is evident that an iterative sequence $\{x_n\}$ which is either almost or half T-stable may fail to be weakly T-stable. Accordingly, it is of important theoretical interest for us to study the weak stability.

Lemma 3.6.3 *Let $\{a\}_{n=0}^{\infty}$, $\{b_n\}_{n=0}^{\infty}$ be two nonnegative real sequence satisfying condition:*

$$a_{n+1} \leq a_n + b_n, \qquad \sum_{n=1}^{\infty} b_n < \infty, \qquad \text{then } \lim_{n\to\infty} a_n \exists.$$

Lemma 3.6.4 *Let be a real banach space. then, for all $x, y \in X$ and $j(x+y) \in J(x+y)$,*

$$\|x+y\|^2 \leq \|x\|^2 + 2\langle y, j(x+y)\rangle.$$

Theorem 3.6.5 *Let X be a real uniformly smooth Banach space, $T : X \to X$ be a continuous and Φ-strong pseudocontraction mapping with a bounded range. Suppose q is the fixed point of the mapping T. Let $\{\alpha_n\}_{n=1}^{\infty}$, $\{\beta_n\}_{n=1}^{\infty}$ be two real sequence satisfying:*

$$0 \leq \alpha_n, \beta_n < 1, \forall n \geq 1$$
$$\lim_{n\to\infty} \alpha_n = \lim_{n\to\infty} \beta_n = 0$$
$$\textstyle\sum_{n=1}^{\infty} \alpha_n = \infty.$$

Let $\{x_n\}_{n=1}^{\infty}$ be the sequence of the Ishikawa iteration [198] generated from an arbitrary $x_1 \in X$ by

$$\begin{aligned} x_{n+1} &= (1-\alpha_n)x_n + \alpha_n T z_n \\ z_n &= (1-\beta_n)x_n + \beta_n T x_n, \quad n \geq 1. \end{aligned}$$

Let $\{y_n\}_{n=1}^{\infty}$ be any sequence in X and define $\{\varepsilon_n\}_{n=1}^{\infty} \subset R^+$ by

$$\begin{aligned} \varepsilon_n &= o(\alpha_n) = \|y_{n+1} - (1-\alpha_n)y_n + \alpha_n T\tau_n\| \\ \tau_n &= (1-\beta_n)y_n + \beta_n T y_n, \quad n \geq 1. \end{aligned}$$

Then the Ishikawa iteration procedure $\{x_n\}_{n=1}^{\infty}$ is half T-stable.

Proof. Let q denote the unique fixed point of T. Let $\varepsilon_n = o(\alpha_n) \to 0$ as $n \to \infty$. It suffices to prove that $y_n \to q$ as $n \to \infty$. Without loss of generality, we may assume that $\varepsilon_n \leq \alpha_n$ for all mn≥ 1. Now set

$$M_0 = \sup_{x\in X}\{\|Tx - q\|\} + \|y_1 - q\|, \quad M = M_0 + 1.$$

For any $n \geq 1$, using induction, we get $\|y_n - q\| \leq M$. Indeed, we have

$$\begin{aligned} \|y_1 - q\| &\leq M_0 \leq M, \\ \|y_2 - q\| &\leq \|y_2 - (1-\alpha_1)y_1 - \alpha_1 T\tau_1\| + (1-\alpha_1)\|y_1 - q\| + \alpha_1 \|T\tau_1 - q\| \\ &\leq \varepsilon_1 + (1-\alpha_1)M_0 + \alpha_1 M_0 \leq \alpha_1 + M_0 \leq M, \\ \|y_3 - q\| &\leq \|y_3 - (1-\alpha_2)y_2 - \alpha_2 T\tau_2\| + (1-\alpha_2)\|y_2 - q\| + \alpha_2 \|T\tau_2 - q\| \\ &\leq \varepsilon_2 + (1-\alpha_2)M + \alpha_2 M_0 \leq \alpha_2 + (1-\alpha_2)M_0 + \alpha_2 M_0 \leq M, \\ &\cdots \end{aligned}$$

$$\|y_3 - q\| \leq \varepsilon_{n-1} + (1-\alpha_{n-1})M + \alpha_{n-1}M_0 \leq \alpha_{n-1} + (1-\alpha_{n-1})M + \alpha_{n-1}M_0 \leq M.$$

Since T is Φ-strong pseudocontraction mapping, then we have

$$\langle Tx - Ty, J(x-y)\rangle \le \|x-y\|^2 - \Phi(\|x-y\|)\,\|x-y\|,$$

for $\forall x, y \in X$.

Let $v_n = (1-\alpha_n)y_n + \alpha_n T\tau_n$, $u_n = y_{n+1} - v_n$. Then we have

$$y_{n+1} = u_n + v_n, \quad \|u_n\| = \varepsilon_n$$

and

$$\|y_{n+1} - q\| \le \|v_n - q\| + \|u_n\| = \|v_n - q\| + \varepsilon_n. \tag{3.6.2}$$

Using Lemma 3.6.4, we have

$$\begin{aligned}
\|v_n - q\|^2 &= \|(1-\alpha_n)y_n + \alpha_n T\tau_n - q\|^2 \\
&= \|(1-\alpha_n)(y_n - q) + \alpha_n(T\tau_n - Tq\|^2 \\
&\le (1-\alpha_n)^2\|y_n - q\|^2 + 2\alpha_n\langle T\tau_n - Tq, J(v_n - q)\rangle \\
&= (1-\alpha_n)^2\|y_n - q\|^2 + 2\alpha_n\langle T\tau_n - Tq, J(\tau_n - q)\rangle \\
&+ 2\alpha_n\langle T\tau_n - Tq, J(v_n - q) - J(\tau_n - q)\rangle \\
&\le (1-\alpha_n)^2\|y_n - q\|^2 + 2\alpha_n\|\tau_n - q\|^2 \qquad (3.6.3)\\
&- 2\alpha_n\Phi(\|\tau_n - q\|)\,\|\tau_n - q\| + 2\alpha_n d_n
\end{aligned}$$

where $d_n = \langle T\tau_n - Tq, J(v_n - q) - J(\tau_n - q)\rangle \to 0$ $(n \to \infty)$ (using J is uniformly continuous on the bounded subsets of X).

Again applying Lemma 3.6.4, we have

$$\begin{aligned}
\|\tau_n - q\|^2 &= \|(1-\beta_n)(y_n - q) + \beta_n(Ty_n - q)\|^2 \\
&\le (1-\beta_n)^2\|y_n - q\|^2 + 2\beta_n\langle Ty_n - Tq, J(\tau_n - q)\rangle \\
&= (1-\beta_n)^2\|y_n - q\|^2 + 2\beta_n\langle Ty_n - Tq, J(y_n - q)\rangle \\
&+ 2\beta_n\langle Ty_n - Tq, J(\tau_n - q) - J(y_n - q)\rangle \\
&\le (1-\beta_n)^2\|y_n - q\|^2 + 2\beta_n\|y_n - q\|^2 \\
&- 2\beta_n\Phi(\|y_n - q\|)\,\|y_n - q\| + 2\beta_n e_n \\
&= (1+\beta_n{}^2)\|y_n - q\|^2 - 2\beta_n\Phi(\|y_n - q\|)\,\|y_n - q\| + 2\beta_n e_n, \qquad (3.6.4)
\end{aligned}$$

where $e_n = \langle T\tau_n - Tq, J(\tau_n - q) - J(y_n - q)\rangle \to 0$ $(n \to \infty)$ (same as above).

Substituting (3.6.4) into (3.6.3) yields

$$\begin{aligned}\|v_n - q\|^2 &\le (1-\alpha_n)^2 \|y_n - q\|^2 + 2\alpha_n(1+\beta_n{}^2)\|y_n - q\|^2 \\ &\quad - 4\alpha_n\beta_n\Phi(\|y_n - q\|)\|y_n - q\| \\ &\quad + 4\alpha_n\beta_n e_n - 2\alpha_n\Phi(\|\tau_n - q\|)\|\tau_n - q\| + 2\alpha_n d_n \\ &= (1+\alpha_n{}^2 + 2\alpha_n\beta_n^2)\|y_n - q\|^2 + 4\alpha_n\beta_n[e_n - \Phi(\|y_n - q\|)\|y_n - q\|] \\ &\quad + 2\alpha_n[d_n - \Phi(\|\tau_n - q\|)\|\tau_n - q\|] \\ &\le \|y_n - q\|^2 + (\alpha_n{}^2 + 2\alpha_n\beta_n^2)M^2 + 4\alpha_n\beta_n[e_n - \Phi(\|y_n - q\|)\|y_n - q\|] \\ &\quad + 2\alpha_n[d_n - \Phi(\|\tau_n - q\|)\|\tau_n - q\|] \\ &= \|y_n - q\|^2 + 4\alpha_n\beta_n(M^2/2\beta_n + e_n - \Phi(\|y_n - q\|)\|y_n - q\|] \\ &\quad + 2\alpha_n[M^2/2\alpha_n + d_n - \Phi(\|\tau_n - q\|)\|\tau_n - q\|].\end{aligned} \tag{3.6.5}$$

We shall prove: $\inf_{n\ge 1}\{\|y_n - q\|\} = 0$. If not, there must exist $\delta > 0$, such that $\inf_{n\ge 1}\{\|y_n - q\|\} = \delta$. This is to say that $\|y_n - q\| \ge \delta$, for all $n \ge 1$, and hence

$$\Phi(\|y_n - q\|)\|y_n - q\| \ge \Phi(\delta)\delta.$$

From (3.6.4), we obtain

$$\begin{aligned}\|\tau_n - q\|^2 &\le \|y_n - q\|^2 + M^2\beta_n^2 + 2e_n\beta_n - 2\beta_n\Phi(\delta)\delta \\ &= \|y_n - q\|^2 + \beta_n(M^2\beta_n + 2e_n - 2\Phi(\delta)\delta).\end{aligned} \tag{3.6.6}$$

Since $\beta_n \to 0$ and $e_n \to 0$ as $n \to \infty$, then there exists some integer N_1 such that $M^2\beta_n + 2e_n - 2\Phi(\delta)\delta < 0$ for all $n > N_1$. Thus we have

$$\|\tau_n - q\| \le \|y_n - q\|$$

and

$$-\Phi(\|\tau_n - q\|)\|\tau_n - q\| \ge -\Phi(\|y_n - q\|)\|y_n - q\| \tag{3.6.7}$$

for all $n > N_1$. Take these facts into consideration in (3.6.5), we obtain

$$\begin{aligned}\|v_n - q\|^2 &\le \|y_n - q\|^2 + 4\alpha_n\beta_n(M^2/2\beta_n + e_n - \Phi(\|\tau_n - q\|)\|\tau_n - q\|] \\ &\quad + 2\alpha_n[M^2/2\alpha_n + d_n - \Phi(\|\tau_n - q\|)\|\tau_n - q\|],\end{aligned} \tag{3.6.8}$$

for all $n > N_1$. Furthermore, we have the following estimate:

$$\begin{aligned}\|\tau_n - q\| &= \|(1-\beta_n)(y_n - q) + \beta_n(Ty_n - Tq)\| \\ &\ge \|y_n - q\| - \beta_n\|y_n - q\| - \beta_n\|Ty_n - Tq\| \\ &\ge \|y_n - q\| - \beta_n(M + M_0).\end{aligned} \tag{3.6.9}$$

Again since $\beta_n \to 0$ as $n \to \infty$, then there exists some integer $N_2 > N_1$ such that $\beta_n < \delta/(2(M + M_0))$ for all $n > N_2$. Hence, from (3.6.8) and (3.6.9) we obtain

$$\delta/2 \leq \|y_n - q\| - \delta/2 \leq \|\tau_n - q\| \tag{3.6.10}$$

for all $n > N_2 > N_1$.

Using (3.6.10) and (3.6.8) we have for all $n > N_3$

$$\begin{aligned} \|v_n - q\|^2 \leq \|y_n - q\|^2 &+ 4\alpha_n\beta_n(M^2/2\beta_n + e_n - \Phi(\delta/2)\delta/2] \\ &+ \alpha_n[M^2\alpha_n + 2d_n - \Phi(\delta/2)\delta/2] - \alpha_n\Phi(\|\tau_n - q\|)\|\tau_n - q\| . \end{aligned} \tag{3.6.11}$$

Owning to $\alpha_n, \beta_n, e_n, d_n \to 0$ as $n \to \infty$, without loss generality, it must be that there exists integer $N_3 > N_2$ such that

$$M^2/2\beta_n + e_n - \Phi(\delta/2)\delta/2 < 0, \qquad M^2\alpha_n + 2d_n - \Phi(\delta/2)\delta/2 < 0$$

for all $n > N_3$. Thus from (3.6.11) we get

$$\begin{aligned} &\|v_n - q\| \leq \|y_n - q\| , \\ &\|v_n - q\|^2 \leq \|y_n - q\|^2 - \alpha_n\Phi(\|\tau_n - q\|)\|\tau_n - q\| \end{aligned} \tag{3.6.12}$$

for all $n > N_3$.

Since $\varepsilon_n = o(\alpha_n)$, we may choose positive integer $N_4 > N_3$ such that

$$\frac{1}{2}\Phi(\delta/2)\delta/2 - 2M\frac{o(\alpha_n)}{\alpha_n} - \frac{[o(\alpha_n)]^2}{\alpha_n} > 0 \tag{3.6.13}$$

for all $n > N_4$. Again from (3.6.2), (3.6.10), (3.6.12) and (3.6.13), we obtain

$$\begin{aligned} \|y_{n+1} - q\|^2 &\leq \|v_n - q\|^2 + 2\|v_n - q\|\varepsilon_n + \varepsilon_n^2 \\ &\leq \|y_n - q\|^2 - \alpha_n\Phi(\|\tau_n - q\|)\|\tau_n - q\| + 2M\varepsilon_n + \varepsilon_n^2 \\ &\leq \|y_n - q\|^2 - \alpha_n\Phi(\|\tau_n - q\|)\|\tau_n - q\| + 2M[o(\alpha_n)] + [o(\alpha_n)]^2 \\ &\leq \|y_n - q\|^2 - \frac{1}{2}\alpha_n\Phi(\delta/2)\delta/2 \\ &\quad - \alpha_n\left[\frac{1}{2}\alpha_n\Phi(\delta/2)\delta/2 - \frac{2M[o(\alpha_n)]}{\alpha_n} - \frac{[o(\alpha_n)]^2}{\alpha_n}\right] \\ &\leq \|y_n - q\|^2 - \frac{1}{2}\alpha_n\Phi(\delta/2)\delta/2 &(3.6.14) \\ &\leq \|y_n - q\|^2 &(3.6.15) \end{aligned}$$

for all $n > N_4$. From above (3.6.15), we get $\lim_{n\to\infty}\|y_n - q\| \exists$.

Thus, (3.6.14) implies that

$$\frac{1}{2}\sum_{n=N_4+1}^{\infty}\alpha_n\Phi(\delta/2)\delta/2 \leq \|y_{N_4+1} - q\|^2 - \lim_{n\to\infty}\|y_{n+1} - q\|^2 < \infty.$$

This is a contradiction. Thus

$$\inf_n \|y_n - q\| = 0. \tag{3.6.16}$$

It follows that we want to prove $\lim_{n\to\infty} \|y_n - q\| = 0$.

From (3.6.16) we get that there exists an infinite subsequence $\{y_{n_j} - q\} \to 0$ as $j \to \infty$.

Since $\|\tau_{n_j} - y_{n_j}\| = \|-\beta_{n_j} y_{n_j} + \beta_{n_j} T y_{n_j}\| \to 0$ as $j \to \infty$, so that $\|\tau_{n_j} - y_{n_j}\| \leq \|\tau_{n_j} - y_{n_j}\| + \|y_{n_j} - q\| \to 0$ as $j \to \infty$, thus we have

$$\begin{aligned}\|y_{n_j+1} - q\| &\leq \|v_{n_j} - q\| + \varepsilon_{n_j} \\ &\leq \Big\{\|y_{n_j} - q\|^2 + 4\alpha_{n_j}\beta_{n_j}[M^2/2\beta_{n_j} + e_{n_j} - \Phi(\|y_{n_j} - q\|)\,\|y_{n_j} - q\|] \\ &\quad +2\alpha_{n_j}[M^2/2\alpha_{n_j} + d_{n_j} - \Phi(\|\tau_{n_j} - q\|)\,\|\tau_{n_j} - q\|]\Big\}^{1/2} + \varepsilon_{n_j} \to 0,\end{aligned}$$

as $j \to \infty$. Using continuous of Φ and induce method, we obtain $\|y_{n_j+m} - q\| \to 0$ as $j \to \infty$ for any nature number m. ■

Theorem 3.6.6 *Let X be a real uniformly smooth Banach space, $T : X \to X$ be a contiuous and Φ-strong accretive operator. Suppose $(I - T)$ has a bounded range. Define $S : X \to X$ by $Sx = f + x - Tx$. Suppose q is the fixed point of the mapping S. Let $\{\alpha_n\}_{n=1}^\infty$, $\{\beta_n\}_{n=1}^\infty$ be two real sequence satisfying:*

$$(1)\ 0 \leq \alpha_n, \beta_n < 1,\ \forall n \geq 1 \qquad (2)\ \lim_{n\to\infty} \alpha_n = \lim_{n\to\infty} \beta_n = 0 \qquad (3)\ \sum_{n=1}^{\infty} \alpha_n = \infty.$$

Let $\{x_n\}_{n=1}^\infty$ be the sequence of the Ishikawa iteration generated from an arbitrary $x_1 \in X$ by

$$\begin{cases} x_{n+1} = (1 - \alpha_n)x_n + \alpha_n S z_n \\ z_n = (1 - \beta_n)x_n + \beta_n S x_n, \quad n \geq 1. \end{cases}$$

Let $\{y_n\}_{n=1}^\infty$ be any sequence in X and define $\{\varepsilon_n\}_{n=1}^\infty \subset R^+$ by

$$\begin{cases} \varepsilon_n = o(\alpha_n) = \|y_{n+1} - (1 - \alpha_n)y_n - \alpha_n S\tau_n\| \\ \tau_n = (1 - \beta_n)y_n + \beta_n S y_n, \quad n \geq 1. \end{cases}$$

Then the Ishikawa iteration procedure $\{x_n\}_{n=1}^\infty$ is half S-stable.

Proof. Since T is Φ-strong accretive operator, then S is Φ-strong pseudocontraction. Now the conclusion follows from Theorem 3.6.5. ■

Theorem 3.6.7 *Let X be a real uniformly smooth Banach space, $T : X \to X$ be a continuous and Φ-strong pseudocontraction mapping with the bounded range.*

Suppose mq is the fixed point of the mapping T. Let $\{\alpha_n\}_{n=1}^{\infty}$, $\{\beta_n\}_{n=1}^{\infty}$ be two real sequence satisfying:

$$(1)\ 0 \le \alpha_n, \beta_n < 1,\ \ \forall n \ge 1 \qquad (2)\ \lim_{n\to\infty} \alpha_n = \lim_{n\to\infty} \beta_n = 0 \qquad (3)\ \sum_{n=1}^{\infty} \alpha_n = \infty.$$

Let $\{x_n\}_{n=1}^{\infty}$ be the sequence of the Ishikawa iteration generated from an arbitrary $x_1 \in X$ by

$$\begin{cases} x_{n+1} = (1-\alpha_n)x_n + \alpha_n T z_n \\ z_n = (1-\beta_n)x_n + \beta_n T x_n, \quad n \ge 1. \end{cases}$$

Let $\{y_n\}_{n=1}^{\infty}$ be any sequence in X and define $\{\varepsilon_n\}_{n=1}^{\infty} \subset R^+$ by

$$\begin{cases} \varepsilon_n = \|y_{n+1} - (1-\alpha_n)y_n - \alpha_n T\tau_n\| \\ \tau_n = (1-\beta_n)y_n + \beta_n T y_n, \quad n \ge 1. \end{cases}$$

Then the Ishikawa iteration procedure $\{x_n\}_{n=1}^{\infty}$ is almost T-stable.

Proof. Set $M_0 = \sup_{x\in X}\{\|Tx - q\|\} + \|y_1 - q\|$, $M = M_0 + \sum_{n=1}^{\infty} \varepsilon_n$.

For any $n \ge 1$, using induction, we get $\|y_n - q\| \le M$. Indeed we have

$$\begin{aligned} \|y_1 - q\| &\le M_0 \le M, \\ \|y_2 - q\| &\le \|y_2 - (1-\alpha_1)y_1 - \alpha_1 T\tau_1\| + (1-\alpha_1)\|y_1 - q\| + \alpha_1 \|T\tau_1 - q\| \\ &\le \varepsilon_1 + (1-\alpha_1)M_0 + \alpha_1 M_0 \le \varepsilon_1 + M_0 \le M, \\ \|y_3 - q\| &\le \varepsilon_2 + \varepsilon_1 + M_0 \le M, \\ &\cdots \\ \|y_3 - q\| &\le \sum_{i=1}^{n-1} \varepsilon_i + M_0 \le M. \end{aligned}$$

Since T is Φ-strong pseudocontraction mapping, for $\forall x, y \in X$ then have

$$\langle Tx - Ty, J(x-y)\rangle \le \|x-y\|^2 - \Phi(\|x-y\|)\,\|x-y\|.$$

Let $v_n = (1-\alpha_n)y_n + \alpha_n T\tau_n$, $u_n = y_{n+1} - v_n$. Then we have

$$y_{n+1} = u_n + v_n,\ \|u_n\| = \varepsilon_n$$

and

$$\|y_{n+1} - q\| \le \|v_n - q\| + \|u_n\| = \|v_n - q\| + \varepsilon_n. \tag{3.6.17}$$

Using Lemma 3.6.4, we have

$$\begin{aligned}
\|v_n - q\|^2 &= \|(1-\alpha_n)y_n + \alpha_n T\tau_n - q\|^2 \\
&= \|(1-\alpha_n)(y_n - q) + \alpha_n(T\tau_n - Tq\|^2 \\
&\le (1-\alpha_n)^2 \|y_n - q\|^2 + 2\alpha_n\langle T\tau_n - Tq, J(v_n - q)\rangle \\
&= (1-\alpha_n)^2 \|y_n - q\|^2 + 2\alpha_n\langle T\tau_n - Tq, J(\tau_n - q)\rangle \\
&+ 2\alpha_n\langle T\tau_n - Tq, J(v_n - q) - J(\tau_n - q)\rangle \\
&\le (1-\alpha_n)^2 \|y_n - q\|^2 + 2\alpha_n \|\tau_n - q\|^2 \\
&- 2\alpha_n\Phi(\|\tau_n - q\|)\|\tau_n - q\| + 2\alpha_n d_n
\end{aligned} \tag{3.6.18}$$

where $d_n = \langle T\tau_n - Tq, J(v_n - q) - J(\tau_n - q)\rangle \to 0$ $(n \to \infty)$ (using J is uniformly continuous on the bounded subsets of X). Again applying Lemma 3.6.4, we have

$$\begin{aligned}
\|\tau_n - q\|^2 &= \|(1-\beta_n)(y_n - q) + \beta_n(Ty_n - q)\|^2 \\
&\le (1-\beta_n)^2 \|y_n - q\|^2 + 2\beta_n\langle Ty_n - Tq, J(\tau_n - q)\rangle \\
&= (1-\beta_n)^2 \|y_n - q\|^2 + 2\beta_n\langle Ty_n - Tq, J(y_n - q)\rangle \\
&+ 2\beta_n\langle Ty_n - Tq, J(\tau_n - q) - J(y_n - q)\rangle \\
&\le (1-\beta_n)^2 \|y_n - q\|^2 + 2\beta_n \|y_n - q\|^2 \\
&- 2\beta_n\Phi(\|y_n - q\|)\|y_n - q\| + 2\beta_n e_n \\
&= (1+\beta_n{}^2)\|y_n - q\|^2 - 2\beta_n\Phi(\|y_n - q\|)\|y_n - q\| + 2\beta_n e_n,
\end{aligned} \tag{3.6.19}$$

where $e_n = \langle T\tau_n - Tq, J(\tau_n - q) - J(y_n - q)\rangle \to 0$ $(n \to \infty)$ (same as above).

Substituting (3.6.19) into (3.6.18) yields

$$\begin{aligned}
\|v_n - q\|^2 &\le (1-\alpha_n)^2 \|y_n - q\|^2 + 2\alpha_n(1+\beta_n{}^2)\|y_n - q\|^2 \\
&- 4\alpha_n\beta_n\Phi(\|y_n - q\|)\|y_n - q\| \\
&+ 4\alpha_n\beta_n e_n - 2\alpha_n\Phi(\|\tau_n - q\|)\|\tau_n - q\| + 2\alpha_n d_n \\
&= (1+\alpha_n{}^2 + 2\alpha_n\beta_n^2)\|y_n - q\|^2 + 4\alpha_n\beta_n[e_n - \Phi(\|y_n - q\|)\|y_n - q\|] \\
&+ 2\alpha_n[d_n - \Phi(\|\tau_n - q\|)\|\tau_n - q\|] \\
&\le \|y_n - q\|^2 + (\alpha_n{}^2 + 2\alpha_n\beta_n^2)M^2 + 4\alpha_n\beta_n[e_n - \Phi(\|y_n - q\|)\|y_n - q\|] \\
&+ 2\alpha_n[d_n - \Phi(\|\tau_n - q\|)\|\tau_n - q\|] \\
&= \|y_n - q\|^2 + 4\alpha_n\beta_n(M^2/2\beta_n + e_n - \Phi(\|y_n - q\|)\|y_n - q\|] \\
&+ 2\alpha_n[M^2/2\alpha_n + d_n - \Phi(\|\tau_n - q\|)\|\tau_n - q\|].
\end{aligned} \tag{3.6.20}$$

We shall prove: $\inf_{n\ge 1}\{\|y_n - q\|\} = 0$. If not, there must exist $\delta > 0$, such that

$\inf_{n\geq 1}\{\|y_n - q\|\} = \delta$. This is to say that $\|y_n - q\| \geq \delta$, for all $n \geq 1$, and hence

$$\Phi(\|y_n - q\|)\,\|y_n - q\| \geq \Phi(\delta)\delta.$$

From (3.6.19), we obtain

$$\begin{aligned}\|\tau_n - q\|^2 &\leq \|y_n - q\|^2 + M^2\beta_n^2 + 2e_n\beta_n - 2\beta_n\Phi(\delta)\delta \\ &= \|y_n - q\|^2 + \beta_n(M^2\beta_n + 2e_n - 2\Phi(\delta)\delta).\end{aligned} \tag{3.6.21}$$

Since $\beta_n \to 0$ and $e_n \to 0$ as $n \to \infty$, then there exists some integer N_1 such that $M^2\beta_n + 2e_n - 2\Phi(\delta)\delta < 0$ for all $n > N_1$. Thus we have

$$\|\tau_n - q\| \leq \|y_n - q\|$$

and

$$-\Phi(\|\tau_n - q\|)\,\|\tau_n - q\| \geq -\Phi(\|y_n - q\|)\,\|y_n - q\| \tag{3.6.22}$$

for all $n > N_1$. Take these facts into consideration in (3.6.20), we obtain

$$\begin{aligned}\|v_n - q\|^2 \leq \|y_n - q\|^2 &+ 4\alpha_n\beta_n(M^2/2\beta_n + e_n - \Phi(\|\tau_n - q\|)\,\|\tau_n - q\|] \\ &+ 2\alpha_n[M^2/2\alpha_n + d_n - \Phi(\|\tau_n - q\|)\,\|\tau_n - q\|],\end{aligned} \tag{3.6.23}$$

for all $n > N_1$. Furthermore, we have the following estimate:

$$\begin{aligned}\|\tau_n - q\| &= \|(1 - \beta_n)(y_n - q) + \beta_n(Ty_n - Tq)\| \\ &\geq \|y_n - q\| - \beta_n\|y_n - q\| - \beta_n\|Ty_n - Tq\| \\ &\geq \|y_n - q\| - \beta_n(M + M_0).\end{aligned} \tag{3.6.24}$$

Again since $\beta_n \to 0$ as $n \to \infty$, then there exists some integer $N_2 > N_1$ such that $\beta_n < \delta/(2(M + M_0))$ for all $n > N_2$. Hence, from (3.6.23) and (3.6.24) we obtain

$$\delta/2 \leq \|y_n - q\| - \delta/2 \leq \|\tau_n - q\| \tag{3.6.25}$$

for all $n > N_2 > N_1$.

Using (3.6.25) and (3.6.23) we have for all $n > N_3$

$$\begin{aligned}\|v_n - q\|^2 \leq \|y_n - q\|^2 &+ 4\alpha_n\beta_n(M^2/2\beta_n + e_n - \Phi(\delta/2)\delta/2] \\ &+ \alpha_n[M^2\alpha_n + 2d_n - \Phi(\delta/2)\delta/2] - \alpha_n\Phi(\|\tau_n - q\|)\,\|\tau_n - q\|.\end{aligned} \tag{3.6.26}$$

Owning to $\alpha_n, \beta_n, e_n, d_n \to 0$ as $n \to \infty$, without loss generality, we may assume that $\exists N_3$ such that

$$M^2/2\beta_n + e_n - \Phi(\delta/2)\delta/2 < 0, \qquad M^2\alpha_n + 2d_n - \Phi(\delta/2)\delta/2 < 0$$

for all $n > N_3$. Thus from (3.6.26) we have $\|v_n - q\| \le \|y_n - q\|$ for all $n > N_3$. It follows that

$$\|y_{n+1} - q\| \le \|y_n - q\| + \varepsilon_n$$

for all $n > N_3$. Using Lemma 3.6.3, we obtain $\lim_{n\to\infty} \|y_n - q\| \exists$.

From (3.6.17) and (3.6.26) we get

$$\begin{aligned}\|y_{n+1} - q\|^2 &\le \|v_n - q\|^2 + 2\|v_n - q\| \varepsilon_n + \varepsilon_n^2 \\ &\le \|y_n - q\|^2 - \alpha_n \Phi(\|\tau_n - q\|) \|\tau_n - q\| + 2M\varepsilon_n + \varepsilon_n^2\end{aligned}$$

for all $n > N_3$. Which implies that

$$\frac{1}{2} \sum_{n=N_3+1}^{\infty} \alpha_n \Phi(\delta/2)\delta/2 \le \|y_{N_3+1} - q\|^2 - \lim_{n\to\infty} \|y_{n+1} - q\|^2 + \sum (2M\varepsilon_n + \varepsilon_n^2) < \infty.$$

This is a contradiction. Thus

$$\inf_n \|y_n - q\| = 0. \tag{3.6.27}$$

It follows that we want to prove $\lim_{n\to\infty} \|y_n - q\| = 0$.

From (3.6.27) we get that there exists an infinite subsequence $\{y_{n_j} - q\} \to 0$ as $j \to \infty$.

Since $\|\tau_{n_j} - y_{n_j}\| = \|-\beta_{n_j} y_{n_j} + \beta_{n_j} T y_{n_j}\| \to 0$ as $j \to \infty$, so that $\|\tau_{n_j} - y_{n_j}\| \le \|\tau_{n_j} - y_{n_j}\| + \|y_{n_j} - q\| \to 0$ as $j \to \infty$, thus we have

$$\begin{aligned}\|y_{n_j+1} - q\| &\le \|v_{n_j} - q\| + \varepsilon_{n_j} \\ &\le \Big\{\|y_{n_j} - q\|^2 + 4\alpha_{n_j}\beta_{n_j}[M^2/2\beta_{n_j} + e_{n_j} - \Phi(\|y_{n_j} - q\|)\|y_{n_j} - q\|] \\ &\quad + 2\alpha_{n_j}[M^2/2\alpha_{n_j} + d_{n_j} - \Phi(\|\tau_{n_j} - q\|)\|\tau_{n_j} - q\|]\Big\}^{1/2} + \varepsilon_{n_j} \to 0,\end{aligned}$$

as $j \to \infty$. Using continuous of Φ and induce method, we obtain $\|y_{n_j+m} - q\| \to 0$ as $j \to \infty$ for any nature number m. ■

Theorem 3.6.8 *Let X be a real uniformly smooth Banach space, $T : X \to X$ be a continuous and Φ-strong accretive operator. Suppose $(I - T)$ has a bounded range. Define $S : X \to X$ by $Sx = f + x - Tx$. Suppose q is the fixed point of the mapping S. Let $\{\alpha_n\}_{n=1}^\infty$, $\{\beta_n\}_{n=1}^\infty$ be two real sequence satisfying:*

$$(1)\ 0 \le \alpha_n, \beta_n < 1,\ \forall n \ge 1 \qquad (2)\ \lim_{n\to\infty} \alpha_n = \lim_{n\to\infty} \beta_n = 0 \qquad (3)\ \sum_{n=1}^{\infty} \alpha_n = \infty.$$

Let $\{x_n\}_{n=1}^\infty$ be the sequence of the Ishikawa iteration generated from an arbitrary $x_1 \in X$ by

$$\begin{cases} x_{n+1} = (1-\alpha_n)x_n + \alpha_n S z_n \\ z_n = (1-\beta_n)x_n + \beta_n S x_n, \quad n \ge 1. \end{cases}$$

Let $\{y_n\}_{n=1}^{\infty}$ *be any sequence in* X *and define* $\{\varepsilon_n\}_{n=1}^{\infty} \subset R^+$ *by*

$$\begin{cases} \varepsilon_n = \|y_{n+1} - (1-\alpha_n)y_n - \alpha_n S\tau_n\| \\ \tau_n = (1-\beta_n)y_n + \beta_n S y_n, \quad n \geq 1. \end{cases}$$

Then the Ishikawa iteration procedure $\{x_n\}_{n=1}^{\infty}$ *is almost* S*-stable.*

Proof. Since T is Φ-strong accretive operator, then S is Φ-strong pseudocontraction. Now the conclusion follows from Theorem 3.6.7. ■

3.7 Exercises

3.1 Consider the problem of approximating a solution $y \in C'[0, t_0]$ of the nonlinear ordinary differential equation

$$\frac{dy}{dt} = K(t, y(t)), \quad 0 \leq t \leq t_0, \ y(0) = y_0.$$

The above equation may be turned into a fixed point problem of the form

$$y(t) = y_0 + \int_0^t K(s, y(s))ds, \quad 0 \leq t \leq t_0.$$

Assume $K(x, y)$ is continuous on $[0, t_0] \times [0, t_0]$ and satisfies the Lipschitz condition

$$\max_{[0,t_0]} \left| K(s, q_1(s)) - K(s, q_2(s)) \right| \leq M\|q_1 - q_2\|, \quad \text{for all } q_1, q_2 \in C[0, t_0].$$

Note that the integral equation above defines an operator P from $C[0, t_0]$ into itself. Find a sufficient condition for P to be a contraction mapping.

3.2. Let F be a contraction mapping on the ball $\bar{U}(x_0, r)$ in a Banach space X, and let

$$\|F(x_0) - x_0\| \leq (1-c)r.$$

Show F has a unique fixed point in $\bar{U}(x_0, r)$.

3.3. Under the assumptions of Theorem 3.1.2, show that the sequence generalized by (3.1.2) minimizes the functional

$$f(x) = \|x - F(x)\|$$

for any x_0 belonging to a closed set A such that $F(A) \subseteq A$.

3.4. Let the equation $F(x) = x$ have a unique solution in a closed subset A of a Banach space X. Assume that there exists an operator F_1 that $F_1(A) \subseteq A$ and F_1 commutes with F on A. Show the equation $x = F_1(x)$ has at least one solution in A.

3.5. Assume that operator F maps a closed set A into a compact subset of itself and satisfies

$$\|F(x) - F(y)\| < \|x - y\| \quad (x \neq y), \quad \text{for all } x, y \in A.$$

Show F has a unique fixed point in A. Apply these results to the mapping $F(x) = x - \frac{x^2}{2}$ of the interval $[0, 1]$ into itself.

3.6. Show that operator F defined by

$$F(x) = x + \tfrac{1}{x}$$

maps the half line $[1, \infty)$ into itself and satisfies

$$\|F(x) - F(y)\| < \|x - y\|$$

but has no fixed point in this set.

3.7. Consider condition

$$\|F(x) - F(y)\| \leq \|x - y\| \quad (x, y \in A).$$

Let A be either an interval $[a, b]$ or a disk $x^2 + y^2 \leq r^2$. Find conditions in both cases under which F has a fixed point.

3.8. Consider the set c_0 of null sequences $x = \{x_1, x_2, \ldots\}$ $(x_n \to 0)$ equipped with the norm $\|x\| = \max_n |x_n|$. Define the operator F by

$$F(x) = \left\{\tfrac{1}{2}(1 + \|x\|), \tfrac{3}{4}x_1, \tfrac{7}{8}x_2, \ldots, \left(1 - \tfrac{1}{2^{n+1}}\right) x_n, \ldots\right\}.$$

Show that $F : \bar{U}(0, 1) \to \bar{U}(0, 1)$ satisfies

$$\|F(x) - F(y)\| < \|x - y\|,$$

but has no fixed points.

3.9. Repeat Exercise 3.8 for the operator F defined in c_0 by $F(x) = \{y_1, \ldots, y_n, \ldots\}$, where $y_n = \frac{n-1}{n} x_n + \frac{1}{n} \sin(n)$ $(n \geq 1)$.

3.10. Repeat Exercise 3.8 for the operator F defined in $C[0, 1]$ by

$$Ax(t) = (1 - t)x(t) + t \sin\left(\tfrac{1}{t}\right).$$

3.11. Let F be a nonlinear operator on a Banach space X which satisfies (3.1.3) on $\bar{U}(0, r)$. Let $F(0) = 0$. Define the resolvent $R(x)$ of F by

$$F(x)f = xFR(x)f + f.$$

Show:

(a) $R(x)$ is defined on the ball $\|f\| \leq (1 - |x|c)r$ if $|x| < c^{-1}$;

(b) $\frac{1}{1+|x|c}\|f - g\| \leq \|R(x)f - R(x)g\| \leq \frac{1}{1-|x|c}\|f - g\|$;

(c) $\|R(x)f - R(y)f\| \leq \frac{c\|f\|\,|x-y|}{(1-|x|c)(1-|y|c)}$.

3.12. Let A be an operator mapping a closed set A of a Banach space X into itself. Assume that there exists a positive integer m such that A^m is a contraction operator. Prove that sequence (3.1.2) converges to a unique fixed point of F in A.

3.13. Let F be an operator mapping a compact set $A \subseteq X$ into itself with $\|F(x) - F(y)\| < \|x - y\|$ ($x \neq y$, all $x, y \in A$). Show that sequence (3.1.2) converges to a fixed point of (3.1.1).

3.14. Let F be a continuous function on $[0, 1]$ with $0 \leq f(x) \leq 1$ for all $x \in [0, 1]$. Define the sequence

$$x_{n+1} = x_n + \tfrac{1}{n+1}(F(x_n) - x_n).$$

Show that for any $x_0 \in [0, 1]$ sequence $\{x_n\}$ ($n \geq 0$) converges to a fixed point of F.

3.15. Show:

(a) A system $x = Ax + b$ of n linear equations in n unknowns $x_1, x_2, \ldots, x_n$ (the components of x) with $A = \{a_{jk}\}$, $j, k = 1, 2, \ldots, n$, b given, has a unique solution x^* if

$$\sum_{k=1}^{n} |a_{jk}| < 1, \qquad j = 1, 2, \ldots, n.$$

(b) The solution x^* can be obtained as the limit of the iteration $(x^{(0)}$, $x^{(1)}$, $x^{(2)}, \ldots\}$, where $x^{(0)}$ is arbitrary and

$$x^{(m+1)} = Ax^{(m)} + b \quad (m \geq 0).$$

(c) The following error bounds hold:

$$\|x^{(m)} - x^*\| \leq \frac{c}{1-c}\|x^{(m-1)} - x^{(m)}\| \leq \frac{c^m}{1-c}\|x^{(0)} - x^{(1)}\|,$$

where

$$c = \max_j \sum_{k=1}^{n} |a_{jk}| \quad \text{and} \quad \|x - z\| = \max_j |x_i - z_i|, \; j = 1, 2, \ldots, n.$$

3.16. (Gershgorin's theorem): If λ is an eigenvalue of a square matrix $A = \{a_{jk}\}$, then for some j, where $1 \leq j \leq n$,

$$|a_{jj} - \lambda| \leq \sum_{\substack{k=1 \\ k \neq j}}^{n} |a_{jk}|.$$

Show that $x = Ax + b$ can be written $Bx = b$, where $B = I - A$, and $\sum_{k=1}^{n} |a_{jk}| < 1$ together with the theorem imply that 0 is not an eigenvalue of B and A has spectral radius less than 1.

3.17. Let (X, d), (X, d_1), (X, d_2) be metric spaces with $d(x, z) = \max_j |x_j - z_j|$, $j = 1, 2, \ldots, n$,

$$d_1(x, z) = \sum_{j=1}^{n} |x_j - z_j| \quad \text{and} \quad d_2(x, z) = \left[\sum_{j=1}^{n} (x_j - z_j)^2\right]^{1/2},$$

respectively. Show that instead of $\sum_{k=1}^{n} |a_{jk}| < 1$, $j = 1, 2, \ldots, n$, we obtain the conditions

$$\sum_{j=1}^{n} |a_{jk}| < 1, \quad k = 1, 2, \ldots, n \quad \text{and} \quad \sum_{j=1}^{n} \sum_{k=1}^{n} a_{jk}^2 < 1.$$

3.18. Let us consider the ordinary differential equation of the first order (ODE)

$$x' = f(t, x), \quad x(t_0) = x_0,$$

where t_0 and x_0 are given real numbers. Assume:

$$|f(t, x)| \leq c_0$$

on $R = \{(t, x) \mid |t - t_0| \leq a, |x - x_0| \leq b\}$,

$$|f(t, x) - f(t, v)| \leq c_1 |x - v|, \qquad \text{for all } (t, x),\ (t, v) \in R.$$

Then show: the (ODE) has a unique solution on $[t_0 - c_2, t_0 + c_2]$, where

$$c_2 < \min\left\{a, \tfrac{b}{c_0}, \tfrac{1}{c_1}\right\}.$$

3.19. Show that f defined by $f(x, y) = |\sin y| + x$ satisfies a Lipschitz condition with respect to the second variable (on the whole xy-plane).

3.20. Does f defined by $f(t, x) = |x|^{1/2}$ satisfy a Lipschitz condition?

3.21. Apply Picard's iteration $x_{n+1}(t) = \int_{t_0}^{t} f(s, x_n(s))ds$ used for the (ODE) $x' = f(t, x)$, $x(t_0) = x_0 + 0$, $x' = 1 + x^2$, $x(0) = 0$. Verify that for x_3, the terms involving $t, t^2, \ldots, t^5$ are the same as those of the exact solution.

3.22. Show that $x' = 3x^{2/3}$, $x(0) = 0$ has infinitely many solutions x.

3.23. Assume that the hypotheses of the contraction mapping principle hold, then show that x^* is accessible from any point $\bar{U}(x_0, r_0)$.

3.24. Define the sequence $\{\bar{x}_n\}$ $(n \geq 0)$ by $\bar{x}_0 = x_0$, $\bar{x}_{n+1} = F(\bar{x}_n) + \varepsilon_n$ $(n \geq 0)$. Assume:

$$\|\varepsilon_n\| \leq \lambda^n \varepsilon \quad (n \geq 0) \quad (0 \leq \lambda < 1);$$

F is a c-contraction operator on $U(x_0, r)$. Then show sequence $\{\bar{x}_n\}$ $(n \geq 0)$ converges to the unique fixed point x^* of F in $\bar{U}(x_0, r)$ provided that

$$r \geq r_0 + \frac{\varepsilon}{1-c}.$$

3.25.

(a) Let $F\colon D \subseteq X \to X$ be an analytic operator. Assume:

- there exists $\alpha \in [0, 1)$ such that

$$\|F'(x)\| \leq \alpha \quad (x \in D); \tag{1}$$

- $$\gamma = \sup_{\substack{k>1 \\ x\in D}} \left\| \tfrac{1}{k!} F^{(k)}(x) \right\|^{\frac{1}{k-1}} \quad \text{is finite;}$$

- there exists $x_0 \in D$ such that

$$\|x_0 - F(x_0)\| \leq \eta \leq \tfrac{3-\alpha-2\sqrt{2-\alpha}}{\gamma}, \quad \gamma \neq 0;$$

- $\bar{U}(x_0, r_1) \subseteq D$, where, r_1, r_2 with $0 \leq r_1 \leq r_2$ are the two zeros of function f, given by

$$f(r) = \gamma(2-\alpha)r^2 - (1 + \eta\gamma - \alpha)r + \eta.$$

Show: method of successive substitutions is well defined, remains in $\bar{U}(x_0, r_1)$ for all $n \geq 0$ and converges to a fixed point $x^* \in \bar{U}(x_0, r_1)$ of operator F.

Moreover, x^* is the unique fixed point of F in $\bar{U}(x_0, r_2)$. Furthermore, the following error bounds hold for all $n \geq 0$:

$$\|x_{n+2} - x_{n+1}\| \leq \beta \|x_{n+1} - x_n\|$$

and

$$\|x_n - x^*\| \leq \frac{\beta^n}{1-\beta}\eta,$$

where

$$\beta = \frac{\gamma\eta}{1-\gamma\eta} + \alpha.$$

The above result is based on the assumption that the sequence

$$\gamma_k = \left\| \frac{1}{k!} F^{(k)}(x) \right\|^{\frac{1}{k-1}} \quad (x \in D), \quad (k > 1)$$

is bounded above by γ. This kind of assumption does not always hold. Let us then not assume sequence $\{\gamma_k\}$ $(k > 1)$ is bounded and define "function" f_1 by

$$f_1(r) = \eta - (1-\alpha)r + \sum_{k=2}^{\infty} \gamma_k^{k-1} r^k.$$

(b) Let $F\colon D \subseteq X \to X$ be an analytic operator. Assume for $x_0 \in D$ function f_1 has a minimum positive zero r_3 such that

$$\bar{U}(x_0, r_3) \subseteq D.$$

Show: method of successive substitutions is well defined, remains in $\bar{U}(x_0, r_3)$ for all $n \geq 0$ and converges to a unique fixed point $x^* \in \bar{U}(x_0, r_3)$ of operator F. Moreover the following error bounds hold for all $n \geq 0$

$$\|x_{n+2} - x_{n+1}\| \leq \beta_1 \|x_{n+1} - x_n\|$$

and

$$\|x_n - x^*\| \leq \frac{\beta_1^n}{1-\beta_1}\eta,$$

where,

$$\beta_1 = \sum_{k=2}^{\infty} \gamma_k^{k-1} \eta^{k-1} + \alpha.$$

3.26.

(a) With F, D as in Exercise 3.25 assume that there exists α such that

$$\|F'(x^*)\| \leq \alpha, \tag{2}$$

and

$$\bar{U}(x^*, r^*) \subseteq D,$$

where,

$$r^* = \begin{cases} \infty, & \text{if } \gamma = 0 \\ \frac{1}{\gamma} \cdot \frac{1-\alpha}{2-\alpha}, & \text{if } \gamma \neq 0. \end{cases}$$

Then, if

$$\beta = \alpha + \frac{\gamma r^*}{1 - \gamma r^*} < 1,$$

show: the method of successive substitutions remains in $\bar{U}(x^*, r^*)$ for all $n \geq 0$ and converges to x^* for any $x_0 \in U(x^*, r^*)$. Moreover, the following error bounds hold for all $n \geq 0$:

$$\|x_{n+1} - x^*\| \leq \beta_n \|x_n - x^*\| \leq \beta \|x_n - x^*\|,$$

where,

$$\beta_0 = 1, \quad \beta_{n+1} = \alpha + \frac{\gamma r^* \beta_n}{1 - \gamma r^* \beta_n} \quad (n \geq 0).$$

The above result was based on the assumption that the sequence

$$\gamma_k = \left\| \frac{1}{k!} F^{(k)}(x^*) \right\|^{\frac{1}{k-1}} \quad (k \geq 2)$$

is bounded γ. In the case where the assumption of boundedness does not necessarily hold, we have the following local alternative.

(b) Let x^* be a fixed point of operator F. Moreover, assume: $\max\limits_{r>0} \sum\limits_{k=2}^{\infty} (\gamma_k r)^{k-1}$ exists and is attained at some $r_0 > 0$. Set

$$p = \sum_{k=2}^{\infty} (\gamma_k r_0)^{k-1};$$

there exist α, δ with $\alpha \in [0, 1)$, $\delta \in (\alpha, 1)$ such that

$$p + \alpha - \delta \leq 0$$

and

$$\bar{U}(x^*, r_0) \subseteq D.$$

Show: the method of successive substitutions $\{x_n\}$ $(n \geq 0)$ remains in $\bar{U}(x^*, r_0)$ for all $n \geq 0$ and converges to x^* for any $x_0 \in \bar{U}(x^*, r_0)$. Moveover the following error bounds hold for all $n \geq 0$:

$$\|x_{n+1} - x^*\| \leq \alpha \|x_n - x^*\| + \sum_{k=2}^{\infty} \gamma_k^{k-1} \|x_n - x^*\|^k \leq \delta \|x_n - x^*\|.$$

Chapter 4

Solving Smooth Equations

In this chapter we are concerned with the problem of approximating a locally unique solution of an equation in a Banach space. The Newton–Kantorovich method is undoubtedly the most popular method for solving such equations.

The theoretical importance of this method should be also emphasized, since using it we can draw conclusions about the existence, uniqueness and domain of distribution of solutions without finding the solution itself, this being sometimes of greater value than the actual knowledge of the solution.

4.1 Linearization of Equations

Let F be a Fréchet-differentiable operator mapping an open and convex subset D of a Banach space X into a Banach space Y. Consider the equation

$$F(x) = 0. \tag{4.1.1}$$

The principal method for constructing approximations x_n to a solution x^* (if it exists) of Equations (4.1.1) is based on successive linearization of the equation.

Assuming that an approximation x_n has been found, we compute the next x_{n+1} by replacing Equation (4.1.1) by

$$F(x_n) + F'(x_n)(x_{n+1} - x_n) = 0. \tag{4.1.2}$$

If $F'(x_n)^{-1} \in L(Y, X)$, then approximation x_{n+1} is given by

$$x_{n+1} = x_n - F'(x_n)^{-1} F(x_n) \quad (n \geq 0). \tag{4.1.3}$$

The iterative procedure generated by (4.1.3) is the famous Newton or Newton–Kantorovich method [205].

Iterates $\{x_n\}$ $(n \geq 0)$ are not always easy to compute. First, x_n may fall outside the set D for a certain n, and secondly, even if this does not happen, $F'(x_n)^{-1}$

may not exist. If sequence $\{x_n\}$ converges to x^* and x_0 is chosen sufficiently close to x^*, the operations $F'(x_n)$ and $F'(x_0)$ will differ only by a small amount due to the continuity of F'. Therefore the modified Newton-Kantorovich method

$$y_{n+1} = y_n - F'(y_0)^{-1} F(y_n) \quad (n \geq 0) \quad (y_0 = x_0) \tag{4.1.4}$$

can also be used to approximate x^*. Method (4.1.4) is easier to use than (4.1.3) since the inverses are computed only once and for all. However the latter method yields worse approximations.

4.2 The Newton-Kantorovich Method

Define the operator P on D by

$$P(x) = x - F'(x)^{-1} F(x) \tag{4.2.1}$$

Then the Newton–Kantorovich method may be regarded as the usual iterative method

$$x_{n+1} = P(x_n) \quad (n \geq 0) \ (x_0 \in D), \tag{4.2.2}$$

for approximating the fixed point x^* of the equation

$$x = P(x) \tag{4.2.3}$$

Consequently, all the results of the previous chapters involving Equation (4.2.3) are applicable.

Suppose that

$$\lim_{n \to \infty} x_n = x^*. \tag{4.2.4}$$

We would like to know under what conditions on F and F' the point x^* is a solution of Equation (4.1.1).

Proposition 4.2.1 *If F' is continuous at $x = x^* \in D$, then we have*

$$F(x^*) = 0. \tag{4.2.5}$$

Proof. The approximations x_n satisfy the equation

$$F'(x_n)(x_{n+1} - x_n) = -F(x_n). \tag{4.2.6}$$

Since the continuity of F at x^* follows from the continuity of F', taking the limit as $n \to \infty$ in (4.2.6) we obtain (4.2.5). ∎

Proposition 4.2.2 *If*

$$\|F'(x)\| \leq b \tag{4.2.7}$$

in some closed ball in D which contains $\{x_n\}$, then x^* is a solution of $F(x) = 0$.

Proof. By (4.2.7) we get

$$\lim_{n\to\infty} F(x_n) = F(x^*), \tag{4.2.8}$$

and since

$$\|F(x_n)\| \leq b\|x_{n+1} - x_n\|, \tag{4.2.9}$$

(4.2.5) is obtained by taking the limit as $n \to \infty$ in (4.2.4). ■

Proposition 4.2.3 *If*

$$\|F''(x)\| \leq K \tag{4.2.10}$$

in some closed ball

$$\bar{U}(x_0, r) = \{x \in X \mid \|x - x_0\| \leq r\} \subseteq D \quad (r \in (0, +\infty)) \tag{4.2.11}$$

which contains $\{x_n\}$*, then* x^* *is a solution of equation* $F(x) = 0$.

Proof. By (4.2.10)

$$\|F'(x) - F'(x_0)\| \leq K\|x - x_0\| \leq Kr, \tag{4.2.12}$$

for all $x \in \bar{U}(x_0, r)$. Moreover, we can write

$$\|F'(x)\| \leq \|F'(x_0)\| + \|F'(x) - F'(x_0)\|, \tag{4.2.13}$$

so the conditions of Proposition 4.2.2 hold with

$$b = \|F'(x_0)\| + Kr. \tag{4.2.14}$$

As in Kantorovich (see, e.g., [207]) consider constants B_0, η_0 such that

$$\left\|[F'(x_0)]^{-1}\right\| \leq B_0, \tag{4.2.15}$$

$$\|x_1 - x_0\| \leq \eta_0 = \eta, \tag{4.2.16}$$

respectively. ■

We state and give a slightly different and improved proof of the celebrated Newton-Kantorovich theorem [206], [207, Th. 6, (I.XVII), p. 708]:

Theorem 4.2.4 *If condition (4.2.10) holds in some closed ball* $\bar{U}(x_0, r)$ *in* D *and*

$$h_0 = B_0\eta_0 K \leq \tfrac{1}{2}, \tag{4.2.17}$$

then the Newton-Kantorovich sequence (4.1.3)*, starting from* x_0*, converges to a solution* x^* *of Equation* (4.1.1) *which exists in* $\bar{U}(x_0, r)$*, provided that*

$$r \geq r_0 = \frac{1 - \sqrt{1 - 2h_0}}{h_0}\eta_0. \tag{4.2.18}$$

Proof. We first show that (4.1.3) is well defined. By (4.2.10) we get

$$\|F'(x_1) - F'(x_0)\| \le K\|x_1 - x_0\| = K\eta_0. \tag{4.2.19}$$

Using (4.2.17), we obtain

$$\|F'(x_1) - F'(x_0)\| \le K\eta_0 \le \frac{1}{2B_0} < \frac{1}{\|[F'(x_0)]^{-1}\|}. \tag{4.2.20}$$

By Theorem 1.3.2, $[F'(x_1)]^{-1}$ exists, and

$$\|[F'(x_1)]^{-1}\| \le \frac{\|[F'(x_0)]^{-1}\|}{1 - \|[F'(x_0)]^{-1}\| \cdot \|F'(x_1) - F'(x_0)\|} \tag{4.2.21}$$

or

$$\|[F'(x_1)]^{-1}\| \le \frac{B_0}{1 - B_0\eta_0 K} = \frac{B_0}{1 - h_0} = B_1. \tag{4.2.22}$$

Hence, x_2 exists. To estimate $\|x_2 - x_1\|$, note that

$$\|x_2 - x_1\| = \|[F'(x_1)]^{-1}F(x_1)\| \tag{4.2.23}$$

and

$$\begin{aligned}[F'(x_1)]^{-1} &= \left(\sum_{n=0}^{\infty}\left\{I - [F'(x_0)]^{-1}F'(x_1)\right\}^n\right)[F'(x_0)]^{-1} \\ &= \left(\sum_{n=0}^{\infty}\left\{[F'(x_0)]^{-1}[F'(x_0) - F'(x_1)]\right\}^n\right)[F'(x_0)]^{-1}. \end{aligned} \tag{4.2.24}$$

Consequently

$$\|[F'(x_1)]^{-1}F(x_1)\| \le \frac{1}{1 - B_0\eta_0 K}\|[F'(x_0)]^{-1}F(x_1)\| \tag{4.2.25}$$

or

$$\|x_2 - x_1\| \le \frac{1}{1 - h_0}\|[F'(x_0)]^{-1}F(x_1)\|. \tag{4.2.26}$$

To estimate $\|[F'(x_0)]^{-1}F(x_1)\|$ consider the operator

$$F_1(x) = x - [F'(x_0)]^{-1}F(x). \tag{4.2.27}$$

Note

$$F_1'(x_0) = I - [F'(x_0)]^{-1}F(x_0) = 0. \tag{4.2.28}$$

By (4.2.27)

$$[F'(x_0)]^{-1} F(x_1) = x_1 - F_1(x_1), \tag{4.2.29}$$

and since

$$x_1 = F_1(x_0) \tag{4.2.30}$$

(4.2.28) may be used to write (4.2.29) as

$$[F'(x_0)]^{-1} F(x_1) = -[F_1(x_1) - F_1(x_0) - F_1'(x_0)(x_1 - x_0)], \tag{4.2.31}$$

and

$$\left\| [F'(x_0)]^{-1} F(x_1) \right\| \leq \sup_{\bar{x} \in L(x_0, x_1)} \left\| F_1''(\bar{x}) \right\| \frac{\|x_1 - x_0\|^2}{2}. \tag{4.2.32}$$

Using (4.2.27)

$$F_1''(x) = -[F'(x_0)]^{-1} F''(x). \tag{4.2.33}$$

Since $x_1 \in \bar{U}(x_0, r)$, if (4.2.28) holds, then

$$\|F_1''(x)\| \leq B_0 K \tag{4.2.34}$$

on $L(x_0, x_1)$. From (4.2.32)

$$\left\| [F'(x_0)]^{-1} F(x_1) \right\| \leq \frac{B_0 K \eta_0^2}{2} = \frac{h_0 \eta_0}{2}, \tag{4.2.35}$$

and from (4.2.26)

$$\|x_2 - x_1\| \leq \frac{1}{2} \frac{h_0}{1 - h_0} \eta_0 = \eta_1 \tag{4.2.36}$$

and obviously $\eta_1 \leq \frac{1}{2}\eta_0$. The constant

$$h_1 = B_1 \eta_1 K = \frac{B_0}{1 - h_0} \frac{1}{2} \frac{h_0}{1 - h_0} \eta_0 K = \frac{1}{2} \frac{h_0^2}{(1 - h_0)^2} \leq 2h_0^2 \leq \frac{1}{2}, \tag{4.2.37}$$

so that (4.2.17) is satisfied if x_0 is replaced by x_1. Now if for

$$r_1 = \frac{1 - \sqrt{1 - 2h}}{h_1} \eta_1 \tag{4.2.38}$$

it can be shown that

$$\bar{U}(x_1, r_1) \subset \bar{U}(x_0, r_0), \tag{4.2.39}$$

then condition (4.2.10) will hold with x_0 replaced by x_1. By direct substitution

$$\sqrt{1-2h_1} = \left(1 - \frac{h_0^2}{(1-h_0)^2}\right)^{1/2} = \frac{1}{1-h_0}\sqrt{1-2h_0} \tag{4.2.40}$$

and

$$\begin{aligned} r_1 &= \frac{1-(1-h_0)^{-1}\sqrt{1-2h_0}}{\frac{1}{2}\left[h_0^2/(1-h_0)^2\right]} \frac{1}{2}\frac{h_0}{1-h_0}\eta_0 \\ &= \left(\frac{1-\sqrt{1-2h_0}}{h_0} - 1\right)\eta_0 = r_0 - \eta_0. \end{aligned} \tag{4.2.41}$$

Consequently, if $x \in \bar{U}(x_1, r_1)$, then $\|x - x_1\| \leq r_1 = r_0 - \eta_0$ and

$$\|x - x_0\| \leq \|x - x_1\| + \|x_1 - x_0\| \leq (r_0 - \eta_0) + \eta_0 = r_0, \tag{4.2.42}$$

so that $x \in \bar{U}(x_0, r_0)$, which establishes (4.2.39). It follows by mathematical induction that the Newton process (4.1.3) generates an infinite sequence $\{x_n\}$, starting from an x_0 at which the hypotheses of the theorem are satisfied. It remains to be shown that this sequence converges to a solution x^* of (4.1.1). Along with $\{x_n\}$ the sequences of numbers $\{B_n\}, \{\eta_n\}$, and $\{h_n\}$ defined by

$$B_n = \frac{B_{n-1}}{1-h_{n-1}}, \tag{4.2.43}$$

$$\eta_n = \frac{1}{2}\frac{h_{n-1}\eta_{n-1}}{1-h_{n-1}}, \tag{4.2.44}$$

$$h_n = \frac{1}{2}\frac{h_{n-1}^2}{(1-h_{n-1})^2}, \tag{4.2.45}$$

respectively, are obtained for $n = 1, 2, \ldots$. We have

$$h_n \leq 2h_{n-1}^2 \leq \tfrac{1}{2}(2h_{n-2})^4 \leq \cdots \leq \tfrac{1}{2}(2h_0)^{2^n} \tag{4.2.46}$$

from (4.2.37) and

$$\begin{aligned} \eta_n &= \tfrac{1}{2}\tfrac{h_{n-1}\eta_{n-1}}{1-h_{n-1}} \leq h_{n-1}\eta_{n-1} \leq h_{n-1}h_{n-2}\eta_{n-2} \leq \cdots \\ &\leq \tfrac{1}{2^n}(2h_0)^{2^{n-1}}(2h_0)^{2^{n-2}}\cdots(2h_0)\eta_0 = \tfrac{1}{2^n}(2h_0)^{2^n-1}\eta_0. \end{aligned} \tag{4.2.47}$$

For any positive integer $p, x_{n+p} \in \bar{U}(x_n, r_n)$, so that

$$\|x_{n+p} - x_n\| \leq r_n = \tfrac{1-\sqrt{1-2h_n}}{h_n}\eta_n \leq 2\eta_n, \tag{4.2.48}$$

and thus from (4.2.47)

$$\|x_{n+p} - x_n\| \leq \tfrac{1}{2^{n-1}}(2h_0)^{2^n-1}\eta_0. \tag{4.2.49}$$

Therefore $\{x_n\}$ is a Cauchy sequence which has a limit $x^* \in \bar{U}(x_0, r_0)$, and x^* is a solution of (4.1.1) by Proposition 4.2.3. ■

Remark 4.2.1 *It is well known (see, e.g., [207]) that conclusions of Theorem 4.2.4 hold, if we assume operator F is only once Fréchet differentiable; conditions (4.2.14), (4.2.16), (4.2.17) are replaced by Lipschitz condition*

$$\|F'(x_0)\left[F'(x) - F'(y)\right]\| \leq \ell \|x - y\|, \quad \text{for all } x, y \in \bar{U}(x_0, r), \tag{4.2.50}$$

and the so-called Newton-Kantorovich hypothesis

$$h \leq \ell\eta \leq \frac{1}{2}. \tag{4.2.51}$$

The advantages of presenting the results in affine invariant form have been explained in [143].

Remark 4.2.2 *Although upper bounds on the Lipschitz K have already been given by severala authors [6], [68], [99], [129], [130], [134], [141], [143], [145], [180], [183], [207], [210], [211], [213], [223], [257], [263], [264], [288], and for large classes of operators F (see also the exercises at the end of the Chapters 1 and 4) here we decided to provide an interesting and nontrivial example where in contrast to the results mentioned in the references above, the actual value of ℓ and not just an upper bound can be obtained.*

Example 4.2.1 *Assume that the second Fréchet derivative of F is a quadratic integral operator Q, where*

$$\left[F''(x)vw\right](s) = Q(x)(s) = \int_0^1 (s - t)x^2(t)v(t)w(t)dt, \tag{4.2.52}$$

in the space $C[0,1]$ of continuous functions on the interval $0 \leq s \leq 1$, then

$$|Q(x)(s)| = \left| s\int_0^1 x^2(t)dt - \int_0^1 tx^2(t)dt \right| \leq \|x\|^2 \left(1 + \tfrac{1}{2}\right)$$

so

$$\max_s \|Q(x)\| \leq \|x\|^2 \tfrac{3}{2}$$

which yields

$$\|Q\| \leq \tfrac{3}{2}.$$

We can do better here. Take a fixed x, with $\|x\| \leq 1$; then

$$\|Q(x)\| = \max_s |sa - b|,$$

where

$$a = \int_0^1 x^2(t)dt \geq 0,$$

and

$$b = \int_0^1 tx^2(t)dt \geq 0.$$

Notice that

$$ab = \int_0^1 (1-t)x^2(t)dt \geq 0$$

and

$$\|Q(x)\| = \max\{b, a-b\}.$$

But

$$tx^2(x) \leq t, \text{ for all } x^2(t) \leq 1,$$

so

$$b \leq \int_0^1 tdt = \tfrac{1}{2}$$

and

$$(1-t)x^2(t) \leq 1-t,$$

so

$$a - b \leq \int_0^1 (1-t)dt = \tfrac{1}{2}, \text{ for all } \|x\| \leq 1.$$

Therefore,

$$\|Q\| = \sup_{\|x\|\leq 1} \|Q(x)\| \leq \max\left\{\tfrac{1}{2}, \tfrac{1}{2}\right\} = \tfrac{1}{2}$$

and

$$\|Q\| \leq \tfrac{1}{2}. \tag{4.2.53}$$

Finally, we can do even better. Take $x(t) = 1$ for all t so

$$\|x\| = 1$$

and

$$\|Q(x)\| = \sup_s \left| s\int_0^1 dt - \int_0^1 tdt \right| = \sup_s \left|s - \tfrac{1}{2}\right| \geq \tfrac{1}{2}. \tag{4.2.54}$$

Now (4.2.53) and (4.2.54) give $\|Q\| = \frac{1}{2} = K$.

Finally we complete this section with the solution of a nonlinear integral equation of mixed Hammerstein type [68], [99]:

$$x(s) = 1 + \int_0^1 G(s,t)\left[x(t)^{\frac{5}{2}} + \frac{x(t)^3}{5}\right] dt,$$

where $x \in C[0,1]$, $s,t \in [0,1]$ and the kernel function G is given by

$$G(s,t) = \begin{cases} (1-s)t, & t \leq s, \\ s(1-t), & s \leq t. \end{cases}$$

Example 4.2.2 *Define operator* $F : \bar{U}(0, \frac{7}{5}) \subseteq C[0,1] \to C[0,1]$ *by*

$$[F(x)](s) = x(s) - 1 - \int_0^1 G(s,t)\left[x(t)^{\frac{5}{2}} + \frac{x(t)^3}{5}\right] dt.$$

Then we get (see Section 1.5)

$$\left[F''(x)yz\right](s) = -\int_0^1 G(s,t)\left[\tfrac{15}{4}x(t)^{\frac{1}{2}} + \tfrac{6}{5}x(t)\right] z(t)y(t)dt.$$

Set $x_0(s) = 1$. *Then we get* $\|I - F'(x_0)\| \leq \frac{31}{80} < 1$, *so* $F'(x_0)^{-1}$ *exists* $\left\|F'(x_0)^{-1}\right\| \leq \beta_0 = \frac{80}{49}$, $\left\|F'(x_0)^{-1}F'(x_0)\right\| \leq \frac{12}{49} = \eta_0$ *and* $\|F''(x_0)\| \leq K = .76463275$. *Therefore condition (4.2.17) becomes*

$$h_0 = .305725714 < .5.$$

That is, the conclusions of Theorem 4.2.4 hold in $\bar{U}(x_0, r_0) \subseteq \bar{U}(0, \frac{7}{5})$ *where for* r_0 *given by (4.2.18) we found*

$$r_0 = .301721716.$$

4.3 Local Convergence

We can now show a local result for (4.1.3).

Theorem 4.3.1 *[288] If* x^* *is a simple zero of* $F, B^* \geq \left\|[F'(x^*)]^{-1}\right\|$, *and*

$$A = \left\{x : \|x - x^*\| < 1/(B^*K)\right\} \subset \bar{U}(x_0, r) \tag{4.3.1}$$

then the hypotheses (i) with $h < \frac{1}{2}$ *and (ii) of theorem 4.2.4 are satisfied at each* $x_0 \in B$, *where*

$$B = \left\{x : \|x - x^*\| < (2-\sqrt{2})/(2B^*K)\right\}. \tag{4.3.2}$$

Proof. For $x_0 \in B$,

$$\|F'(x_0) - F'(x^*)\| \leq K\|x_0 - x^*\| < \left(2-\sqrt{2}\right)/\left(2B^*\right) < \left\|[F'(x^*)]^{-1}\right\|^{-1}, \quad (4.3.3)$$

so that $[F'(x_0)]^{-1}$ exists, and

$$B = \frac{B^*}{1 - B^*K\|x_0 - x^*\|} \geq \left\|[F'(x_0]^{-1}\right\| \quad (4.3.4)$$

by Theorem 1.3.2. By the fundamental theorem of calculus we obtain

$$\begin{aligned} F(x^*) - F(x_0) =& \int_0^1 F'(x_0 + \theta(x^* - x_0))(x^* - x_0)\,d\theta \\ =& F'(x_0)(x^* - x_0) + \\ &+ \int_0^1 [F'(x_0 + \theta(x^* - x_0)) - F'(x_0)](x^* - x_0)\,d\theta. \end{aligned} \quad (4.3.5)$$

Since $F(x^*) = 0$,

$$\begin{aligned} -[F'(x_0)]^{-1} F(x_0) =& x^* - x_0 + [F'(x_0)]^{-1} \cdot \\ &\cdot \int_0^1 [F'(x_0 + \theta(x^* - x_0)) - F'(x_0)](x^* - x_0)\,d\theta, \end{aligned} \quad (4.3.6)$$

and hence

$$\left\|[F'(x_0)]^{-1}F(x_0)\right\| \leq \left\{1 + BK\|x^* - x_0\| \int_0^1 \theta d\theta\right\} \|x^* - x_0\|. \quad (4.3.7)$$

By (4.3.4) we get

$$\eta = \frac{1 - \frac{1}{2}B^*K\|x^* - x_0\|}{1 - B^*K\|x^* - x_0\|}\|x^* - x_0\| \geq \left\|[F'(x_0)]^{-1}F(x_0)\right\|. \quad (4.3.8)$$

It follows that for $x_0 \in A$,

$$h = BK\eta = \frac{1 - \frac{1}{2}B^*K\|x^* - x_0\|}{\left(1 - B^*K\|x^* - x_0\|\right)^2}B^*K\|x^* - x_0\| < \tfrac{1}{2}, \quad (4.3.9)$$

hence (4.2.17) holds with $h < \frac{1}{2}$. Using the values of h and η given above,

$$\left(1 - \sqrt{1 - 2h}\right)\frac{\eta}{h} < \frac{\eta}{h} = \frac{1}{B^*K} - \|x^* - x_0\|, \quad (4.3.10)$$

and thus $U(x_0, r_0) \subseteq A \subseteq U(x_0, r) \subset \Omega_* \subset \Omega$ satisfying (4.1.2). ∎

Remark 4.3.1 *The value* $\left(2-\sqrt{2}\right)(2B^*K)$ *for the radius of* Ω^* *is best possible, as the following scalar example shows. Take* $x^* > 0$ *real, and consider the quadratic*

$$F(x) = \tfrac{1}{2}K\left(x^2 - x^{*2}\right),$$

where $K > 0$. *One has* $\left(2-\sqrt{2}\right)/\left(2B^*K\right) = \left(1-\tfrac{1}{2}\sqrt{2}\right)x^*$, *and for*

$$x_0 = x^* - \left(1-\tfrac{1}{2}\sqrt{2}\right)x^* = \tfrac{1}{2}\sqrt{2}x^*,$$

it follows that

$$B = \left(\tfrac{1}{2}\sqrt{2}Kx^*\right)^{-1}, \quad \eta = \tfrac{1}{4}\sqrt{2}x^*,$$

from which $h = BK\eta = \tfrac{1}{2}$.

4.4 Convergence under Twice-Fréchet Differentiability Only at a Point

Recently Hu and Ding in [196] being motivated by our work in [91] provided a semilocal convergence analysis for method (4.1.3), assuming that operator F is continuously twice Fréchet-differentiable only at the initial guess x_0. The order of convergence according to their analysis is only linear. Here we show that the order of convergence is quadratic and under weaker hypotheses. The local case not considered in [196] is also covered here.

Finally a numerical example is provided to show how our results can be applied.

4.4.1 Semilocal Convergence Analysis for Comparison Reasons

We can state the following semilocal convergence result for Newton's method whose proof can be found in [196] (see also our proof in Theorem 4.2.4):

Theorem 4.4.1 *Let* X, Y *be Banach spaces, and* $F\colon D \subseteq X \to Y$ *be a twice continuously Fréchet-differentiable operator at some* $x_0 \in D$. *The hypothesis on the operator* F *implies: for all* $\varepsilon > 0$ *there exists* $\delta = \delta(\varepsilon) > 0$ *such that*

$$\|F'(x_0)^{-1}[F'(x) - F'(x_0)]\| < \varepsilon$$

and

$$\|F'(x_0)^{-1}[F''(x) - F''(x_0)]\| < \varepsilon$$

for all $x \in \overline{U}(x_0, \delta)$. *Assume:*

(a) $F'(x_0)^{-1} \in L(Y, X)$;

(b) there exist parameters η, b *such that*

$$0 < \|F'(x_0)^{-1}F(x_0)\| \leq \eta < 1, \quad \|F'(x_0)^{-1}F''(x_0)\| \leq b;$$

(c) set

$$c = \frac{b+\varepsilon}{2(1-\varepsilon)}, \quad \frac{1}{1-c}\eta < \delta$$

for $b+\varepsilon \in \left(0, \frac{2}{3}\right)$*;*

(d) $\overline{U}(x_0, \delta) \subseteq D$.

Then sequence $\{x_n\}$ $(n \geq 0)$ *generated by (4.1.3) is well defined, remains in* $\overline{U}(x_0, \delta)$*, for all* $n \geq 0$*, and converges to a solution* $x^* \in \overline{U}(x_0, \delta)$ *of equation* $F(x) = 0$*. Moreover the following estimates hold for all* $n \geq 0$*:*

$$\|x_{n+1} - x_n\| \leq c^{\eta}\|x_1 - x_0\|, \tag{4.4.1}$$

and

$$\|x_n - x^*\| \leq \frac{c^{\eta}}{1-c}\|x_1 - x_0\|. \tag{4.4.2}$$

Next we show how to improve on the above result.

We can obtain the following local result for method (4.1.3) corresponding to Theorem 4.4.1.

Theorem 4.4.2 *Let* $F\colon D \subseteq X \to Y$ *be a twice Fréchet-differentiable operator. Assume:*

(a) there exists $x_0 \in D$ *such that* $F'(x_0)^{-1} \in L(Y, X)$*;*

(b) for all $\varepsilon_0 > 0$*,* $\varepsilon > 0$*,* $t \in [0,1]$ *there exist* $\delta = \delta(\varepsilon, \varepsilon_0, t)$ *and* $b \geq 0$ *such that*

$$\|F'(x_0)^{-1}[F''(x_0 + t(x - x_0)) - F''(x_0)] < \varepsilon_0, \tag{4.4.3}$$

$$\|F'(x_0)^{-1}[F''(x + t(y - x)) - F''(x_0)] < \varepsilon, \tag{4.4.4}$$

and

$$\|F'(x_0)^{-1}F''(x_0)\| \leq b; \tag{4.4.5}$$

(c) for

$$a_0 = \frac{b+\varepsilon}{2[1 - (\varepsilon_0 + b)\delta]}, \quad a = a_0\eta \tag{4.4.6}$$

the following conditions hold

$$a < 1, \quad a_0(1 - a^2) \leq 1, \tag{4.4.7}$$

$$\frac{1}{a_0}\frac{\eta}{1 - a^2} \leq \delta, \tag{4.4.8}$$

$$(\varepsilon_0 + b)\delta < 1, \tag{4.4.9}$$

and

$$\overline{U}(x_0, \delta) \subseteq D. \tag{4.4.10}$$

Then, sequence $\{x_n\}$ $(n \geq 0)$ generated by method (4.1.3) is well defined, remains in $\overline{U}(x_0, \delta)$ for all $n \geq 0$ and converges to a solution x^* of equation $F(x) = 0$ with $x^* \in \overline{U}(x_0, \delta)$. Moreover the following estimates hold for all $n \geq 0$

$$\|x_{n+1} - x_n\| \leq \frac{1}{a_0} a^{2^n} \leq \frac{1}{a_0} a^{2n} \tag{4.4.11}$$

and

$$\|x^* - x_n\| \leq \frac{1}{a_0} \lim_{m\to\infty} \left[a^{2^{n+m}} + a^{2^{n+m-1}} + \cdots + a^{2^n}\right] \tag{4.4.12}$$

$$\leq \frac{1}{a_0} \frac{a^{2n}}{1 - a^2} \eta. \tag{4.4.13}$$

Furthermore the solution x^* is unique in $\overline{U}(x_0, \delta)$ if $a_0 \delta < 1$.

Proof. We first show operator $F'(x)^{-1}$ is invertible for all $x \in \overline{U}(x_0, \delta)$. Indeed we can obtain the identity:

$$F'(x_0)^{-1}[F'(x) - F'(x_0)] = \int_0^1 F'(x_0)^{-1}[F''(x_0 + \theta(x - x_0)) \tag{4.4.14}$$
$$- F''(x_0)]d\theta(x - x_0) + F'(x_0)^{-1}F''(x_0)(x - x_0).$$

By hypotheses we get

$$\|F'(x_0)^{-1}[F'(x) - F'(x_0)]\| \leq (\varepsilon_0 + b)\|x - x_0\| \leq (\varepsilon_0 + b)\delta < 1. \tag{4.4.15}$$

It follows from (*4.4.15*) and the Banach Lemma on invertible operators 1.3.2 that $F'(x)^{-1} \in L(Y, X)$ and

$$\|F'(x)^{-1}F'(x_0)\| \leq \frac{1}{1 - (\varepsilon_0 + b)\delta}. \tag{4.4.16}$$

Note that:

$$\|x + \theta(y - x) - x_0\| \leq (1 - \theta)\|x - x_0\| + \theta\|y - x_0\|$$
$$\leq (1 - \theta)\delta + \theta\delta = \delta.$$

Hence $x + \theta(y - x) - x_0 \in U(x_0, \delta)$ for all $x, y \in U(x_0, \delta)$. By (4.1.3) for $n = 0$, we get

$$\|x_1 - x_0\| = \|F'(x_0)^{-1}F(x_0)\| \leq \eta \leq \frac{\eta}{a_0(1 - a^2)} \leq \delta.$$

That is $x_1 \in U(x_0, \delta)$. Assume $x_k \in U(x_0, \delta)$ for all $k = 0, 1, \ldots, m-1$. Using (4.1.3) we can obtain the identity

$$F(x_{k+1}) = F(x_{k+1}) - F(x_k) - F'(x_k)(x_{k+1} - x_k) \tag{4.4.17}$$

$$= \int_0^1 \{F''[x_k + \theta(x_{k+1} - x_k)] - F''(x_0)\}(1-\theta)(x_{k+1} - x_k)^2 d\theta$$

$$+ \frac{1}{2}F''(x_0)(x_{k+1} - x_k)^2. \tag{4.4.18}$$

By (*4.4.4*), (*4.4.5*) and (*4.4.17*) we get

$$\|F'(x_0)^{-1}F(x_{k+1})\| \leq \frac{b+\varepsilon}{2}\|x_{k+1} - x_k\|^2. \tag{4.4.19}$$

Moreover by (*4.4.16*) for $x = x_k$, (4.1.3) and (*4.4.18*) we deduce

$$\|x_{k+1} - x_k\| \leq \|F'(x_k)^{-1}F'(x_0)\| \cdot \|F'(x_0)^{-1}F(x_k)\|$$

$$\leq a_0\|x_k - x_{k-1}\|^2$$

$$\Rightarrow$$

$$a_0\|x_{k+1} - x_k\| \leq (a_0\|x_k - x_{k-1}\|)^{2^1} \leq \cdots \leq (a_0\|x_1 - x_0\|)^{2^k}$$

$$\Rightarrow$$

$$\|x_{k+1} - x_k\| \leq \frac{1}{a_0}a^{2^k} \leq \frac{1}{a_0}a^{2k}, \tag{4.4.20}$$

which shows (*4.4.11*). For all $m \geq 0$ we get

$$\|x_{k+m} - x_k\| \leq$$

$$\leq \|x_{k+m} - x_{k+m-1}\| + \|x_{k+m-1} - x_{k+m-2}\| + \cdots + \|x_{k+1} - x_k\|$$

$$\leq \frac{1}{a_0}\left[a^{2^{k+m}} + a^{2^{k+m-1}} + \cdots + a^{2^k}\right]$$

$$\leq \frac{1}{a_0}\left[(a^{2(k+m)} + a^{2(k+m-1)} + \cdots + a^{2k}\right]$$

$$\leq \frac{1}{a_0}\frac{1-a^{2m}}{1-a^2}a^{2k}. \tag{4.4.21}$$

In particular for $k = 0$ it follows from (4.4.20) and (4.4.8) that $x_n \in \overline{U}(x_0, \delta)$ for all $n \geq 0$. Sequence $\{x_n\}$ is Cauchy in a Banach space X and as such it converges to some $x^* \in \overline{U}(x_0, \delta)$. Estimate (4.4.12) follows from (4.4.20) by letting $m \to \infty$. Finally, to show uniqueness let $y^* \in \overline{U}(x_0, \delta)$ be a solution of equation $F(x) = 0$. Since,

$$x_{k+1} - y^* = [F'(x_k)^{-1}F'(x_0)]F'(x_0)^{-1}\int_0^1 \{F''[x_k + \theta(x_k - y^*)]$$

$$- F''(x_0)\}(1-\theta)(x_k - y^*)^2 d\theta$$

$$+ [F'(x_k)^{-1}F'(x_0)]\frac{1}{2}F'(x_0)^{-1}F''(x_0)(x_k - y^*)^2, \tag{4.4.22}$$

we obtain

$$\|x_{k+1} - y^*\| \le a_0 \|x_k - y^*\|^2, \tag{4.4.23}$$

from which it follows $\lim_{k\to\infty} x_k = y^*$. However, we showed above $\lim_{k\to\infty} x_k = x^*$. Hence, we deduce

$$x^* = y^*.$$

■

Remark 4.4.1 *(a) In general*

$$\varepsilon_0 \le \varepsilon \tag{4.4.24}$$

holds. We can certainly choose $\varepsilon = \varepsilon_0$ in (4.4.3). However in case strict inequality holds in (4.4.23) more precise error bounds can be obtained. Conditions (4.4.3) and (4.4.4) certainly hold if operator F is twice continuously Fréchet-differentiable.

(b) If condition (*4.4.3*) is replaced by

$$\|F'(x_0)^{-1}[F'(x) - F'(x_0)]\| < \varepsilon \tag{4.4.22'}$$

then it can easily be seen by (*4.4.15*) that conclusions of Theorem 4.4.2 hold for $\overline{a}_0$, $\overline{a}$ replacing a_0, a where

$$\overline{a}_0 = \frac{b+\varepsilon}{2(1-\varepsilon)} \quad \text{and} \quad \overline{a} = \overline{a}_0 \eta. \tag{4.4.25}$$

This case was considered in [196] (see Theorem 1). However the stronger condition that operator F is actually twice continuously Fréchet-differentiable was imposed and the convergence of the sequence $\{x_n\}$ is only linear.

(c) Note that even in the case of Theorem 4.4.2 using (*4.4.3*), (*4.4.4*) instead of (*4.4.22*)$'$ and (*4.4.4*) essentially used there the results can be weaker provided that

$$(\varepsilon_0 + b)\delta < \varepsilon. \tag{4.4.26}$$

We can prove the following local convergence theorem for Newton's method:

Theorem 4.4.3 *Let $F\colon D \subseteq X \to Y$ be a twice-Fréchet differentiable operator. Assume:*

(a) there exists a solution x^* of equation $F(x) = 0$ such that $F'(x^*)^{-1} \in L(Y, X)$;

(b) for all $\varepsilon_0 > 0$, $\varepsilon > 0$, $t \in [0,1]$ there exist $\delta = \delta(\varepsilon, \varepsilon_0, t)$ and $b \ge 0$ such that:

$$\|F'(x^*)^{-1}[F''(x^* + \theta(x - x^*)) - F''(x^*)]\| < \varepsilon_0, \tag{4.4.27}$$

$$\|F'(x^*)^{-1}[F''(x + \theta(x^* - x)) - F''(x^*)]\| < \varepsilon, \tag{4.4.28}$$

and

$$\|F'(x^*)^{-1}F''(x^*)\| \le b; \tag{4.4.29}$$

(c)

$$r^* \le \delta \tag{4.4.30}$$

where,

$$r^* = \frac{2}{2\varepsilon_0 + \varepsilon + 3b}. \tag{4.4.31}$$

Then, sequence $\{x_n\}$ generated by method (4.1.3) is well defined, remains in $U(x^*, r^*)$ for all $n \ge 0$ and converges to x^* provided that $x_0 \in U(x^*, r^*)$. Moreover the following estimates hold for all $n \ge 0$

$$\|x_{n+1} - x^*\| \le \alpha \|x_n - x^*\|^2, \tag{4.4.32}$$

where,

$$\alpha = \frac{b + \varepsilon}{2[1 - (\varepsilon_0 + b)r^*]}. \tag{4.4.33}$$

Proof. It follows immediately from (*4.4.21*) by using x^*, (*4.4.26*), (*4.4.27*) instead of x_0, (*4.4.3*) and (*4.4.4*) respectively. ■

Remark 4.4.2 *Local results were not given in [196]. If they were the following assumptions would have been made*

$$\|F'(x^*)^{-1}[F'(x) - F'(x^*)]\| < \overline{\varepsilon} \tag{4.4.34}$$

$$\|F'(x^*)^{-1}[F''(x) - F''(x^*)]\| < \overline{\varepsilon} \tag{4.4.35}$$

with the corresponding radius r_1^* given by

$$r_1^* = \frac{2(1 - \overline{\varepsilon})}{b + \overline{\varepsilon}}. \tag{4.4.36}$$

Moreover according to the analog of the proof of Theorem 4.4.1 the order of convergence would have been only linear.

We provide an example for Theorem 4.4.3.

Example 4.4.1 *Let $X = Y = \mathbf{R}$, $D = U(0, 1)$ and define function F on D by*

$$F(x) = e^x - 1. \tag{4.4.37}$$

According to Theorem 4.4.3 for $\varepsilon_0 = \varepsilon = 1$ we can have by (4.4.26)–(4.4.28), (4.4.30) $b = 1$, $r^* = \frac{1}{2}$. Choose $\delta = \frac{1}{2}$. Then all hypotheses of Theorem 4.4.3 hold.

Moreover if (4.4.33), (4.4.34) are used instead of (4.4.26) and (4.4.27) respectively, then for $\overline{\varepsilon} = .999$ we obtain using (4.4.35) that $r_1^* = .0010005$. Choosing $\overline{\delta} < .368$ all conditions hold but

$$r_1^* < r^*. \tag{4.4.38}$$

Hence in this case our Theorem 4.4.3 provides a wider range of initial guesses x_0 than the corresponding one mentioned in Remark 4.4.2. Furthermore the order of convergence is quadratic instead of only linear.

Remark 4.4.3 *If only (4.4.33) is used it can easily be seen by (4.4.21) that*

$$r_2^* = \frac{1-\overline{\varepsilon}}{\overline{\varepsilon}}.$$

In the case of Example 4.4.1 for $\overline{\varepsilon} = .999$ *we get*

$$r_2^* = \overline{.001} < \overline{r}_1 < r^*.$$

Another more interesting example is given by the following.

Example 4.4.2 *Let* $D = X = Y = \mathbf{R}$, $x_0 = 0$ *and define function* F *on* D *by*

$$F(x) = \frac{4}{3\pi}\int_0^x \tanh y^{\frac{5}{4}}\,dy + x + 1.$$

In [91] we used the weaker condition

$$\left\|F'(x_0)^{-1}[F'(x) - F'(x_0)]\right\| < \varepsilon$$

and hypotheses

$$\left[1 + c_0 + \frac{c_1^2}{1-c_1}\right]\eta < \delta, \tag{4.4.39}$$

where,

$$c_0 = (1-\varepsilon)^{-1}\varepsilon \ \textit{and} \ c_1 = 2c_0.$$

It can easily be seen that

$$\left[1 + c_0 + \frac{c_1^2}{1-c_1}\right]\eta > 1 \ \ (\textit{since}\ \eta = 1)$$

for all $\varepsilon \in (0, \frac{1}{3})$. *On the other hand, for* $\varepsilon \in (0, \frac{1}{3})$ *we can get*

$$\delta = \left(\tan\frac{3}{4}\pi\varepsilon\right)^{\frac{4}{5}} < 1.$$

Therefore (4.3.9) is violated. However the conditions of Theorem 4.4.1 are satisfied, since $b = 0, \eta = 1$ *and if we choose* $\varepsilon = \frac{1}{2}$ *and take* $\delta = \left(1+\sqrt{2}\right)^{\frac{4}{5}}$. *That is, Theorem 4.4.1 guarantees the convergence of method (4.1.3) (starting at* $x_0 = 0$*) to the solution* x^* *of equation* $F(x) = 0$.

Remark 4.4.4 *The results obtained here can be extended to hold for m-times Fréchet differentiable operators* ($m > 2$) *[79].*

4.5 Exercises

4.1. If $h_0 < \frac{1}{2}$ holds in $U(x_0, r^*)$, where $r^* = \frac{1+\sqrt{1-2h_0}}{h_0}\eta_0$, then show: Equation (4.1.1) has a unique solution x^* in $U(x_0, r^*)$.

4.2. If the hypotheses of Theorem 4.2.4 are satisfied and $h_0 = \frac{1}{2}$, then show: There exists a unique solution x^* of Equation (4.1.1) in $\bar{U}(x_0, r_0) = \bar{U}(x_0, 2\eta_0)$

4.3. Let x^* be a solution of Equation (4.1.1). If the linear operator $F'(x^*)$ has a bounded inverse, and $\lim_{\|x-x^*\|\to 0} \|F'(x) - F'(x^*)\| = 0$, then show Newton's method (4.1.3) converges to x^* if x_0 is sufficiently close to x^* and

$$\|x_n - x^*\| \leq d\varepsilon^n \quad (n \leq 0),$$

where ε is any positive number; d is a constant depending on x_0 and ε.

4.4. The above result cannot be strengthened, in the sense that for every sequence of positive numbers c_n such that: $\lim_{n\to\infty} \frac{c_{n+1}}{c_n} = 0$, there is an equation for which (4.1.3) converges less rapidly than c_n. Define

$$s_n = \begin{cases} c_{n/2}, & \text{if } n \text{ is even} \\ \sqrt{c_{(n-1)/2}c_{(n+1)/2}}, & \text{if } n \text{ is odd.} \end{cases}$$

Show: $s_n \to 0$, $\frac{s_{n+1}}{s_n} \to 0$, and $\lim_{n\to\infty} \frac{c_n}{s_{n+k}} = 0$, $(k \geq 1)$.

4.5. Assume operator $F'(x)$ satisfies a Hölder condition

$$\|F'(x) - F'(y)\| \leq a\|x - y\|^b,$$

with $0 < b < 1$ and $U(x_0, R)$. Define $h_0 = b_0 a \eta_0^b \leq c_0$, where c_0 is a root of

$$\left(\frac{c}{1+b}\right)^b = (1-c)^{1+b} \quad (0 \leq c \leq 1)$$

and let $R \geq \frac{\eta_0}{1-d_0} = r_0$, where $d_0 = \frac{h_0}{(1+b)(1-h_0)}$. Show that method (4.1.3) converges to a solution x^* of Equation (4.1.1) in $U(x_0, r_0)$.

4.6. Let K, b_0, η_0 be as in Theorem 4.2.4. If $h_0 = b_0\eta_0 K < \frac{1}{2}$, and

$$r_0 = \frac{1 - \sqrt{1 - 2h_0}}{h_0}\eta_0 \leq r.$$

Then show: modified Newton's method (4.1.4)

$$x_{n+1} = x_n - F'(x_0)^{-1} F(x_n) \quad (n \geq 0)$$

converges to a solution $x^* \in U(x_0, r_0)$ of Equation (4.1.1). Moreover, if

$$r_0 \leq r < \frac{1+\sqrt{1-2h_0}}{h_0}\eta_0,$$

then show: Equation (4.1.1) has a unique solution x^* in $U(x_0, r)$. Furthermore show: method (4.1.4) converges to a solution x^* of Equation (4.1.1) for any initial guess $x_0 \in U(x_0, r)$.

4.7. Under the hypotheses of Theorem 4.2.4, let us introduce $\bar{U} = \bar{U}(x_1, r_0 - \eta)$, sequence $\{t_n\}$ $(n \geq 0)$, $t_0 = 0$, $t_{n+1} = t_n - \frac{f(t_n)}{f'(t_n)}$, $f(t) = \frac{1}{2}kt^2 - t + \eta$, $\Delta = r^* - r_0$, $\theta = \frac{r_0}{r^*}$, $\nabla t_{n+1} = t_{n+1} - t_n$, $d_n = \|x_{n+1} - x_n\|$, $\Delta_n = \|x_n - x_0\|$, $\bar{U}_0 = \bar{U}, \bar{U}_n = \bar{U}(x_n, r_0 - t_n)$ $(n \geq 1)$, $K_0 = L_0 = K$,

$$K_n = \sup_{\substack{x,y \in \bar{U}_n \\ x \neq y}} \frac{\|F'(x_n)^{-1}(F'(x) - F'(y))\|}{\|x-y\|} \,(n \geq 1),$$

$$L_n = \sup_{\substack{x,y \in \bar{U} \\ x \neq y}} \frac{\|F'(x_n)^{-1}(F'(x) - F'(y))\|}{\|x-y\|} \,(n \geq 1),$$

$$\underline{\lambda}_n = \frac{2d_n}{1+\sqrt{1+2L_n d_n}} \,(n \geq 0),$$

$$\lambda_n = \frac{2d_n}{1+\sqrt{1-2L_n d_n}} \,(n \geq 0), \quad \underline{\kappa}_n = \frac{2d_n}{1+\sqrt{1+2K_n d_n}},$$

$$k_n = \frac{2d_n}{1+\sqrt{1-2K_n d_n}} \,(n \geq 0),$$

$$s_0 = 1, \; s_n = \frac{s_{n-1}^2}{2^{n-1}\sqrt{1-2h} + s_{n-1}\left(1-\sqrt{1-2h}\right)^{2^n-1}} \,(n \geq 0).$$

With the notation introduced above show (Yamamoto [280]):

$$\begin{aligned}\|x^* - x_n\| &\leq \mathcal{K}_n \,(n \geq 0) \leq \lambda_n \,(n \geq 0) \\ &\leq \frac{2d_n}{1+\sqrt{1-2K(1-K\Delta_n)^{-1} d_n}} \,(n \geq 0)\end{aligned}$$

$$\begin{aligned}
&\le \frac{2d_n}{1+\sqrt{1-2K\left(1-Kt_n\right)^{-1}d_n}} \ (n\ge 0)\\
&= \frac{2d_n}{1+\sqrt{1-2KB_nd_n}} \ (n\ge 0)\\
&= \begin{cases} \frac{2d_n}{1+\sqrt{1-\frac{4}{\Delta}\cdot\frac{1-\theta^{2^n}}{1+\theta^{2^n}}d_n}} & (2h<1)\\ \frac{2d_n}{1+\sqrt{1-\frac{2^n}{\eta}d_n}} & (2h=1)\ (n\ge 0)\end{cases}\\
&\le \frac{r_0-t_n}{\nabla t_{n+1}}d_n \quad (n\ge 0)\\
&= \frac{2d_n}{1+\sqrt{1-2h_n}} \quad (n\ge 0)\\
&\le \frac{KB_nd_{n-1}^2}{1+\sqrt{1-2h_n}} \quad (n\ge 0)\\
&= \frac{r_0-t_n}{\left(\nabla t_n\right)^2}d_{n-1}^2 \quad (n\ge 0)\\
&= \begin{cases} \frac{1-\theta^{2^n}}{\Delta}d_{n-1}^2 & (2h<1)\\ \frac{2^{n-1}}{\eta}d_{n-1}^2 & (2h=1)\,(n\ge 1)\end{cases}\\
&\le \frac{Kd_{n-1}^2}{\sqrt{1-2h}+\sqrt{1-2h+\left(Kd_{n-1}\right)^2}} \quad (n\ge 1)\\
&\le \frac{K\eta_{n-1}d_{n-1}}{\sqrt{1-2h}+\sqrt{1-2h+\left(K\eta_{n-1}\right)^2}} \quad (n\ge 1)\\
&= e^{-2^{n-1}\varphi}d_{n-1} \quad (n\ge 1)\\
&= \theta^{2^{n-1}}d_{n-1} \quad (n\ge 1)\\
&= \frac{r_0-t_n}{\nabla t_n}d_{n-1} \quad (n\ge 1)\\
&\le r_0-t_n \quad (n\ge 0)\\
&= \frac{2\eta_n}{1+\sqrt{1-2h_n}} \quad (n\ge 0)\\
&= \begin{cases} e^{-2^{n-1}\varphi}\frac{\sinh\varphi}{\sinh 2^{n-1}\varphi}\eta & (2h<1)\\ 2^{1-n}\eta & (2h<1)\end{cases}\\
&= \begin{cases} \frac{\Delta\theta^{2^n}}{1-\theta^{2^n}} & (2h<1)\\ 2^{1-n}\eta & (2h=1)\,(n\ge 0)\end{cases}
\end{aligned}$$

$$= \frac{s_n}{2^n K}\left(\frac{2h}{1+\sqrt{1-2h}}\right)^{2^n} \quad (n \geq 0)$$
$$\leq \frac{1}{2^n} K \left(\frac{2h}{1+\sqrt{1-2h}}\right)^{2^n} \quad (n \geq 0)$$
$$\leq \frac{1}{2^{n-1}} (2h)^{2n-1} \eta \quad (n \geq 0),$$

$$\begin{aligned}
\|x^* - x_n\| &\leq \lambda_n \quad (n \geq 0) \\
&\leq \frac{L_n d_{n-1}^2}{1+\sqrt{1-(L_n d_{n-1})^2}} \quad (n \geq 1) \\
&\leq \frac{L_{n-1} d_{n-1}^2}{1 - L_{n-1} d_{n-1} + \sqrt{1 - 2L_{n-1} d_{n-1}}} \quad (n \geq 1) \\
&\leq \frac{L_{n-1} d_{n-1}^2}{1 - L_{n-1} d_{n-1}} \quad (n \geq 1),
\end{aligned}$$

$$\begin{aligned}
\|x^* - x_n\| &\leq \lambda_n \\
&\leq \frac{2d_n}{1+\sqrt{1-2L_0 (1-L_0\Delta_n)^{-1} d_n}} \\
&\leq \frac{2\|F'(x_0)^{-1} F(x_n)\|}{1 - L_0\Delta_n + \sqrt{(1-L_0\Delta_n)^2 - 2L_0\|F'(x_0)^{-1} F(x_n)\|}} \\
&\leq \frac{L_0 d_{n-1}^2}{1 - L_0\Delta_n + \sqrt{(1-L_0\Delta_n)^2 - (L_0 d_{n-1})^2}},
\end{aligned}$$

$$\begin{aligned}
\|x^* - x_n\| &\geq \underline{\kappa}_n \, (n \geq 0) \geq \underline{\lambda}_n \, (n \geq 0) \\
&\geq \frac{2d_n}{1+\sqrt{1+2K(1-K\Delta_n)^{-1} d_n}} \quad (n \geq 0) \\
&\geq \frac{2d_n}{1+\sqrt{1+2K(1-Kt_n)^{-1} d_n}} \quad (n \geq 0) \\
&= \frac{2d_n}{1+\sqrt{1+2KB_n d_n}} \quad (n \geq 0)
\end{aligned}$$

$$
\begin{aligned}
&= \frac{2d_n}{1+\sqrt{1+4\cdot\frac{r_0-t_{n+1}}{(r_0-t_n)^2}d_n}} \quad (n \geq 0) \\
&= \frac{2d_n}{1+\sqrt{1+4\cdot\frac{\nabla t_{n+1}}{(\nabla t_n)^2}d_n}} \quad (n \geq 0) \\
&= \frac{2d_n}{1+\sqrt{1+\frac{2Kd_n}{\sqrt{1-2h+(K\eta_{n-1})^2}}}} \quad (n \geq 0) \\
&\geq \frac{2d_n}{1+\sqrt{1+\frac{2Kd_n}{\sqrt{1-2h+(Kd_{n-1})^2}}}} \quad (n \geq 0) \\
&= \frac{2d_n}{1+\sqrt{1+\frac{2d_n}{\sqrt{a^2+d_{n-1}^2}}}} \quad \left(a = \sqrt{1-2h}/K, n \geq 1\right) \\
&\geq \frac{2d_n}{1+\sqrt{1+\frac{2d_n}{d_n+\sqrt{a^2+d_n^2}}}} \quad (n \geq 0) \\
&\geq \frac{2d_n}{1+\sqrt{1+2h_n}} \quad (n \geq 0) \\
&= \frac{2d_n}{1+\sqrt{1+\frac{4\theta^{2^n}}{(1+\theta^{2^n})^2}}} \quad (n \geq 0),
\end{aligned}
$$

$$
\|x^* - x_{n+1}\| \leq \kappa_{n+1} \leq \kappa_n - d_n \leq \frac{r_0 - t_{n+1}}{\nabla t_{n+1}} d_n,
$$

$$
\|x^* - x_{n+1}\| \leq \lambda_{n+1} \leq \lambda_n - d_n \leq \frac{r_0 - t_{n+1}}{\nabla t_{n+1}} d_n,
$$

$$
\begin{aligned}
d_n &\leq \tfrac{1}{2} K_n d_{n-1}^2 \\
&\leq \tfrac{1}{2} L_n d_{n-1}^2 \\
&\leq \tfrac{1}{2} K \left(1 - K\Delta_n\right)^{-1} d_{n-1}^2 \\
&\leq \tfrac{1}{2} K \left(1 - K\Delta_{n-1} - Kd_{n-1}\right)^{-1} d_{n-1}^2 \\
&\leq \tfrac{1}{2} K \left(1 - Kt_n\right)^{-1} d_{n-1}^2 \\
&= \tfrac{1}{2} K B_n d_{n-1}^2
\end{aligned}
$$

$$
\begin{aligned}
&= \frac{r_0 - t_{n+1}}{(r_0 - t_n)^2} d_{n-1}^2 \\
&= \frac{\nabla t_{n+1}}{(\nabla t_n)^2} d_{n-1}^2 \\
&= \frac{d_{n-1}^2}{2\sqrt{a^2 + \eta_{n-1}^2}} \\
&\leq \frac{d_{n-1}^2}{2\sqrt{a^2 + d_{n-1}^2}} \\
&\leq \frac{\nabla t_{n+1}}{\nabla t_n} d_{n-1} \\
&= \frac{\eta_n}{\eta_{n-1}} d_{n-1} \\
&= \frac{1}{2\cosh 2^{n-1}\varphi} d_{n-1} \\
&\leq \tfrac{1}{2} d_{n-1} \\
&\leq \tfrac{1}{2}\eta_{n-1} = \tfrac{1}{2} \nabla t_n,
\end{aligned}
$$

and

$$
\begin{aligned}
d_n &\leq \eta_n \\
&= \nabla t_{n+1} \\
&= (r_0 - t_{n+1})\, \theta^{-2^n} \\
&= (r_0 - t_{n+1})\, e^{2^n} \varphi \\
&= \begin{cases} \frac{\sinh \varphi}{\sinh 2^n \varphi} \eta & (2h < 1) \\ 2^{-n}\eta & (2h = 1) \end{cases} \\
&= \frac{\Delta \theta^{2^n}}{1 - \theta^{2^{n+1}}} \quad (2h < 1).
\end{aligned}
$$

4.8. Let $F : D \subseteq X \to Y$ be m-times Fréchet-differentiable in D ($m \geq 2$ an integer). Assume that for some $x_0 \in D$ and parameters $\eta > 0$, $\alpha_i \geq 0$ ($i = 2, \ldots, m$), $a > 0$, $F'(x_0)^{-1}$ exists,

$$
\begin{aligned}
\|F'(x_0)^{-1} F(x_0)\| &\leq \eta, \\
\|F'(x_0)^{-1}[F'(x) - F'(x_0)]\| &\leq a\|x - x_0\|, \\
\|F'(x_0)^{-1} F^{(i)}(x_0)\| &\leq \alpha_i, \quad i = 2, \ldots, m, \\
\|F'(x_0)[F^{(m)}(x) - F^{(m)}(x_0)]\| &\leq \alpha_{m+1}\|x - x_0\|
\end{aligned}
$$

for all $x \in D$, and

$$2b\eta \leq 1,$$

where,

$$b = \max\left\{a, \alpha_2 + \tfrac{2}{3}\alpha_3\eta + \cdots + \tfrac{2^m \alpha_{m+1}}{(m+1)!}\eta^{m-1}\right\},$$

and

$$U\left(x_0, \tfrac{1}{a}\right) \subseteq D.$$

Show:

(a) the real polynomial

$$p(t) = \tfrac{b}{2}t^2 - t + \eta$$

has two positive roots r_1, r_2 with $r_1 \le r_2$ and iteration $\{t_n\}$ $(n \ge 0)$ generated by

$$t_0 = 0, \quad t_{n+1} = t_n - \frac{p(t_n)}{p'(t_n)} \quad (n \ge 0)$$

is monotonically increasing with $\lim_{n\to\infty} t_n = r_1$.

(b) Sequence $\{x_n\}$ $(n \ge 0)$ generated by method (4.1.3) is well defined, remains in $\bar{U}(x_0, r_1)$ for all $n \ge 0$, and converges to a solution $x^* \in \bar{U}(x_0, r_1)$ of equation $F(x) = 0$, which is unique in $U(x_0, r_2)$ if $r_1 < r_2$. If $r_1 = r_2$ the solution x^* is unique in $\bar{U}(x_0, r_1)$. Moreover, the following estimates hold for all $n \ge 0$

$$\|x_{n+1} - x_n\| \le t_{n+1} - t_n,$$

and

$$\|x_n - x^*\| \le r_1 - t_n = \left(\frac{r_1}{r_2}\right)^{2^n} (r_2 - t_n).$$

Let

$$F(x) = \tfrac{1}{6}x^3 - (\tfrac{2}{6}^{3/2} + .23), \quad x \in [\sqrt{2} - 1, \sqrt{2} + 1]$$

show that the Newton–Kantorovich hypothesis method (4.2.51) is violated, but the conditions in this result hold.

4.9. Let F be a Fréchet-differentiable operator defined on some closed convex subset D of a Banach space X with values in a Banach space Y. Assume: there exists $x_0 \in D$ such that $F'(x_0) \in L(X, Y)$, and $F'(x_0)^{-1} \in L(Y, X)$.

Show:

(a) for all $\varepsilon_0 > 0$ there exists $\delta_0 > 0$ such that

$$\|F'(x_0)^{-1}(F'(x_0) - F'(x))\| < \varepsilon_0, \qquad \text{for all } x \in U(x_0, \delta_0).$$

(b) for all $\varepsilon_1 > 0$ there exists $\delta_1 > 0$ such that

$$\|(F'(x_0)^{-1} - F'(x)^{-1})F'(x_0)\| < \varepsilon_1, \qquad \text{for all } x \in U(x_0, \delta_1).$$

Set $\delta = \min\{\delta_0, \delta_1\}$ and $\varepsilon = \max\{\varepsilon_0, \varepsilon_1\}$.

(c) for $\varepsilon > 0$ there exists $\delta > 0$ as defined above such that

$$\|F'(x_0)^{-1}(F'(x_0) - F'(x))\| < \varepsilon,$$
$$\|(F'(x_0)^{-1} - F'(x)^{-1})F'(x_0)\| < \varepsilon,$$

for all $x \in U(x_0, \delta)$.
Define parameters b, c, η by

$$b \geq \tfrac{-1+\sqrt{3}}{2}, \quad c \geq 2\varepsilon^2 + 2\varepsilon, \text{ for a fixed } \varepsilon \in [0, b), \quad \|F'(x_0)^{-1}F(x_0)\| \leq \eta.$$

(d) Assume further:

$$\tfrac{\eta}{1-c} \leq \delta, \quad \varepsilon \in (0, b), \quad \text{and} \quad \bar{U}(x_0, \delta) \subseteq D.$$

Show: Sequence $\{x_n\}$ $(n \geq 0)$ generated by method method (4.1.3) is well defined, remains in $U(x_0, \delta)$ for all $n \geq 0$, and converges to a solution $x^* \in \bar{U}(x_0, \delta)$ of equation $F(x) = 0$. Moreover, x^* is the unique solution of equation $F(x) = 0$ in $\bar{U}(x_0, \delta)$. Furthermore, the following estimates hold for all $n \geq 0$

$$\|x_{n+1} - x_n\| \leq c^n \|x_1 - x_0\| \leq c^n \eta$$

and

$$\|x_n - x^*\| \leq \tfrac{c^n}{1-c}\|x_1 - x_0\|.$$

(e) Assume: hypotheses on F, F' hold, with x_0 being replaced by a simple solution x^* of equation $F(x) = 0$; $\varepsilon \in (0, b)$.
Show: sequence $\{x_n\}$ $(n \geq 0)$ generated by method (4.1.3) is well defined, remains in $U(x^*, \delta)$ for all $n \geq 0$ and converges to x^*, provided that $x_0 \in \bar{U}(x^*, \delta) \subseteq D$. Moreover, the following estimates hold for all $n \geq 0$

$$\|x_{n+1} - x^*\| \leq c\|x_n - x^*\| \leq c^n \|x_0 - x^*\|.$$

4.10. Let $m \geq 2$ be an integer and F an m-times Fréchet-differentiable operator defined on a convex subset D of a Banach space X with values in a Banach space Y. Assume:

(a) there exists $x_0 \in D$ such that $F'(x_0)^{-1} \in L(Y, X)$;

(b) there exist parameters $\eta > 0$, $\alpha_i \geq 0$, $i = 2, \ldots, m$ such that

$$\|F'(x_0)^{-1}F(x_0)\| \leq \eta,$$
$$\|F'(x_0)^{-1}F^{(i)}(x_0)\| \leq \alpha_i, \quad i = 2, \ldots, m$$

Since F is m-times Fréchet-differentiable for all $\varepsilon > 0$ there exists $\delta_0 > 0$ such that

$$\left\|F'(x_0)^{-1}[F^{(m)}(x) - F^{(m)}(x_0)]\right\| < \varepsilon$$

for all $x \in U(x_0, \delta_0)$;

(c) the positive zeros of p' is such that

$$p(s) \leq 0,$$

where

$$p(t) = \eta - t + \tfrac{\alpha_2}{2!}t^2 + \cdots + \tfrac{(\varepsilon+\alpha_m)}{m!}t^m;$$

Then polynomial p has only two positive zeros denoted by δ_1, δ_2 ($\delta_1 \leq \delta_2$).

(d) $\bar{U}(x_0, \delta) \subseteq D$;

(e) $\delta_0 \in [\delta_1, \delta_2]$ or $\delta_0 > \delta_2$, where

$$\delta = \max\{\delta_0, \delta_1, \delta_2\}.$$

Show: sequence $\{x_n\}$ $(n \geq 0)$ generated by method (4.1.3) is well defined, remains in $U(x_0, \delta_1)$ for all $n \geq 0$ and converges to a solution $x^* \in \bar{U}(x_0, \delta_1)$ of equation $F(x) = 0$. The solution x^* is unique in $U(x_0, \delta_0)$ if $\delta_0 \in [\delta_1, \delta_2]$ or x^* is unique in $U(x_0, \delta_2)$ if $\delta_0 > \delta_2$. Moreover, the following estimates hold for all $n \geq 0$

$$\|x_{n+1} - x_n\| \leq t_{n+1} - t_n$$

$$\|x_n - x^*\| \leq t^* - t_n,$$

where $\{t_n\}$ $(n \geq 0)$ is a monotonically increasing sequence converging to t^*, generated by

$$t_0 = 0, \quad t_{n+1} = t_n - \tfrac{p(t_n)}{p'(t_n)} \quad (n \geq 0).$$

4.11. (a) Let F be a twice Fréchet-differentiable operator defined on a convex subset D of a Banach space X with values in a Banach space Y. Assume that the equation $F(x) = 0$ has a simple zero $x^* \in D$, in the sense that

$F'(x^*)$ has an inverse $F'(x^*)^{-1} \in L(Y, X)$. Then for all $\ell_1 > 0$ there exists $r^* > 0$ such that

$$\left\| F'(x^*)^{-1}[F''(x) - F''(y)] \right\| \leq \ell_1, \quad \text{for all } x, y \in U(x^*, r^*).$$

Moreover, assume there exists $b_1 > 0$ such that

$$\|F'(x^*)^{-1} F''(x^*)\| \leq b_1.$$

Then if

$$r^* < (3c_1)^{-1}, \quad c_1 = \tfrac{\ell_1 + b_1}{2}$$

show: sequence $\{x_n\}$ $(n \geq 0)$ generated by method (4.1.3) is well defined, remains in $U(x^*, r^*)$ for all $n \geq 0$ and converges to x^* provided that $x_0 \in U(x^*, r^*)$. Furthermore, the following estimates hold for all $n \geq 0$:

$$\|x_{n+1} - x^*\| \leq \tfrac{c_1}{1 - 2c_1 \|x_n - x^*\|} \|x_n - x^*\|^2.$$

(b) Assume:

(i) there exists $\eta \geq 0$, $x_0 \in D$ such that $F'(x_0)^{-1} \in L(Y, X)$

$$\|F'(x_0)^{-1} F(x_0)\| \leq \eta;$$

Then, for all $\ell \geq 0$ there exists $r > 0$ such that

$$\left\| F'(x_0)^{-1}[F''(x) - F''(y)] \right\| < \ell_0, \qquad \forall x, y \in \bar{U}(x_0, r) \subseteq D;$$

(ii) there exists $b_0 \geq 0$ such that

$$\|F'(x_0)^{-1} F''(x_0)\| \leq b_0;$$

(iii) $c_0 \eta \leq \frac{2 - \sqrt{3}}{2}, \quad c_0 = \frac{1}{2}(\ell_0 + b_0);$

(iv) r is the smallest positive zero of equation

$$3c_0 s^2 - (1 + 2c_0 \eta)s + \eta = 0.$$

Show: sequence $\{x_n\}$ $(n \geq 0)$ generated by Newton's method is well defined, remains in $U(x_0, r)$ for all $n \geq 0$ and converges to a unique solution $x^* \in \bar{U}(x_0, r)$ of equation $F(x) = 0$. Moreover, the following error bounds hold for all $n \geq 0$

$$\|x_{n+2} - x_{n+1}\| \leq \tfrac{c_0}{1 - 2c_0 \|x_{n+1} - x_n\|} \|x_{n+1} - x_n\|^2$$

and

$$\|x_{n+1} - x^*\| \leq \tfrac{c_0^n}{1 - c_0} \|x_n - x^*\|.$$

4.12. (a) Let : $D \subseteq X \to Y$ be a continuously Fréchet-differentiable operator defined on an open convex subset D of a Banach space X with values in a Banach space Y.

Assume:

- there exists $x_0 \in D$ such that $F(x_0) \neq 0$;
- $F'(x)^{-1} \in L(Y,X)$ $(x \in D)$, and there exists $b > 0$ such that:

$$\|F'(x)^{-1} F'(x_0)\| \leq b \quad (x \in D;)$$

- for each fixed $p \in D$ with $F(p) \neq 0$, there exists

$$\delta = \delta(p,t) \geq t\|F'(p)^{-1} F(p)\|, \quad t \in [0,1]$$

such that

$$\left\|F'(x_0)^{-1}[F'(p) - F'(y_t)]\right\| < t\|F'(p)^{-1} F(p)\|$$

for all

$$y_t \in U(p,\delta) \subseteq D,$$

and y_t collinear to p;

- there exists $\eta > 0$ such that:

$$\|F'(x_0)^{-1} F(x_0)\| \leq \eta,$$

$$c = \tfrac{b\eta}{2} \in (0,1);$$

and $U = U(x_0, \delta^*) \subseteq D$, where

$$\delta^* \geq \tfrac{\eta}{1-c}.$$

Show: sequence $\{x_n\}$ $(n \geq 0)$ generated by method (4.1.3) is well defined, remains in U for all $n \geq 0$ and converges to a unique solution $x^* \in \bar{U}$ of equation $F(x) = 0$. Moreover the following estimates hold:

$$\|x_{n+1} - x_n\| < \tfrac{b}{2}\|x_n - x_{n-1}\|^2 \quad (n \geq 1),$$

and

$$\|x_n - x^*\| \leq a_n\|x_n - x_{n-1}\|^2 \quad (n \geq 1),$$

where

$$a_n = \tfrac{b}{2}\sum_{j=0}^{\infty}\left(c^{2^n}\right)^{2^j-1} \leq \tfrac{b}{2(1-c^{2n})}.$$

(b) Assume:

- there exists a simple zero $x^* \in D$ of F in the sense that $F'(x^*)^{-1} \in L(Y, X)$;
- $F'(x)^{-1} \in L(Y, X)$ $(x \in D)$ and there exists $q > 0$ such that:

$$\|F'(x)^{-1} F'(x^*)\| \leq q \quad (x \in D);$$

- for each fixed $p \in D$ with $F(p) \neq 0$ there exists $\delta \geq t\|p - x^*\|$, $t \in (0, 1]$ such that:

$$\|F'(x^*)^{-1} [F'(p) - F'(p_t)]\| < t\|p - x^*\|$$

for all $p_t = x + t(x^* - x) \in U(p, \delta) \subseteq D$,

- $U^* = U(x^*, r^*) \subseteq D$, where

$$r^* > \tfrac{2}{q}.$$

Show: sequence $\{x_n\}$ $(n \geq 0)$ generated by method (4.1.3) is well defined, remains in U^* for all $n \geq 0$ and converges to x^* provided that $x_0 \in U^*$. Moreover, the following estimates hold for all $n \geq 0$:

$$\|x_{n+1} - x^*\| \leq \tfrac{q}{2} \|x_n - x^*\|^2.$$

4.13. (a) Suppose that F', F'' are uniformly bounded by non-negative constants $\alpha < 1$, K respectively, on a convex subset D of X and the ball

$$\bar{U}\left(x_0, r_0 \equiv \tfrac{2\|x_0 - F(x_0)\|}{1-\alpha}\right) \subseteq D.$$

Moreover, if

$$h_S = \tfrac{K}{2} \tfrac{1+2\alpha}{1-\alpha} \tfrac{\|x_0 - F(x_0)\|}{1-\alpha} < 1,$$

holds, then Stirling's method

$$x_{n+1} = x_n - \big(I - F'(F(x_n))\big)(x_n - F(x_n)) \quad (n \geq 0) \tag{4.5.1}$$

converges to the unique fixed point x^* of F in $\bar{U}(x_0, r_0)$. Moreover, the following estimates hold for all $n \geq 0$:

$$\|x_n - x^*\| \leq h_S^{2^n - 1} \frac{\|x_0 - F(x_0)\|}{1 - \alpha}. \tag{4.5.2}$$

(b) Let $F\colon D \subseteq X \to Y$ be analytic. Assume:

$$\|F'(x)\| \leq \alpha < 1, \quad \text{for all } x \in D,$$
$$x_0 \neq F(x_0), \quad x_0 \in D,$$
$$\tfrac{\gamma(1+2\alpha)\|x_0 - F(x_0)\|}{(1-\alpha)^2} < 1,$$
$$r_0 < r_1,$$
$$U(x_0, r_1) \subseteq D,$$

and

$$0 \neq \gamma \equiv \sup_{\substack{k>1 \\ x\in D}} \left\| \tfrac{1}{k!} F^{(k)}(x_0) \right\|^{\frac{1}{k-1}} < \infty,$$

where,

$$r_1 = \tfrac{1}{\gamma}\left[1 - \sqrt[3]{\tfrac{\gamma(1+2\alpha)\|x_0 - F(x_0)\|}{(1-\alpha)^2}}\right].$$

Show: sequence $\{x_n\}$ $(n \geq 0)$ generated by Stirling's method (4.5.1) is well defined, remains in $U(x_0, r_0)$ for all $n \geq 0$ and converges to a unique fixed point x^* of operator F at the rate given by (4.5.2) with

$$K = \tfrac{2\gamma}{(1-\gamma r_0)^3}\,.$$

(c) Let $X = D = \mathbb{R}$ and define function F on D by

$$F(x) = \begin{cases} -\frac{1}{3}x, & x \leq 3 \\ \frac{1}{45}(x^2 - 7x - 33), & 3 \leq x \leq 4 \\ \frac{1}{3}(x-7), & x > 4. \end{cases}$$

Using Stirling's method for $x_0 = 3$ we obtain the fixed point $x^* = 0$ of F in one iteration, since $x_1 = 3 - (1 + \frac{1}{3})^{-1}(3+1) = 0$.
Show: method (4.1.3) fails to converge.

4.14. It is convenient for us to define certain parameters, sequences and functions. Let $\{t_n\}$ $(n \geq 0)$ be a Fibonacci sequence given by

$$t_0 = t_1 = 1, \quad t_{n+1} = t_n + t_{n-1} \quad (n \geq 1).$$

Let also c, ℓ, η be non-negative parameters and define:

- the real function f by

$$f(x) = \tfrac{1}{1-x}, \quad x \in [0, 1),$$

- sequences $\{s_n\}$ $(n \geq -1)$, $\{a_n\}$ $(n \geq -1)$, $\{A_n\}$ $(n \geq -1)$ by

$$s_{-1} = \tfrac{\eta}{c+\eta}, \quad s_0 = \ell(c+\eta), \quad s_n = f^2(s_{n-1})a_{n-1}a_{n-2} \quad (n \geq 1),$$

$$a_{-1} = a_{-2} = 0, \quad a_{n-2} = \sum_{j=0}^{n-1} c_j, \quad c_j = t_0 + t_1 + \cdots + t_{j+1},$$

$$A_n = [x_n, x_{n-1}; F],$$

for $x_n \in X$, and

- parameters b, d, r_0 by

$$b = \max\left\{\frac{\ell \eta}{(1-s_0)^2 s_0}, \frac{s_0}{(1-s_0)^2}\right\}, \quad d = \frac{s_0}{1-s_0}, \quad r_0 = \frac{\eta}{1-d}.$$

Let $F\colon D \subseteq X \to Y$ be a nonlinear operator. Assume there exist $x_{-1}, x_0 \in D$ and non-negative parameters c, ℓ, η such that:

$$\begin{aligned}
A_0^{-1} &\text{ exists},\\
\|x_0 - x_{-1}\| &\le c,\\
\|A_0^{-1}F(x_0)\| &\le \eta,\\
\big\|A_0^{-1}\big([x,y;F] - [z,w;F]\big)\big\| &\le \ell\big(\|x-z\| + \|y-w\|\big),\\
&\quad \forall x,y,z \in D, x \ne y, w \ne z,\\
s_0 &< \tfrac{3-\sqrt{5}}{2},\\
\ell\eta &< (1-s_0)^2 s_0 = \alpha,
\end{aligned}$$

and

$$\bar{U}(x_0, r_0) \subseteq D.$$

Show: sequence $\{x_n\}$ $(n \ge -1)$ generated by the Secant method

$$x_{n+1} = x_n - [x_n, x_{n-1}; F]^{-1} F(x_n) \quad (n \ge 0)\ (x_{-1}, x_0 \in D)$$

is well defined, remains in $U(x_0, r_0)$ for all $n \ge 0$ and converges to a unique solution $x^* \in \bar{U}(x_0, r_0)$ of equation $F(x) = 0$. Moreover the following estimates hold for all $n \ge 1$:

$$\|x_n - x^*\| \le \tfrac{d^n}{1-d} b^{a_{n-2}} \|x_1 - x_0\|,$$

and

$$\|L_0^{-1}F(x_{n+1})\| \le b^{c_{n-1}} s_0.$$

Furthermore, let

$$r^* = \tfrac{1}{\ell} - r_0 - \eta.$$

Then $r^* > r_0$ and the solution x^* is unique in $U(x_0, r^*)$.

4.15. (1) Let F be a twice Fréchet differentiable operator defined on a convex subset D of a Hilbert space H with values in H; x_0 be a point in D. Assume:

(a) there exist constants a, b, c, d and η such that

$$\begin{aligned} \|F''(x) - F''(x_0)\| &\leq a\|x - x_0\|, \\ \|F''(x_0)\| &\leq b, \\ \tfrac{1}{c}\|y\|^2 &\leq \big|\langle F'(x)(y), y\rangle\big|, \\ \|F'(x)\| &\leq d, \end{aligned}$$

and

$$\|F(x_0)\| \leq \eta,$$

for all $x \in D$, $y \in H$;

(b) $p = \frac{a}{6}c^3\eta^2 + \frac{b}{2}c^2\eta + \sqrt{1 - \frac{1}{d^2c^2}} \in [0, 1)$
and

(c) $U(x_0, r^*) \subseteq D$,
where

$$r^* = \tfrac{c\eta}{1-p}.$$

Show: iteration $\{x_n\}$ $(n \geq 0)$ generated by

$$x_{n+1} = x_n - \frac{\langle F'(x_n)F(x_n), F(x_n)\rangle}{\|F'(x_n)F(x_n)\|^2} F(x_n) \ (n \geq 0)$$

is well defined, remains in $U(x_0, r^*)$ for all $n \geq 0$ and converges to a unique solution x^* of equation $F(x) = 0$ in $U(x_0, r^*)$. Moreover, the following estimates hold for all $n \geq 0$

$$\|F(x_n)\| \leq p^n\eta$$

and

$$\|x_n - x^*\| \leq \tfrac{c\eta}{1-p}p^n$$

If we let $a = 0$, then (1) reduces to Theorem 13.2 in [213, p. 161].

(2) Under the hypothesis of (1), show the same conclusions of (1) hold for iteration

$$x_{n+1} = x_n - \frac{\|F(x_n)\|^2}{\big\|[F'(x_n)]^*F(x_n)\big\|^2}[F'(x_n)]^*F(x_n) \ (n \geq 0)$$

but with p replaced by

$$p_0 = \sqrt{c^2\left[d^2 + b\eta + \tfrac{ac\eta^2}{3}\right] - 1}\,.$$

(3) Under the hypotheses of (2), show that the conclusions of (1) hold for iteration

$$x_{n+1} = x_n - \frac{\|F(x_n)\|^2}{\langle F'(x_n)F(x_n), F(x_n)\rangle} F(x_n) \ (n \geq 0).$$

Results (2) and (3) reduce to the corresponding ones in [213, p. 163] for $a = 0$. We give a numerical example to show our results apply, whereas the corresponding ones in [213, pp. 161–163] do not.

(4) Let $H = \mathbb{R}$, $D = [-1, 1]$, $x_0 = 0$ and consider equation

$$F(x) = -\tfrac{1}{6}x^3 - \tfrac{1}{6}x^2 + \tfrac{5}{6}x - \tfrac{1}{3} = 0.$$

The convergence condition in [187, p. 161] using our notation is

$$q = \sqrt{1 - \tfrac{1}{d^2c^2}} + \tfrac{c^2 L\eta}{2} \in [0, 1) \tag{i}$$

where L is the Lipschitz constant such that

$$\|F'(x) - F'(y)\| \leq L\|x - y\|, \quad x, y \in D.$$

Show:

$$\eta = \tfrac{1}{3},\ d = \tfrac{5}{6},\ c = \tfrac{3}{2},\ L = \tfrac{4}{3} \quad \text{and} \quad q = \tfrac{11}{10} > 1.$$

Hence, (i) is violated.
Moreover, show: the condition corresponding to iterations in (2) and (3) in [213, p. 163] is also violated since

$$q_0 = \sqrt{c^2(d^2 + L\eta) - 1} \in [0, 1),$$

gives

$$q_0 = \tfrac{5}{4} > 1.$$

Furthermore show: our conditions hold since for $a = 1$, $b = \frac{1}{3}$, we get

$$p = .7875 < 1$$

and

$$p_0 = \sqrt{\tfrac{15}{16}} < 1.$$

Conclude: there is no guarantee that iterations converge to a solution x^* of equation $F(x) = 0$ under the conditions in [213]. But our results (1)–(3) guarantee convergence for the same iterations.

(5) The results obtained here can easily be extended to include m-times Fréchet differentiable operators, where $m \geq 2$ an integer. Indeed, let us assume there exist constants $a_2, \ldots, a_{m+1}$ such that

$$\|F^{(m)}(x) - F^{(m)}(x_0)\| \leq a_{m+1}\|x - x_0\|$$
$$\|F^{(i)}(x_0)\| \leq a_i, \quad i = 2, \ldots, m.$$

Show:

$$\|F(x_{k+1}) - F(x_k) - F'(x_k)(x_{k+1} - x_k)\| \leq$$
$$\leq \tfrac{a_{m+1}}{(m+1)!}\|x_{k+1} - x_k\|^{m+1}\| + \cdots + \tfrac{a_2}{2!}\|x_{k+1} - x_k\|^2.$$

The results (1)–(3) hold if we replace p, p_0 by

$$p = \sqrt{1 - \tfrac{1}{d^2c^2}} + \tfrac{a_{m+1}}{(m+1)!}c^{m+1}\eta^n + \cdots + \tfrac{a_2}{2!}c^2\eta,$$

and

$$p_0 = \sqrt{c^2\left[d^2 + \left(\tfrac{2a_{m+1}}{(m+1)!}c^{m-1}\eta^{m-1} + \cdots + a_2\right)\eta\right] - 1}.$$

Chapter 5

Newton-like methods

We provide a local as well as a semilocal convergence analysis for Newton-like methods in a Banach space setting under very relaxed conditions.

5.1 Two-point methods

In this section we are concerned with the problem of approximating a locally unique solution x^* of the nonlinear equation

$$F(x) + G(x) = 0, \tag{5.1.1}$$

where F, G are operators defined on a closed ball $\overline{U}(w, R)$ centered at point w and of radius $R > 0$, which is a subset of a Banach space X with values in a Banach space Y. F is Fréchet-differentiable on $\overline{U}(w, R)$, while the differentiability of operator G is not assumed. However we note that the actual condition (4) (see below) used here implies the continuity (but not necessarily the differentiability) of operator G on $\overline{U}(w, R)$.

We use the two-point Newton-like method

$$y_{-1}, y_0 \in \overline{U}(w, R), \; y_{n+1} = y_n - A(y_{n-1}, y_n)^{-1}[F(y_n) + G(y_n)] \quad (n \geq 0) \tag{5.1.2}$$

to generate a sequence converging to x^*. Here $A(x, y) \in L(X, Y)$, for each fixed $x, y \in \overline{U}(w, R)$. Special cases of method (5.1.2) have already been studied. For example: if $A(x, y) = F'(y)$ and $G(x) = 0$ we obtain the Newton–Kantorovich method (4.1.3); if $A(x, y) = [x, y; F]$, and $G(x) = 0$ method (5.1.2) reduces to the Secant method; if $A(x, y) = F'(y)$, we obtain the Zincenko iteration, and for $A(x, y) = A(y)$ $(A(x) \in L(X, Y))$ we obtain the Newton-like method studied.

Our motivation originates from the following observation: Consider the scalar equation $F(d) = d^3 - p = 0$ on interval $[p, 2-p]$ $(p \in [0, \frac{1}{2}))$. Then for the initial guess $d_0 = 1$, the famous Newton–Kantorovich hypothesis, which is the sufficient convergence condition for the convergence of Newton's method to x^* is violated.

Therefore under the Newton–Kantorovich theorem 4.2.4 there is no guarantee that method (4.1.3) converges to x^* in this case. However one can easily see that Newton's method indeed converges for several values of p, say $p \in [0, .46)$ (see also Example 5.1.3). Therefore one expects that the Newton–Kantorovich hypothesis (4.2.51) can be weakened. In order for us to cover a wider range of problems and present a finer unified convergence analysis for iterative methods we decided to introduce method (5.1.2).

We provide a local as well as a semilocal convergence analysis for method (5.1.2) under very relaxed hypotheses (see (5.1.3), (5.1.4)). Since we are further motivated by optimization considerations, our new idea is to use center-Lipschitz conditions instead of Lipschitz conditions for the upper bounds on the inverses of the linear operators involved. It turns out that this way we obtain more precise majorizing sequences [263]. Moreover, despite the fact that our conditions are more general than related ones already in the literature, we can provide weaker sufficient convergence conditions, and finer estimates on the distances involved.

Several applications are provided: e.g., in the semilocal case we show that the famous for its simplicity and transparency Newton–Kantorovich hypothesis (see (4.2.51)) is weakened (see (5.1.57)), whereas in the local case we can provide a larger convergence radius using the same information (see (5.1.131) and (5.1.132)).

Assume there exists fixed $v \in X$ such that $v \in \overline{U}(w, R)$, $A(v, w)^{-1} \in L(Y, X)$, and for any $x, y, z \in \overline{U}(w, r) \subseteq \overline{U}(w, R)$, $r \in [0, R]$, $t \in [0, 1]$, the following hold:

$$\|A(v, w)^{-1}[A(x, y) - A(v, w)]\| \leq \theta_0(\|x - v\|, \|y - w\|), \tag{5.1.3}$$

and

$$\begin{aligned} &\|A(v, w)^{-1}\{[F'(y + t(z - y)) - A(x, y)](z - y) + G(z) - G(y)\}\| \\ &\quad \leq \theta(\|y - w\|, \|z - y\|, \|z - x\|, t)\|z - y\| \end{aligned} \tag{5.1.4}$$

where $\theta_0 : [0, +\infty)^2 \to [0, +\infty)$, $\theta : [0, +\infty)^3 \times [0, 1] \to [0, +\infty)$ are continuous, nondecreasing functions on $[0, +\infty)^2$, $[0, +\infty)^3 \times [0, 1]$ respectively. Define parameters c_{-1}, c, c_1 by

$$\|y_{-1} - v\| \leq c_{-1}, \quad \|y_{-1} - y_0\| \leq c, \quad \|v - w\| \leq c_1. \tag{5.1.5}$$

With the above choices, we show the following result on majorizing sequences for method (5.1.2).

Lemma 5.1.1 *Assume: there exist parameters* $\eta \geq 0$, $\delta \in [0, 2)$, $r_0 \in [0, r]$ *such that:*

$$h_\delta = 2\int_0^1 \theta(r_0, \eta, c + \eta, t)dt + \delta\theta_0(c + c_{-1}, \eta + r_0) \leq \delta, \tag{5.1.6}$$

$$\frac{2\eta}{2 - \delta} + r_0 + c \leq r, \tag{5.1.7}$$

$$\theta_0\left[\frac{1 - \left(\frac{\delta}{2}\right)^{n+1}}{1 - \frac{\delta}{2}}\eta + c + c_{-1}, \frac{1 - \left(\frac{\delta}{2}\right)^{n+2}}{1 - \frac{\delta}{2}}\eta + r_0\right] < 1, \tag{5.1.8}$$

and

$$2\int_0^1 \theta\left[\frac{1-\left(\frac{\delta}{2}\right)^{n+1}}{1-\frac{\delta}{2}}\eta + r_0, \left(\frac{\delta}{2}\right)^{n+1}\eta, \frac{1-\left(\frac{\delta}{2}\right)^{n+1}}{1-\frac{\delta}{2}}\eta + r_0, t\right] dt$$
$$+ \ \delta\theta_0\left[\frac{1-\left(\frac{\delta}{2}\right)^{n+1}}{1-\frac{\delta}{2}}\eta + c + c_{-1}, \frac{1-\left(\frac{\delta}{2}\right)^{n+2}}{1-\frac{\delta}{2}}\eta + r_0\right] \leq \delta \tag{5.1.9}$$

for all $n \geq 0$. Then iteration $\{t_n\}$ $(n \geq -1)$ given by

$$\begin{aligned} t_{-1} &= r_0, \quad t_0 = c + r_0, \quad t_1 = c + r_0 + \eta, \\ t_{n+2} &= t_{n+1} + p_n(t_{n+1} - t_n) \\ &\equiv t_{n+1} + \frac{\left\{\int_0^1 \theta(t_n - c, t_{n+1} - t_n, t_{n+1} - t_{n-1}, t)dt\right\}(t_{n+1} - t_n)}{1 - \theta_0(t_n - r_0 + c_{-1}, t_{n+1} - c)} \end{aligned} \tag{5.1.10}$$

is nondecreasing, bounded above by

$$t^{**} = \frac{2\eta}{2-\delta} + r_0 + c \tag{5.1.11}$$

and converges to some t^* such that

$$0 \leq t^* \leq t^{**} \leq r. \tag{5.1.12}$$

Moreover the following estimates hold for all $n \geq 0$:

$$0 \leq t_{n+2} - t_{n+1} \leq \frac{\delta}{2}(t_{n+1} - t_n) \leq \left(\frac{\delta}{2}\right)^{n+1}\eta. \tag{5.1.13}$$

Note that for $\delta = 0$ by (5.1.6) we require $h_0 = 0$ provided $\theta = 0$. However (5.1.8) is really used to show (5.1.16).

Proof. Using induction we show:

$$2\int_0^1 \theta(t_k - c, t_{k+1} - t_k, t_{k+1} - t_{k-1}, t)dt$$
$$+ \ \delta\theta_0(t_k - r_0 + c_{-1}, t_{k+1} - c) \leq \delta, \tag{5.1.14}$$

$$0 \leq t_{k+1} - t_k, \tag{5.1.15}$$

and

$$\theta_0(t_k - t_{-1} + c_{-1}, t_{k+1} - t_0 + r_0) < 1 \tag{5.1.16}$$

for all $k \geq 0$. Estimate (5.1.13) can then follow from (5.1.14)–(5.1.16) and (5.1.10). For $k = 0$ estimates (5.1.14)–(5.1.16) follow by the initial conditions. By (5.1.10) we then get

$$0 \leq t_2 - t_1 \leq \frac{\delta}{2}(t_1 - t_0). \tag{5.1.17}$$

Assume (5.1.14)–(5.1.16) hold for all $k \leq n+1$. We obtain in turn

$$\begin{aligned}
&2\int_0^1 \theta(t_{k+1}-t_0+r_0, t_{k+2}-t_{k+1}, t_{k+2}-t_k, t)dt \\
&\quad + \delta\theta_0(t_{k+1}-t_{-1}+c_{-1}, t_{k+2}-t_0+r_0) \\
&\leq 2\int_0^1 \theta\left[\frac{1-\left(\frac{\delta}{2}\right)^{k+1}}{1-\frac{\delta}{2}}\eta+r_0, \left(\frac{\delta}{2}\right)^{k+1}\eta, \frac{1-\left(\frac{\delta}{2}\right)^{k+1}}{1-\frac{\delta}{2}}\eta+r_0, t\right]dt \\
&\quad + \delta\theta_0\left[\frac{1-\left(\frac{\delta}{2}\right)^{k+1}}{1-\frac{\delta}{2}}\eta+c+c_{-1}, \frac{1-\left(\frac{\delta}{2}\right)^{k+2}}{1-\frac{\delta}{2}}\eta+r_0\right] \\
&\leq \delta \quad \text{by (9) and (13).}
\end{aligned} \tag{5.1.18}$$

Hence we showed (5.1.14) holds for $k=n+2$. Moreover we show

$$t_k \leq t^{**}. \tag{5.1.19}$$

Indeed we have:

$$\begin{aligned}
&t_{-1} = r_0 \leq t^{**}, \quad t_0 = r_0 + c \leq t^{**}, \quad t_1 = c + r_0 + \eta \leq t^{**}, \\
&t_2 \leq c + r_0 + \eta + \frac{\delta}{2}\eta = \frac{2+\delta}{2}\eta + r_0 + c \leq t^{**}.
\end{aligned} \tag{5.1.20}$$

Assume (5.1.19) holds for all $k \leq n+1$. It follows from (5.1.13)–(5.1.16):

$$\begin{aligned}
t_{k+2} &\leq t_{k+1} + \frac{\delta}{2}(t_{k+1}-t_k) \leq t_k + \frac{\delta}{2}(t_k - t_{k-1}) + \frac{\delta}{2}(t_{k+1}-t_k) \\
&\leq \cdots \leq c + r_0 + \eta + \frac{\delta}{2}\eta + \left(\frac{\delta}{2}\right)^2\eta + \cdots + \left(\frac{\delta}{2}\right)^{k+1}\eta \\
&= \frac{1-\left(\frac{\delta}{2}\right)^{k+2}}{1-\frac{\delta}{2}}\eta + r_0 + c \leq \frac{2\eta}{2-\delta} + r_0 + c = t^{**}.
\end{aligned} \tag{5.1.21}$$

Hence, sequence $\{t_n\}$ $(n \geq -1)$ is bounded above by t^{**}. Estimate (5.1.15) holds for $k=n+2$ by (5.1.14) and (5.1.21). Furthermore (5.1.16) holds for $k=n+2$ by (5.1.14) and (5.1.21). Finally, sequence $\{t_n\}$ $(n \geq -1)$ is monotonically increasing by (5.1.15), bounded above by (5.1.21) and as such it converges to some t^* satisfying (5.1.12). ∎

Remark 5.1.1 *It follows from (5.1.10) that t^{**} can be replaced in (5.1.11) by the finer t_0^{**} given by*

$$t_0^{**} = r_0 + c + \left[1 + \frac{p_0}{2} + \frac{1}{1-\frac{\delta}{2}}\frac{\delta}{2}\frac{p_0}{2}\right]\eta,$$

since $p_0 \leq \delta$.

We provide the main result on the semilocal convergence of method (5.1.2) using majorizing sequence (5.1.10).

Theorem 5.1.2 *Under the hypotheses of Lemma 5.1.1 assume further that for each fixed pair* (r, r_0)*:*

$$y_{-1} \in \overline{U}(w,r), \;\; y_0 \in \overline{U}(w,r_0) \tag{5.1.22}$$

and

$$\|A(y_{-1},y_0)^{-1}[F(y_0)+G(y_0)]\| \leq \eta. \tag{5.1.23}$$

Note that the existence of $A(y_{-1},y_0)^{-1}$ *follows from* $\theta_0(c_{-1},c_0) < 1$*. Then, sequence* $\{y_n\}$ $(n \geq -1)$ *generated by Newton-like method (5.1.2) is well defined, remains in* $\overline{U}(w,t^*)$ *for all* $n \geq -1$*, and converges to a solution* x^* *of equation* $F(x)+G(x)=0$*. Moreover the following estimates hold for all* $n \geq -1$*:*

$$\|y_{n+1}-y_n\| \leq t_{n+1}-t_n \tag{5.1.24}$$

and

$$\|y_n - x^*\| \leq t^* - t_n. \tag{5.1.25}$$

Furthermore the solution x^* *is unique in* $\overline{U}(w,t^*)$ *if*

$$\int_0^1 \theta(t^*,2t^*,2t^*,t)dt + \theta_0(t^*+c_1,t^*) < 1, \tag{5.1.26}$$

and in $\overline{U}(w,R_0)$ *for* $R_0 \in (t^*,r]$ *if*

$$\int_0^1 \theta(t^*,t^*+R_0,t^*+R_0,t)dt + \theta_0(t^*+c_1,t^*) < 1. \tag{5.1.27}$$

Proof. We first show estimate (5.1.24), and $y_n \in \overline{U}(w,t^*)$ for all $n \geq -1$. For $n=-1,0$, (5.1.24) follows from (5.1.5), (5.1.10) and (5.1.23). Suppose (5.1.24) holds for all $n=0,1,\ldots,k+1$; this implies in particular (using (5.1.5), (5.1.22))

$$\begin{aligned}\|y_{k+1}-w\| &\leq \|y_{k+1}-y_k\| + \|y_k-y_{k-1}\| + \cdots + \|y_1-y_0\| + \|y_0-w\| \\ &\leq (t_{k+1}-t_k)+(t_k-t_{k-1})+\cdots+(t_1-t_0)+r_0 \\ &= t_{k+1}-t_0+r_0 \leq t_{k+1} \leq t^*.\end{aligned}$$

That is, $y_{k+1} \in \overline{U}(w,t^*)$. We show (5.1.24) holds for $n=k+2$. By (5.1.3) and (5.1.10) we obtain for all $x,y \in \overline{U}(w,t^*)$

$$\|A(v,w)^{-1}[A(x,y)-A(v,w)]\| \leq \theta_0(\|x-v\|,\|y-w\|). \tag{5.1.28}$$

In particular for $x = y_k$ and $y = y_{k+1}$ we get using (5.1.3), (5.1.5),

$$\begin{aligned}
&\|A(v,w)^{-1}[A(y_k,y_{k+1}) - A(v,w)]\| \le \theta_0(\|y_k - v\|, \|y_{k+1} - w\|) \\
&\quad\le \theta_0(\|y_k - y_{-1}\| + \|y_{-1} - v\|, \|y_{k+1} - y_0\| + \|y_0 - w\|) \\
&\le \theta_0(t_k - t_{-1} + c_{-1}, t_{k+1} - t_0 + r_0) \\
&\le \theta_0\left[\frac{1-\left(\frac{\delta}{2}\right)^k}{1-\frac{\delta}{2}}\eta + c + c_{-1}, \frac{1-\left(\frac{\delta}{2}\right)^{k+1}}{1-\frac{\delta}{2}}\eta + r_0\right] < 1, \quad \text{(by (8))}. \qquad (5.1.29)
\end{aligned}$$

It also follows from (5.1.29) and the Banach Lemma on invertible operators 1.3.2 that $A(y_k, y_{k+1})^{-1}$ exists, and

$$\|A(y_k,y_{k+1})^{-1}A(v,w)\| \le [1-\theta_0(t_k - t_{-1} + c_{-1}, t_{k+1} - t_0 + r_0)]^{-1} \le b_0, \qquad (5.1.30)$$

where,

$$b_0 = [1 - \theta_0(R - t_{-1} + c_{-1}, R - t_0 + r_0)]^{-1}.$$

Using (5.1.2), (5.1.4), (5.1.10), (5.1.30) we obtain in turn

$$\begin{aligned}
&\|y_{k+2} - y_{k+1}\| = \|A(y_k,y_{k+1})^{-1}[F(y_{k+1}) + G(y_{k+1})]\| \\
&= \|A(y_k,y_{k+1})^{-1}[F(y_{k+1}) + G(y_{k+1}) \\
&\quad - A(y_{k-1},y_k)(y_{k+1} - y_k) - F(y_k) - G(y_k)]\| \\
&\le \|A(y_k,y_{k+1})^{-1}A(v,w)\|\,\|A(v,w)^{-1}[F(y_{k+1}) - F(y_k) \\
&\quad - A(y_{k-1},y_k)(y_{k+1} - y_k) + G(y_{k+1}) - G(y_k)]\| \\
&\le \frac{\int_0^1 \theta(\|y_k - w\|, \|y_{k+1} - y_k\|, \|y_{k+1} - y_{k-1}\|, t)dt\|y_{k+1} - y_k\|}{1 - \theta_0(t_k - t_{-1} + c_{-1}, t_{k+1} - t_0 + r_0)} \\
&\le \frac{\{\int_0^1 \theta(t_k - c, t_{k+1} - t_k, t_{k+1} - t_{k-1}, t)dt\}(t_{k+1} - t_k)}{1 - \theta_0(t_k - t_{-1} + c_{-1}, t_{k+1} - t_0 + r_0)} \\
&= t_{k+2} - t_{k+1}, \qquad (5.1.31)
\end{aligned}$$

which shows (5.1.24) for all $n \ge 0$. Note also that

$$\begin{aligned}
\|y_{k+2} - w\| &\le \|y_{k+2} - y_{k+1}\| + \|y_{k+1} - w\| \qquad (5.1.32)\\
&\le t_{k+2} - t_{k+1} + t_{k+1} - t_0 + r_0 = t_{k+2} - t_0 + r_0 \le t_{k+2} \le t^*.
\end{aligned}$$

That is, $y_{k+2} \in \overline{U}(w, t^*)$. It follows from (5.1.24) that $\{y_n\}$ $(n \ge -1)$ is a Cauchy sequence in a Banach space X, and as such it converges to some $x^* \in \overline{U}(w, t^*)$. We can have as above

$$\|y_{k+2} - y_{k+1}\| \le b_0\|A(v,w)^{-1}[F(y_{k+1}) + G(y_{k+1})]\| \le b\|y_{k+1} - y_k\| \qquad (5.1.33)$$

where,

$$b = b_0 b_1, \qquad (5.1.34)$$

and

$$b_1 = \int_0^1 \theta(R - c, \eta, 2\eta, t)\, dt. \tag{5.1.35}$$

By letting $k \to \infty$ in (5.1.33), using (5.1.30), and the continuity of operators F, G we get $b_0 \| A(v,w)^{-1}(F(x^*) + G(x^*)) \| = 0$ from which we obtain $F(x^*) + G(x^*) = 0$ (since $b_0 > 0$). Estimate (5.1.25) follows from (5.1.24) by using standard majorization techniques [99], [263]. To show uniqueness in $\overline{U}(w, t^*)$, let y^* be a solution of equation (5.1.1) in $\overline{U}(w, t^*)$. Then as in (5.1.31) we obtain the identity:

$$\begin{aligned} y^* - y_{k+1} &= y^* - y_k + A(y_{k-1}, y_k)^{-1}(F(y_k) + G(y_k)) \\ &- A(y_{k-1}, y_k)^{-1}(F(y^*) + G(y^*)) \\ &= -[A(y_{k-1}, y_k)^{-1} A(v,w)] A(v,w)^{-1} [F(y^*) - F(y_k) \\ &- A(y_{k-1}, y_k)(y^* - y_k) + G(y^*) - G(y_k)]. \end{aligned} \tag{5.1.36}$$

Using (5.1.36) we obtain in turn:

$$\begin{aligned} \|y^* - y_{k+1}\| &\leq \frac{\int_0^1 \theta(\|y_k - w\|, \|y^* - y_k\|, \|y^* - y_{k-1}\|, t) dt \|y^* - y_k\|}{1 - \theta_0(\|y_{k-1} - v\|, \|y_k - w\|)} \\ &\leq \frac{\int_0^1 \theta[t^*, 2t^*, 2t^*, t] dt}{1 - \theta_0(t^* + c_1, t^*)} \|y^* - y_k\| < \|y^* - y_k\|. \end{aligned} \tag{5.1.37}$$

That is, $x^* = y^*$, since $\lim_{k\to\infty} y_k = y^*$. If $y^* \in \overline{U}(w, R_0)$ then as in (5.1.37) we get

$$\|y^* - y_{k+1}\| \leq \frac{\int_0^1 \theta(t^*, R_0 + t^*, R_0 + t^*, t) dt}{1 - \theta_0(t^* + c_1, t^*)} \|y^* - y_k\| < \|y^* - y_k\|. \tag{5.1.38}$$

Hence, again we get $x^* = y*$. ■

Remark 5.1.2 *Conditions (5.1.8), (5.1.9) can be replaced by the stronger but easier to check*

$$\theta_0 \left[\frac{2\eta}{2-\delta} + c + c_{-1}, \frac{2\eta}{2-\delta} + r_0 \right] < 1, \tag{5.1.39}$$

and

$$\begin{aligned} 2 \int_0^1 \theta &\left[\frac{2\eta}{2-\delta} + r_0, \frac{\delta}{2}\eta, \frac{2\eta}{2-\delta} + r_0, t \right] dt \\ &+ \delta\theta_0 \left[\frac{2\eta}{2-\delta} + c + c_{-1}, \frac{2\eta}{2-\delta} + r_0 \right] \leq \delta \end{aligned} \tag{5.1.40}$$

respectively. Note also that conditions (5.1.6)–(5.1.9), (5.1.39), (5.1.40) are of the Newton–Kantorovich-type hypotheses (see also (5.1.58)) which are always present in the study of Newton-like methods.

Remark 5.1.3 *(a) Conditions similar to (5.1.3), (5.1.4) but less flexible were considered by Chen and Yamamoto in [121] in the special case when $A(x,y) = A(x)$ for all $x, y \in \overline{U}(w, R)$ $(A(x) \in L(X,Y))$ (see also Theorem 5.1.6). Operator $A(x)$ is intended there to be an approximation to the Fréchet-derivative $F'(x)$ of F. However we also want the choice of the operator A to be more flexible, and be related to the difference $G(z) - G(y)$ for all $y, z \in \overline{U}(w, R)$ (see Application 5.1.1 and Example 5.1.1). (b) If we choose*

$$\begin{aligned} &A(x,y) = F'(y), \ G(x) = 0, \ \theta_0(q,q) = \ell_0 q, \\ &\theta(q_1,q_2,q_3,t) = \ell t q_2 \quad (\gamma_0 \geq 0, \gamma_1 \geq 0) \end{aligned} \tag{5.1.41}$$

for all $x, y \in \overline{U}(w,r)$, q, q_1, q_2, $q_3 \in [0,r]$, $t \in [0,1]$ we obtain the Newton–Kantorovich method (4.1.3) (see also Application 5.1.4). (c) Assume:

$$\|A(v,w)^{-1}\| \leq d. \tag{5.1.42}$$

Define $F = 0$, $A(x,y) = [y,x;G]$, $G(z) - G(y) = [z,y;G](z-y)$, $\theta(q_1,q_2,q_3,t) = d\overline{\overline{\theta}}(q_2,q_3)q_2$ where $[z,y;G]$ is a divided difference of order one for G (see Chapter 2), and $\overline{\overline{\theta}}$ is a continuous nondecreasing function in two variables. Then note that function $\overline{\overline{\theta}}$ was used in [190] in the case of the secant method. Note also that obviously (5.1.4) implies again the continuity of operator G for this special choice of function θ.

Application 5.1.3 *Let us consider some special cases. Define*

$$A(x,y) = F'(y) + [x,y;G], \tag{5.1.43}$$

$$v = y_{-1}, \quad w = y_0, \tag{5.1.44}$$

and set

$$r_0 = 0, \tag{5.1.45}$$

where F', $[\cdot,\cdot;G]$ denote the Fréchet-derivative of F and the divided difference of order one for operator G (see Chapter 1). Hence, we consider Newton-like method (5.1.2) in the form

$$y_{n+1} = y_n - (F'(y_n) + [y_{n-1}, y_n; G])^{-1}(F(y_n) + G(y_n)) \quad (n \geq 0). \tag{5.1.46}$$

The method was studied in [117]. It is shown to be of order $\frac{1+\sqrt{5}}{2} \approx 1.618\ldots$ (same as the order of Chord), but higher than the order of

$$z_{n+1} = z_n - F'(z_n)^{-1}(F(z_n) + G(z_n)) \quad (n \geq 0) \tag{5.1.47}$$

and

$$w_{n+1} = w_n - A(w_n)^{-1}(F(w_n) + G(w_n)) \quad (n \geq 0), \tag{5.1.48}$$

where $A(\cdot)$ is an operator approximating F'. Assume:

$$\|A(y_{-1},y_0)^{-1}[F'(y)-F'(y_0)]\| \le \ell_1\|y-y_0\|, \quad (5.1.49)$$

$$\|A(y_{-1},y_0)^{-1}[F'(x)-F'(y)]\| \le \ell_2\|x-y\|, \quad (5.1.50)$$

$$\|A(y_{-1},y_0)^{-1}([x,y;G]-[y_{-1},y_0;G])\| \le \ell_3(\|x-y_{-1}\|+\|y-y_0\|), \quad (5.1.51)$$

and

$$\|A(y_{-1},y_0)^{-1}([x,y;G]-[z,x;G])\| \le \ell_4\|z-y\| \quad (5.1.52)$$

for some non-negative parameters ℓ_i, $i=1,2,3,4$ and all $x,y\in\overline{U}(y_0,r)\subseteq\overline{U}(y_0,R)$. Then we can define:

$$w_1=w_2,\quad w_1(q)=\ell_2 q,\quad w_3(q)=\ell_4 q,\ \theta_0(q_1,q_2)=\ell_3 q_1+(\ell_1+\ell_3)q_2 \quad (5.1.53)$$

and

$$\theta(q_1,q_2,q_3,t)=w_1(q_1+tq_2)-w_2(q_1)+w_3(q_3).$$

If the hypotheses of Theorem 5.1.2 hold for the above choices, the conclusions follow.

Note that conditions (5.1.49)–(5.1.52) are weaker than the corresponding ones in [117, pp. 48–49], [99]. Indeed, conditions $\|F'(x)-F'(y)\|\le\ell_5\|x-y\|$, $\|A(x,y)^{-1}\|\le\ell_6$, $\|[x,y,z;G]\|\le\ell_7$, and

$$\|[x,y;G]-[z,w;G]\|\le\ell_8(\|x-z\|+\|y-w\|)$$

for all $x,y,z,w\in\overline{U}(y_0,r)$ are used there instead of (5.1.49)–(5.1.52), where $[x,y,z;G]$ denotes a second order divided difference of G at (x,y,z), and ℓ_i, $i=5,6,7,8$ are non-negative parameters.

Let us provide an example for this case:

Example 5.1.1 *Let $X=Y=(\mathbf{R}^2,\|\cdot\|_\infty)$. Consider the system*

$$3x^2y+y^2-1+|x-1|=0$$
$$x^4+xy^3-1+|y|=0.$$

Set $\|x\|_\infty=\|(x',x'')\|_\infty=\max\{|x'|,|x''|\}$, $F=(F_1,F_2)$, $G=(G_1,G_2)$. For $x=(x',x'')\in\mathbf{R}^2$ we take $F_1(x',x'')=3(x')^2x''+(x'')^2-1$, $F_2(x',x'')=(x')^4+x'(x'')^3-1$, $G_1(x',x'')=|x'-1|$, $G_2(x',x'')=|x''|$. We shall take $[x,y;G]\in M_{2\times2}(\mathbf{R})$ as $[x,y;G]_{i,1}=\frac{G_i(y',y'')-G_i(x',y'')}{y'-x'}$, $[x,y;G]_{i,2}=\frac{G_i(x',y'')-G_i(x',x'')}{y''-x''}$, $i=1,2$, provided that $y'\ne x'$ and $y''\ne x''$. Otherwise define $[x,y;G]$ to be the zero matrix in $M_{2\times2}(\mathbf{R})$. Using method (5.1.47) with $z_0=(1,0)$ we obtain the results in Table 5.1.

Using the method of chord with $w_{-1}=(5,5)$, and $w_0=(1,0)$, we obtain the results in Table 5.2.

n	$z_n^{(1)}$	$z_n^{(2)}$	$\|z_n - z_{n-1}\|$
0	1	0	
1	1	0.333333333333333	3.333E–1
2	0.906550218340611	0.354002911208151	9.344E–2
3	0.885328400663412	0.338027276361322	2.122E–2
4	0.891329556832800	0.326613976593566	1.141E–2
5	0.895238815463844	0.326406852843625	3.909E–3
6	0.895154671372635	0.327730334045043	1.323E–3
7	0.894673743471137	0.327979154372032	4.809E–4
8	0.894598908977448	0.327865059348755	1.140E–4
9	0.894643228355865	0.327815039208286	5.002E–5
10	0.894659993615645	0.327819889264891	1.676E–5
11	0.894657640195329	0.327826728208560	6.838E–6
12	0.894655219565091	0.327827351826856	2.420E–6
13	0.894655074977661	0.327826643198819	7.086E–7
...			
39	0.894655373334687	0.327826521746298	5.149E–19

Table 5.1: Method (5.1.47) with $z_0 = (1, 0)$.

n	$w_n^{(1)}$	$w_n^{(2)}$	$\|w_n - w_{n-1}\|$
−1	5	5	
0	1	0	5.000E+00
1	0.989800874210782	0.012627489072365	1.262E–02
2	0.921814765493287	0.307939916152262	2.953E–01
3	0.900073765669214	0.325927010697792	2.174E–02
4	0.894939851625105	0.327725437396226	5.133E–03
5	0.894658420586013	0.327825363500783	2.814E–04
6	0.894655375077418	0.327826521051833	3.045E–04
7	0.894655373334698	0.327826521746293	1.742E–09
8	0.894655373334687	0.327826521746298	1.076E–14
9	0.894655373334687	0.327826521746298	5.421E–20

Table 5.2: Method of chord with $w_{-1} = (5, 5)$, and $w_0 = (1, 0)$.

Using our method (5.1.46) with $y_{-1} = (5, 5)$, $y_0 = (1, 0)$, *we obtain the results in Table 5.3.*

We did not verify the hypotheses of Theorem 5.1.2 for the above starting points. However, it is clear that the hypotheses of Theorem 5.1.2 are satisfied for all three

n	$y_n^{(1)}$	$y_n^{(2)}$	$\|y_n - y_{n-1}\|$
−1	5	5	
0	1	0	5
1	0.909090909090909	0.363636363636364	3.636E–01
2	0.894886945874111	0.329098638203090	3.453E–02
3	0.894655531991499	0.327827544745569	1.271E–03
4	0.894655373334793	0.327826521746906	1.022E–06
5	0.894655373334687	0.327826521746298	6.089E–13
6	0.894655373334687	0.327826521746298	2.710E–20

Table 5.3: Our method (5.1.46) with $y_{-1} = (5,5)$, $y_0 = (1,0)$.

methods for starting points closer to the solution

$$x^* = (.894655373334687, .327826521746298)$$

chosen from the lists of the tables displayed above. Hence method (5.1.2) (i.e. method (5.1.46) in this case) converges faster than (5.1.47) suggested in Chen and Yamamoto [121], Zabrejko and Nguen [363] in this case and the method of chord or secant method.

In the application that follows we show that the famous Newton–Kantorovich hypothesis (see (5.1.58)) is weakened under the same hypotheses/information.

Application 5.1.4 *Returning back to Remark 5.1.3 and (5.1.41), iteration (5.1.2) reduces to the famous Newton–Kantorovich method (4.1.3). Condition (5.1.6) reduces to:*

$$h_\delta = (\ell + \delta \ell_0)\eta \leq \delta. \tag{5.1.54}$$

Case 1. *Let us restrict $\delta \in [0,1]$. Hypothesis (5.1.9) now becomes*

$$\begin{aligned} 2\int_0^1 \ell \left[\frac{2\eta}{2-\delta}\left(1-\left(\frac{\delta}{2}\right)^{k+1}\right) + t\left(\frac{\delta}{2}\right)^{k+1}\eta\right] dt \\ - 2\ell\left[\frac{2\eta}{2-\delta}\left(1-\left(\frac{\delta}{2}\right)^{k+1}\right)\right] + \delta\ell_0\left[\frac{2\eta}{2-\delta}\left(1-\left(\frac{\delta}{2}\right)^{k+1}\right)\right] \\ \leq 2\ell\int_0^1 t\eta\, dt + \delta\ell_0\eta \end{aligned}$$

or

$$\left[\frac{\ell_0\delta^2}{2-\delta} - \ell\right]\left[1-\left(\frac{\delta}{2}\right)^{k+1}\right] \leq 1,$$

which is true for all $k \geq 0$ by the choice of δ. Furthermore (5.1.8) gives

$$\frac{2\ell_0\eta}{2-\delta} \leq 2\ell_0\eta < 1. \tag{5.1.55}$$

Hence in this case conditions (5.1.6), (5.1.8) and (5.1.9) reduce only to (5.1.56) provided $\delta \in [0,1]$. Condition (5.1.54) for say $\delta = 1$ reduces to:

$$h_1 \leq 1 \tag{5.1.56}$$

which weakens the famous Newton–Kantorovich hypothesis (4.2.51) since

$$\ell_0 \leq \ell \tag{5.1.57}$$

in general.

Case 2. *It follows from Case 1 that (5.1.6), (5.1.8) and (5.1.9) reduce to (5.1.54),*

$$\frac{2\ell_0\eta}{2-\delta} \leq 1 \tag{5.1.58}$$

and

$$\frac{\ell_0\delta^2}{2-\delta} \leq \ell_1 \tag{5.1.59}$$

respectively provided $\delta \in [0,2)$.

Case 3. *It turns out that the range for δ can be extended (see also Example 5.1.3). Introduce conditions:*

$$\ell_0\eta \leq 1 - \frac{1}{2}\delta, \quad \textit{for } \delta \in [\ell_0, 2),$$

where,

$$\delta_0 = \frac{-b+\sqrt{b^2+8b}}{2}, \quad b = \frac{\ell}{\ell_0} \textit{ and } \ell_0 \neq 0.$$

Indeed the proof of Lemma 5.1.1 goes through if instead of (5.1.18) we show the weaker condition in this case

$$\ell\left(\frac{\delta}{2}\right)^{k+1} + \frac{2\ell_0\delta}{2-\delta}\left[1-\left(\frac{\delta}{2}\right)^{k+2}\right] \leq \delta,$$

or

$$\left(b - \frac{b\delta}{2} - \frac{\delta^2}{2}\right)\left(\frac{\delta}{2}\right)^{k+1} \leq 0,$$

or

$$\delta \geq \delta^0,$$

which is true by the choice of δ^0.

In the example that follows we show that $\frac{\ell}{\ell_0}$ can be arbitrarily large. Indeed:

Example 5.1.2 *Let $X = Y = \mathbf{R}$, $d_0 = 0$ and define function F on $\mathbf{R}$ by*

$$F(x) = \alpha_0 x + \alpha_1 + \alpha_2 \sin e^{\alpha_3 x} \tag{5.1.60}$$

where α_i, $i = 0, 1, 2, 3$ are given parameters. Then with the rest of the choices of operators and functions given by (5.1.41) we see from (5.1.60) that for α_3 large and α_2 sufficiently small $\frac{\ell}{\ell_0}$ can be arbitrarily large.

Next we provide an example also mentioned in the introduction to show that (5.1.57) holds but not (5.1.58).

Example 5.1.3 *Let $X = Y = \mathbf{R}$, $d_0 = 1$, $D = [p, 2-p]$, $p \in \left[0, \frac{1}{2}\right)$, and define F on D by*

$$F(d) = d^3 - p. \tag{5.1.61}$$

Using (5.1.23), (5.1.41) and (5.1.61) we get

$$\eta = \frac{1}{3}(1-p), \quad \ell = 2(2-p), \tag{5.1.62}$$

which imply

$$h = \frac{4}{3}(1-p)(2-p) > 1 \quad \textit{for all } p \in \left[0, \frac{1}{2}\right). \tag{5.1.63}$$

That is there is no guarantee that method (5.1.42) converges since (4.2.51) is violated. However since $\ell_0 = 3 - p$ we find that (5.1.56) holds if $p \in \left[\frac{5-\sqrt{13}}{3}, \frac{1}{2}\right)$, which improves (4.2.51) under the same computational cost.

Finally using Case 3 we can do better. Indeed, choose

$$p = p^0 = .4505 < \frac{5-\sqrt{13}}{3} = .464816242\ldots.$$

Then we get

$$\eta = .183166\ldots, \quad \ell_0 = 2.5495, \quad \ell = 3.099, \quad \text{and} \quad \delta^0 = 1.0656867.$$

Choose $\delta = \delta^0$. Then we get

$$\ell_0 \eta = .466983415 < 1 - \frac{\delta^0}{2} = .46715665.$$

That is the interval $\left[\frac{5-\sqrt{13}}{3}, \frac{1}{2}\right)$ can be extended to $\left[p^0, \frac{1}{2}\right)$.

In order to compare with earlier results we consider the case of single step methods by setting $x = y$ and $v = w$. We can then prove along the same lines to Lemma 5.1.1 and Theorem 5.1.2 respectively the following results by assuming:

there exists $w \in X$ such that $A(w)^{-1} \in L(Y, X)$, for any $x, y \in \overline{U}(w, r) \subseteq \overline{U}(w, R)$, $t \in [0, 1]$:

$$\|A(w)^{-1}[A(x) - A(w)]\| \leq m_0(\|x - w\|) \tag{5.1.64}$$

and

$$\begin{aligned}&\|A(w)^{-1}\{[F'(x + t(y - x)) - A(x)](y - x) + G(y) - G(x)\}\| \\ &\qquad\leq [m(\|x - w\|, \|y - x\|, t) + m_1(r)]\|y - x\|\end{aligned} \tag{5.1.65}$$

where m_0, m_1, m are as θ_0, θ_0 (one variable), θ (three variables), respectively.

Then as in Lemma 5.1.1 we can show the following result on majorizing sequences:

Lemma 5.1.5 *Assume:*

$$\overline{h}_\delta = 2\int_0^1 m(r_0, \eta, t)dt + 2m_1(r_0 + \eta) + \delta m_0(r_0 + \eta) \leq \delta, \tag{5.1.66}$$

$$\frac{2\eta}{2 - \delta} + r_0 \leq r, \tag{5.1.67}$$

$$m_0\left[\frac{2\eta}{2 - \delta}\left(1 - \left(\frac{\delta}{2}\right)^{n+1}\right) + r_0\right] < 1, \tag{5.1.68}$$

and

$$\begin{aligned}&2\int_0^1 m\left[\frac{2\eta}{2 - \delta}\left(1 - \left(\frac{\delta}{2}\right)^{n+1}\right) + r_0, \left(\frac{\delta}{2}\right)^{n+1}\eta, t\right] dt \\ &\quad + 2m_1\left[\frac{2\eta}{2 - \delta}\left(1 - \left(\frac{\delta}{2}\right)^{n+1}\right) + r_0\right] \\ &\quad + \delta m_0\left[\frac{2\eta}{2 - \delta}\left(1 - \left(\frac{\delta}{2}\right)^{n+1}\right) + r_0\right] \leq \delta\end{aligned} \tag{5.1.69}$$

for all $n \geq 0$. Then, iteration $\{s_n\}$ $(n \geq 0)$ given by

$$\begin{aligned}&s_0 = r_0, \qquad s_1 = r_0 + \eta, \\ &s_{n+2} = s_{n+1} + \frac{\int_0^1 m(s_n, s_{n+1} - s_n, t)dt(s_{n+1} - s_n) + \int_{s_n}^{s_{n+1}} m_1(t)dt}{1 - m_0(s_{n+1})}\end{aligned} \tag{5.1.70}$$

is nondecreasing, bounded above by

$$s^{**} = \frac{2\eta}{2-\delta} + r_0, \tag{5.1.71}$$

and converges to some s^* *such that*

$$0 \le s^* \le s^{**}. \tag{5.1.72}$$

Moreover the following estimates hold for all $n \ge 0$

$$0 \le s_{n+2} - s_{n+1} \le \frac{\delta}{2}(s_{n+1} - s_n) \le \left(\frac{\delta}{2}\right)^{n+1} \eta. \tag{5.1.73}$$

Remark 5.1.4 *Comments similar to Remark 5.1.2 can now follow for Lemma 5.1.5.*

Using m_0, $m(\cdot,\cdot,\cdot) + m_1$, $A(\cdot), w$ instead of θ_0, $\theta(\cdot,\cdot,\cdot,\cdot)$, $A(\cdot,\cdot)$, (v,w) respectively and going through the proof of Theorem 5.1.2 step by step we can show:

Theorem 5.1.6 *Under the hypotheses of Lemma 5.1.5 hold assume further that*

$$\|A(y_0)^{-1}[F(y_0) + G(y_0)]\| \le \eta. \tag{5.1.74}$$

Then, sequence $\{w_n\}$ $(n \ge 0)$ *generated by Newton-like method (5.1.48) is well-defined, remains in* $\overline{U}(w, s^*)$ *for all* $n \ge 0$, *and converges to a solution* x^* *of equation* $F(x) + G(x) = 0$. *Moreover the following estimates hold for all* $n \ge 0$

$$\|w_{n+1} - w_n\| \le s_{n+1} - s_n \tag{5.1.75}$$

and

$$\|w_n - x^*\| \le s^* - s_n. \tag{5.1.76}$$

Furthermore the solution x^* *is unique in* $\overline{U}(w, s^*)$ *if*

$$\int_0^1 m(s^*, 2s^*, t)dt + m_1(s^*) + m_0(s^*) < 1, \tag{5.1.77}$$

or in $\overline{U}(w, R_0)$, *if* $s^* < R_0 \le r$, *and*

$$\int_0^1 m(s^*, R_0 + s^*, t)dt + m_1(s^*) + m_0(s^*) < 1. \tag{5.1.78}$$

Remark 5.1.5 *We state the relevant results due to Chen and Yamamoto [15, pp. 40]. We assume: $A(w)^{-1}$ exists, and for any $x, y \in \overline{U}(w,r) \subseteq \overline{U}(w,R)$:*

$$0 < \|A(w)^{-1}(F(w) + G(w)\| \leq \overline{\eta}, \tag{5.1.79}$$

$$\|A(w)^{-1}(A(x) - A(w))\| \leq \overline{g}_0(\|x - w\|) + \overline{\alpha}, \tag{5.1.80}$$

$$\|A(w)^{-1}[F'(x + t(y - x)) - A(x)]\| \leq \overline{g}_1(\|x - w\|) + t\|y - x\|) - \overline{g}_0(\|x - w\|) + \overline{\beta}, \quad t \in [0,1], \tag{5.1.81}$$

$$\|A(w)^{-1}[G(x) - G(y)]\| \leq g_3(r)\|x - y\|, \tag{5.1.82}$$

where $\overline{g}_0$, $\overline{g}_1$, are as m_0, m_1, respectively, $\overline{\alpha} \geq 0$, $\overline{\beta} \geq 0$ but $\overline{g}_0$ is also differentiable with $\overline{g}_0'(r) > 0$, $r \in [0,R]$, and $\overline{\alpha} + \overline{\beta} < 1$. Note that conditions (5.1.79)–(5.1.82) generalize the ones considered in [263] even in the case $G = 0$.

As in [121] set:

$$\varphi(r) = \overline{\eta} - r + \int_0^r \overline{g}_1(t)dt, \quad \psi(r) = \int_0^r g_3(t)dt, \tag{5.1.83}$$

$$\chi(r) = \phi(r) + \psi(r) + (\overline{\alpha} + \overline{\beta})r. \tag{5.1.84}$$

Denote the minimal value of $\chi(r)$ on $[0,R]$ by χ^*, and the minimal point by r^*. If $\chi(R) \leq 0$, denote the unique zero of χ by $r_0^* \in (0, r^*]$. Define scalar sequence $\{r_n\}$ $(n \geq 0)$ by

$$r_0 \in [0,R], \quad r_{n+1} = r_n + \frac{u(r_n)}{g(r_n)} \quad (n \geq 0) \tag{5.1.85}$$

where

$$u(r) = \chi(r) - x^* \tag{5.1.86}$$

and

$$g(r) = 1 - \overline{g}_0(r) - \overline{\alpha}. \tag{5.1.87}$$

With the above notation they showed:

Theorem 5.1.7 *Suppose $\chi(R) \leq 0$. Then equation (5.1.1) has a solution $x^* \in \overline{U}(w, r_0^*)$, which is unique in*

$$\tilde{U} = \begin{cases} \overline{U}(w,R) & \text{if } \chi(R) = 0 \text{ or } \psi(R) = 0, \text{ and } r_0^* < R. \\ U(w,R) & \text{if } \chi(R) = 0 \text{ and } r_0^* < R. \end{cases} \tag{5.1.88}$$

Let

$$D^* = \overline{U}_{r \in [0,r^*)} \left\{ y \in \overline{U}(w,r) \mid \|A(y)^{-1}[F(y) + G(y)]\| \leq \frac{u(r)}{g(r)} \right\}. \tag{5.1.89}$$

Then, for any $y_0 \in D$, sequence $\{y_n\}$ $(n \geq 0)$ generated by Newton-like method (5.1.48) is well defined, remains in $\overline{U}(w, r^*)$ and satisfies

$$\|y_{n+1} - y_n\| \leq r_{n+1} - r_n, \tag{5.1.90}$$

and

$$\|y_n - x^*\| \leq r^* - r_n \tag{5.1.91}$$

provided that r_0 is chosen as in (5.1.85) so that $r_0 \in Ry_0$, where for $y \in D^*$

$$R_y = \left\{ r \in [0, r^*) \mid \|A(y)^{-1}(F(y) + G(y))\| \leq \frac{u(r)}{y(r)}, \|y - z\| \leq r \right\}. \tag{5.1.92}$$

Remark 5.1.6 *(a) Hypothesis on $\overline{g}_0$ is stronger than the corresponding one on v_0.*

(b) Iteration (5.1.85) converges to r^* (even if $r_0 = 0$) not r_0^*.

(c) Choices of y_{-1}, y_0 other than the ones in Theorems 5.1.2, 5.1.6 can be given by (5.1.89) and (5.1.90).

From now on in this section we compare our Theorem 5.1.6 with Theorem 5.1.7 for some special choices of functions m_0, m_1 and m. Indeed we define:

$$m_0(q) = g_0(q) + \alpha, \tag{5.1.93}$$

$$m_1(q) = g_3(q), \tag{5.1.94}$$

and

$$m(q_1, q_2, t) = g_1(q_1 + tq_2) - g_2(q_1) + \beta, \tag{5.1.95}$$

where $g_0(r)$, $g_1(r+\overline{r}) - g_2(r)$ $(\overline{r} \geq 0)$, $g_3(r)$ are continuous nondecreasing functions for $r \in [0, R]$, and constants α, β satisfy $\alpha \geq 0$ and $\beta \geq 0$.

Remark 5.1.7 *The conclusions of Theorem 5.1.7 hold if the more general conditions (5.1.64), (5.1.65) replace (5.1.80)–(5.1.82), and*

$$\overline{g}_0(r) \leq g_2(r), \quad r \in [0, R], \tag{5.1.96}$$

is satisfied. Moreover if strict inequality holds in (5.1.96) we obtain more precise estimates. Indeed, define the sequence $\{\overline{r}_n\}$ $(n \geq 0)$, using (5.1.65), g_2 instead of (5.1.81), $\overline{g}_0$ respectively (with $\overline{g}_1 = g_1$, $\alpha = \overline{\alpha}$, $\beta = \overline{\beta}$) by

$$\overline{r}_0 = r_0, \quad \overline{r}_1 = r_1,$$

$$\overline{r}_{n+1} - \overline{r}_n = \frac{u(\overline{r}_n) - u(\overline{r}_{n-1}) + (1 - g_2(\overline{r}_{n-1}) - \overline{\alpha})(\overline{r}_n - \overline{r}_{n-1})}{g(\overline{r}_n)} \quad (n \geq 1) \tag{5.1.97}$$

It can easily be seen using induction on n (see also the proof of Proposition 5.1.8 that follows) that

$$\overline{r}_{n+1} - \overline{r}_n < r_{n+1} - r_n, \tag{5.1.98}$$

$$\overline{r}_n < r_n, \tag{5.1.99}$$

$$\overline{r}^* - \overline{r}_n \leq r^* - r_n, \quad \overline{r}^* = \lim_{n\to\infty} \overline{r}_n, \tag{5.1.100}$$

and

$$\overline{r}^* \leq r^*. \tag{5.1.101}$$

Furthermore condition (5.1.65) allows us more flexibility in choosing functions and constants.

Remark 5.1.8 *Returning back to method (4.1.3) (see also (5.1.41)), the iterations corresponding to (5.1.85) and (5.1.97) are:*

$$\overline{d}_0 = 0, \ \overline{d}_1 = \eta, \ \overline{d}_{n+2} = \overline{d}_{n+1} + \frac{\gamma_1(\overline{d}_{n+1} - \overline{d}_n)^2}{2(1 - \gamma_1 \overline{d}_{n+1})}$$

and

$$\overline{\overline{d}}_0 = 0, \ \overline{\overline{d}}_1 = \eta, \ \overline{\overline{d}}_{n+2} = \overline{\overline{d}}_{n+1} + \frac{\gamma_1(\overline{\overline{d}}_{n+1} - \overline{\overline{d}}_n)^2}{2(1 - \gamma_0 \overline{\overline{d}}_{n+1})},$$

respectively. Moreover condition (5.1.96) reduces to (5.1.59), and in case $\gamma_0 < \gamma_1$, estimates:

$$\|d_{n+1} - d_n\| \leq \overline{\overline{d}}_{n+1} - \overline{\overline{d}}_n < \overline{d}_{n+1} - \overline{d}_n, \tag{5.1.102}$$

$$\|d_n - d^*\| \leq d^3 - \overline{\overline{d}}_n \leq d_1 - \overline{d}_n, \tag{5.1.103}$$

$$d^3 = \lim_{n\to\infty} \overline{\overline{d}}_n \leq d_1 = \lim_{n\to\infty} \overline{d}_n \tag{5.1.104}$$

hold.

Remark 5.1.9 *Our estimates (5.1.75), (5.1.76) are finer than the corresponding ones (5.1.90) and (5.1.91) respectively in many interesting cases. Let us choose:*

$$\alpha = \overline{\alpha}, \quad \beta = \overline{\beta}, \quad g_0(r) = \overline{g}_0(r), \quad g_1(r) = g_2(r) = \overline{g}_1(r), \quad \textit{and} \tag{5.1.105}$$

$$\overline{g}_3(r) = g_3(r) \qquad \textit{for all } r \in [0, R]. \tag{5.1.106}$$

Then we can show:

Proposition 5.1.8 *Under the hypotheses of Theorems 5.1.6 and 5.1.7, further assume:*

$$s_1 < r_1. \tag{5.1.107}$$

Then, the following hold:

$$s_n < r_n \quad (n \geq 1), \tag{5.1.108}$$

$$s_{n+1} - s_n < r_{n+1} - r_n \quad (n \geq 0), \tag{5.1.109}$$

$$s^* - s_n \leq r^* - r_n \quad (n \geq 0), \tag{5.1.110}$$

and

$$s^* \leq r^*. \tag{5.1.111}$$

Proof. It suffices to show (5.1.108) and (5.1.109), since then (5.1.110) and (5.1.111) respectively can easily follow from them. Inequality (5.1.108) holds for $n = 1$ by (5.1.107). By (5.1.70) and (5.1.85) we get in turn

$$\begin{aligned} s_2 - s_1 &= \frac{\int_0^1 \{g_1(s_0 + t(s_1 - s_0))dt - g_2(s_0) + \alpha\}(s_1 - s_0) + \int_{s_0}^{s_1} g_3(t)dt}{1 - \beta - g_0(s_1)} \\ &< \frac{\int_0^1 \{\overline{g}_1(r_0 + t(r_1 - r_0))dt - \overline{g}_2(r_0) + \overline{\alpha}\}(r_1 - r_0) + \int_{r_0}^{r_1} \overline{g}_3(t)dt}{1 - \overline{\beta} - \overline{g}_0(r_1)} \\ &= \frac{u(r_1) - u(r_0) + g(r_0)(r_1 - r_0)}{1 - \overline{\beta} - \overline{g}_0(r_1)} = \frac{u(r_1)}{g(r_1)} = r_2 - r_1. \end{aligned} \tag{5.1.112}$$

Assume:

$$s_{k+1} < r_{k+1}, \tag{5.1.113}$$

and

$$s_{k+1} - s_k < r_{k+1} - r_k \tag{5.1.114}$$

hold for all $k \leq n$. Using (5.1.70), (5.1.76), and (5.1.114) we obtain

$$\begin{aligned} & s_{k+2} - s_{k+1} \\ &= \frac{\int_0^1 \{g_1[s_k + t(s_{k+1} - s_k)]dt - g_2(s_k) + \alpha\}(s_{k+1} - s_k) + \int_{s_k}^{s_{k+1}} g_3(t)dt}{1 - \beta - g_0(s_{k+1})} \\ &< \frac{\int_0^1 \{\overline{g}_1[r_k + t(r_{k+1} - r_k)]dt - \overline{g}_2(r_k) + \overline{\alpha}\}(r_{k+1} - r_k) + \int_{r_k}^{r_{k+1}} \overline{g}_3(t)dt}{1 - \overline{\beta} - \overline{g}_0(r_{k+1})} \\ &= \frac{u(r_{k+1}) - u(r_k) + g(r_k)(r_{k+1} - r_k)}{g(r_{k+1})} = \frac{u(r_{k+1})}{g(r_{k+1})} = r_{k+2} - r_{k+1}. \end{aligned} \tag{5.1.115}$$

■

In order for us to include a case where operator G is nontrivial, we consider the following example:

Example 5.1.4 *Let $X = Y = C[0,1]$ the space of continuous functions on $[0,1]$ equipped with the sup-norm. Consider the integral equation on $\overline{U}\left(x_0, \frac{r}{2}\right)$, $(r \in [0, R])$ given by*

$$x(t) = \int_0^1 k(t, s, x(s))ds, \tag{5.1.116}$$

where the kernel $k(t, s, x(s))$ with $(t, s) \in [0,1] \times [0,1]$ is a nondifferentiable operator on $\overline{U}\left(x_0, \frac{r}{2}\right)$. Define operators F, G on $\overline{U}\left(x_0, \frac{r}{2}\right)$ by

$$F(x)(t) = Ix(t) \quad (I \text{ the identity operator}) \tag{5.1.117}$$

$$G(x)(t) = -\int_0^1 k(t, s, x(s))ds. \tag{5.1.118}$$

Choose $x_0 = 0$, and assume there exists a constant $k_0 \in [0, 1)$, a real function $k_1(t, s)$ such that

$$\|k(t, s, x) - k(t, s, y)\| \leq k_1(t, s)\|x - y\| \tag{5.1.119}$$

and

$$\sup_{t \in [0,1]} \int_0^1 k_1(t, s)ds \leq k_0 \tag{5.1.120}$$

for all $t, s \in [0,1]$, $x, y \in \overline{U}\left(x_0, \frac{r}{2}\right)$. Moreover choose: $r_0 = 0$, $y_0 = y_{-1}$, $A(x) = I(x)$, I the identity operator on X, $v_0(r) = r$, $a = b = 0$, $v = 0$, and $v_1(r) = k_0$ for all $x, y \in \overline{U}\left(x_0, \frac{r}{2}\right)$, $r, s \in [0,1]$. It can easily be seen that the conditions of Theorem 5.1.6 hold if

$$t^* = \frac{\eta}{1 - k_0} \leq \frac{r}{2}. \tag{5.1.121}$$

In order to cover the local case, let us assume x^* is a zero of equation (5.1.1), $A(x^*, x^*)^{-1}$ exists and for any $x, y \in \overline{U}(x^*, r) \subseteq \overline{U}(x^*, R)$, $r \in [0, R]$, $t \in [0,1]$:

$$\|A(x^*, x^*)^{-1}[A(x, y) - A(x^*, x^*)]\| \leq \overline{\theta}_0(\|x - x^*\|, \|y - x^*\|), \tag{5.1.122}$$

and

$$\begin{aligned} \|A(x^*, x^*)^{-1}[F'(x^* + t(y - x^*))(y - x^*) \\ - A(x, y)(y - x^*) + G(y) - G(x^*)]\| \\ \leq \overline{\theta}(\|y - x^*\|, \|x - x^*\|, t)\|y - x^*\|, \end{aligned} \tag{5.1.123}$$

where, $\overline{\theta}_0$, $\overline{\theta}_1$ are as θ_0, θ (in three variables), respectively. In order for us to compare our results with earlier ones, we only consider the case $r_0 = 0$, $x_{-1} = v$, $x_0 = w$ in (5.1.2), and call the corresponding sequence $\{x_n\}$ instead of $\{y_n\}$. Then as in the identity (5.1.36) and estimate (5.1.37) with y^*, v, w, replaced by x^*, x^*, x^* respectively but using (5.1.122), (5.1.123), instead of (5.1.4), (5.1.5) we can show along the lines of the proof of the uniqueness part of Theorem 5.1.2 the following local result for method (5.1.2).

Theorem 5.1.9 *Assume: there exist solutions of equation*

$$f(\lambda) = 0, \tag{5.1.124}$$

in $[0, R]$ *where*

$$f(\lambda) = \int_0^1 \overline{\theta}(\lambda, \lambda, t)dt + \overline{\theta}_0(\lambda, \lambda) - 1. \tag{5.1.125}$$

Denote by λ^* *the smallest of the solutions in* $[0, r]$. *Then, sequence* $\{x_n\}$ $(n \geq -1)$ *generated by Newton-like method (5.1.2) is well defined, remains in* $\overline{U}(x^*, \lambda^*)$ *for all* $n \geq 0$ *and converges to* x^* *provided that* $x_{-1}, x_0 \in \overline{U}(x^*, \lambda^*)$. *Moreover the following estimates hold for all* $n \geq 0$:

$$\|x^* - x_{n+1}\| \leq p_n, \tag{5.1.126}$$

where,

$$p_n = \frac{\{\int_0^1 \overline{\theta}(\|x_n - x^*\|, \|x_{n-1} - x^*\|, t)dt}{1 - \overline{\theta}_0(\|x_n - x^*\|)} \|x_n - x^*\|. \tag{5.1.127}$$

Application 5.1.10 *Let us again consider Newton's method, i.e.,* $F'(x) = A(x, y)$, $G(x) = 0$, *and assume:*

$$\|F'(x^*)^{-1}[F'(x) - F'(x^*)]\| \leq \lambda_0 \|x - x^*\|, \tag{5.1.128}$$

and

$$\|F'(x^*)^{-1}[F'(x) - F'(y)]\| \leq \lambda \|x - y\| \tag{5.1.129}$$

for all $x, y \in \overline{U}(x^*, r) \subseteq \overline{U}(x^*, R)$. *Then we can define:*

$$\overline{\theta}(r_1, r_2, t) = \overline{h}_1((1+t)r_1) - \overline{h}_2(r_2) + \overline{h}_3(r_2) \quad \textit{and } \overline{\theta}_0(r, r) = \lambda_0 r$$

where

$$\overline{h}_3 = 0, \quad \overline{h}_1(r) = \overline{h}_2(r) = \lambda r, \quad \textit{for all } r \in [0, R]. \tag{5.1.130}$$

Using (5.1.128), (5.1.129) we get:

$$\lambda^* = \frac{2}{2\lambda_0 + \lambda}. \tag{5.1.131}$$

Local results were not given in [121], [138], [351], [363]. However Rheinboldt in [295] showed that under only (129) (call the corresponding radius λ_{RH} *in this case) the convergence radius is given by*

$$\lambda_{RH} = \frac{2}{3\lambda}. \tag{5.1.132}$$

But in general

$$\lambda_0 \leq \lambda. \tag{5.1.133}$$

Hence we conclude:

$$\lambda_{RH} \leq \lambda^*. \tag{5.1.134}$$

The corresponding estimates become:

$$\|x_{n+1} - x^*\| \leq e_n, \tag{5.1.135}$$

$$\|x_{n+1} - x^*\| \leq e_n^1, \tag{5.1.136}$$

where,

$$e_n = \frac{\lambda \|x_n - x^*\|^2}{2[1 - \lambda_0 \|x_n - x^*\|]} \tag{5.1.137}$$

and

$$e_n^1 = \frac{\lambda \|x_n - x^*\|^2}{2[1 - \lambda \|x_n - x^*\|]}. \tag{5.1.138}$$

That is

$$e_n \leq e_n^1 \quad (n \geq 0). \tag{5.1.139}$$

If strict inequality holds in (5.1.133) then (5.1.134) and (5.1.139) hold as strict inequalities also (see also Example 5.1.4).

Remark 5.1.10 *As noted in [68], [112], [99], [142], [143], [355]–[359], the local results obtained here can be used for projection methods such as Arnoldi's, the generalized minimum residual method (GMRES), the generalized conjugate residual method (GCR), for combined Newton/finite projection methods and in connection with the mesh independence principle to develop the cheapest and most efficient mesh refinement strategies.*

Remark 5.1.11 *The local results can also be used to solve equations of the form $F(x) = 0$, where F' satisfies the autonomous differential equation [68]:*

$$F'(x) = P(F(x)), \tag{5.1.140}$$

where, $P\colon Y \to X$ is a known continuous operator. Since $F'(x^*) = P(F(x^*)) = P(0)$, we can apply our results without actually knowing the solution x^* of equation (5.1.1).

Example 5.1.5 *Let* $X = Y = \mathbf{R}$, $\overline{U}(x^*, R) = \overline{U}(0,1)$, $G = 0$, $A(x,y) = F'(x)$, *and define function* F *on* $\overline{U}(0,1)$ *by*

$$F(x) = e^x - 1. \tag{5.1.141}$$

Then we can set $P(x) = x + 1$ *in (5.1.140). Using (5.1.129) we get* $\lambda_2 = e$. *Moreover by (5.1.141) we get*

$$\begin{aligned} F'(x) - F'(x^*) &= e^x - 1 = x + \frac{x^2}{2!} + \cdots + \frac{x^n}{n!} + \cdots \\ &= \left(1 + \frac{x}{2!} + \cdots + \frac{x^{n-1}}{n!} + \cdots\right)(x - x^*) \end{aligned} \tag{5.1.142}$$

and

$$F'(x^*)^{-1}[\|F'(x) - F'(x^*)\|] \leq (e-1)\|x - x^*\|.$$

That is $\lambda_0 = e - 1$. *By (5.1.131) and (5.1.132) we get*

$$\lambda_{RH} = .245252961 \tag{5.1.143}$$

and

$$\lambda^* = .324947231. \tag{5.1.144}$$

That is our convergence radius λ^* *is larger than the corresponding one* λ_{RH} *due to Rheinboldt [295] and our error bounds (5.1.137) are also finer than (5.1.138) so that (5.1.139) holds as a strict inequality. If we define*

$$F(x) = \frac{1}{5}e^{x^5} - x - \frac{1}{5}, \quad A(x,x) = A(x), \quad G(x) = 0 \tag{5.1.145}$$

we can choose

$$\overline{\theta}_0(r,r) = er^4 \tag{5.1.146}$$

and

$$\overline{\theta}(r,r,t) = e[(1-t)r]^4. \tag{5.1.147}$$

Solving equation (5.1.124) with the above choices we obtain

$$\lambda^* = .639470787, \tag{5.1.148}$$

whereas (5.1.132) for $\lambda = 9e$ *gives*

$$\lambda_{RH} = .027250329. \tag{5.1.149}$$

Hence again our convergence radius λ^* *constitutes a significant improvement over* λ_{RH}.

Finally note that all these improvements are made using the same hypotheses/information as in the earlier results. This observation is important in computational mathematics, since a wider choice of initial guesses x_0 becomes available (see also Remark 5.1.10).

Remark 5.1.12 *The results obtained here can be extended to m-point methods ($m > 2$ an integer) [85], and can be used in the solution of variational inequalities (see chapter 10). However we leave the details to the motivated reader.*

5.2 Relaxing the convergence conditions

In this section we further relax conditions (5.1.64) and (5.1.65). For simplicity we set $G(x) = 0$ for all $x \in D$. We consider Newton-like methods of the form:

$$x_{n+1} = x_n - A(x_n)^{-1}F(x_n) \quad (x_0 \in D), \ (n \geq 0) \tag{5.2.1}$$

where $A(x) \in L(X, Y)$ $(x \in D)$ to approximate x^* solving equation (5.1.1).

The conditions used so far (see section 5.1) can be summarized as follows:

For all $x, y \in \overline{U}(x_0, R)$, $A(x_0)^{-1} \in L(Y, X)$, $R > 0$,

$$\|A(x_0)^{-1}[F'(x) - F'(y)]\| \leq v(\|x - y\|), \tag{5.2.2}$$

$$\|A(x_0)^{-1}[A(x) - F'(x_0)]\| \leq v_0(\|x - x_0\|), \tag{5.2.3}$$

$$\|A(x_0)^{-1}[A(x) - A(y)]\| \leq v_1(\|x - y\|) \tag{5.2.4}$$

and R is such that

$$\overline{U}(x_0, R) \subseteq D, \tag{5.2.5}$$

where v, v_0, v_1 are nondecreasing nonnegative functions defined on $[0, R]$.

Our motivation for introducing conditions weaker than (5.2.2)–(5.2.4) can be seen from the following example in the case of Newton's method. Let $X = Y = \mathbf{R}$, and $D = [0, \infty)$. Define F on D by

$$F(x) = \frac{x^{1+\frac{1}{i}}}{1+\frac{1}{i}} + c_1 x + c_2 \tag{5.2.6}$$

where c_1, c_2 are real parameters and $i > 2$ an integer. Set $v_0(r) = v(r) = v_1(r) = r$. Then $F'(x) = x^{1/i} + c_1$ is not Lipschitz on D. That is conditions (5.2.2)–(5.2.4) do not hold in this case. Therefore we need to introduce conditions that will cover such cases.

Indeed we assume instead of (5.2.2)–(5.2.4) that the following weaker center-Lipschitz conditions hold:

$$\|A(x_0)^{-1}[F'(x) - F'(x_0)]\| \leq w_0(\|x - x_0\|), \tag{5.2.7}$$

$$\|A(x_0)^{-1}[A(x) - A(x_0)]\| \leq w_1(\|x - x_0\|), \tag{5.2.8}$$

where functions w_0, w_1 are as v_0, v_1. Further introduce the conditions:

$$\|A(x_0)^{-1}(F'(x_0) - A(x_0))\| \leq a < 1, \tag{5.2.9}$$

and

$$\|A(x_0)^{-1}F(x_0)\| \leq \eta. \tag{5.2.10}$$

We call by (C) conditions (5.2.5), (5.2.7)–(5.2.10).

Clearly

$$w_0(r) \leq v(r), \quad w_1(r) \leq v_1(r) \quad \text{and} \quad a \leq v(R) \ \text{ for all } r \in [0, R], \tag{5.2.11}$$

hold in general. Moreover $\frac{v(r)}{w_0(r)}$, $\frac{v_1(r)}{w_1(r)}$ and $\frac{v(R)}{a}$ can be arbitrarily large. Moreover w_0, w_1 and a are easier to compute than v and v_1.

Note that returning back to the example given by (5.2.6) and setting $w_0(r) = w_1(r) = r$ and for say $x_0 = 1$, we obtain

$$\begin{aligned}\|F'(x) - F'(x_0)\| &= \left|x^{\frac{1}{i}} - x_0^{\frac{1}{i}}\right| \\ &= \frac{|x - x_0|}{x_0^{\frac{i-1}{i}} + \cdots + x^{\frac{i-1}{i}}} \leq \|x - x_0\|.\end{aligned} \tag{5.2.12}$$

That is conditions (C) can be used in this case.

Here motivated by the above we provide a local as well as a semilocal convergence analysis for method (5.2.1). Certain examples are also provided.

It is convenient to define functions q and f on $(0, R]$ by

$$q(r) = \frac{w_0(r) + w_1(r) + a}{1 - w_1(r)}, \quad \text{provided that } w_1(r) \neq 1, \tag{5.2.13}$$

$$f(r) = \left[1 + \frac{w_0(r) + a}{1 - w_1(r)} + \frac{q^2(r)}{1 - q(r)}\right]\eta, \ \text{ provided that } w_1(r) \neq 1 \text{ and } q(r) \neq 1. \tag{5.2.14}$$

We can now state and prove the main semilocal convergence theorem for Newton-like method (5.2.1):

Theorem 5.2.1 *Under conditions (C) assume: (a) $\eta < R$ and function f has fixed points in $[\eta, R]$; Denote by r_0 the smallest; and (b)*

$$w_1(r_0) < 1 \quad \textit{and} \quad q(r_0) < 1. \tag{5.2.15}$$

Then sequence $\{x_n\}$ $(n \geq 0)$ generated by Newton-like method (5.2.1) is well defined, remains in $\overline{U}(x_0, r_0)$ and converges to a unique solution $x^* \in \overline{U}(x_0, r_0)$ of equation $F(x) = 0$. Moreover the following estimates hold for all $n \geq 0$:

$$\|x_{n+2} - x_{n+1}\| \leq q\|x_{n+1} - x_n\| \leq q^{n+1}\eta, \quad q = q(r_0), \tag{5.2.16}$$

and

$$\|x_n - x^*\| \le \frac{\eta}{1-q} q^n. \tag{5.2.17}$$

Proof. Let $x \in \overline{U}(x_0, r_0)$. Using condition (5.2.8) we get

$$\|A(x_0)^{-1}[A(x) - A(x_0)]\| \le w_1(\|x - x_0\|) \le w_1(r_0) < 1. \tag{5.2.18}$$

By (5.2.18) and the Banach Lemma on invertible operators 1.3.2 we deduce $A(x)^{-1} \in L(Y, X)$,
and

$$\|A(x)^{-1}A(x_0)\| \le [1 - w_1(\|x - x_0\|)]^{-1} \le [1 - w_1(r_0)]^{-1}. \tag{5.2.19}$$

In particular $x_1 \in \overline{U}(x_0, r_0)$, since we have $\eta \le r_0$. We assume $x_k \in \overline{U}(x_0, r_0)$, $k = 0, 1, \ldots, n$. Using (5.2.1) we obtain the identity

$$\begin{aligned} F(x_n) &= F(x_n) - F(x_{n-1}) - A(x_{n-1})(x_n - x_{n-1}) \\ &= \left\{ \int_0^1 [F'(x_{n-1} + t(x_n - x_{n-1})) - F'(x_0)]dt \right. \\ &\quad \left. + \ [F'(x_0) - A(x_0)] + [A(x_0) - A(x_{n-1})] \right\} (x_n - x_{n-1}). \end{aligned} \tag{5.2.20}$$

By (5.2.7)–(5.2.9) we get after composing by $A(x_0)^{-1}$ and taking norms

$$\begin{aligned} \|A(x_0)^{-1}F(x_n)\| &\le \left\{ \int_0^1 \|A(x_0)^{-1}[F'(x_{n-1} + t(x_n - x_{n-1})) - F'(x_0)]\|dt \right. \\ &\quad + \ \|A(x_0)^{-1}[F'(x_0) - A(x_0)]\| \\ &\quad \left. + \ \|A(x_0)^{-1}[A(x_0) - A(x_{n-1})]\| \right\} \|x_n - x_{n-1}\| \\ &\le \left\{ \int_0^1 w_0[(1-t)\|x_{n-1} - x_0\| + t\|x_n - x_0\|]dt \right. \\ &\quad \left. + \ a + w_1(\|x_{n-1} - x_0\|) \right\} \|x_n - x_{n-1}\| \\ &\le \ [w_0(r_0) + w_1(r_0) + a]\|x_n - x_{n-1}\|. \end{aligned} \tag{5.2.21}$$

Moreover by (5.2.1), (5.2.19) and (5.2.21) we get

$$\begin{aligned} \|x_2 - x_1\| &\le \frac{w_0(r) + a}{1 - w_1(r)}\|x_1 - x_0\|, \\ \|x_{n+1} - x_n\| &\le \|A(x_n)^{-1}A(x_0)\| \cdot \|A(x_0)^{-1}F(x_n)\| \\ &\le q\|x_n - x_{n-1}\| \quad \text{for } n > 1, \end{aligned} \tag{5.2.22}$$

which shows (5.2.16). Note that $x_{n-1}+t(x_n-x_{n-1})\in\overline{U}(x_0,r_0)$, since

$$\begin{aligned}\|x_{n-1}-x_0+t(x_n-x_{n-1})\| &\leq (1-t)\|x_{n-1}-x_0\|+t\|x_n-x_0\|\\ &\leq (1-t)r_0+tr_0=r_0.\end{aligned}$$

We must show $x_{n+1}\in\overline{U}(x_0,r_0)$. But

$$\begin{aligned}\|x_{n+1}-x_0\| &\leq \|x_{n+1}-x_n\|+\|x_n-x_{n-1}\|+\cdots+\|x_1-x_0\|\\ &\leq \left(q^n+q^{n-1}+\cdots+q^2+\frac{w_0(r_0)+a}{1-w_1(r)}+1\right)\eta\\ &= f(r_0)=r_0,\end{aligned}\tag{5.2.23}$$

which shows $x_n\in\overline{U}(x_0,r_0)$ for all $n\geq 0$. Using (5.2.16) we get for $m\geq 0$, $n\geq 2$

$$\begin{aligned}\|x_{n+m}-x_n\| &\leq \|x_{n+m}-x_{n+m-1}\|+\cdots+\|x_{n+1}-x_n\|\\ &\leq (q^{n+m-1}+\cdots+q^n)\eta=q^n\frac{1-q^m}{1-q}\eta.\end{aligned}\tag{5.2.24}$$

By letting $m\to\infty$ we see $\{x_n\}$ is a Cauchy sequence in a Banach space X and as such it converges to some $x^*\in\overline{U}(x_0,r_0)$. Moreover for $n\to\infty$ in (5.2.21) we get $F(x^*)=0$. To show uniqueness, let $y^*\in\overline{U}(x_0,r_0)$ then as in (5.2.20) we obtain

$$\begin{aligned}x_n-y^* = A(x_{n-1})^{-1}\Bigg\{&\int_0^1[F'(x_{n-1}+t(y^*-x_{n-1}))-F'(x_0)]dt\\ &+[F'(x_0)-A(x_0)]+[A(x_0)-A(x_{n-1})]\Bigg\}(y^*-x_{n-1}),\end{aligned}\tag{5.2.25}$$

and consequently

$$\|x_n-y^*\|\leq q\|x_{n-1}-y^*\|<\|x_{n-1}-y^*\|.\tag{5.2.26}$$

We get $\lim_{n\to\infty}x_n=y^*$. But we showed $\lim_{n\to\infty}x_n=x^*$. Hence, we deduce $x^*=y^*$.. ■

In order for us to cover the local convergence case we introduce conditions similar to (5.2.7) and (5.2.8) as: assume there exists a solution $x^*\in D$ of equation (5.2.1), functions $\overline{w}_0$, $\overline{w}_1$ as w_0 and w_1 respectively and a parameter $\overline{a}$ such that for all $x\in\overline{U}(x^*,R)\subseteq D$:

$$A(x^*)^{-1}\in L(Y,X),$$

$$\|A(x^*)^{-1}[F'(x)-F'(x^*)]\|\leq\overline{w}_0(\|x-x^*\|),\tag{5.2.27}$$

$$\|A(x^*)^{-1}[A(x)-A(x^*)]\|\leq\overline{w}_1(\|x-x^*\|),\tag{5.2.28}$$

$$\|A(x^*)^{-1}[F'(x^*)-A(x^*)]\|\leq\overline{a}<1\tag{5.2.29}$$

and

$$\overline{U}(x^*, R) \subseteq D. \tag{5.2.30}$$

Call the above hypotheses conditions $(\overline{C})$.

Define functions $\overline{q}$, $\overline{f}$ on $(0, R]$ by

$$\overline{q}(r) = \frac{\overline{w}_0(r) + \overline{w}_1(r) + \overline{a}}{1 - \overline{w}_1(r)} \tag{5.2.31}$$

provided that $\overline{w}_1(r) \neq 1$, and

$$\overline{f}(r) = \overline{w}_0(r) + 2\overline{w}_1(r) + \overline{a} - 1. \tag{5.2.32}$$

Then we can show the following local convergence theorem for Newton-like method (5.2.1):

Theorem 5.2.2 *Under condition $(\overline{C})$ assume: (a) function $\overline{f}$ has fixed points in $(0, R]$; Denote by r^* the smallest. and (b) $\overline{w}_1(r^*) < 1$. Then sequence $\{x_n\}$ $(n \geq 0)$ generated by Newton-like method (5.2.1) is well defined, remains in $\overline{U}(x^*, r^*)$ for all $n \geq 0$, and converges to x^* provided that $x_0 \in U(x^*, r^*)$. Moreover the following estimates hold for all $n \geq 0$:*

$$\|x_{n+1} - x^*\| \leq \overline{q}(r^*)\|x_n - x^*\| \leq \overline{q}^{n+1}(r^*)\|x_0 - x^*\|.$$

Proof. By simply replacing y^*, x_0, w_0, w_1, a, q, f by x^*, x^*, $\overline{w}_0$, $\overline{w}_1$, $\overline{a}$, $\overline{q}$, $\overline{f}$ respectively in (5.2.25) and using (5.2.27)–(5.2.32) instead of (5.2.7)–(5.2.9), (5.2.12), (5.2.13), we obtain

$$\|x_n - x^*\| \leq \overline{q}(r^*)\|x_{n-1} - x^*\| < \|x_{n-1} - x^*\|,$$

from which it follows $x_n \in \overline{U}(x^*, r^*)$, and $\lim_{n\to\infty} x_n = x^*$. ∎

Remark 5.2.1 *(a) Conditions for the existence of fixed points of even more general functions than f and $\overline{f}$ are available in the literature [99], [149].*

(b) It follows from (5.2.19), (5.2.21) and (5.2.22) that scalar $\{t_n\}$ $(n \geq 0)$ given by

$$t_{n+2} = t_{n+1} + \frac{\int_0^1 w_0(t_n + t(t_{n+1} - t_n))dt + a + w_1(t_n)}{1 - w_0(t_{n+1})}(t_{n+1} - t_n),$$
$$t_0 = 0, \quad t_1 = b \tag{5.2.33}$$

is a majorizing sequence for $\{x_n\}$ $(n \geq 0)$. In case of convergence we have the following error bounds replacing (5.2.16) and (5.2.17) respectively:

$$\|x_{n+2} - x_{n+1}\| \leq t_{n+2} - t_{n+1} \tag{5.2.34}$$

and

$$\|x_n - x^*\| \leq t^* - t_n, \quad t^* = \lim_{n\to\infty} t_n. \tag{5.2.35}$$

Sufficient convergence conditions for scalar sequence even more general than (5.2.33) have already been given by us in Section 5.1. Therefore under such conditions the conclusions of Theorem 5.2.1 will hold but using scalar sequence $\{t_n\}$.

(c) In the special case of method (4.1.3) we set $A(x) = F'(x)$, $x \in D$, $w_0(r) = w_1(r) = \ell_0 r$ on $[0, R]$ and $a = 0$. It can easily be seen that f has a positive zero r_0 if

$$h_0 = \ell_0 b \leq \frac{2-\sqrt{3}}{2} = .133974596\ldots, \tag{5.2.36}$$

and

$$r_0 = \frac{1 + 2\ell_0 b - \sqrt{(1+2\ell_0 b)^2 - 12\ell_0 b}}{6\ell_0}. \tag{5.2.37}$$

Note also that center Lipschitz conditions (5.2.7) and (5.2.8) reduce to

$$\|F'(x_0)^{-1}[F'(x) - F'(x_0)]\| \leq \ell_0 \|x - x_0\|. \tag{5.2.38}$$

By comparing (5.2.36) with (5.2.2) we see that if

$$\ell_0 \leq (2-\sqrt{3})\ell,$$

then

$$h \leq .5 \Rightarrow h_0 \leq \frac{2-\sqrt{3}}{2}.$$

That is not only (5.2.38) is weaker than (5.2.2) but (5.2.36) may hold but not

$$h = \ell b \leq .5.$$

(d) In the local case for Newton's method, by setting $A(x) = F'(x)$, $\overline{w}_0(r) = \overline{w}_1(r) = r\bar{\lambda}_0$ and $\overline{a} = 0$, we get

$$r^* = \frac{1}{3\bar{\lambda}_0}. \tag{5.2.39}$$

The corresponding radius of convergence r_1^* due to Rheinboldt [295] (using stronger condition (5.1.129)) instead of (5.1.27) is given by (5.1.132). Clearly if

$$2\bar{\lambda}_0 < \lambda, \tag{5.2.40}$$

we get

$$\lambda_{RH} < r^*. \tag{5.2.41}$$

In this case our Theorem 5.2.2 provides a wider choice of initial guesses x_0.

We complete this study with an example where (5.2.3) holds but (5.2.2) fails. For simplicity we take $A(x) = F'(x)$ $(x \in D)$. Let us illustrate the previous results with an example.

Example 5.2.1 *We consider the integral equation*

$$u(s) = f(s) + \lambda \int_a^b G(s,t)u(t)^{1+1/n}dt, \quad n \in \mathbb{N}. \tag{5.2.42}$$

Here, f is a given continuous function satisfying $f(s) > 0$, $s \in [a,b]$, λ is a real number, and the kernel G is continuous and positive in $[a,b] \times [a,b]$. For instance, when $G(s,t)$ is the Green kernel, the corresponding integral equation is equivalent to the boundary value problem

$$\begin{cases} u'' = \lambda u^{1+1/n} \\ u(a) = f(a), \quad u(b) = f(b). \end{cases}$$

These kinds of problems have been also treated in [134]. Equations like (5.2.42) are a type of generalization of equation of the form

$$u(s) = \lambda \int_a^b G(s,t)u(t)^n dt, \quad n \in \mathbb{N},$$

studied by Davis [134]. Instead of (5.2.42) we can try to solve the equation $F(u) = 0$ where

$$F : \Omega \subseteq C[a,b] \to C[a,b],$$
$$\Omega = \{u \in C[a,b] : u(s) \geq 0, s \in [a,b]\},$$

and

$$F(u)(s) = u(s) - \lambda \int_a^b G(s,t)u(t)^{1+1/n}dt.$$

The norm we consider is the max-norm. The derivative F' is defined by

$$F'(u)v(s) = v(s) - \lambda(1+1/n) \int_a^b G(s,t)u(t)^{1/n}v(t)dt, \quad v \in \Omega.$$

First of all, we notice that F' does not satisfy a Lipschitz-type condition in Ω. Let us consider, for instance, $[a,b] = [0,1]$, $G(s,t) = 1$ and $y(t) = 0$. Then $F'(y)v(s) = v(s)$ and

$$\|F'(x) - F'(y)\| = |\lambda| (1+1/n) \int_0^1 x(t)^{1/n}dt.$$

If F' were a Lipschitz function, then

$$\|F'(x) - F'(y)\| \leq L_1 \|x - y\|$$

or equivalently, the inequality

$$\int_0^1 x(t)^{1/n}dt \le L_2 \max_{t\in[0,1]} x(s) \tag{5.2.43}$$

would hold for all $x \in \Omega$ and for a constant L_2. But this is not true. Consider, for example, the function

$$x_j(t) = \frac{t}{j}, \quad j \ge 1, \ t \in [0,1].$$

If these are substituted into (5.2.43),

$$\frac{1}{j^{1/n}(1+1/n)} \le \frac{L_2}{j} \Longleftrightarrow j^{1-1/n} \le L_2(1+1/n), \quad \forall j \ge 1.$$

This inequality is not true when $j \to \infty$. Therefore, condition (5.2.2) fails in this case. However, condition (5.2.3) holds. To show this, let $x_0(t) = f(t)$ and $\alpha = \min_{s\in[a,b]} f(s)$, $\alpha > 0$. Then, for $v \in \Omega$,

$$\begin{aligned}
&\|[F'(x) - F'(x_0)]\, v\| = \\
&= |\lambda|\,(1+1/n) \max_{s\in[a,b]} \left| \int_a^b G(s,t)\left(x(t)^{1/n} f(t)^{1/n}\right) v(t)dt \right| \\
&\le |\lambda|\,(1+1/n) \max_{s\in[a,b]} \int_a^b \frac{G(s,t)|x(t)-f(t)|}{x(t)^{(n-1)/n}+x(t)^{(n-2)/n}f(t)^{1/n}+\ldots+f(t)^{(n-1)/n}} df\, \|v\|.
\end{aligned}$$

Hence,

$$\begin{aligned}
\|F'(x) - F'(x_0)\| &\le \frac{|\lambda|\,(1+1/n)}{\alpha^{(n-1)/n}} \max_{s\in[a,b]} \int_a^b G(s,t)dt\, \|x - x_0\| \\
&\le K\, \|x - x_0\|,
\end{aligned}$$

where $K = \frac{|\lambda|(1+1/n)}{\alpha^{(n-1)/n}} N$ and $N = \max_{s\in[a,b]} \int_a^b G(s,t)dt$. Set $\ell_0 = \left\|F'(x_0)^{-1}\right\| K$.

5.3 A fast method

It turns out that special choices of linear operator A introduced in Section 5.2 can increase the order of convergence.

In this section we are concerned with the problem of approximating a solution x^* of the nonlinear equation

$$F(x) = 0, \tag{5.3.1}$$

where F is a Fréchet-differentiable operator defined on an open subset D of a Banach space X with values in a Banach space Y.

The secant method is the most popular iterative procedure using two previous iterates and divided differences of order one for approximating x^*. The order of the Secant method is 1.618... . In the elegant paper of F.A. Potra [277] a three point method was used and divided differences of order one (see also (8.1.24) and (8.1.25)). This method is of order 1.839.... More recently Secant-like methods of order between 1.618... and 1.839... were introduced in the works of Hernandez, Gutierrez et al. [186]–[191]. The question arises if it is then possible to realize an iterative method using two previous iterates and divided differences of only order one with at least quadratic convergence.

It turns out that this is possible. Indeed we introduce the method

$$x_{n+1} = x_n - [2x_n - x_{n-1}, x_{n-1}]^{-1} F(x_n) \quad (x_{-1}, x_0 \in D) \quad (n \geq 0) \tag{5.3.2}$$

for approximating x^*. Here, a linear operator from X into Y, denoted by $[x, y; F]$ or simply $[x, y]$ which satisfies the condition

$$[x, y](x - y) = F(x) - F(y), \tag{5.3.3}$$

is called a divided difference of order one. Iteration (5.3.2) has a geometrical interpretation similar to the Secant method in the scalar case.

We provide a local and semilocal convergence analysis for method (5.3.2) using Lipschitz-type conditions and the majorant principle. The monotone convergence of method (5.3.2) is also examined on partially ordered topological spaces.

We can show the following local convergence result for method (5.3.2).

Theorem 5.3.1 *Let F be a nonlinear operator defined on an open subset D of a Banach space X with values in a Banach space Y. Assume: equation $F(x) = 0$ has a solution $x^* \in D$ at which the Fréchet derivative $F'(x^*)$ exists, and is invertible; operator F is Fréchet-differentiable with divided difference of order one on $D_0 \subseteq D$ satisfying the Lipschitz conditions:*

$$\|F'(x^*)^{-1}[F'(x) - F'(x^*)]\| \leq a\|x - x^*\|, \tag{5.3.4}$$

$$\|F'(x^*)^{-1}([x, y] - [x, x^*])\| \leq b\|y - x^*\| \tag{5.3.5}$$

and

$$\|F'(x^*)^{-1}([y, y] - [2y - x, x])\| \leq c\|y - x\|^2; \tag{5.3.6}$$

the ball

$$U^* = U(x^*, r^*) \subseteq D_0, \tag{5.3.7}$$

where,

$$r^* = \frac{4}{a + b + \sqrt{(a+b)^2 + 32c}}; \tag{5.3.8}$$

$$F'(x) = [x, x] \quad \text{for all} \quad x \in D; \tag{5.3.9}$$

$$\text{for all } x, y \in D_0 \Rightarrow 2y - x \in D_0. \tag{5.3.10}$$

Then, sequence $\{x_n\}$ $(n \geq 0)$ generated by method (5.3.2) is well defined, remains in $U(x^*, r^*)$ for all $n \geq 0$ and converges to x^* provided that x_{-1}, x_0 belong in $U(x^*, r^*)$. Moreover the following estimates hold for all $n \geq 0$:

$$\|x_{n+1} - x^*\| \leq \frac{b\|x_n - x^*\| + c\|x_{n-1} - x_n\|^2}{1 - a\|x_n - x^*\| - c\|x_{n-1} - x_n\|^2}\|x_n - x^*\|. \tag{5.3.11}$$

Proof. Let us denote by $L = L(x, y)$ the linear operator

$$L = [2y - x, x]. \tag{5.3.12}$$

Assume $x, y \in U(x^*, r^*)$. We shall show L is invertible on $U(x^*, r^*)$, and

$$\|L^{-1}F'(x^*)\| \leq [1 - a\|y - x^*\| - c\|x - y\|^2]^{-1} \leq [1 - ar^* - 4c(r^*)^2]^{-1}. \tag{5.3.13}$$

Using (5.3.2), (5.3.4)–(5.3.10), we obtain in turn:

$$\begin{aligned}
&\|F'(x^*)^{-1}[F'(x^*) - L]\| = \\
&= \|F'(x^*)^{-1}[([x^*, x^*] - [y, y]) + ([y, y] - [2y - x, x])] \\
&\leq a\|y - x^*\| + c\|y - x\|^2 \\
&\leq ar^* + c[\|y - x^*\| + \|x^* - x\|]^2 \\
&\leq ar^* + 4c(r^*)^2 < 1
\end{aligned} \tag{5.3.14}$$

by the choice of r^*. It follows from the Banach lemma on invertible operators 1.3.2 and (5.3.14) that L^{-1} exists on $U(x^*, r^*)$, so that estimate (5.3.13) holds. We can also have:

$$\begin{aligned}
&\|F'(x^*)^{-1}([y, x^*] - L)\| = \\
&= \|F'(x^*)^{-1}\left[([y, x^*] - [y, y]) + ([y, y] - L)\right]\| \\
&\leq \|F'(x^*)^{-1}([y, x^*] - [y, y])\| + \|F'(x^*)^{-1}([y, y] - L)\| \\
&\leq b\|y - x^*\| + c\|y - x\|^2 \\
&\leq br^* + 4c(r^*)^2.
\end{aligned} \tag{5.3.15}$$

Moreover by (5.3.2) we get for $y = x_n$, $x = x_{n-1}$

$$\begin{aligned}
\|x_{n+1} - x^*\| &= \| - L_n^{-1}([x_n, x^*] - L_n)(x_n - x^*)\| \\
&\leq \|L_n^{-1}F'(x^*)\| \cdot \|F'(x^*)^{-1}([x_n, x^*] - L_n)\| \cdot \|x_n - x^*\|.
\end{aligned} \tag{5.3.16}$$

Estimate (5.3.11) now follows from (5.3.14)–(5.3.16). Furthermore from (5.3.11), (5.3.14) and (5.3.15) we get

$$\|x_{n+1} - x^*\| < \|x_n - x^*\| < r^* \quad (n \geq 0). \tag{5.3.17}$$

Hence, sequence $\{x_n\}$ $(n \geq -1)$ is well defined, remains in $U(x^*, r^*)$ for all $n \geq -1$ and converges to x^*. ■

Let $x, y, z \in D_0$, and define the divided differences of order two of operator F at the points x, y and z denoted by $[x, y, z]$ by

$$[x, y, z](y - z) = [x, y] - [x, z]. \tag{5.3.18}$$

Remark 5.3.1 *In order for us to compare method* (5.3.2) *with others [236] using divided differences of order one, consider the condition*

$$\|F'(x^*)^{-1}([x, y] - [u, v])\| \leq \overline{a}(\|x - u\| + \|y - v\|) \tag{5.3.19}$$

instead of (5.3.4) *and* (5.3.5). *Note that* (5.3.19) *implies* (5.3.4) *and* (5.3.5). *Moreover we have:*

$$a \leq 2\overline{a} \tag{5.3.20}$$

and

$$b \leq 2\overline{a}. \tag{5.3.21}$$

Therefore stronger condition (5.3.19) *can replace* (5.3.4), (5.3.6) *and* (5.3.9) *in Theorem 5.3.1.*

Assuming F has divided differences of order two, condition (5.3.6) can be replaced by the stronger

$$\|F'(x^*)^{-1}([y, x, y] - [2y - x, x, y])(y - x)\| \leq \overline{c}\|y - x\|, \tag{5.3.22}$$

or the even stronger

$$\|F'(x^*)^{-1}([u, x, y] - [v, x, y])(y - x)\| \leq \overline{\overline{c}}\|u - v\|. \tag{5.3.23}$$

Note also that

$$\overline{c} \leq \overline{\overline{c}} \tag{5.3.24}$$

and we can set

$$c = \overline{c} \tag{5.3.25}$$

despite the fact that $\overline{c}$ is more difficult to compute since we use divided differences of order two (instead of one). Conditions (5.3.19) and (5.3.23) were used in [236] to show method

$$y_{n+1} = y_n - ([y_n, y_{n-1}] + [y_{n-2}, y_n] - [y_{n-2}, y_{n-1}])^{-1}F(y_n) \quad (n \geq 0) \tag{5.3.26}$$

converges to x^* with order 1.839... which is the solution of the scalar equation

$$t^3 - t^2 - t - 1 = 0. \tag{5.3.27}$$

Potra in [236] has also shown how to compute the Lipschitz constants appearing here in some cases. It follows from (5.3.11) that there exist c_0, N sufficiently large such that:

$$\|x_{n+1} - x^*\| \leq c_0 \|x_n - x^*\|^2 \quad \text{for all } n \geq N. \tag{5.3.28}$$

Hence the order of convergence for method (5.3.2) is essentially at least two, which is higher than 1.839.... Note also that the radius of convergence r^* given by (5.3.9) is larger than the corresponding one given in [[236], estimate (5.3.22)]. This observation is very important since it allows a wider choice of initial guesses x_{-1} and x_0.

It turns out that our convergence radius r^* given by (5.3.9) can even be larger than the one given by Rheinboldt [295] (see, e.g., Remark 4.2 in [277]) for Newton's method. Indeed under condition (5.3.19) radius r_R^* is given by

$$r_R^* = \frac{1}{3\overline{a}}. \tag{5.3.29}$$

We showed in Section 5.1 that $\frac{\overline{a}}{a}$ (or $\frac{\overline{a}}{b}$) can be arbitrarily large. Hence we can have:

$$r_R^* < r^*. \tag{5.3.30}$$

In 5.1 we also showed that r_R^* is enlarged under the same hypotheses and computational cost as in [247].

We note that condition (5.3.10) suffices to hold only for x, y being iterates of method (5.3.2) (see, e.g., Example 5.3.2).

Condition (5.3.24) can be removed if $D = X$. In this case (5.3.7) is also satisfied.

Delicate condition (5.3.9) can also be replaced by stronger but more practical one which we decided not to introduce originally in Theorem 5.3.1, so we can leave the result as uncluttered-general as possible.

Indeed, define ball U_1 by

$$U_1 = U(x^*, R^*) \quad \text{with} \quad R^* = 3r^*.$$

If $x_{n-1}, x_n \in U^*$ $(n \geq 0)$ then we conclude $2x_n - x_{n-1} \in U_1$ $(n \geq 0)$. This is true since it follows from the estimates

$$\begin{aligned} \|2x_n - x_{n-1} - x^*\| &\leq \|x_n - x^*\| + \|x_n - x_{n-1}\| \\ &\leq 2\|x_n - x^*\| + \|x_{n-1} - x^*\| \\ &< 3r^* = R^* \ (n \geq 0). \end{aligned}$$

Hence the proof of Theorem 5.3.1 goes through if both conditions (5.3.7), (5.3.9) are replaced by

$$U_1 \subseteq D_0. \tag{(5.3.7')}$$

We provide an example to justify estimate (5.3.30).

Example 5.3.1 *Returning back to Example 5.1.5 we obtain* $a = b = e - 1$, $c = e$ *and* $\bar{a} = \frac{e}{2}$. *In view of (5.3.8) and (5.3.29), we get*

$$r_R^* = .24525296 < .299040145 = r^*.$$

We can also set $R^* = 3r^* = .897120435$.

We can show the following result for the semilocal convergence of method (5.3.2).

Theorem 5.3.2 *Let* F *be a nonlinear operator defined on an open set* D *of a Banach space* X *with values in a Banach space* Y. *Assume: operator* F *has divided differences of order one and two on* $D_0 \subseteq D$; *there exist points* x_{-1}, x_0 *in* D_0 *such that* $2x_0 - x_{-1} \in D_0$ *and* $A_0 = [2x_0 - x_{-1}, x_{-1}]$ *is invertible on* D; *Set* $A_n = [2x_n - x_{n-1}, x_{n-1}]$ $(n \geq 0)$. *there exist constants* α, β *such that:*

$$\|A_0^{-1}([x,y] - [u,v])\| \leq \alpha(\|x-u\| + \|y-v\|), \tag{5.3.31}$$

$$\|A_0^{-1}([y,x,y] - [2y-x,x,y])\| \leq \beta\|x-y\| \tag{5.3.32}$$

for all $x, y, u, v \in D$, *and condition (5.3.10) holds; Define constants* γ, δ *by*

$$\|x_0 - x_{-1}\| \leq \gamma, \tag{5.3.33}$$

$$\|A_0^{-1}F(x_0)\| \leq \delta, \tag{5.3.34}$$

$$2\beta\gamma^2 \leq 1; \tag{5.3.35}$$

define θ, r, h *by*

$$\theta = \left\{(\alpha + \beta\gamma)^2 + 3\beta(1 - \beta\gamma^2)\right\}^{1/2}, \tag{5.3.36}$$

$$r = \frac{1 - \beta\gamma^2}{\alpha + \beta\gamma + \theta}, \tag{5.3.37}$$

and

$$h(t) = -\beta t^3 - (\alpha + \beta\gamma)t^2 + (1 - \beta\gamma^2)t, \tag{5.3.38}$$

$$\delta \leq h(r) = \frac{1}{3}\,\frac{\alpha + \beta\gamma + 2\theta}{1 - 2\beta\gamma^2}r^2; \tag{5.3.39}$$

$$U_0 = U(x_0, r_0) \subseteq D_0, \tag{5.3.40}$$

where $r_0 \in (0, r]$ *is the unique solution of equation*

$$h(t) = (1 - 2\beta\gamma^2)\delta \tag{5.3.41}$$

on interval $(0, r]$. *Then sequence* $\{x_n\}$ $(n \geq -1)$ *generated by method (5.3.2) is well defined, remains in* $U(x_0, r_0)$ *for all* $n \geq -1$ *and converges to a solution* x^* *of equation* $F(x) = 0$. *Moreover the following estimates hold for all* $n \geq -1$

$$\|x_{n+1} - x_n\| \leq t_n - t_{n+1}, \tag{5.3.42}$$

and

$$\|x_n - x^*\| \leq t_n, \tag{5.3.43}$$

where,

$$t_{-1} = r_0 + \gamma, \quad t_0 = r_0, \tag{5.3.44}$$

$$\gamma_0 = \alpha + 3\beta r_0 + \beta\gamma, \quad \gamma_1 = 3\beta r_0^2 - 2\gamma_0 r_0 - \beta\gamma^2 + 1, \tag{5.3.45}$$

and for $n \geq 0$

$$t_{n+1} = \frac{\gamma_0 t_n - (t_n - t_{n-1})^2\beta - 2\beta t_n^2}{\gamma_1 + 2\gamma_0 t_n - (t_n - t_{n-1})^2 - 3\beta t_n^2} \cdot t_n. \tag{5.3.46}$$

Furthermore if $r_0 \leq r_1$ *and*

$$\alpha(2\gamma + r_0 + r_1) < 1, \tag{5.3.47}$$

x^* *is the unique solution of equation* (5.3.1) *in* $\overline{U}(x_0, r_0)$.

Proof. Sequence $\{t_n\}$ $(n \geq -1)$ generated by (5.3.44)–(5.3.46) can be obtained if method (5.3.2) is applied to the scalar polynomial

$$f(t) = -\beta t^3 + \gamma_0 t^2 + \gamma_1 t. \tag{5.3.48}$$

It is simple calculus to show sequence $\{t_n\}$ $(n \geq -1)$ converges monotonically to zero (decreasingly). We can have:

$$\begin{aligned} & t_{n+1} - t_{n+2} = \\ & = \frac{2(t_{n+1} - t_n)}{f(2t_{n+1} - t_n) - f(t_n)} f(t_{n+1}) \qquad (5.3.49) \\ & = \frac{\{[\gamma_0 - (2t_n + t_{n+1})\beta](t_n - t_{n+1}) + (t_n - t_{n-1})^2\beta\}(t_n - t_{n+1})}{1 - \beta\gamma^2 - 2(t_0 - t_{n+1})\alpha - [3(t_0 - t_{n+1})(3t_0 + t_{n+1}) - (t_n - t_{n+1})^2]\beta} \\ & \geq \frac{(t_n - t_{n+1})\alpha + (t_{n-1} - t_n)^2\beta}{1 - 2(t_0 - t_{n+1})\alpha - \beta\gamma^2}(t_n - t_{n+1}). \qquad (5.3.50) \end{aligned}$$

We show (5.3.42) holds for all $n \geq -1$. Using (5.3.33)–(5.3.38) and

$$t_0 - t_1 = \left[1 - \frac{\gamma_0 t_0 - (t_0 - t_{-1})^2\beta - 2\beta t_0^2}{\gamma_1 + 2\gamma_0 t_0 - (t_0 - t_{-1})^2\beta - 3\beta t_0^2}\right] t_0 = \frac{h(r_0)}{1 - 2\beta\gamma^2} = c \tag{5.3.51}$$

we conclude that (5.3.42) holds for $n = -1, 0$. Assume (5.3.42) holds for all $n \leq k$ and $x_k \in U(x_0, r_0)$. By (5.3.10) and (5.3.42) $x_{k+1} \in U(x_0, r_0)$. By (5.3.31), (5.3.32) and (5.3.42)

$$\begin{aligned}
&\|A_0^{-1}(A_0 - A_{k+1})\| = \\
&= \|A_0^{-1}([2x_0 - x_{-1}, x_{-1}] - [x_0, x_{-1}] + [x_0, x_{-1}] - [x_0, x_0] \\
&\quad + [x_0, x_0] - [x_{k+1}, x_0] + [x_{k+1}, x_0] - [x_{k+1}, x_k] \\
&\quad + [x_{k+1}, x_k] - [2x_{k+1} - x_k, x_k])\| \\
&= \|A_0^{-1}(([2x_0 - x_{-1}, x_{-1}, x_0] - [x_0, x_{-1}, x_0])(x_0 - x_{-1}) \\
&\quad + ([x_0, x_0] - [x_{k+1}, x_0]) \\
&\quad + ([x_{k+1}, x_0] - [x_{k+1}, x_k]) + ([x_{k+1}, x_k] - [2x_{k+1} - x_k, x_k]))\| \\
&\le \beta\gamma^2 + (\|x_0 - x_{k+1}\| + \|x_0 - x_k\| + \|x_k - x_{k+1}\|)\alpha \\
&\le \beta\gamma^2 + 2(t_0 - t_{k+1})\alpha < 2\beta\gamma^2 + 2\alpha r \le 1. \qquad (5.3.52)
\end{aligned}$$

It follows by the Banach lemma on invertible operators 1.3.2 and (5.3.52) that A_{k+1}^{-1} exists, so that

$$\|A_{k+1}^{-1}A_0\| \le [1 - \beta\gamma^2 - (\|x_0 - x_{k+1}\| + \|x_0 - x_k\| + \|x_k - x_{k+1}\|)\alpha]^{-1}. \qquad (5.3.53)$$

We can also obtain

$$\begin{aligned}
&\|A_0^{-1}([x_{k+1}, x_k] - A_k)\| = \\
&= \|A_0^{-1}([x_{k+1}, x_k] - [x_k, x_k] \\
&\quad + [x_k, x_k] - [x_k, x_{k-1}] + [x_k, x_{k-1}] - [2x_k - x_{k-1}, x_{k-1}])\| \\
&= \|A_0^{-1}(([x_{k+1}, x_k] - [x_k, x_k]) \\
&\quad + ([x_k, x_{k-1}, x_k] - [2x_k - x_{k-1}, x_{k-1}, x_k])(x_k - x_{k-1}))\| \\
&\le \alpha\|x_k - x_{k+1}\| + \beta\|x_{k-1} - x_k\|^2. \qquad (5.3.54)
\end{aligned}$$

Using (5.3.2), (5.3.53) and (5.3.54) we get

$$\begin{aligned}
&\|x_{k+2} - x_{k+1}\| = \\
&= \|A_{k+1}^{-1}F(x_{k+1})\| = \|A_{k+1}^{-1}(F(x_{k+1}) - F(x_k) - A_k(x_{k+1} - x_k))\| \\
&\le \|A_{k+1}^{-1}A_0\| \, \|A_0^{-1}([x_{k+1}, x_k] - A_k)\| \cdot \|x_k - x_{k+1}\| \\
&\le \frac{\alpha\|x_k - x_{k+1}\| + \beta\|x_{k-1} - x_k\|^2}{1 - \beta\gamma^2 - \alpha(\|x_0 - x_{k+1}\| + \|x_0 - x_k\| + \|x_k - x_{k+1}\|)}\|x_k - x_{k+1}\| \\
&\le \frac{[\alpha(t_k - t_{k+1}) + \beta(t_{k-1} - t_k)^2](t_k - t_{k+1})}{1 - \beta\gamma^2 - 2(t_0 - t_{k+1})\alpha} \le t_{k+1} - t_{k+2}, \qquad (5.3.55)
\end{aligned}$$

which together with (5.3.41) completes the induction. It follows from (5.3.42) that sequence $\{x_n\}$ $(n \ge -1)$ is Cauchy in a Banach space X and as such it converges to some $x^* \in \overline{U}(x_0, r_0)$. By letting $k \to \infty$ in (5.3.55) we obtain $F(x^*) = 0$. Finally to show uniqueness, define operator

$$B_0 = [y^*, x^*] \qquad (5.3.56)$$

where y^* is a solution of equation (5.3.1) in $\overline{U}(x_0, r_1)$. We can have

$$\begin{aligned}
&\|A_0^{-1}(A_0 - B_0)\| \leq \\
&\leq \alpha\,[\|y^* - (2x_0 - x_{-1})\| + \|x^* - x_{-1}\|] \\
&\leq \alpha\,[\|(y^* - x_0) - (x_0 - x_{-1})\| + \|(x^* - x_0) + (x_0 - x_{-1})\|] \\
&\leq \alpha\,[\|y^* - x_0\| + 2\,\|x_0 - x_{-1}\| + \|x^* - x_0\|] \\
&\leq \alpha(2\gamma + r_0 + r_1) < 1.
\end{aligned} \tag{5.3.57}$$

It follows from the Banach lemma on invertible operators and (5.3.57) that linear operator B_0 is invertible. We deduce from (5.3.56) and the identity

$$F(x^*) - F(y^*) = B_0(x^* - y^*) \tag{5.3.58}$$

that

$$x^* = y^*. \tag{5.3.59}$$

■

Remark 5.3.2 *(a) It follows from* (5.3.42), (5.3.43), (5.3.50) *and* (5.3.55) *that the order of convergence of scalar sequence* $\{t_n\}$ *and iteration* $\{x_n\}$ *is quadratic.*

(b) The conclusions of Theorem 5.3.2 hold in a weaker setting. Indeed assume:

$$\|A_0^{-1}([x_0, x_0] - [x, x_0])\| \leq \alpha_0\|x - x_0\|, \tag{5.3.60}$$

$$\|A_0^{-1}([x, x_0] - [x, y])\| \leq \alpha_1\|y - x_0\|, \tag{5.3.61}$$

$$\|A_0^{-1}([y, x] - [2y - x, x])\| \leq \alpha_2\|y - x\|, \tag{5.3.62}$$

$$\|A_0^{-1}([y, x] - [x, x])\| \leq \alpha_3\|y - x\|, \tag{5.3.63}$$

$$\|A_0^{-1}([2x_0 - x_{-1}, x_0] - [x, y])\| \leq \alpha_4(\|2x_0 - x_{-1} - x\| + \|x_0 - y\|) \tag{5.3.64}$$

and

$$\|A_0^{-1}([2x_0 - x_{-1}, x_{-1}, x_0] - [x_0, x_{-1}, x_0])\| \leq \beta_0\|x_0 - x_{-1}\| \tag{5.3.65}$$

for all $x, y \in D_0$. *It follows from* (5.3.31), (5.3.32) *and* (5.3.60) – (5.3.65) *that*

$$\alpha_i \leq 2\alpha, \quad i = 1, 2, 3, 4 \tag{5.3.66}$$

and

$$\beta_0 \leq \beta. \tag{5.3.67}$$

For the derivation of: (5.3.53), *we can use* (5.3.60) – (5.3.62) *and* (5.3.65) *instead of* (5.3.31) *and* (5.3.32), *respectively;* (5.3.54), *we can use* (5.3.63) *instead of* (5.3.31; (5.3.57), *we can use* (5.3.65) *instead of* (5.3.31). *The resulting majorizing sequence call it* $\{s_n\}$ *is also converging to zero and is finer than* $\{t_n\}$ *because of* (5.3.66) *and* (5.3.67). *Therefore if* (5.3.10), (5.3.60) – (5.3.65) *are used in Theorem 5.3.2 instead of* (5.3.31) *we draw the same conclusions but with weaker conditions, and corresponding estimates are such that:*

$$\|x_{n+1} - x_n\| \leq s_n - s_{n+1} \leq t_{n+1} - t_n \tag{5.3.68}$$

and

$$\|x_n - x^*\| \leq s_n \leq t_n \tag{5.3.69}$$

for all $n \geq 0$. *(c) Condition* (5.3.32) *can be replaced by the stronger (not really needed in the proof) but more popular,*

$$\|A_0^{-1}([v, x, y] - [u, x, y])\| \leq \beta_1 \|u - v\| \tag{5.3.70}$$

for all $v, u, x, y \in D$. *(d) As already noted at the end of Remark 5.3.1, conditions* (5.3.10) *and* (5.3.40) *can be replaced by*

$$U_2 = U(x_0, R_0) \subseteq D_0 \quad \text{with} \quad R_0 = 3r_0 \tag{(5.3.40')}$$

provided that $x_{-1} \in U_2$. *Indeed if* $x_{n-1}, x_n \in U_0$ $(n \geq 0)$ *then*

$$\|2x_n - x_{n-1}\| \leq 2\|x_n - x_0\| + \|x_{n-1} - x_0\| < 3r_0.$$

That is $2x_n - x_{n-1} \in U_2$ $(n \geq 0)$.

We can also provide a posteriori estimates for method (5.3.2):

Proposition 5.3.3 *Assume hypotheses of Theorem 5.3.3 hold. Define scalar sequences* $\{p_n\}$ *and* $\{q_n\}$ *for all* $n \geq 1$ *by:*

$$p_n = \alpha\|x_{n-1} - x_n\|^2 + \beta\|x_{n-1} - x_{n-2}\|^2\|x_{n-1} - x_n\| \tag{5.3.71}$$

and

$$q_n = 1 - 2\alpha\|x_n - x_0\| + \beta\gamma^2. \tag{5.3.72}$$

Then the following error bounds hold for all $n \geq 1$*:*

$$\|x_n - x^*\| \leq \varepsilon_n \leq t_n, \tag{5.3.73}$$

where,

$$\varepsilon_n = 2\{q_n + (q_n^2 - 4\alpha p_n)^{1/2}\}^{-1} p_n. \tag{5.3.74}$$

Proof. As in (5.3.52) we can have in turn:

$$\begin{aligned}\|A_0^{-1}(A_0-[x_n,x^*])\| &= \|A_0^{-1}(A_0-[x_0,x_0]+[x_0,x_0]-[x_n,x^*])\| \\ &\leq \beta\gamma^2+\alpha(\|x_0-x_n\|+\|x_0-x^*\|) \\ &\leq \beta\gamma^2+\alpha(2t_0-t_n) \\ &< \beta\gamma^2+2\alpha r_0 \leq 1. \end{aligned} \tag{5.3.75}$$

It follows from (5.3.71) and the Banach lemma on invertible operators 1.3.2 that linear operator $[x_n,x^*]$ is invertible, and

$$\|[x_n,x^*]^{-1}A_0\| \leq (q_n-\alpha\|x_n-x^*\|)^{-1}. \tag{5.3.76}$$

Using (5.3.2) we obtain the approximation:

$$x_n-x^* = ([x_n,x^*]^{-1}A_0)(A_0^{-1}F(x_n)). \tag{5.3.77}$$

By (5.3.54), (5.3.76) and (5.3.77) we obtain the left-hand side estimate of (5.3.73). Moreover we can have in turn:

$$\begin{aligned} &\|x_n-x^*\| \leq \\ &\leq \frac{p_n}{q_n-\alpha\|x_n-x^*\|} \\ &\leq \frac{\alpha(t_{n-1}-t_n)^2+\beta(t_{n-1}-t_n)(t_{n-1}-t_{n-2})^2}{1-\beta\gamma^2-2\alpha(t_0-t_n)-\alpha t_n} \\ &\leq \frac{\{[\gamma_0-\beta(2t_{n-1}+t_n)](t_{n-1}-t_n)+\beta(t_{n-2}-t_{n-1})^2\}(t_{n-1}-t_n)}{1-\beta\gamma^2-2\alpha(t_0-t_n)-\alpha t_n-(3r_0+\gamma)(t_0-t_n)\beta-\beta t_n^2-\beta\gamma r_0} \\ &\leq \frac{-\beta t_n^3+\alpha t_n^2+\gamma_1}{-\beta r_n^2+\alpha t_n+\gamma_1} = t_n. \end{aligned} \tag{5.3.78}$$

■

A simple numerical example follows to show:

(a) how to choose divided difference in method (5.3.2);

(b) method (5.3.2) is faster than the Secant method [68]

$$x_{n+1} = x_n-[x_n,x_{n-1}]^{-1}F(x_n) \quad (n \geq 0) \tag{5.3.79}$$

(c) method (5.3.2) can be as fast as Newton-Kantorovich method (4.1.3)

Note that the analytical representation of $F'(x_n)$ may be complicated which makes the use of method (5.3.2) very attractive.

Example 5.3.2 *Let $X = Y = \mathbf{R}$, and define function F on $D_0 = D = (.4, 1.5)$ by*

$$F(x) = x^2 - 6x + 5. \tag{5.3.80}$$

Moreover define divided difference of order one appearing in method (5.3.2) by

$$[2y - x, x] = \frac{F(2y - x) - F(x)}{2(y - x)}. \tag{5.3.81}$$

In this case method (5.3.2) becomes

$$x_{n+1} = \frac{x_n^2 - 5}{2(x_n - 3)}, \tag{5.3.82}$$

and coincides with method (4.1.3) applied to F. Furthermore Secant method (5.3.79) becomes:

$$x_{n+1} = \frac{x_{n-1}x_n - 5}{x_{n-1} + x_n - 6}. \tag{5.3.83}$$

Choose $x_{-1} = .6$ and $x_0 = .7$. Then we obtain:

n	Method 2	Secant method (5.3.83)
1	.980434783	.96875
2	.999905228	.997835498
3	.999999998	.99998323
4	$1 = x^*$	.999999991
5	–	1

We conclude this section with an example involving a nonlinear integral equation:

Example 5.3.3 *Let $H(x, t, x(t))$ be a continuous function of its arguments which is sufficiently many times differentiable with respect to x. It can easily be seen that if operator F in (5.3.1) is given by*

$$F(x(s)) = x(s) - \int_0^1 H(s, t, x(t))dt, \tag{5.3.84}$$

then divided difference of order one appearing in (5.3.2) can be defined as

$$h_n(s, t) = \frac{H(s, t, 2x_n(t) - x_{n-1}(t)) - H(s, t, x_{n-1}(t))}{2(x_n(t) - x_{n-1}(t))}, \tag{5.3.85}$$

provided that if for $t = t_m$ we get $x_n(t) = x_{n-1}(t)$, then the above function equals $H'_x(s, t_m, x_n(t_m))$. Note that this way $h_n(s, t)$ is continuous for all $t \in [0, 1]$.

We refer the reader to Chapter 2 for the concepts concerning partially ordered topological spaces (POTL-spaces).

The monotone convergence of method (5.3.2) is examined in the next result.

Theorem 5.3.4 *Let F be a nonlinear operator defined on an open subset of a regular POTL-space X with values in a POTL-space Y. Let x_0, y_0, y_{-1} be points of D such that:*

$$x_0 \leq y_0 \leq y_{-1}, \quad D_0 = \langle x_0, y_{-1} \rangle \subseteq D, \quad F(x_0) \leq 0 \leq F(y_0). \tag{5.3.86}$$

Moreover assume: there exists a divided difference $[\cdot,\cdot]\colon D \to L(X,Y)$ such that for all $(x,y) \in D_0^2$ with $x \leq y$

$$2y - x \in D_0, \tag{5.3.87}$$

and

$$F(y) - F(x) \leq [x, 2y-x](y-x). \tag{5.3.88}$$

Furthermore, assume that for any $(x,y) \in D_0^2$ with $x \leq y$, and $(x, 2y-x) \in D_0^2$ the linear operator $[x, 2y-x]$ has a continuous non-singular, non-negative left subinverse. Then there exist two sequences $\{x_n\}$ $(n \geq 1)$, $\{y_n\}$ $(n \geq 1)$, and two points x^*, y^* of X such that for all $n \geq 0$:

$$F(y_n) + [y_{n-1}, 2y_n - y_{n-1}](y_{n+1} - y_n) = 0, \tag{5.3.89}$$

$$F(x_n) + [y_{n-1}, 2y_n - y_{n-1}](x_{n+1} - x_n) = 0, \tag{5.3.90}$$

$$F(x_n) \leq 0 \leq F(y_n), \tag{5.3.91}$$

$$x_0 \leq x_1 \leq \cdots \leq x_n \leq x_{n+1} \leq y_{n+1} \leq y_n \leq \cdots \leq y_1 \leq y_0, \tag{5.3.92}$$

$$\lim_{n\to\infty} x_n = x^*, \qquad \lim_{n\to\infty} y_n = y^*. \tag{5.3.93}$$

Finally, if linear operators $A_n = [y_{n-1}, 2y_n - y_{n-1}]$ are inverse non-negative, then any solution of the equation $F(x) = 0$ from the interval D_0 belongs to the interval $\langle x^*, y^* \rangle$ (i.e., $x_0 \leq v \leq y_0$ and $F(v) = 0$ imply $x^* \leq v \leq y^*$).

Proof. Let $\overline{A}_0$ be a continuous non-singular, non-negative left subinverse of A_0. Define the operator $Q\colon \langle 0, y_0 - x_0 \rangle \to X$ by

$$Q(x) = x - \overline{A}_0[F(x_0) + A_0(x)].$$

It is easy to see that Q is isotone and continuous. We also have:

$$\begin{aligned} Q(0) &= -\overline{A}_0 F(x_0) \geq 0, \\ Q(y_0 - x_0) &= y_0 - x_0 - \overline{A}_0(F(y_0)) + \overline{A}_0(F(y_0) - F(x_0) - A_0(y_0 - x_0)) \\ &\leq y_0 - x_0 - \overline{A}_0(F(y_0)) \leq y_0 - x_0. \end{aligned}$$

According to Kantorovich's theorem 2.2.2 on POTL-spaces [99], [272], operator Q has a fixed point $w \in \langle 0, y_0 - x_0 \rangle$. Set $x_1 = x_0 + w$. Then we get

$$F(x_0) + A_0(x_1 - x_0) = 0, \quad x_0 \le x_1 \le y_0. \tag{5.3.94}$$

By (5.3.88) and (5.3.94) we deduce:

$$F(x_1) = F(x_1) - F(x_0) + A_0(x_0 - x_1) \le 0.$$

Consider the operator $H\colon \langle 0, y_0 - x_1 \rangle \to X$ given by

$$H(x) = x + \overline{A}_0(F(y_0) - A_0(x)).$$

Operator H is clearly continuous, isotone and we have:

$$\begin{aligned} H(0) &= \overline{A}_0 F(y_0) \ge 0, \\ H(y_0 - x_1) &= y_0 - x_1 + \overline{A}_0 F(x_1) + \overline{A}_0[F(y_0) - F(x_1) - A_0(y_0 - x_1)] \\ &\le y_0 - x_1 + \overline{A}_0 F(x_1) \le y_0 - x_1. \end{aligned}$$

By Kantorovich's theorem there exists a point $z \in \langle 0, y_0 - x_1 \rangle$ such that $H(z) = z$. Set $y_1 = y_0 - z$ to obtain

$$F(y_0) + A_0(y_1 - y_0) = 0, \quad x_1 \le y_1 \le y_0. \tag{5.3.95}$$

Using (5.3.88), (5.3.95) we get:

$$F(y_1) = F(y_1) - F(y_0) - A_0(y_1 - y_0) \ge 0.$$

Proceeding by induction we can show that there exist two sequences $\{x_n\}$ $(n \ge 1)$, $\{y_n\}$ $(n \ge 1)$ satisfying (5.3.89)–(5.3.92) in a regular space X, and as such they converge to points $x^*, y^* \in X$ respectively. We obviously have $x^* \le y^*$. If $x_0 \le u \le y_0$ and $F(u) = 0$, then we can write

$$A_0(y_1 - u) = A_0(y_0) - F(y_0) - A_0(u) = A_0(y_0 - u) - (F(y_0) - F(u)) \ge 0$$

and

$$A_0(x_1 - u) = A_0(x_0) - F(x_0) - A_0(u) = A_0(x_0 - u) - (F(x_0) - F(u)) \le 0.$$

If the operator A_0 is inverse non-negative then it follows that $x_1 \le u \le y_1$. Proceeding by induction we deduce that $x_n \le u \le y_n$ holds for all $n \ge 0$. Hence we conclude

$$x^* \le u \le y^*.$$

■

In what follows we give some natural conditions under which the points x^* and y^* are solutions of equation $F(x) = 0$.

Proposition 5.3.5 *Under the hypotheses of Theorem 5.3.4, assume that F is continuous at x^* and y^* if one of the following conditions is satisfied:*

(a) $x^* = y^*$;

(b) *X is normal, and there exists an operator $T\colon X \to Y$ ($T(0) = 0$) which has an isotone inverse continuous at the origin and such that $A_n \leq T$ for sufficiently large n;*

(c) *Y is normal and there exists an operator $Q\colon X \to Y$ ($Q(0) = 0$) continuous at the origin and such that $A_n \leq Q$ for sufficiently large n;*

(d) *operators A_n ($n \geq 0$) are equicontinuous.*

Then we deduce

$$F(x^*) = F(y^*) = 0. \tag{5.3.96}$$

Proof. (a) Using the continuity of F and (92) we get

$$F(x^*) \leq 0 \leq F(y^*).$$

Hence, we conclude

$$F(x^*) = 0.$$

(b) Using (90)–(93) we get

$$0 \geq F(x_n) = A_n(x_n - x_{n+1}) \geq T(x_n - x_{n+1}),$$
$$0 \leq F(y_n) = A_n(y_n - y_{n+1}) \leq T(y_n - y_{n+1}).$$

Therefore, it follows:

$$0 \geq T^{-1}F(x_n) \geq x_n - x_{n+1}, \quad 0 \leq T^{-1}F(y_n) \leq y_n - y_{n+1}.$$

By the normality of X, and

$$\lim_{n\to\infty}(x_n - x_{n+1}) = \lim_{n\to\infty}(y_n - y_{n+1}) = 0,$$

we get $\lim_{n\to\infty} T^{-1}F(x_n)) = \lim_{n\to\infty} T^{-1}(F(y_n)) = 0$. Using the continuity of F we obtain (5.3.96). (c) As before for sufficiently large n

$$0 \geq F(x_n) \geq Q(x_n - x_{n+1}), \quad 0 \leq F(y_n) \leq Q(y_n - y_{n+1}).$$

By the normality of Y and the continuity of F and Q we obtain (5.3.96). (d) It follows from the equicontinuity of operator A_n that $\lim_{n\to\infty} A_n v_n = 0$ whenever $\lim_{n\to\infty} v_n = 0$. Therefore, we get $\lim_{n\to\infty} A_n(x_n - x_{n+1}) = \lim_{n\to\infty} A_n(y_n - y_{n+1}) = 0$. By (5.3.89), (5.3.90), and the continuity of F at x^* and y^* we obtain (5.3.96). That completes the proof of Proposition 5.3.5. ∎

Remark 5.3.3 *Hypotheses of Theorem 5.3.4 can be weakened along the lines of Remarks 5.3.1 and 5.3.2 above and the works in [277, pp.102-105], or Chapter 2 on the monotone convergence of Newton-like methods. However, we leave the details to the motivated reader.*

Remark 5.3.4 *We finally note that (5.3.2) is a special case of the class of methods of he form:*

$$x_{n+1} = x_n - [(1+\lambda_n)x_n - \lambda_n x_{n-1}, x_{n-1}]^{-1} F(x_n) \quad (n \geq 0) \tag{5.3.97}$$

where λ_n *are real numbers depending on* x_{n-1} *and* x_n*, i.e.*

$$\lambda_n = \lambda(x_{n-1}, x_n) \quad (n \geq 0), \quad \lambda\colon X^2 \to R, \tag{5.3.98}$$

and are chosen so that in practice, e.g.,

$$\textit{for all } x, y \in D \Rightarrow (1+\lambda(x,y))y - \lambda(x,y)x \in D. \tag{5.3.99}$$

Note that setting $\lambda(x,y) = 1$ *for all* $x, y \in D$ *in (5.3.97) we obtain (5.3.2).*

Using (5.3.97) instead of (5.3.2) all the results obtained here can immediately be reproduced in this more general setting.

5.4 Exercise

5.1. Let F be a nonlinear operator defined on an open convex subset D of a Banach space X with values in a Banach space Y and let $A(x) \in L(X,Y)$ $(x \in D)$. Assume:

- there exists $x_0 \in D$ such that $A(x_0)^{-1} \in L(Y,X)$;
- there exist non-decreasing, non-negative functions a, b such that:

$$\|A(x_0)^{-1}[A(x) - A(x_0)]\| \leq a(\|x - x_0\|),$$
$$\|A(x_0)^{-1}[F(y) - F(x) - A(x)(y-x)]\| \leq b(\|x-y\|)\|x-y\|,$$
$$\text{for all } x, y \in D;$$

- there exist $\eta \geq 0$, $r_0 > \eta$ such that

$$\|A(x_0)^{-1}F(x_0)\| \leq \eta,$$
$$a(r) < 1,$$

and

$$d(r) < 1, \quad \text{for all } r \in (0, r_0],$$

where

$$c(r) = (1 - a(r))^{-1},$$

and

$$d(r) = c(r)b(r);$$

- r_0 is the minimum positive root of equation $h(r) = 0$ on $(0, r_0]$, where

$$h(r) = \frac{\eta}{1-d(r)} - r;$$

- $\bar{U}(x_0, r_0) \subseteq D$.

Show: sequence $\{x_n\}$ $(n \geq 0)$ generated by Newton-like method (5.2.1) is well defined, remains in $U(x_0, r_0)$ for all $n \geq 0$ and converges to a solution $x^* \in \bar{U}(x_0, r_0)$ of equation $F(x) = 0$.

5.2. (a) Let F be a Fréchet-differentiable operator defined on some closed convex subset D of a Banach space X with values in a Banach space Y; let $A(x) \in L(X, Y)$ $(x \in D)$. Assume: there exists $x_0 \in D$ such that $A(x_0) \in L(X, Y)$, $A(x_0)^{-1} \in L(Y, X)$, and

$$\left\|A(x_0)^{-1}[F'(y) - A(x)]\right\| < \varepsilon_0, \quad \text{for all } x, y \in U(x_0, \delta_0).$$

Then, show:

(1) for all $\varepsilon_1 > 0$ there exists $\delta_1 > 0$ such that

$$\left\|[A(x)^{-1} - A(x_0)^{-1}]A(x_0)\right\| < \varepsilon_1, \quad \text{for all } x \in U(x_0, \delta_1).$$

Set $\delta = \min\{\delta_0, \delta_1\}$ and $\varepsilon = \max\{\varepsilon_0, \varepsilon_1\}$.

(2) for $\varepsilon > 0$ there exist $\delta > 0$ as defined above such that

$$\left\|A(x_0)^{-1}[F'(y) - A(x)]\right\| < \varepsilon$$

and

$$\left\|[A(x)^{-1} - A(x_0)^{-1}]A(x_0)\right\| < \varepsilon,$$

for all $x, y \in U(x_0, \delta)$.

(b) Let operators F, A, point $x_0 \in D$, and parameters ε, δ be as in (1). Assume there exist $\eta \geq 0$, $c \geq 0$ such that

$$\|A(x_0)^{-1}F(x_0)\| \leq \eta,$$
$$(1 + \varepsilon)\varepsilon \leq c < 1,$$
$$\frac{\eta}{1-c} \leq \delta,$$

and

$$\bar{U}(x_0, \delta) \subseteq D.$$

Show: sequence $\{x_n\}$ $(n \geq 0)$ generated by Newton-like method (5.2.1) is well defined, remains in $U(x_0, \delta)$ for all $n \geq 0$, and converges to a solution $x^* \in \bar{U}(x_0, \delta)$ of equation $F(x) = 0$. Moreover, if linear operator

$$L = \int_0^1 F'\big(x + t(y - x)\big)dt$$

is invertible for all $x, y \in D$, then x^* is the unique solution of equation $F(x) = 0$ in $\bar{U}(x_0, \delta)$. Furthermore, the following estimates hold for all $n \geq 0$

$$\|x_{n+1} - x_n\| \leq c^n \|x_1 - x_0\| \leq c^n \eta$$

and

$$\|x_n - x^*\| \leq \tfrac{c^n}{1-c} \|x_1 - x_0\|.$$

(c) Let $X = Y = \mathbb{R}$, $D \supseteq U(0, .3)$, $x_0 = 0$,

$$F(x) = \tfrac{x^2}{2} + x - .04.$$

Set $A(x) = F'(x)$ $(x \in D)$, $\delta_3 = \delta_4 = \varepsilon_3 = \varepsilon_4 = .3$.
Then we obtain

$$c_3 = \tfrac{6}{7} < 1,$$
$$\tfrac{\eta}{1-c_3} = .28 < \delta = \delta_3.$$

The conclusions of (b) hold and

$$x^* = .039230485 \in U(x_0, \delta).$$

(d) Let $x^* \in D$ be a simple zero of equation $F(x) = 0$. Assume:

$$\big\|A(x^*)^{-1}[F'(y) - A(x)]\big\| < \varepsilon_{11}, \quad \text{for all } \ x \in U(x^*, \delta_{11}),$$
$$\big\|[A(x)^{-1} - A(x^*)^{-1}]A(x^*)\big\| < \varepsilon_{12}, \quad \text{for all } \ x \in U(x^*, \delta_{12}).$$

Set

$$\delta_{13} = \min\{\delta_{11}, \delta_{12}\}, \quad c_8 = (1 + \varepsilon_{12})\varepsilon_{11}.$$

Further, assume:

$$0 < c_8 < 1$$
$$x_0 \in \bar{U}(x^*, \delta_{13}),$$

and

$$\bar{U}(x^*, \delta_{13}) \subseteq D.$$

Show: sequence, $\{x_n\}$ $(n \geq 0)$ generated by Newton-like method (5.2.1) is well defined, remains in $U(x^*, \delta_{13})$ for all $n \geq 0$ and converges to x^* with

$$\|x_{n+1} - x^*\| \leq c_8 \|x_n - x^*\|, \qquad \text{for all } n \geq 0.$$

5.3. Consider equation (5.1.1) and iteration (5.1.48). Assume:

(a) $\left\| A(x_n)^{-1} (A(x) - A(x_n)) \right\| \leq v_n(r) + b_n,$

$$\left\| A(x_n)^{-1} (F'(x + t(y - x)) - A(x)) \right\|$$
$$\leq w_n(r + t\|y - x\|) - v_n(r) + c_n$$

and

$$\left\| A(x_n)^{-1} (G(x) - G(y)) \right\| \leq e_n(r) \|x - y\|$$

for all $x_n, x, y \in \bar{U}(x_0, r) \subseteq \overline{U}(x_0, R)$, $t \in [0, 1]$, where $w_n(r+t) - v_n(r)\, t \geq 0$ and $e_n(r)$ $(n \geq 0)$ are nondecreasing non-negative functions with $w_n(0) = v_n(0) = e_n(0) = 0$ $(n \geq 0)$, $v_n(r)$ are differentiable, $v_n'(r) > 0$ $(n \geq 0)$ for all $r \in [0, R]$, and the constants b_n, c_n satisfy $b_n \geq 0$, $c_n \geq 0$ and $b_n + c_n < 1$ for all $n \geq 0$. Introduce for all $n, i \geq 0$, $a_n = \|A(x_n)^{-1}(F(x_n) + G(x_n))\|$, $\varphi_{n,i}(r) = a_i - r + c_{n,i} \int_0^r w_n(t)\, dt$, $z_n(r) = 1 - v_n(r) - b_n$, $\psi_n(r) = c_{n,i} \int_0^r e_n(t)\, dt$, $c_{n,i} = z_n(r_i)^{-1}$, $h_{n,i}(r) = \varphi_{n,i}(r) + \psi_{n,i}(r)$, $r_n = \|x_n - x_0\|$, $a_n = \|x_{n+1} - x_n\|$, the equations

$$r = a_n + c_{0,n} \left(\int_0^r (w_0(r_n + t) + e_n(r_n + t))\, dt + (b_n + c_n - 1)\, r \right) \tag{5.4.1}$$

$$r = a_n + c_{n,n} \left(\int_0^r (w_n(r_n + t) + e_n(r_n + t))\, dt + (b_n - c_n - 1)\, r \right) \tag{5.4.2}$$

$$a_n = r + c_{0,n} \left(\int_0^r (w_0(t_n + t) + e_n(r_n + t))\, dt + (b_n - c_n - 1)\, r \right) \tag{5.4.3}$$

$$a_n = r + c_{n,n} \left(\int_0^r (w_n(r + t) + e_n(r_n + t))\, dt + (b_n - c_n - 1)\, r \right) \tag{5.4.4}$$

and the scalar iterations

$$s_{0,n} = s_{n,n}^0 = 0, \; s_{k+1,n}^0 = s_{k,n}^0 + \frac{h_{0,n}(s_{k,n}^0 + r_n)}{c_{0,n} z_0(s_{k,n}^0 + r_n)} \quad (k \geq 0)$$

$$s_{k+1,n} = s_{k,n} + \frac{h_{n,n}(s_{k,n}+r_n)}{c_{n,n}z_{n,n}(s_{k,n}+r_n)} \quad (k \geq n);$$

(b) The function $h_{0,0}(r)$ has a unique zero s_0^* in the interval $[0, R]$ and $h_{0,0}(R) \leq 0$;

(c) The following estimates are true:

$$\frac{h_{n,n}(r+r_n)}{c_{n,n}z_{n,n}(r+r_n)} \leq \frac{h_{0,n}(r+r_n)}{c_{0,n}z_n(r+r_n)}$$

for all $r \in [0, R - r_n]$ and for each fixed $n \geq 0$.

Then show:

i. the scalar iterations $\left\{s_{k+1,n}^0\right\}$ and $\{s_{k+1,n}\}$ for $k \geq 0$ are monotonically increasing and convergence to s_n^* and s_n^{**} for each fixed $n \geq 0$, which are the unique solutions of equations (5.4.1) and (5.4.2) in $[0, R - s_n]$ respectively with $s_n^{**} \leq s_n^* \ (n \geq 0)$.

ii. Iteration (5.1.48) is well defined, remains in $\bar{U}(x_0, s^*)$ for all $n \geq 0$ and converges to a solution of equation (5.1.1), which is unique in $\bar{U}(x_0, R)$.

iii. The following estimates are true:

$$\|x_{n+1} - x_n\| \leq s_{n+1,n+1} - s_{n,n} \leq s_{n+1,n+1}^0 - s_{n,n}^0,$$
$$\|x^* - x_n\| \leq s_n^{**} - s_{n,n} \leq s_n^* - s_{n,n}^0 \leq s_0^* - s_{n,0}^0$$
$$\|x^* - x_n\| \geq I_n^*, \quad \|x^* - x_n\| \geq I_n^{**}$$

and

$$I_n^{**} \leq I_n^* \quad (n \geq 0)$$

where I_n^* and I_n^{**} are the solutions of the equations (5.4.3) and (5.4.4) respectively for all $n \geq 0.i$
The above approach shows how to improve upon the results given in [121] by Yamamoto and Chen.

iv. Define the sequence $\left\{s_n^1\right\}$ $(n \geq 0)$ by

$$s_0^1 = 0, \ s_{n+1}^1 + \frac{h_{n,n}(s_n^1+r_n)}{c_{n,n}} \quad (n \geq 0).$$

Then under the hypotheses (5.4.1)-(5.4.3) above show that:

$$\|x_{n+1} - x_n\| \leq s_{n+1}^1 - s_n^1 \leq s_{n+1,n+1} - s_{n,n} \leq s_{n+1,n+1}^0 - s_{n,n}^0$$

and

$$\|x^* - x_n\| \leq t^* - s_n^1 \leq s_n^{**} - s_{n,n} \leq s_n^* - s_{n,n}^0 \leq s_0^* - s_{n,0}^0 \quad (n \geq 0),$$

where

$$t^* = \lim_{n\to\infty} s_n^1.$$

5.4. Consider the Newton-like method (5.2.1). Let $A : D \to L(X,Y)$, $x_0 \in D$, $M_{-1} \in L(X,Y)$, $X \subset Y$, $L_{-1} \in L(X,X)$. For $n \geq 0$ choose $N_n \in L(X,X)$ and define $M_n = M_{n-1}N_n + A(x_n)L_{n-1}$, $L_n = L_{n-1} + L_{n-1}N_n$, $x_{n+1} = x_n + L_n(y_n)$, y_n being a solution of $M_n(y_n) = -[F(x_n) + z_n]$ for a suitable $z_n \in y$.

Assume:

(a) F is Fréchet-differentiable on D.

(b) There exist non-negative numbers α, α_0 and nondecreasing functions w, w_0 : $\mathbb{R}^+ \to \mathbb{R}^+$ with $w(0) = w_0(0) = 0$ such that

$$\|F(x_0)\| \leq \alpha_0, \quad \|R_0(y_0)\| \leq \alpha,$$
$$\|A(x) - A(x_0)\| \leq w_0(\|x - x_0\|)$$

and

$$\|F'(x + t(y - x)) - A(x)\| \leq w(\|x - x_0\| + t\|x - y\|)$$

for all $x, y \in \bar{U}(x_0, R)$ and $t \in [0,1]$.

(c) Let M_{-1} and L_{-1} be such that M_{-1} is invertible,

$$\left\|M_{-1}^{-1}\right\| \leq \beta, \ \|L_{-1}\| \leq \gamma \text{ and } \|M_{-1} - A(x_0)L_{-1}\| \leq \delta.$$

(d) There exist non-negative sequence $\{a_n\}$, $\{\bar{a}_n\}$, $\{b_n\}$ and $\{c_n\}$ such that for all $n \geq 0$

$$\|N_n\| \leq a_n, \ \|I + N_n\| \leq \bar{a}_n, \ \left\|M_{-1}^{-1}\right\| \cdot \|M_{-1} - M_n\| \leq b_n < 1$$

and

$$\|z_n\| \leq c_n \|F(x_n)\|.$$

(e) The scalar sequence $\{t_n\}$ $(n \geq 0)$ given by

$$t_{n+1} = t_{n+1} + e_{n+1}d_{n+1}(1 + c_{n+1}) \left[I_n + \sum_{i=1}^{n} h_i w(l_i)(l_i - l_{i-1}) + w(t_{n+1})(t_{n+1} - t_n)\right]$$

$(n \geq)$, $t_0 = 0$, $t_1 = \alpha$ is bounded above by a t_0^* with $0 < t_0^* \leq R$, where

$$e_0 = \gamma\bar{a}_0, \ e_n = I_{n-1}\bar{a}_n \ (n \geq 1), \ d_n = \tfrac{\beta}{1-d_n} \ (n \geq 0)$$
$$I_n = \varepsilon_n\varepsilon_{n-1}\ldots\varepsilon_0\alpha_0 \ (n \geq 0) \ \ \varepsilon_n = p_n d_n(1 + c_n) + c_n,$$
$$p_n = q_{n-1}a_n \ (n \geq 1), \ p_0 = \delta a_0,$$
$$q_n = p_n + w_0(t_{n+1})e_n \ (n \geq 1)$$

and

$$h_i = \prod_{m=i}^{n} \varepsilon_m \quad (i \leq n)\,.$$

(f) The following estimate is true $\varepsilon_n \leq \varepsilon < 1 \;\; (n \geq 0)$.

Then show:

i. scalar sequence $\{t_n\}$ $(n \geq 0)$ is nondecreasing and convergence to a t^* with $0 < t^* \leq t_0^*$ as $n \to \infty$.

ii. Newton-like method (5.2.1) is well defined, remains in $\bar{U}\,(x_0, t^*)$ and converges to a solution x^* of equation $F\,(x) = 0$.

iii. The following estimates are true:

$$\|x_{n+1} - x_n\| \leq t_{n+1} - t_n$$

and

$$\|x_n - x^*\| \leq t^* - t_n \quad (n \geq 0)\,.$$

5.5. (a) Let $F\colon D \subseteq X \to Y$ be a Fréchet differentiable operator and $A(x) \in L(X,Y)$ $(x \in D)$. Assume there exists a point $x_0 \in D$, $\eta \geq 0$ and nonnegative continuous functions a, b, c such that

$$\begin{gathered}
A(x_0)^{-1} \in L(Y,X), \\
\|A(x_0)^{-1}F(x_0)\| \leq \eta, \\
\|A(x_0)^{-1}[F'(x) - F'(x_0)]\| \leq a(\|x - x_0\|), \\
\|A(x_0)^{-1}[F'(x_0) - A(x)]\| \leq b(\|x - x_0\|), \\
\|A(x_0)^{-1}[A(x) - A(x_0)]\| \leq c(\|x - x_0\|)
\end{gathered}$$

for all $x \in D$;

equation

$$\int_0^1 a[(1-t)r]r\,dt + [b(r) + c(r) - 1]r + \eta = 0$$

has nonnegative solutions. Denote by r_0 the smallest.

Point r_0 satisfies

$$a(r_0) + b(r_0) + c(r_0) < 1,$$

and

$$\bar{U}(x_0, r_0) = \subseteq D.$$

Then show sequence $\{x_n\}$ $(n \geq 0)$ generated by Newton-like method (5.2.1) is well defined, remains in $\bar{U}(x_0, r_0)$ for all $n \geq 0$ and converges to a unique solution $x^* \in \bar{U}(x_0, r_0)$ of equation $F(x) = 0$. Moroever the following estimates hold for all $n \geq 0$

$$\|x_{n+2} - x_{n+1}\| \leq q\|x_{n+1} - x_n\|$$

and

$$\|x_n - x^*\| \leq \frac{\eta}{1-q} q^{n+1}$$

where,

$$q = \frac{a(r_0) + b(r_0)}{1 - c(r_0)}.$$

Furthermore x^* is unique in $U(x_0, R)$ for $R > t^*$ and

$$U(x_0, R) \subseteq D$$

if

$$\int_0^1 a[(1-t)r_0 + tR]dt + b(0) \leq 1.$$

(b) Let $F\colon D \subseteq X \subseteq Y$ be a Fréchet-differentiable operator and $A(x) \in L(X, Y)$. Assume: there exist a simple zero x^* of F and nonnegative continuous functions α, β, γ such that

$$A(x^*)^{-1} \in L(Y, X),$$

$$\|A(x^*)^{-1}[F'(x) - F'(x^*)]\| \leq \alpha(\|x - x^*\|),$$
$$\|A(x^*)^{-1}[F'(x^*) - A(x)]\| \leq \beta(\|x - x^*\|),$$
$$\|A(x^*)^{-1}[A(x) - A(x^*)]\| \leq \gamma(\|x - x^*\|)$$

for all $x \in D$;

equation

$$\int_0^1 \alpha[(1-t)r]dt + \beta(r) + \gamma(r) = 1$$

has nonnegative solutions. Denote by r^* the smallest; and

$$\bar{U}(x^*, r^*) \subseteq D.$$

Show: Under the above stated hypotheses: sequence $\{x_n\}$ $(n \geq 0)$ generated by the Newton-like method (5.2.1) is well defined, remains in $\bar{U}(x^*, r^*)$ for all $n \geq 0$ and converges to x^*, provided $x_0 \in U(x^*, r^*)$.

Moreover the following estimates hold for all $n \geq 0$:

$$\|x_{n+1} - x^*\| \leq \delta_n \|x_n - x^*\|,$$

where

$$\delta_n = \frac{\int_0^1 \alpha[(1-t)\|x_n - x^*\|]dt + \beta(\|x_n - x^*\|)}{1-\gamma(\|x_n - x^*\|)}.$$

5.6. (a) Assume:

there exist parameters $K \geq 0$, $M \geq 0$, $L \geq 0$, $\ell \geq 0$, $\mu \geq 0$, $\eta \geq 0$, λ_1, λ_2, $\lambda_3 \in [0, 1]$, $\delta \in [0, 2)$ such that:

$$h_q = K\eta^{\lambda_1} + (1 + \lambda_1)\left[M\left(\frac{\eta}{1-q}\right)^{\lambda_2} + \mu\right] + \left[\ell + L\left(\frac{\eta}{1-q}\right)^{\lambda_3}\right]\delta \leq \delta,$$

and

$$\ell + L\left(\frac{\eta}{1-q}\right)^{\lambda_3} \leq 1,$$

where,

$$q = \frac{\delta}{1+\lambda_1}.$$

Then, show: iteration $\{t_n\}$ $(n \geq 0)$ given by

$$t_0 = 0,\ t_1 = \eta,\ t_{n+2} = t_{n+1} + \frac{K(t_{n+1}-t_n)^{\lambda_1} + (1+\lambda_1)[Mt_n^{\lambda_2} + \mu]}{(1+\lambda_1)[1-\ell-Lt_{n+1}^{\lambda_3}]} \cdot (t_{n+1} - t_n)\ (n \geq 0)$$

is non-decreasing, bounded above by

$$t^{**} = \frac{\eta}{1-q},$$

and converges to some t^* such that

$$0 \leq t^* \leq t^{**}.$$

Moreover, the following estimates hold for all $n \geq 0$

$$0 \leq t_{n+2} - t_{n+1} \leq q(t_{n+1} - t_n) \leq q^{n+1}\eta.$$

(b) Let $\lambda_1 = \lambda_2 = \lambda_3 = 1$. Assume:

there exist parameters $K \geq 0$, $M \geq 0$, $L \geq 0$, $\ell \geq 0$, $\mu \geq 0$, $\eta \geq 0$, $\delta \in [0, 1]$ such that:

$$h_\delta = \left(K + L\delta + \frac{4M}{2-\delta}\right)\eta + \delta\ell + 2\mu \leq \delta,$$

$$\ell + \frac{2L\eta}{2-\delta} \leq 1,$$

$$L \leq K,$$

and

$$\ell + 2\mu < 1,$$

then, show: iteration $\{t_n\}$ $(n \geq 0)$ is non-decreasing, bounded above

$$t^{**} = \frac{2\eta}{2-\delta}$$

and converges to some t^* such that

$$0 \leq t^* \leq t^{**}.$$

Moreover the following estimates hold for all $n \geq 0$

$$0 \leq t_{n+2} - t_{n+1} \leq \frac{\delta}{2}(t_{n+1} - t_n) \leq \left(\frac{\delta}{2}\right)^{n+1} \eta.$$

(c) Let $F\colon D \subseteq X \to Y$ be a Fréchet-differentiable operator. Assume:

(1) there exist an approximation $A(x) \in L(X,Y)$ of $F'(x)$, an open convex subset D_0 of D, $x_0 \in D_0$, parameters $\eta \geq 0$, $K \geq 0$, $M \geq 0$, $L \geq 0$, $\mu \geq 0$, $\ell \geq 0$, $\lambda_1 \in [0,1]$, $\lambda_2 \in [0,1]$, $\lambda_3 \in [0,1]$ such that:

$$A(x_0)^{-1} \in L(Y,X), \quad \|A(x_0)^{-1}F(x_0)\| \leq \eta,$$
$$\|A(x_0)^{-1}[F'(x) - F'(y)]\| \leq K\|x - y\|^{\lambda_1},$$
$$\|A(x_0)^{-1}[F'(x) - A(x)]\| \leq M\|x - x_0\|^{\lambda_2} + \mu,$$

and

$$\|A(x_0)^{-1}[A(x) - A(x_0)]\| \leq L\|x - x_0\|^{\lambda_3} + \ell \quad \text{for all } x, y \in D_0;$$

(2) hypotheses of (a) or (b) hold;

(3)

$$\bar{U}(x_0, t^*) \subseteq D_0.$$

Then, show sequence $\{x_n\}$ $(n \geq 0)$ generated by Newton-like method (5.2.1) is well defined, remains in $\bar{U}(x_0, t^*)$ for all $n \geq 0$ and converges to a solution $x^* \in \bar{U}(x_0, t^*)$ of equation $F(x) = 0$.

Moreover the following estimates hold for all $n \geq 0$:

$$\|x_{n+1} - x_n\| \leq t_{n+1} - t_n$$

and

$$\|x_n - x^*\| \leq t^* - t_n.$$

Furthermore the solution x^* is unique in $\bar{U}(x_0, t^*)$ if

$$\frac{1}{1-\ell-L(t^*)^{\lambda_3}}\left[\frac{K}{1+\lambda_1}(t^*)^{1+\lambda_1}+M(t^*)^{\lambda_2}+\mu\right]<1,$$

or in $U(x_0, R_0)$ if $R_0 > t^*$, $U(x_0, R_0) \subseteq D_0$, and

$$\frac{1}{1-\ell-L(t^*)^{\lambda_3}}\left[\frac{K}{1+\lambda_1}(R+t^*)^{1+\lambda_1}+M(t^*)^{\lambda_2}+\mu\right]\leq 1.$$

(d) Let $F\colon D \subseteq X \to Y$ be a Fréchet-differentiable operator. Assume:

(a) there exist an approximation $A(x) \in L(X, Y)$ of $F'(x)$, a simple solution $x^* \in D$ of equation $F(x) = 0$, a bounded inverse $A(x^*)$ and parameters $\bar{K}$, $\bar{L}$, $\bar{M}$, $\bar{\mu}$, $\bar{\ell} \geq 0$, λ_4, λ_5, $\lambda_6 \in [0,1]$ such that:

$$\|A(x^*)^{-1}[F'(x)-F'(y)]\| \leq \overline{K}\|x-y\|^{\lambda_4},$$
$$\|A(x^*)^{-1}[F'(x)-A(x)]\| \leq \bar{M}\|x-x^*\|^{\lambda_5}+\bar{\mu},$$

and

$$\|A(x^*)^{-1}[A(x)-A(x^*)]\| \leq \bar{L}\|x-x^*\|^{\lambda_6}+\bar{\ell}$$

for all $x, y \in D$;

(b) equation

$$\frac{\bar{K}}{1+\lambda_4}r^{\lambda_4}+\bar{L}r^{\lambda_6}+\bar{M}r^{\lambda_5}+\bar{\mu}+\bar{\ell}-1=0$$

has a minimal positive zero r_0 which also satisfies:

$$\bar{L}r_0^{\lambda_6}+\bar{\ell}<1$$

and

$$U(x^*, r_0) \subseteq D.$$

Then, show: sequence $\{x_n\}$ $(n \geq 0)$ generated by Newton-like method (5.2.1) is well defined, remains in $U(x^*, r_0)$ for all $n \geq 0$, and converges to x^* provided that $x_0 \in U(x^*, r_0)$. Moreover the following estimates hold for all $n \geq 0$:

$$\|x_{n+1}-x^*\| \leq$$
$$\leq \frac{1}{1-\bar{L}\|x_n-x^*\|^{\lambda_6}-\bar{\ell}}\left[\frac{\bar{K}}{1+\lambda_4}\|x_n-x^*\|^{\lambda_4}+\bar{M}\|x_n-x^*\|^{\lambda_5}+\bar{\mu}\right]\|x_n-x^*\|.$$

5.7. Let F be a nonlinear operator defined on an open convex subset D of a Banach space X with values in a Banach space Y and let $A(x) \in L(X, Y)$ $(x \in D)$. Assume:

(a) there exists $x_0 \in D$ such that $A(x_0)^{-1} \in L(Y, X)$;

(b) there exist non-decreasing, non-negative functions a, b such that:

$$\|A(x_0)^{-1}[A(x) - A(x_0)]\| \le a(\|x - x_0\|),$$
$$\|A(x_0)^{-1}[F(y) - F(x) - A(x)(y - x)]\| \le b(\|x - y\|)\|x - y\|$$

for all $x, y \in D$;

(c) there exist $\eta \ge 0$, $r_0 > \eta$ such that

$$\|A(x_0)^{-1}F(x_0)\| \le \eta,$$
$$a(r) < 1,$$

and

$$d(r) < 1 \quad \text{for all} \quad r \in (0, r_0],$$

where

$$c(r) = (1 - a(r))^{-1},$$

and

$$d(r) = c^2(r)b(r);$$

(d) r_0 is the minimum positive root of equation $h(r) = 0$ on $(0, r_0]$ where,

$$h(r) = \tfrac{\eta}{1-d(r)} - r.$$

(e) $\bar{U}(x_0, r_0) \subseteq D$.

Then show: sequence $\{x_n\}$ $(n \ge 0)$ generated by Newton-like method (5.2.1) is well defined, remains in $U(x_0, r_0)$ for all $n \ge 0$ and converges to a solution $x^* \in \bar{U}(x_0, r_0)$ of equation $F(x) = 0$.

5.8. (a) Let $F\colon U(z, R) \subseteq X \to Y$ be a Fréchet-differentiable operator for some $z \in X$, $R > 0$, and $A(x) \in L(X, Y)$. Assume:

$A(z)^{-1} \in L(Y, X)$ and for any $x, y \in \bar{U}(z, r) \subseteq \bar{U}(z, R)$

$$\|A(z)^{-1}[A(x) - A(x_0)]\| \le w_0(\|x - x_0\|) + a,$$
$$\|A(z)^{-1}[F'(x + t(y - x)) - A(x)]\| \le w(\|x - z\| + t\|x - y\|) - w_1(\|x - z\|) + b, \ t \in [0, 1],$$
$$\|A(z)^{-1}[G(x) - G(y)]\| \le w_2(r)\|x - y\|,$$
$$0 < \|A(z)^{-1}[F(z) + G(z)]\| \le \eta,$$

where $w(r+t) - w_1(r)$ $(t \geq 0)$, $w_1(r)$ and $w_2(r)$ are non-decreasing, non-negative functions with $w(0) = w_0(0) = w_1(0) = w_2(0) = 0$, w_0 is differentiable, $w_0'(r) > 0$, $r \in [0, R]$,

$$w_0(r) \leq w_1(r) \quad r \in [0, R] \tag{5.4.5}$$

and parameters a, b satisfy

$$a \geq 0, \quad b \geq 0, \quad a + b < 1.$$

Define functions φ_1, φ_2, φ by

$$\varphi_1(r) = \eta - r + \int_0^r w(t)dt,$$

$$\varphi_2(r) = \int_0^r w_2(t)dt,$$

$$\varphi(r) = \varphi_1(r) + \varphi_2(r) + (a+b)r,$$

iteration $\{r_n\}$ $(n \geq 0)$ by

$$r_0 \in [0, R], \quad r_{n+1} = r_n + \frac{\varphi(r_n) - \varphi_0}{1 - a - w_0(r_n)} \quad (n \geq 0),$$

where φ_0 is the minimal value of φ on $[0, R]$;

(b)

$$\varphi(R) \leq 0.$$

Then show iteration $\{x_n\}$ $(n \geq 0)$ generated by Newton-like method (5.1.48) is well defined, remains in $\bar{U}(z, r^*)$ for any

$$x_0 \in D(r^*) = \bigcup_{r \in [0, r^*)} \left\{ x \in \overline{U}(z, r) \mid \|A(x)^{-1}(F(x) + G(x))\| \leq \tfrac{\varphi(r) - \varphi_0}{1 - a - w_0(r)} \right\},$$

and converges to a solution $x^* \in \bar{U}(z, r_0^*)$, which is unique in

$$\widetilde{U} = \begin{cases} \bar{U}(z, R) & \text{if } \varphi(R) < 0 \text{ or } \varphi(R) = 0 \text{ and } r_0^* = R, \\ U(z, R) & \text{if } \varphi(R) = 0 \text{ and } r_0^* < R. \end{cases}$$

where r^* is the minimal point, r_0^* is the unique zero on $(0, r^*]$, and

$$r^* = \lim_{n \to \infty} r_n.$$

Moreover sequence $\{r_n\}$ $(n \geq 0)$ is monotonically increasing and converges to r^*. Furthermore the following estimates hold for all $n \geq 0$

$$\|x_{n+1} - x_n\| \leq r_{n+1} - r_n,$$

$$\|x_n - x^*\| \leq r^* - r_n,$$

provided that $r_0 \in R_{x_0}$, where for $x \in D(r^*)$

$$R_x = \left\{ r \in [0, r^*) : \|A(x)^{-1}[F(x) + G(x)]\| \leq \tfrac{\varphi(r) - \varphi_0}{1 - a - w_0(r)}, \|x - z\| \leq r \right\},$$

and

$$\bar{U}\left(z, \tfrac{|\varphi_0|}{2-a}\right) \subseteq D(r^*).$$

In the next result we show how to improve on the error bounds.

(b) Under the hypotheses of (a) show the conclusions hold with $\{r_n\}$ and $D(r^*)$ replaced by $\{t_n\}$ $(n \geq 0)$, $D(t^*)$ given by

$$t_0 = r_0, \quad t_1 = r_1,$$

$$t_{n+1} = t_n + \frac{\int_0^{t_{n-1}} [w(t_{n-1} + t(t_n - t_{n-1})) - w_1(t_{n-1})] dt (t_n - t_{n-1}) + b(t_n - t_{n-1}) + \int_{t_{n-1}}^{t_n} w_2(t) dt}{1 - a - w_0(t_n)}$$

$$(n \geq 1),$$

$$t^* = \lim_{n \to \infty} t_n.$$

Moreover iteration $\{t_n\}$ is monotonically increasing and converges to t^*.

Furthermore the following hold for all $n \geq 1$:

$$t_{n+1} - t_n \leq r_{n+1} - r_n, \tag{5.4.6}$$

$$t_n \leq r_n, \tag{5.4.7}$$

$$t^* - t_n \leq r^* - r_n, \tag{5.4.8}$$

and

$$t^* \leq r^*.$$

If strict inequality holds in (5.4.5) so does in (5.4.6-5.4.8).

If

$$w_0(r) = w_1(r) \quad r \in [0, R]$$

our result reduces to Theorem 1 in [121]. If (5.4.5) holds our error bounds are at least as fine as the ones given by Chen and Yamamoto (under the same hypotheses and a more general condition. Moreover our error bounds are finer.

5.9. (a) Let $F : D \subseteq X \to Y$ be differentiable. Assume:

There exist functions $f_1 : [0,1] \times [0,\infty)^2 \to [0,\infty), f_2, f_3 : [0,\infty) \to [0,\infty)$, nondecreasing on $[0,\infty)^2, [0,\infty), [0,\infty)$ such that

$$\begin{aligned}
&\left\| A(x_0)^{-1} \left[F'(x + t(y-x)) - F'(x)\right]\right\| \le \\
&\le f_1(t, \|x-y\|, \|x-x_0\|, \|y-x_0\|), \\
&\left\| A(x_0)^{-1} \left[F'(x) - A(x)\right]\right\| \le f_2(\|x-x_0\|), \\
&\left\| A(x_0)^{-1} \left[A(x) - A(x_0)\right]\right\| \le f_3(\|x-x_0\|),
\end{aligned}$$

hold for all $t \in [0,1]$ and $x, y \in D$;

For $\left\| A(x_0)^{-1} F(x_0)\right\| \le \eta$, equation

$$\eta + b_0\eta + \tfrac{b_0 b_1 \eta}{1-b(r)} = r$$

has non-negative solutions, and denote by r_0 the smallest one. In addition, r_0 satisfies:

$$\bar{U}(x_0, r_0) \subseteq D,$$

and

$$\int_0^1 f_1(t, b_1 b_0 r_0, r_0, r_0)\,dt + f_2(r_0) + f_3(r_0) < 1,$$

where

$$\begin{aligned}
b_0 &= \tfrac{\int_0^1 f_1(t,\eta,0,\eta)dt + f_2(0)}{1-f_3(\eta)}, \\
b_1 &= \tfrac{\int_0^1 f_1(t,b_0\eta,\eta,\eta+b_0\eta)dt + f_2(\eta)}{1-f_3(\eta+b_0\eta)},
\end{aligned}$$

and

$$b = b(r) = \tfrac{\int_0^1 f_1(t,b_1 b_0\eta,r,r)dt + f_2(r)}{1-f_3(r)},$$

Then, show iteration $\{x_n\}$ $(n \ge 0)$ generated by Newton-like method (5.2.1) is well defined, remains in $\bar{U}(x_0, r_0)$ for all $n \ge 0$ and converges to a solution $x^* \in \bar{U}(x_0, r_0)$ of equation. Moreover, the following estimates hold

$$\begin{aligned}
\|x_2 - x_1\| &\le b_0\eta \\
\|x_3 - x_2\| &\le b_1 \|x_2 - x_1\|. \\
\|x_{n+1} - x^*\| &\le b_2 \|x_n - x_{n-1}\|, \ (n \ge 3)
\end{aligned}$$

and

$$\begin{aligned}
&\|x^* - x_n\| \le \tfrac{b_0 b_1 b_2^{n-2}\eta}{1-b_2}, \ (n \ge 3), \\
&\|x^* - x_n\| \le \tfrac{b_2^n}{1-b_2}, \ (n \ge 0)
\end{aligned}$$

where $b_2 = b(r_0)$.

Furthermore if r_0 satisfies

$$\int_0^1 f_1(t, 2r_0, r_0, r_0)\, dt + f_2(r_0) + f_3(r_0) < 1,$$

x^* is the unique solution of equation $F(x) = 0$ in $\overline{U}(x_0, r_0)$.

Finally if there exists a minimum non-negative number R satisfying equation

$$\int_0^1 f_1(t, r + r_0, r_0, r)\, dt + f_2(r_0) + f_3(r_0) = 1,$$

such that $U(x_0, R) \subseteq D$, then the solution x^* is unique in $U(x_0, R)$.

(b) There exist a simple zero x^* of F and continuous functions $f_4 : [0,1] \times [0,\infty) \to [0,\infty)$, $f_5, f_6 : [0,\infty) \to [0,\infty)$, non-decreasing on $[0,\infty)$ such that

$$\left\| A(x^*)^{-1} \left[F'(x + t(x^* - x)) - F'(x) \right] \right\| \le f_4(t, \|x^* - x\|),$$
$$\left\| A(x^*)^{-1} \left[F'(x) - A(x) \right] \right\| \le f_5(\|x^* - x\|),$$

and

$$\left\| A(x^*)^{-1} \left[A(x) - A(x^*) \right] \right\| \le f_6(\|x^* - x\|),$$

hold for all $t \in [0,1]$ and $x \in D$;

Equation

$$\int_0^1 f_4(t, r)\, dt + f_5(r) + f_6(r) = 1$$

has a minimum positive zero r^*.

$$U(x^*, r^*) \subseteq D.$$

Then, show iteration $\{x_n\}$ $(n \ge 0)$ generated by Newton-like sequence (5.2.1) is well defined, remains in $U(x^*, r^*)$ for all $n \ge 0$ and converges to x^* provided that $x_0 \in U(x^*, r^*)$. Moreover, the following estimates hold for all $n \ge 0$

$$\|x_{n+1} - x^*\| \le \frac{\int_0^1 f_4(t, \|x_n - x^*\|)dt + f_5(\|x_n - x^*\|)}{1 - f_6(\|x_n - x^*\|)} \|x_n - x^*\|$$
$$\le \gamma \|x_n - x^*\|,$$

where

$$\gamma = \frac{\int_0^1 f_4(t, \|x_0 - x^*\|)dt + f_5(\|x_0 - x^*\|)}{1 - f_6(\|x_0 - x^*\|)} < 1.$$

(c) Let $X = Y = \mathbb{R}$, $D = (-1,1)$, $x^* = 0$ and define function F on D by

$$F(x) = x + \tfrac{x^{p+1}}{p+1},\ p > 1.$$

For the case $A = F'$ show

$$r_R = \tfrac{2}{3p} < r^* = \tfrac{2}{2+p},$$

where r_R stands for Rheinboldt's radius (see section 4.1) [295] where r_R is the convergence radius given by Rheinboldt's [295].

(d) Let $X = Y = C[0,1]$, the space of continuous functions defined on $[0,1]$ equipped with the max-norm. Let $D = \{\phi \in C[0,1]; \|\phi\| \leq 1\}$ and F defined on D by

$$F(\phi)(x) = \phi(x) - 5\int_0^1 xt\phi(t)^3\,dt$$

with a solution $\phi^*(x) = 0$ for all $x \in [0,1]$.

In this case, for each $\phi \in D, F'(\phi)$ is a linear operator defined on D by the following expression:

$$F'(\phi)[\nu](x) = \nu(x) - 15\int_0^1 xt\phi(t)^2\,\nu(t)\,dt,\ \nu \in D.$$

In this case and by considering again $A = F'$,

$$r_R = \tfrac{2}{45} < r^* = \tfrac{1}{15}.$$

5.10. Consider inexact Newton methods for solving equation $F(x) = 0$, $F : D \subseteq \mathbb{R}^N \to \mathbb{R}^N$ of the general form

$$x_{n+1} = x_n + s_n \quad (n \geq 0),$$

where $s_n \in \mathbb{R}^N$ satisfies the equation

$$F'(x_n)\,s_n = -F(x_n) + r_n \quad (n \geq 0)$$

for some sequence $\{r_n\} \subseteq \mathbb{R}^N$. Let $F \in F_\lambda(\sigma) \equiv \{F : U(x^*,\sigma) \to \mathbb{R}^N$ with $U(x^*,\sigma) \subseteq D$, where $F(x^*) = 0$, F is Fréchet differentiable on $U(x^*,\sigma)$ and $F'(x)^{-1}$ exists for all $x \in U(x^*,\sigma)$, F' is continous on $U(x^*,\sigma)$ and there exists $\mu_\lambda \geq 0$ such that for all $y, z \in U(x^*,\sigma)$

$$\left\|F'(x^*)^{-1}[F'(y) - F'(z)]\right\| \leq \mu_\lambda \|y - z\|^\lambda\}.$$

Suppose the inexact Newton method $\{x_n\}$ $(n \geq 0)$ satisfy:

$$\frac{\left\| s_n + F'(x_n)^{-1} F(x_n) \right\|}{\left\| F'(x_n)^{-1} F(x_n) \right\|} \equiv \frac{\left\| F'(x_n)^{-1} r_n \right\|}{\left\| F'(x_n)^{-1} F(x_n) \right\|} \leq v_n \ (n \geq 0)$$

for some $\{v_n\} \subseteq \mathbb{R}$ $(n \geq 0)$. If $x_i \in U(x^*, \sigma)$, then show:
(a) $\|x_{n+1} - x^*\| \leq w_n \|x_n - x^*\|$,

$$w_n \equiv v_n + \frac{(1+v_n)\mu_\lambda \|x_n - x^*\|^\lambda}{(1-\lambda)\left[1-\mu_\lambda \|x_n - x^*\|^\lambda\right]} \ (n \geq 0);$$

(b) if $v_n \leq v < 1$ $(n \geq 0)$, then there exists $w \in (0,1)$ given by

$$w \equiv v + \frac{(1+v)\mu_\lambda \|x_0 - x^*\|^\lambda}{(1+\lambda)\left[1-\mu_\lambda \|x_0 - x^*\|^\lambda\right]}$$

such that $\|x_{n+1} - x^*\| \leq w \|x_n - x^*\|$ and $\lim_{n\to\infty} x_n = x^*$;
(c) if $\{x_n\}$ $(n \geq 0)$ converges to x^* and $\lim_{n\to\infty} v_n = 0$, then $\{x_n\}$ converges Q-superlinearly;
(d) if $\{x_n\}$ $(n \geq 0)$ converges to x^* and $\lim_{n\to\infty} v_n^{(1+\lambda)^{-1}} < 1$, then $\{x_n\}$ converges with R-order at least $1 + \lambda$.

5.11 Consider the two-point method [144]:

$$\begin{aligned} y_n &= x_n - F'(x_n)^{-1} F(x_n), \\ H_n &= \tfrac{1}{p} F'(x_n)^{-1} \left[F'(x_n + p(y_n - x_n)) - F'(x_n)\right], \qquad p \in (0,1], \\ x_{n+1} &= y_n - \tfrac{1}{2} H_n [I + H_n]^{-1} (y_n - x_n) \qquad (n \geq 0), \end{aligned}$$

for approximating a solution x^* of equation $F(x) = 0$. Let $F : \Omega \subseteq X \to Y$ be a twice-Fréchet-differentiable operator defined on an open convex subset Ω of a Banach space X with values in a Banach space Y. Assume:

(1) $\Gamma_0 = F'(x_0)^{-1} \in L(Y, X)$ exists for some $x_0 \in \Omega$ with $\|\Gamma_0\| \leq \beta$;

(2) $\|\Gamma_0 F(x_0)\| \leq \eta$;

(3) $\|F''(x)\| \leq M$ $(x \in \Omega)$;

(4) $\|F''(x) - F''(y)\| \leq K \|x - y\|, \qquad x, y \in \Omega.$

Denote by $a_0 = M\beta\eta$, $b_0 = K\beta\eta^2$. Define sequences

$$\begin{aligned} a_{n+1} &= a_n f(a_n)^2 g_p(a_n, b_n), \\ b_{n+1} &= b_n f(a_n)^3 g_p(a_n, b_n)^2, \end{aligned}$$

where $f(x) = \frac{2(1-x)}{x^2-4x+2}$ and $g_p(x,y) = \frac{3x^3+2y(1-x)[(1-6p)x+(2+3p)]}{24(1-x)^2}$. If $a_0 \in \left(0, \frac{1}{2}\right)$, $b_0 < h_p(h_0)$, where

$$h_p(x) = \frac{3(2x-1)(x-2)\left(x-3+\sqrt{5}\right)\left(x-3-\sqrt{5}\right)}{2(1-x)[(1-6p)x+2+3p]}, \qquad \bar{U}(x_0, \tfrac{\eta}{a_0}) \subseteq \Omega$$

then show: iteration $\{x_n\}$ $(n \geq 0)$ is well defined, remains in $\bar{U}(x_0, \frac{\eta}{a_0})$ for all $n \geq 0$ and converges to a solution x^* of equation $F(x) = 0$, which is unique in $U(x_0, \frac{\eta}{a_0})$. Moreover, the following estimates hold for all $n \geq 0$:

$$\|x^* - x_n\| \leq \left[1 + \frac{a_0\gamma^{\frac{3^n-1}{2}}}{2(1-a_0)}\right]\gamma^{\frac{3^n-1}{2}} \frac{\Delta^n}{1-\Delta}\eta,$$

where $\gamma = \frac{a_1}{a_0}$ and $\Delta = f(a_0)^{-1}$.

5.12. Consider the two-step method:

$$\begin{aligned} y_n &= x_n - F'(x_n)^{-1} F(x_n) \\ x_{n+1} &= y_n - F'(x_n)^{-1} F(y_n) \quad (n \geq 0); \end{aligned}$$

for approximating a solution x^* of equation $F(x) = 0$. Let $F : \Omega \subseteq X \subseteq Y$ be a Fréchet-differentiable operator defined on an open convex subset Ω of a Banach space X with values in a Banach space Y. Assume:

(1) $\Gamma_0 = f'(X_0)^{-1} \in L(Y, X)$ for some $x_0 \in \Omega$, $\|\Gamma_0\| \leq \beta$;

(2) $\|\Gamma_0 F(x_0)\| \leq \eta$;

(3) $\|F'(x) - F'(y)\| \leq K\|x - y\|$ $(x, y \in \Omega)$.

Denote $a_0 = k\beta\eta$ and define the sequence $a_{n+1} = f(a_n)^2 g(a_n) a_n$ $(n \geq 0)$, where $f(x) = \frac{2}{2-2x-x^2}$ and $g(x) = x^2(x+4)/8$. If $a_0 \in \left(0, \frac{1}{2}\right)$, $\bar{U}(x_0, R\eta) \subseteq \Omega$, $R = \frac{1+\frac{a_0}{2}}{1-\gamma\Delta}$, $\gamma = \frac{a_1}{a_0}$ and $\Delta = f(a_0)^{-1}$, then show: iteration $\{x_n\}$ $(n \geq 0)$ is well defined, remains in $\bar{U}(x_0, R\eta)$ for all $n \geq 0$ and converges to a solution x^* of equation $F(x) = 0$, which is unique in $U(x_0, \frac{2}{K\beta} - R\eta) \cap \Omega$. Moreover, the following estimates hold for all $n \geq 0$

$$\|x_n - x^*\| \leq \left[1 + \frac{a_0}{2}\gamma^{\frac{3^n-1}{2}}\right]\gamma^{\frac{3^n-1}{2}} \frac{\Delta^n}{1-\gamma^{3^n}\Delta}\eta \, (n \geq 0).$$

5.13. Let X be a Banach space, and let Y be a closed subspace. Assume F is a completely continuous operator defined on $\overline{D} \subseteq X$, D an open set, and assume the values $F(x) \in Y$ for all $x \in \overline{D}$. Let X_n *be* a sequence of finite dimensional subspace of X *such* that

$$\inf_{x \in X_n} \|y - x\| \to 0 \text{ as } n \to \infty \text{ for all } y \in Y.$$

Let F_n be a sequence of projections associated with X_n :

$$F_n : X \to X_n \quad (n \geq 1).$$

Assume that when restricted to Y, the projections are uniformly bounded:

$$\sup_n \|F_n \mid Y\| \leq a < \infty.$$

Then projection method for solving

$$x = F(x)$$

becomes

$$x_n = F_n F(x_n)$$

Suppose that $x^* \in D$ if a fixed point of nonzero index for F. Then show for all sufficiently large n equation $F(x) = x$ has at least one solution $x_n \in X_n \cap D$ such that

$$\lim_{n\to\infty} \|x_n - x^*\| = 0.$$

Let $F : U(x_0, r) \subseteq X \to X$ be differentiable and continuous on $\overline{U}(x_0, r)$ such that $I - P'(x)$ is compact. Suppose $L \in L(X)$ is such that

$$\begin{aligned} \|LF(x_0)\| &\leq a \\ \|I - LF'(x_0)\| &\leq b < 1, \\ \|L[F'(x) - F'(y)]\| &\leq c\|x - y\| \text{ for all } x \in U(x_0, r) \\ h &= \tfrac{ac}{(1-b)^2} \leq \tfrac{1}{2} \\ r_0 &= \tfrac{a}{1-b} f(h) \leq r, \text{ where } f(h) = \tfrac{1-\sqrt{1-2h}}{h};\ f(0) = 1. \end{aligned}$$

Then show:
(a) equation $F(x) = x$ has a solution x^* in $U(x_0, r_0)$; x^* is unique in $\overline{U}(x_0, r)$ if, for $r_1 = \frac{a}{1-b} f_1(h)$, where $f_1(h) = \frac{1+\sqrt{1-2h}}{h}$

$$\begin{aligned} r &< r_1 \quad \text{for } h < \tfrac{1}{2}; \\ r &\leq r_1 \quad \text{for } h = \tfrac{1}{2}. \end{aligned}$$

(b) Furthermore Newton's method is well defined, remains in $U(x_0, r)$ for all $n \geq 0$ and converges to x^*.

5.14. Let Fibonacci sequence $\{a_n\}$ be defined by $aa_n + ba_{n-1} + ca_{n-2} = 0$ where $a_0 = 0$ and $a_1 = 1$. If the characteristice polynomial $p(x) = ax^2 + bx + c$ has zeros r_1 and r_2 with $|r_1| > |r_2|$. Then show:

(a) $a_n \neq 0 \quad (n > 0)$;
(b) $\lim\limits_{n\to\infty} \frac{a_{n+1}}{a_n} = r_1$;
(c) Newton $\left(\frac{a_{n+1}}{a_n}\right) = \frac{a_{2n+1}}{a_{2n}}$
and
(d) secant $\left(\frac{a_{m+1}}{a_m}, \frac{a_{n+1}}{a_n}\right) = \frac{a_{m+n+1}}{a_{m+n}}$,
where,

$$x_n = \text{Newton}\,(x_{n-1}) = x_{n-1} - \frac{F(x_{n-1})}{F'(x_{n-1})} \quad (n \geq 1)$$

and

$$x_n = \text{secant}\ \ (x_{n-1}, x_{n-2}) = x_{n-1} - \frac{F(x_{n-1})(x_{n-1}-x_{n-2})}{F(x_{n-1})-F(x_{n-2})}. \quad (n \geq 1)$$

Chapter 6

More Results on Newton's Method

We exploit the weaker conditions introduced in Section 5 to improve on recent results on the local and semilocal convergence analysis of Newton's method.

6.1 Convergence radii for the Newton-Kantorovich method

Motivated by the works in Wang [341], [343] and using the same type of weak Lipschitz conditions we show that a larger convergence radius can be obtained for Newton-Kantorovich method (4.1.3).

Recently in [341], [343] a Lipschitz-type condition

$$\|F'(x^*)^{-1}[F'(x) - F'(x^\theta)]\| \leq \int_{\theta s(x)}^{s(x)} L(u)du \tag{6.1.1}$$

for all $x \in D$, where L is an integrable, nondecreasing, nonnegative function, $s(x) = \|x - x^*\|$ and $x^\theta = x^* + \theta(x - x^*)$ $\theta \in [0, 1]$, was used to obtain larger convergence radii than the ones given by the relevant works in [136], [340], [312].

Condition (6.1.1) is weaker than the famous Lipschitz condition

$$\|F'(x^*)^{-1}[F'(x) - F'(y)]\| \leq w(\|x - y\|), \quad \text{for all } x, y \in D, \tag{6.1.2}$$

where w is also a nondecreasing, nonnegative function. Condition (6.1.2) has been used e.g. in [295] (when w is a constant function).

It turns out that (6.1.1) instead of the stronger (6.1.2) is really needed to show the local convergence of method (6.1.1) to x^* [341], [343]. Based on this observation it became possible to obtain convergence radii larger than the ones obtained in [136], [295], [340], [312].

Here we show that all the above mentioned results can be further improved under weaker or the same hypotheses and under less or the same computational cost.

A numerical example is also provided where our results compare favorably with all earlier ones. Several areas of application are also suggested.

We provide the following local convergence theorems:

Theorem 6.1.1 *Let $F\colon D \subseteq X \to Y$ be a Fréchet-differentiable operator. Assume: (a) there exists a simple zero $x^* \in D$ of equation $F(x) = 0$ so that $F'(x^*)^{-1} \in L(Y, X)$, positive integrable nondecreasing functions L, L_0 such that the radius Lipschitz condition with L average and the center radius Lipschitz condition with L_0 average*

$$\|F'(x^*)^{-1}[F'(x) - F'(x^\theta)]\| \leq \int_{\theta s(x)}^{s(x)} L(u)du \tag{6.1.3}$$

and

$$\|F'(x^*)^{-1}[F'(x) - F'(x^*)]\| \leq \int_0^{s(x)} L_0(u)du \tag{6.1.4}$$

are satisfied respectively for all $x \in D$, $\theta \in [0, 1]$; (b) equation

$$\int_0^r L(u)u\,du + r \int_0^r L_0(u)du - r = 0 \tag{6.1.5}$$

has positive zeros. Denote by r^* the smallest such zero; and

$$U(x^*, r^*) \subseteq D. \tag{6.1.6}$$

Then, sequence $\{x_n\}$ $(n \geq 0)$ generated by method (4.1.3) is well defined, remains in $U(x^*, r^*)$ for all $n \geq 0$, and converges to x^* provided that $x_0 \in U(x^*, r^*)$. Moreover the following estimates hold for all $n \geq 1$:

$$\|x_n - x^*\| \leq a^{2^n - 1}\|x_0 - x^*\| \tag{6.1.7}$$

where,

$$a = \frac{\int_0^{s(x_0)} L(u)u\,du}{s(x_0)\left[1 - \int_0^{s(x_0)} L_0(u)du\right]}, \tag{6.1.8}$$

and

$$a \in [0, 1). \tag{6.1.9}$$

Proof. We first show $a \in [0, 1)$. Indeed we can have in turn for $0 < u_1 < u_2$ and the monotonicity of L:

$$\begin{aligned}
&\left(\frac{1}{u_2^2}\int_0^{u_2} - \frac{1}{u_1^2}\int_0^{u_1}\right) uL(u)du \\
&\quad = \left[\frac{1}{u_2^2}\int_{u_1}^{u_2} + \left(\frac{1}{u_2^2} - \frac{1}{u_1^2}\right)\int_0^{u_1}\right] L(u)u\,du \\
&\quad \geq L(u_1)\left[\frac{1}{u_2^2}\int_{u_1}^{u_2} + \left(\frac{1}{u_2^2} - \frac{1}{u_1^2}\right)\int_0^{u_1}\right] u\,du \\
&\quad = L(u_1)\left[\frac{1}{u_2^2}\int_0^{u_2} - \frac{1}{u_2^2}\int_0^{u_1}\right] u\,du = 0.
\end{aligned}$$

Therefore $\frac{1}{u^2}\int_0^u L(v)dv$ is a nondecreasing function with respect to u. Then, we can have

$$\begin{aligned}
0 &= \frac{\int_0^{s(x_0)} L(u)u\,du}{s(x_0)^2\left[1-\int_0^{s(x_0)} L_0(u)du\right]}s(x_0)\\
&\le \frac{\int_0^{r^*} L(u)u\,du}{(r^*)^2\left[1-\int_0^{r^*} L_0(u)du\right]}s(x_0)\\
&\le \frac{\|x_0-x^*\|}{r^*} < 1.
\end{aligned}$$

Let $x \in U(x^*, r^*)$. Using (6.1.4) and (6.1.6) we get

$$\|F'(x^*)^{-1}[F'(x)-F'(x^*)]\| \le \int_0^{s(x)} L_0(u)du \le \int_0^{r^*} L_0(u)du < 1. \tag{6.1.10}$$

It follows from (6.1.10) and the Banach Lemma on invertible operators 1.3.2 that $F'(x)^{-1} \in L(Y, X)$ and

$$\|F'(x)^{-1}F'(x^*)\| \le \frac{1}{1-\int_0^{s(x)} L_0(u)du} \le \frac{1}{1-\int_0^{r^*} L_0(u)du}. \tag{6.1.11}$$

In particular $x_0 \in U(x^*, r^*)$ by hypothesis. Let us assume $x_n \in U(x^*, r^*)$. We obtain the identity

$$\begin{aligned}
x_{n+1}-x^* &= x_n - F'(x_n)^{-1}F(x_n) - x^*\\
&= F'(x_n)^{-1}[F(x^*)-F(x_n)-F'(x_n)(x^*-x_n)] \qquad (6.1.12)\\
&= [F'(x_n)^{-1}F'(x^*)]\left[\int_0^1 F'(x^*)^{-1}(F'(x_n)-F'(x^\theta))(x_n-x^*)d\theta\right].
\end{aligned}$$

By (6.1.3), (6.1.4) and (6.1.11) we get

$$\begin{aligned}
&\|x_{n+1}-x^*\| \le\\
&\le \|F'(x_n)^{-1}F'(x^*)\| \int_0^1 \|F'(x^*)^{-1}(F'(x_n)-F'(x^\theta))\|\,\|x_n-x^*\|d\theta\\
&\le \frac{\int_0^1\int_{\theta s(x_n)}^{s(x_n)} L(u)du s(x_n)d\theta}{1-\int_0^{s(x_n)} L_0(u)du}\\
&= \frac{\int_0^{s(x_n)} L(u)u\,du}{1-\int_0^{s(x_n)} L_0(u)du} \le a\|x_n-x^*\| < \|x_n-x^*\|, \qquad (6.1.13)
\end{aligned}$$

which shows the convergence of x_n to x^* and $x_{n+1} \in U(x^*, r^*)$. Hence sequence $\{x_n\}$ $(n \ge 0)$ remains in $U(x^*, r^*)$ by the principle of mathematical induction.

Moreover $s(x_n)$ decreases monotonically. Furthermore by (6.1.13) we obtain more precisely

$$\begin{aligned} \|x_{n+1} - x^*\| &\leq \frac{\int_0^{s(x_n)} L(u)u\,du}{s(x_n)^2\left[1 - \int_0^{s(x_n)} L_0(u)du\right]} s(x_n)^2 \\ &\leq \frac{\int_0^{s(x_0)} L(u)u\,du}{s(x_0)^2\left[1 - \int_0^{s(x_0)} L_0(u)du\right]} s(x_n)^2 \\ &= \frac{a}{s(x_0)} \|x_n - x^*\|^2, \end{aligned} \tag{6.1.14}$$

from which estimate (6.1.7) follows. ■

Remark 6.1.1 *In general*

$$L_0(u) \leq L(u) \tag{6.1.15}$$

holds and $\frac{L(u)}{L_0(u)}$ *can be arbitrarily large (see Section 5.1). If equality holds in (6.1.1518) then our theorem coincides with the corresponding Theorem 3.1 in [341]. However if strict inequality holds in (6.1.15) (see Example 5.1.5), then our convergence radius* r^* *is greater than the corresponding radius* r *in Theorem 3.1. Moreover our convergence rate* a *is smaller than the corresponding* q *in Theorem 3.1. Summarizing, if*

$$L_0(u) < L(u) \tag{6.1.16}$$

then

$$r < r^* \tag{6.1.17}$$

$$a < q. \tag{6.1.18}$$

That is our theorem provides a wider range of initial guesses x_0 *and more precise error estimates on the distances* $\|x_n - x^*\|$ $(n \geq 0)$ *than Theorem 3.1 and under the same computational cost since in practice the evaluation of function* L *requires that of* L_0.

In case L_0, L are constants see section 5.1.

Theorem 6.1.2 *Let* $F\colon D \subseteq X \to Y$ *be a Fréchet-differentiable operator. Assume:*

(a) there exists a simple zero $x^* \in D$ *of equation* $F(x) = 0$*, so that* $F'(x^*)^{-1} \in L(Y, X)$*, and a positive integrable function* L_0 *such that condition (6.1.4) holds;*

(b) equation

$$\frac{1}{r}\int_0^r L_0(u)(r-u)du - 1 = 0 \tag{6.1.19}$$

has a minimum positive zero denoted by r_0^**; and*

(c)

$$U(x^*, r_0^*) \subseteq D. \tag{6.1.20}$$

Then sequence $\{y_n\}$ $(n \geq 0)$ *generated by*

$$y_{n+1} = y_n - F'(x^*)^{-1}F(y_n) \quad (n \geq 0) \tag{6.1.21}$$

remains in $U(x^*, r_0^*)$ *for all* $n \geq 0$ *and converges to a unique solution* x^* *of equation (4.1.1) so that*

$$\|x_{n+1} - x^*\| \leq a_0\|x_n - x^*\| \tag{6.1.22}$$

where $a_0 \in [0, 1)$ *and*

$$a_0 = \frac{1}{s(x_0)} \int_0^{s(x_0)} L_0(u)(s(x_0) - u)du. \tag{6.1.23}$$

Proof. Simply use function L_0 instead of L in the proof of Theorem 4.1 in [341]. ■

Remark 6.1.2 *If* $L_0 = L$ *then Theorem 6.1.2 reduces to Theorem 4.1 in [341]. Moreover if (19) holds then*

$$r_0 < r_0^* \tag{6.1.24}$$

and

$$q_0 < a_0 \tag{6.1.25}$$

where by r_0, q_0 *we denote the radius and ratio found in Theorem 4.1.*

In the case when L, L_0 are constants as expected our results improve Corollaries 6.3 and 6.4 in [342] which in turn compare favorably with the results obtained by Smale [312] and Debieu [139] respectively when F is analytic and satisfies

$$\|F'(x^*)^{-1}F^{(k)}(x^*)\| \leq k!\gamma^{k-1} \quad (k \geq 2) \tag{6.1.26}$$

provided that

$$L(u) = \frac{2\gamma}{(1 - \gamma u)^3} \tag{6.1.27}$$

and

$$L_0(u) = \frac{2\gamma_0}{(1 - \gamma_0 u)^3} \tag{6.1.28}$$

for some $\gamma_0 \geq 0$, $\gamma \geq 0$ with

$$\gamma_0 \leq \gamma. \tag{6.1.29}$$

For example our result corresponding to Corollary 6.4 gives

$$r_0^* \in \left[0, \frac{1}{\gamma_0}\right) \tag{6.1.30}$$

whereas the corresponding radius in [342] is given by

$$r_0 \in \left[0, \frac{1}{\gamma}\right). \tag{6.1.31}$$

Therefore (6.1.24) and (6.1.25) hold (if strict inequality holds in (6.1.29)).

We can provide the local convergence result for twice Fréchet-differentiable operators as an alternative to Theorem 6.1.1.

Theorem 6.1.3 *Let $F\colon D \subseteq X \to Y$ be a twice Fréchet-differentiable operator. Assume: (a) there exists a simple zero $x^* \in D$ of equation $F(x) = 0$, so that $F'(x^*)^{-1} \in L(Y, X)$ and nonnegative constants ℓ, b, ℓ_0 such that*

$$\|F'(x^*)^{-1}[F''(x + \theta(x^* - x)) - F''(x^*)]\| \leq \ell(1 - \theta)\|x - x^*\|, \tag{6.1.32}$$

$$\|F'(x^*)^{-1}F''(x^*)\| \leq b, \tag{6.1.33}$$

and

$$\|F'(x^*)^{-1}[F''(x^\theta) - F''(x^*)]\| \leq \ell_0\theta\|x - x^*\| \tag{6.1.34}$$

for all $x \in D$, $\theta \in [0, 1]$; (b) $U(x^*, R^*) \subseteq D$, where,

$$R^* = \frac{2}{d + \sqrt{d^2 + 4c}}, \tag{6.1.35}$$

$$d = \frac{3b}{2} \quad \text{and} \quad c = \frac{1}{3}\ell + \frac{1}{2}\ell_0. \tag{6.1.36}$$

Then, sequence $\{x_n\}$ generated by method (4.1.3) is well defined, remains in $U(x^*, R^*)$ for all $n \geq 0$ and converges to x^* provided that $x_0 \in U(x^*, R^*)$. Moreover the following error bounds hold for all $n \geq 0$

$$\|x_{n+1} - x^*\| \leq h\|x_n - x^*\|^2 \tag{6.1.37}$$

and

$$\|x_n - x^*\| \leq h^{2^n - 1}\|x_0 - x^*\|. \tag{6.1.38}$$

Proof. Let $x \in U(x^*, R^*)$. Using (6.1.33), (6.1.34) and the definition of R^* we obtain in turn

$$\begin{aligned}F'(x^*)^{-1}[F'(x^*) - F'(x)] &= -F'(x^*)^{-1}[F'(x) - F'(x^*)\\ &- F''(x^*)(x - x^*) + F''(x^*)(x - x^*)]\\ &= -\int_0^1 F'(x^*)^{-1}\{F''[x^* + \theta(x - x^*)] \qquad (6.1.39)\\ &- F''(x^*)\}(x - x^*) - F'(x^*)F''(x^*)(x - x^*)\end{aligned}$$

and

$$\begin{aligned}\|F'(x^*)^{-1}[F'(x^*) - F'(x)]\| &\leq \int_0^1 \ell_0\theta\|x - x^*\|^2 d\theta + b\|x - x^*\|\\ &\leq \frac{1}{2}\ell_0(R^*)^2 + bR^* < 1. \qquad (6.1.40)\end{aligned}$$

It follows from (6.1.40) and the Banach Lemma on invertible operators that $F'(x)^{-1}F'(x^*)$ exists and

$$\begin{aligned}\|F'(x)^{-1}F'(x^*)\| &\leq \frac{1}{1 - \frac{\ell_0}{2}\|x - x^*\|^2 - b\|x - x^*\|}\\ &\leq \frac{1}{1 - \frac{\ell_0}{2}(R^*)^2 - bR^*}. \qquad (6.1.41)\end{aligned}$$

In particular by hypothesis $x_0 \in U(x^*, R^*)$. Assume $x_n \in U(x^*, R^*)$. Using $F(x^*) = 0$ we obtain the identity

$$\begin{aligned}x_{n+1} - x^* &= x_n - F'(x_n)^{-1}F(x_n) - x^*\\ &= F'(x_n)^{-1}[F(x^*) - F(x_n) - F'(x_n)(x^* - x_n)]\\ &= [F'(x_n)^{-1}F'(x^*)]F'(x^*)^{-1}\int_0^1 F''[x_n\\ &+ \theta(x^* - x_n)](1 - \theta)d\theta(x_n - x^*)^2\\ &= [F'(x_n)^{-1}F'(x^*)]F'(x^*)^{-1}\left\{\int_0^1 [F''(x_n + \theta(x^* - x_n))\right.\\ &- F''(x^*)](1 - \theta)d\theta(x_n - x^*)^2\\ &\left.+ \int_0^1 F''(x^*)(1 - \theta)d\theta(x_n - x^*)^2\right\}. \qquad (6.1.42)\end{aligned}$$

By (6.1.32), (6.1.33), (6.1.41), (6.1.42) and the triangle inequality we get

$$\|x_{n+1} - x^*\| \leq \frac{\frac{1}{3}\ell\|x_n - x^*\| + \frac{b}{2}}{1 - \frac{\ell_0}{2}\|x_n - x^*\|^2 - b\|x_n - x^*\|}\|x_n - x^*\|^2 \qquad (6.1.43)$$

$$\leq h\|x_n - x^*\|^2 < \|x_n - x^*\|. \qquad (6.1.44)$$

Estimate (6.1.44) implies $x_{n+1} \in U(x^*, R^*)$ and $\lim_{n\to\infty} x_n = x^*$. ■

The analog of Theorem 6.1.2 is given by

Theorem 6.1.4 *Under the hypotheses of Theorem 6.1.3 (excluding (6.1.46)) and replacing R^*, h by R_0^*, h_0 respectively given by*

$$R_0^* = \frac{2}{d_0 + \sqrt{d_0^2 + 4c_0}} \tag{6.1.45}$$

$$h_0 = c_0 R_0^* + d_0, \tag{6.1.46}$$

where,

$$c_0 = \frac{\ell_0}{6} \quad \text{and} \quad d_0 = \frac{b}{2} \tag{6.1.47}$$

the conclusions of Theorem 6.1.3 hold, for iteration (6.1.21).

Proof. Using iteration (6.1.21) we obtain the identity

$$\begin{aligned}
& y_{n+1} - x^* = \\
& = -F'(x^*)^{-1}[F(y_n) - F(x^*) - F'(x^*)(y_n - x^*)] \\
& = -F'(x^*)^{-1}\Bigg[\int_0^1 F''[x^* + \theta(y_n - x^*)](1-\theta)d\theta(y_n - x^*)^2 \\
& - \int_0^1 F''(x^*)(1-\theta)d\theta(y_n - x^*)^2 + \int_0^1 F''(x^*)(1-\theta)d\theta(y_n - x^*)^2 \\
& = -F'(x^*)^{-1}\Bigg\{\int_0^1 [F''(x^* + \theta(y_n - x^*) - F''(x^*)](1-\theta)d\theta(y_n - x^*)^2 \\
& + \frac{1}{2}F''(x^*)(y_n - x^*)^2\Bigg\}.
\end{aligned} \tag{6.1.48}$$

By (6.1.33), (6.1.34), (6.1.48) and the definition of R_0^*, h_0 we obtain (as in (6.1.43))

$$\begin{aligned}
\|y_{n+1} - x^*\| & \le \left[\frac{1}{6}\ell_0\|y_n - x^*\| + \frac{b_0}{2}\right]\|y_n - x^*\|^2 \\
& \le h_0\|y_n - x^*\|^2 < \|y_n - x^*\|,
\end{aligned} \tag{6.1.49}$$

which completes the proof of Theorem 6.1.4. ■

Remark 6.1.3 *(a) Returning back to Example 5.1.5 and using (6.1.32)–(6.1.36), (6.1.45)–(6.1.47) we obtain* $\ell = \ell_0 = e - 1$, $b = 1$, $d = 1.5$, $c = \frac{5}{6}(e-1)$, $c_0 = \frac{e-1}{6}$, $d_0 = \frac{1}{2}$,

$$R^* = .462484887$$

and we can set

$$R_0^* = 1.$$

Comparing with the earlier results mentioned above (see Example 5.1.5) we see that the convergence radius is significantly enlarged under our new approach. Note also that the order of convergence of iteration (6.1.21) is quadratic whereas it is only linear in the corresponding Theorem 4.1 in [341]. (b) The results can be extended easily along the same lines as above to the case when ℓ, ℓ_0 are not constants but functions. (c) The results obtained here can also be extended to hold for m-Fréchet-differentiable operators ($m \geq 2$) along the lines of our works in [79], [68]. However we leave the details of the works for (b) and (c) to the motivated reader.

We present a new convergence theorem where the hypotheses (see Theorem 6.1.1) that functions L, L_0 are nondecreasing are not assumed:

Theorem 6.1.5 *Let F, x^*, L, L_0 be as in Theorem 6.1.1 but the nondecreasiness of functions L, L_0 is not assumed. Moreover assume: (a) equation*

$$\int_0^r (L(u) + L_0(u))du - 1 = 0 \tag{6.1.50}$$

has positive solutions. Denote by R^* the smallest one; (b)

$$U(x^*, R^*) \subseteq D. \tag{6.1.51}$$

Then sequence $\{x_n\}$ $(n \geq 0)$ generated by method (4.1.3) is well defined, remains in $U(x^*, R^*)$ for all $n \geq 0$ and converges to x^* provided that $x_0 \in U(x^*, R^*)$, so that:

$$\|x_n - x^*\| \leq b^n \|x_0 - x^*\| \quad (n \geq 0) \tag{6.1.52}$$

where,

$$b = \frac{\int_0^{s(x_0)} L(u)du}{1 - \int_0^{s(x_0)} L_0(u)du} \tag{6.1.53}$$

and

$$b \in [0, 1).$$

Furthermore, if function L_α given by

$$L_\alpha(t) = t^{1-\alpha} L(t) \tag{6.1.54}$$

is nondecreasing for some $\alpha \in [0, 1]$, and R^* satisfies

$$\frac{1}{R^*} \int_0^{R^*} L(u)u\, du + \int_0^{R^*} L_0(u)du \leq 1. \tag{6.1.55}$$

Then method (4.1.3) converges to x^* for all $x_0 \in U(x^*, R^*)$ and

$$\|x_n - x^*\| \leq b^{h_{n-1}} \|x_0 - x^*\| \quad (n \geq 0) \tag{6.1.56}$$

where,

$$h_k = \sum_{i=0}^{k} (1+\alpha)^i \quad (k \geq 1). \tag{6.1.57}$$

Proof. Exactly as in the proof of Theorem 6.1.1 we arrive at

$$\|x_{n+1} - x^*\| \leq \frac{\int_0^{s(x_n)} L(u)u\,du}{1 - \int_0^{s(x_n)} L_0(u)du} \tag{6.1.58}$$

$$\leq \frac{\int_0^{s(x_n)} L(u)du\, s(x_n)}{1 - \int_0^{s(x_n)} L_0(u)du} \leq b\|x_n - x^*\| < \|x_n - x^*\|, \tag{6.1.59}$$

which show that if $x_n \in U(x^*, R^*)$ then $x_{n+1} \in U(x^*, R^*)$ and $\lim_{n\to\infty} x_n = x^*$. Estimate (6.1.13) follows from (6.1.20). Furthermore, if function L_α is given by (6.1.15) and R^* satisfies (6.1.16) then by Lemma 2.2 in [343, p. 408] for nondecreasing function

$$\varphi_{\beta,\alpha}(t) = \frac{1}{t^{\alpha+\beta}} \int_0^t u^\beta L(u)du, \tag{6.1.60}$$

we get

$$\|x_{n+1} - x^*\| \leq \frac{\varphi_{1,\alpha}(s(x_n))}{1 - \int_0^{s(x_n)} L_0(u)du} s(x_n)^{1+\alpha}, \tag{6.1.61}$$

$$\leq \frac{\varphi_{1,\alpha}(s(x_0))}{1 - \int_0^{s(x_n)} L_0(u)du} s(x_n)^{1+\alpha}$$

$$= \frac{b}{s(x_0)^\alpha} \|x_n - x^*\|^{1+\alpha} < \|x_n - x^*\|, \tag{6.1.62}$$

which completes the proof of Theorem 6.1.5. ∎

Theorem 6.1.6 *Let F, x^*, L_0 be as in Theorem 6.1.2 but the nondecreasness of function L_0 is not assumed. Moreover assume: (a) equation*

$$3\int_0^r L_0(u)du - 1 = 0, \tag{6.1.63}$$

has positive solutions. Denote by R_0^* the smallest one; (b) $U(x^*, R_0^*) \subseteq D$. Then, sequence $\{x_n\}$ $(n \geq 0)$ generated by method (4.1.3) is well defined, remains in

$U(x^*, R_0^*)$ for all $n \geq 0$ and converges to x^* provided that $x_0 \in U(x^*, R_0^*)$. Moreover error bound (6.1.13) hold for all $n \geq 0$; where

$$b = \frac{2\int_0^{s(x_0)} L_0(u)du}{1 - \int_0^{s(x_0)} L_0(u)du}. \tag{6.1.64}$$

Proof. Simply use L_0 instead of L in the corresponding proof of Theorem 1.2 in [343, p. 409]. ■

Remark 6.1.4 *(a) In general (6.1.16) holds. If equality holds in (6.1.16) then Theorems 6.1.4 and 6.1.5 reduce to Theorems 1.1 and 1.2 in [343] respectively. However if strict inequality holds in (6.1.16) then clearly: our convergence conditions (6.1.11), (6.1.24) are weaker than the corresponding ones (9) and (15) in [343]; convergence radii R^* and R_0^* are also larger than the corresponding ones in [343] denoted here by p and p_0; and finally our convergence rates are smaller.*

For example in case functions L_0, L are constants we obtain

$$p = \frac{1}{2L}, \quad R^* = \frac{1}{L_0 + L}, \quad q = \frac{Lp}{1 - Lp},$$
$$b = \frac{LR^*}{1 - L_0 R^*}, \quad p_0 = \frac{1}{3L}, \quad \text{and} \quad R_0^* = \frac{1}{3L_0}. \tag{6.1.65}$$

Therefore we have:

$$p \leq R^* \leq 2p, \tag{6.1.66}$$
$$p_0 \leq R_0^*, \tag{6.1.67}$$

and

$$b \leq q, \tag{6.1.68}$$

where q is the corresponding ratio in Theorem 1.1 in [343]. Note that if strict inequality holds in (6.1.26) so does in (6.1.28)–(6.1.30). Moreover the computational cost for finding L_0, L in practice is the same as of that of finding just L. As an example, as in Section 5.1 consider real function F given by (5.1.40). Then using (6.1.3) and (6.1.4) we obtain $L_0 = e - 1 < e = L$,

$$R^* = .225399674, \quad p = .183939721, \quad R_0^* = .193992236,$$

and

$$p_0 = .12262648.$$

Hence, we have:

$$p < R^*$$

and

$$p_0 < R_0^*.$$

(c) As already noted in [343, p. 411] and for the same reasons Theorem 6.1.5 is an essential improvement over Theorem 6.1.4 if the radius of convergence is ignored (see Example 3.1 in [343]).

6.2 Convergence under weak continuous derivative

We provide a semilocal convergence analysis of Newton's method (4.1.3) under weaker Lipschitz conditions for approximating a locally unique solution of equation (4.1.1) in a Banach space.

For $p \in [0, 1]$ consider the following Hölder condition:

$$\|F'(x_0)^{-1}[F'(x) - F'(y)]\| \leq \ell\|x - y\|^p. \tag{6.2.1}$$

In Rokne [303], Argyros [11] the Hölder continuity condition replaced (6.2.1) or in Argyros [93] the more general

$$\|F'(x_0)^{-1}[F'(x) - F'(y)]\| \leq v(\|x - y\|) \tag{6.2.2}$$

was used, where v is a nondecreasing function on $[0, R]$ for some given R. In order to combine the Kantorovich condition with Smale α-criterion [296], [312] for the convergence of method (4.1.3) (when F is an analytic operator), Wang in [341], [343] used the weak Lipschitz condition

$$\|F'(x_0)^{-1}[F'(x) - F'(y)]\| \leq \int_{\|x-x_0\|}^{\|x-y\|+\|x-x_0\|} L(u)du \tag{6.2.3}$$

instead of (6.2.1). Huang in [195] noticed that (6.2.3) is a special case of the Generalized Kantorovich condition

$$\|F'(x_0)^{-1}[F'(x) - F'(y)]\| \leq \int_{s(\|x-y\|)}^{s(\|x-y\|+\|x-z\|)} L(u)du \tag{6.2.4}$$

for all $x \in U_1$, $y \in U_2$ where $U_1 = \overline{U}(x_0, t^*)$, $U_2 = \overline{U}(x_0, t^* - \|x - x_0\|)$, and $t^* \in (0, R)$ is the minimum positive simple zero of

$$f(t) = \int_0^{s(t)} L(u)(t - s^{-1}(u))du - t + \eta, \tag{6.2.5}$$

with functions L, s' continuously increasing satisfying $L(u) \geq 0$, on $[0, R]$, $s(0) = 0$ and $s'(t) \geq 0$ on $[0, R)$, respectively. Under condition (6.2.4) a semilocal convergence analysis was provided and examples were given where the Generalized

Kantorovich condition (6.2.4) can be used to show convergence of method (4.3.1) but (6.2.1) cannot.

Here, we provide a semilocal convergence analysis based on a combination of condition (6.2.6) (which is weaker than (6.2.4)) and on the center generalized Kantorovich condition (6.2.7). The advantages of such an approach have already been explained in Section 5.1.

We need the definitions:

Definition 6.2.1 *(Weak Generalized Kantorovich Condition). We say x_0 satisfies the weak generalized Kantorovich condition if for all $\theta \in [0,1]$, $x \in U_1$, $y \in U_2$*

$$\|F'(x_0)^{-1}[F'(x) - F'(x + \theta(y-x))]\| \leq \int_{s_1(\|x-x_0\|)}^{s_1(\|x-x_0\|+\theta\|x-y\|)} L_1(u)du \quad (6.2.6)$$

holds, where functions s_1, L_1 are as s, L respectively.

Definition 6.2.2 *(Center Generalized Kantorovich Condition). We say x_0 satisfies the center generalized Kantorovich condition if for all $x \in U_1$*

$$\|F'(x_0)^{-1}[F'(x) - F'(x_0)]\| \leq \int_0^{s_0(\|x-x_0\|)} L_0(u)du \quad (6.2.7)$$

holds, where functions s_0, L_0 are as s and L respectively.

Note that in general

$$s_0(u) \leq s_1(u) \leq s(u) \quad (6.2.8)$$

and

$$L_0(u) \leq L_1(u) \leq L(u) \quad (6.2.9)$$

hold.

In case strict inequality holds in (6.2.8) or (6.2.9) then the results obtained in [195] can be improved under (6.2.6) or even more under the weaker condition (6.2.7).

Indeed we can show the following semilocal convergence theorem for method (4.1.3). We first define functions f_0, f_1, by

$$f_0(t) = \int_0^{s_0(t)} L_0(u)(t - s_0^{-1}(u))du - t + \eta, \quad (6.2.10)$$

$$f_1(t) = \int_0^{s_1(t)} L_1(u)(t - s_1^{-1}(u))du - t + \eta \quad (6.2.11)$$

and sequences a_n, t_n by

$$a_{n+1} = a_n - \frac{f_1(t_n)}{f_0'(t_n)} \quad (n \geq 0), \quad (t_0 = 0), \quad (6.2.12)$$

$$t_{n+1} = t_n - \frac{f(t_n)}{f'(t_n)} \quad (n \geq 0), \quad (t_0 = 0). \quad (6.2.13)$$

Theorem 6.2.3 *Let $F\colon D \subseteq X \to Y$ be a Fréchet-differentiable operator. Assume: there exists $x_0 \in D$ such that $F'(x_0)^{-1} \in L(Y,X)$; and condition (6.2.5) holds. Then sequence $\{x_n\}$ generated by method (4.1.3) is well defined, remains in $\overline{U}(x_0, a^*)$ for all $n \geq 0$ and converges to a unique solution x^* of equation $F(x) = 0$ in $\overline{U}(x_0, a^*)$, where a^* is the limit of the monotonically increasing sequence $\{a_n\}$ given by (6.2.12). Moreover the following estimates hold:*

$$\|x_{n+1} - x_n\| \leq a_{n+1} - a_n \leq t_{n+1} - t_n, \tag{6.2.14}$$

$$\|x_n - x^*\| \leq a^* - a_n \leq t^* - t_n, \tag{6.2.15}$$

$$a_n \leq t_n, \tag{6.2.16}$$

$$a^* \leq t^*, \tag{6.2.17}$$

$$\|x_n - x^*\| \leq \frac{(\tau - a^*)q_0(n)}{1 - q_0(n)\delta_0}\delta_0 \leq \frac{1 - \delta_0^{2^n - 1}}{1 - \delta_0^{2^n}}a^*, \tag{6.2.18}$$

and

$$\frac{\|F'(x_0)^{-1}F(x_{n+1})\|}{\|F'(x_0)^{-1}F(x_n)\|} \leq \delta_0^{2^n}, \tag{6.2.19}$$

where $\tau \in [a^*, t^{**}]$ satisfies $f(\tau) \leq 0$,

$$t^{**} = \max_{a^* \leq t < R}\{t, f(t) \leq 0\}, \quad \delta_0 = \frac{a^*}{\tau}, \tag{6.2.20}$$

$$q_0(n) = \prod_{i=0}^{n-1}(\lambda_i^0\delta)^{2^{n-i-1}}, \tag{6.2.21}$$

and

$$\lambda_i^0 = \frac{(\tau - a_i)\int_0^1 \int_{s(a_i)}^{s(a_i + \theta(a^* - a_i))} L(u)du\,d\theta}{(a^* - a_i)\left[\int_0^1 \int_{s(a_i)}^{s(a_i + \theta(a^* - a_i))} L_0(u)du\,du - f_0(\tau)\right]} < 1. \tag{6.2.22}$$

Furthermore if U_2 in weak generalized Kantorovich condition (6.2.6) is $U(x, t^{**} - \|x - x_0\|) \cap D$, then equation (4.1.1) has a unique solution in the larger ball $U(x_0, t^{**}) \cap D$.

Proof. We have noticed that in the proof of Theorem 1 in [195, p. 117] weaker condition (6.2.6) can be used instead of (6.2.4) to obtain estimates on $\|F'(x_0)^{-1}F(x_n)\|$. Moreover center generalized Kantorovich condition can be used instead of (6.2.7) to obtain the more precise error bounds on $\|F'(x_n)^{-1}F'(x_0)\|$. The rest of the proof goes along the lines of Theorem 1 in [195, p. 117] by simply replacing L, s, t_n, t^*, $q(n)$, δ, λ_i, f' by L_1, s_1, a_n, a^*, $q_0(n)$, δ_0, λ_i^0, f_0' respectively and using (6.2.6), (6.2.7). ■

Remark 6.2.1 *If equality holds in (6.2.8), (6.2.9) and (6.2.6) is replaced by (6.2.4) our theorem reduces to Theorem 1 in [195, p. 117]. Otherwise our results are more precise under the same hypotheses/computational cost since in practice computing s, L requires the computation of s_1, L_1, s_0, L_0.*

We need the following result on majorizing sequences.
For simplicity we set

$$w_0 = f_0' \quad \text{and} \quad w = f'. \tag{6.2.23}$$

Lemma 6.2.4 *Assume there exist parameters $\eta \geq 0$, $\delta \in [0, 2)$ and functions w_0, w such that:*

$$h_\delta = 2\left[\int_0^1 w(\theta\eta)d\theta - w(0)\right] + \delta w_0(\eta) \leq \delta, \tag{6.2.24}$$

$$w_0\left[\frac{2\eta}{2-\delta}\left(1 - \left(\frac{\delta}{2}\right)^{n+1}\right)\right] < 1, \tag{6.2.25}$$

and

$$\begin{aligned} 2\left\{\int_0^1 w\left[\frac{2\eta}{2-\delta}\left(1 - \left(\frac{\delta}{2}\right)^{n+1}\right) + \theta\left(\frac{\delta}{2}\right)^{n+1}\eta\right] d\theta \right. \\ \left. - w\left[\frac{2\eta}{2-\delta}\left(1 - \left(\frac{\delta}{2}\right)^{n+1}\right)\right]\right\} \\ + \delta w_0\left[\frac{2\eta}{2-\delta}\left(1 - \left(\frac{\delta}{2}\right)^{n+1}\right)\right] \leq \delta, \end{aligned} \tag{6.2.26}$$

hold for all $\eta \geq 0$. Then, iteration $\{b_n\}$ $(n \geq 0)$ given by

$$\begin{aligned} b_0 = 0, \quad b_1 = \eta, \quad b_{n+2} = b_{n+1} \\ + \frac{\int_0^1 \{[w(b_n + \theta(b_{n+1} - b_n)]d\theta - w(b_n)]\}(b_{n+1} - b_n)}{1 - w_0(b_{n+1})} \end{aligned} \tag{6.2.27}$$

is monotonically increasing, bounded above by

$$t^{**} = \frac{2\eta}{2-\delta}, \tag{6.2.28}$$

and converges to some b^* such that

$$0 \leq b^* \leq b^{**}. \tag{6.2.29}$$

Moreover, the following error bounds hold for all $n \geq 0$:

$$0 \leq b_{n+2} - b_{n+1} \leq \frac{\delta}{2}(b_{n+1} - b_n) \leq \left(\frac{\delta}{2}\right)^{n+1} \eta. \tag{6.2.30}$$

The proof is omitted as it follows along the lines of Lemma 5.1.1

We can show the main semilocal convergence result for method (4.1.3):

Theorem 6.2.5 *Assume: hypotheses of Lemma 6.2.4; conditions (6.2.6), (6.2.7); and*

$$\overline{U}(x_0, b^*) \subseteq D. \tag{6.2.31}$$

Then sequence $\{x_n\}$ $(n \geq 0)$ generated by method (4.1.3) is well defined, remains in $\overline{U}(x_0, b^*)$ for all $n \geq 0$ and converges to a solution $x^* \in \overline{U}(x_0, b^*)$ of equation $F(x) = 0$. Moreover, the following estimates hold for all $n \geq 0$:

$$\|x_{n+1} - x_n\| \leq b_{n+1} - b_n \tag{6.2.32}$$

and

$$\|x_n - x^*\| \leq b^* - b_n, \tag{6.2.33}$$

where iteration $\{b_n\}$ $(n \geq 0)$ is given by (6.2.27). The solution x^* is unique in $\overline{U}(x_0, b^*)$ if

$$\int_0^1 [w((1+\theta)b^*) - w(b^*)]d\theta + w_0(b^*) < 1. \tag{6.2.34}$$

Furthermore if there exists R_0 such that

$$b^* < R_0, \quad U(x_0, R_0) \subseteq D \tag{6.2.35}$$

and

$$\int_0^1 [w(b^* + \theta(b^* + R_0)) - w(b^*)]d\theta + w_0(b^*) \leq 1, \tag{6.2.36}$$

then the solution x^* is unique in $U(x_0, R_0)$.

Proof. Using induction we show $x_k \in \overline{U}(x_0, b^*)$ and estimate (6.2.32) hold for all $k \geq 0$. For $n = 0$, (6.2.32) holds, since

$$\|x_1 - x_0\| = \|F'(x_0)^{-1}F(x_0)\| \leq \eta = b_1 - b_0, \quad \text{(by (33))}.$$

Suppose (6.2.32) holds for all $n = 0, 1, \ldots, k+1$; this implies in particular that

$$\|x_{k+1} - x_0\| \leq \sum_{i=0}^{k} \|x_{i+1} - x_i\| \leq \sum_{i=0}^{k} (b_{i+1} - b_i) = b_{k+1} - b_0 = b_{k+1}.$$

We show (6.2.32) holds for $n = k+2$. By (6.2.7) and (6.2.25) we get

$$\|F'(x_0)^{-1}[F'(x_{k+1}) - F'(x_0)]\| \leq w_0(\|x_{k+1} - x_0\|) \leq w_0(b_{k+1}) < 1. \tag{6.2.37}$$

It follows from (6.2.37) and the Banach Lemma on invertible operators 1.3.2 that $F'(x_{k+1})^{-1}$ exists and

$$\|F'(x_{k+1})^{-1}F'(x_0)\| \leq [1 - w_0(\|x_{k+1} - x_0\|)]^{-1} \leq [1 - w_0(b_{k+1})]^{-1}. \quad (6.2.38)$$

By (4.1.3), (6.2.6), and (6.2.23) we obtain

$$\begin{aligned}
&\|F'(x_0)^{-1}F(x_{k+1})\|\|F'(x_0)^{-1}\{F(x_{k+1}) - F'(x_k)(x_{k+1} - x_k) - F(x_k)\}\| \\
&\leq \int_0^1 \|F'(x_0)^{-1}[F'(x_k + \theta(x_{k+1} - x_k)) - F'(x_k)]\| \, \|x_{k+1} - x_k\| d\theta \\
&\leq \int_0^1 [w(\|x_k - x_0\| + \theta\|x_{k+1} - x_k\|) - w(\|x_k - x_0\|)]d\theta\|x_{k+1} - x_k\| \\
&\leq \int_0^1 [w(b_k + \theta(b_{k+1} - b_k)) - w(b_k)]d\theta(b_{k+1} - b_k). \qquad (6.2.39)
\end{aligned}$$

Using (4.1.3), (6.2.38) and (6.2.39) we get

$$\begin{aligned}
\|x_{k+2} - x_{k+1}\| &\leq \|F'(x_{k+1})^{-1}F'(x_0)\| \cdot \|F'(x_0)^{-1}F(x_{k+1})\| \\
&\leq \frac{\int_0^1 [w(b_k + \theta(b_{k+1} - b_k)) - w(b_k)]d\theta(b_{k+1} - b_k)}{1 - w_0(b_{k+1})} \\
&= b_{k+2} - b_{k+1}. \qquad (6.2.40)
\end{aligned}$$

Note also that:

$$\|x_{k+2}-x_0\| \leq \|x_{k+2}-x_{k+1}\|+\|x_{k+1}-x_0\| \leq b_{k+2}-b_{k+1}+b_{k+1}-b_0 = b_{k+2} \leq b^*.$$

That is $x_{k+2} \in \overline{U}(x_0, b^*)$. It follows from (6.2.40) that $\{x_n\}$ $(n \geq 0)$ is a Cauchy sequence in a Banach space X, and as such it converges to some $x^* \in \overline{U}(x_0, b^*)$. By letting $k \to \infty$ in (6.2.39) we obtain $F(x^*) = 0$. Moreover, estimate (6.2.33) follows from (6.2.32) by using standard majorization techniques. Furthermore, to show uniqueness, let y^* be a solution of equation $F(x) = 0$ in $U(x_0, R_0)$. It follows from (4.1.3)

$$\begin{aligned}
\|y^* - x_{k+1}\| &\leq \|F'(x_k)^{-1}F'(x_0)\| \\
&\quad \cdot \left\| \int_0^1 \|F'(x_0)^{-1}[F'(x_k + \theta(y^* - x_k)) - F'(x_k)\right\| \|y^* - x_k\|d\theta \\
&\leq \frac{\int_0^1 [w(\|x_k - x_0\| + \theta\|y^* - x_k\|) - w(\|x_k - x_0\|)]d\theta}{1 - w_0(\|x_k - x_0\|)}\|y^* - x_k\| \\
&< \frac{\int_0^1 [w(b^* + \theta(b^* + R_0)) - w(b^*)]d\theta}{1 - w_0(b^*)}\|y^* - x_k\| \qquad (6.2.41) \\
&< \|y^* - x_k\| \quad \text{(by (46))}.
\end{aligned}$$

Similarly if $y^* \in \overline{U}(x_0, b^*)$ we arrive at

$$\|y^* - x_{k+1}\| \leq \frac{\int_0^1 [w(b^* + \theta b^*) - w(b^*)]d\theta}{1 - w_0(b^*)} \|y^* - x_k\|, \tag{6.2.42}$$

$$< \|y^* - x_k\| \quad \text{(by (44))}. \tag{6.2.43}$$

In both cases above $\lim_{k \to \infty} x_k = y^*$. Hence, we conclude

$$x^* = y^*.$$

■

We can now compare majorizing sequences $\{t_n\}$, $\{a_n\}$, $\{b_n\}$ under the hypotheses of Theorem 1 in [195, p. 117], and Theorems 6.2.3, 6.2.5 above:

Lemma 6.2.6 *Under the hypotheses stated above and if strict inequality holds in (6.2.7) or (6.2.8) then the following estimates*

$$b_{n+1} < a_{n+1} < t_{n+1} \tag{6.2.44}$$

$$b_{n+1} - b_n < a_{n+1} - a_n < t_{n+1} - t_n \tag{6.2.45}$$

$$b^* - b_n \leq a^* - a_n \leq t^* - t_n \tag{6.2.46}$$

and

$$b^* \leq a^* \leq t^* \tag{6.2.47}$$

hold for all $n \geq 0$.

Proof. It follows easily by using induction, definitions of sequences $\{t_n\}$, $\{a_n\}$, $\{b_n\}$ given by (6.2.12), (6.2.11), (6.2.27) respectively and hypotheses (6.2.8), (6.2.9). ■

Remark 6.2.2 *The above lemma justifies the advantages of our approach claimed at the beginning of the section.*

6.3 Convergence and universal constants

We improve the elegant work by Wang, Li and Lai in [342], where they gave a unified convergence theory for several Newton-type methods by using different Lipschitz conditions and a finer "universal constant".

It is convenient to define majorizing function f and majorizing sequence $\{t_n\}$ by

$$f(t) = \eta - t + \frac{1}{2}\ell t^2 + \frac{1}{6}Lt^3 \tag{6.3.1}$$

and

$$t_{n+1} = t_n - \frac{f(t_n)}{f'(t_n)}, \quad t_0 = 0 \tag{6.3.2}$$

respectively where η, ℓ, L are nonnegative constants.

Proposition 6.3.1 *[342, p. 207] Let*

$$r = \frac{2}{\ell + \sqrt{\ell^2 + 2L}}, \tag{6.3.3}$$

and

$$b = \frac{2(\ell + 2\sqrt{\ell^2 + 2L})}{3(\ell + \sqrt{\ell^2 + 2L})^2}. \tag{6.3.4}$$

Then, (a) function f is monotonically decreasing in $[0, r]$, while it is increasing monotonically in $[r, +\infty)$. Moreover, if $\eta \leq b$ then

$$f(\eta) > 0, \quad f(r) = \eta - b \leq 0, \quad f(+\infty) > \eta > 0.$$

Therefore f has a unique zero in two intervals denoted by t^* and t^{**} respectively, satisfying

$$\eta < t^* < \frac{r}{b}\eta < r < t^{**}, \tag{6.3.5}$$

when $\eta < b$ and $t^* = t^{**}$, when $\eta = b$; (b) Iteration $\{t_n\}$ is monotonically increasing and converges to t^*.

From now on we set $D = \overline{U}(x_0, r)$.

We can define a more precise majorizing sequence $\{s_n\}$ $(n \geq 0)$ by

$$s_{n+2} = s_{n+1} + \frac{\frac{L}{3}(s_{n+1} - s_n) + Ls_n + \ell}{2(1 - \ell_0 s_{n+1})}(s_{n+1} - s_n)^2, \tag{6.3.6}$$

$s_0 = 0$, $s_1 = \eta$, for some constant $\ell_0 \in [0, \ell]$.

Iteration $\{t_n\}$ $(n \geq 0)$ can also be written as

$$t_{n+2} = t_{n+1} + \frac{\frac{L}{3}(t_{n+1} - t_n) + Lt_n + \ell}{2\left[1 - \ell t_{n+1} - \frac{L}{2}t_{n+1}^2\right]}, \quad t_0 = 0, t_1 = \eta. \tag{6.3.7}$$

Then under the hypotheses of Proposition 6.3.1, using (6.3.6), (6.3.7) and induction on n we can easily show:

Proposition 6.3.2 *Under the hypotheses of Proposition 6.3.1, sequence $\{s_n\}$ is monotonically increasing and converges to some $s^* \in (\eta, t^*]$. Moreover the following estimates hold:*

$$s_n < t_n \quad (n \geq 1) \tag{6.3.8}$$

$$0 \leq s_{n+1} - s_n < t_{n+1} - t_n \quad (n \geq 0) \tag{6.3.9}$$

$$s^* - s_n \leq t^* - t_n \tag{6.3.10}$$

and

$$s^* \leq t^* \tag{6.3.11}$$

provided that $L > 0$ if $\ell_0 = \ell$.

Let $x_0 \in D$ such that $F'(x_0)^{-1} \in L(Y, X)$.

We now need to introduce some Lipschitz conditions:

$$\|F'(x_0)^{-1}[F'(x) - F'(x_0)]\| \leq \ell_0 \|x - x_0\|, \quad \text{for all } x \in \overline{U}(x_0, r) \tag{6.3.12}$$

$$\|F'(x_0)^{-1}[F'(y) - F'(x)]\| \leq \left[\ell + L\|x - x_0\| + \frac{1}{2} L \|y - x\|\right] \|y - x\|, \tag{6.3.13}$$
$$\text{for all } x, y\colon \|y - x\| + \|x - x_0\| \leq r,$$

$$\|F'(x_0)^{-1}[F'(x) - F'(x_0)]\| \leq \left(\ell + \frac{1}{2} L \|x - x_0\|\right) \|x - x_0\|, \tag{6.3.14}$$
$$\text{for all } x \in \overline{U}(x_0, r),$$

$$\|F'(x_0)^{-1}[F'(x) - F'(x_0)]\| \leq \ell \|x - x_0\|, \quad \text{for all } x \in \overline{U}(x_0, r) \tag{6.3.15}$$

$$\|F'(x_0)^{-1} F''(x_0)\| \leq \ell \tag{6.3.16}$$

$$\|F'(x_0)^{-1} F''(x_0)\| \leq \ell_0 \tag{6.3.17}$$

$$\|F'(x_0)^{-1}[F''(y) - F''(x)]\| \leq L\|y - x\|, \text{ for all } y, x\colon \|y - x\| + \|x - x_0\| \leq r, \tag{6.3.18}$$

$$\|F'(x_0)^{-1}[F''(x) - F''(x_0)]\| \leq L\|x - x_0\|, \text{ for all } x \in \overline{U}(x_0, r) \tag{6.3.19}$$

and

$$\|F'(x_0)^{-1}[F''(x) - F''(x_0)]\| \leq L_0 \|x - x_0\|, \quad \text{for all } x \in \overline{U}(x_0, r). \tag{6.3.20}$$

Note that we will use (6.3.12), (6.3.17), (6.3.20) instead of (6.3.15), (6.3.16) and (6.3.19) respectively in our proofs.

We define as in [342]

$$\begin{aligned}
K^{(1)}(x_0, \ell) &= \{F \in C^1(x_0, r)\colon (6.3.13) \text{ holds for } L = 0\},\\
K^{(1)}_{cent}(x_0, \ell) &= \{F \in C^1(x_0, r)\colon (6.3.15) \text{ holds}\},\\
K^{(1)}(x_0, \ell, L) &= \{F \in C^1(x_0, r)\colon (6.3.13) \text{ holds}\},\\
K^{(1)}_{cent}(x_0, \ell, L) &= \{F \in C^1(x_0, r)\colon (6.3.14) \text{ holds}\},\\
K^{(2)}(x_0, \ell, L) &= \{F \in C^2(x_0, r)\colon (6.3.16) \text{ and } (6.3.18) \text{ hold}\},\\
K^{(2)}_{cent}(x_0, \ell, L) &= \{F \in C^2(x_0, r)\colon (6.3.16) \text{ and } (6.3.20) \text{ hold}\}.
\end{aligned}$$

Moreover for our study we need to define:

$$\begin{aligned}
K^{(1)}(x_0, \ell_0, \ell, L) &= \{F \in C^1(x_0, r)\colon (6.3.12) \text{ and } (6.3.13) \text{ hold}\},\\
K^{(1)}_{cent}(x_0, \ell_0) &= \{F \in C^1(x_0, r)\colon (6.3.12) \text{ holds}\},\\
K^{(2)}(x_0, \ell_0, \ell, L) &= \{F \in C^2(x_0, r)\colon (6.3.12), (6.3.17) \text{ and } (6.3.18) \text{ hold}\\
&\qquad \text{or } (6.3.12), (6.3.16) \text{ and } (6.3.18) \text{ hold}\}.
\end{aligned}$$

We can provide the main semilocal convergence result for method (4.1.3).

Theorem 6.3.3 *Suppose that $F \in K^{(1)}(x_0, \ell_0, \ell, L)$. If $\eta \leq b$, then sequence $\{x_n\}$ generated by method (4.1.3) is well defined, remains in $\overline{U}(x_0, s^*)$ for all $n \geq 0$, and converges to a solution $x^* \in \overline{U}(x_0, r)$ of equation $F(x) = 0$.*

Proof. Simply use (6.3.12) instead of (6.3.15) in Lemma 2.3 in [342] to obtain $F'(x)^{-1}$ exists for all $x \in \overline{U}(x_0, r)$ and satisfies

$$\|F'(x)^{-1}F'(x_0)\| \leq \frac{1}{1 - \ell_0\|x - x_0\|}. \tag{6.3.21}$$

The rest follows exactly as in Theorem 3.1 in [342] but by replacing $\{t_n\}$, $\{t^*\}$, (6.3.11) by $\{s_n\}$, $\{s^*\}$, (6.3.21) respectively. ■

Theorem 6.3.4 *Suppose that $F \in K^1_{cent}(x_0, \ell_0)$. If $\eta \leq b$, then the equation $F(x) = 0$ has a unique solution x^* in $\overline{U}(x_0, s_0^*)$, where $s_0^* = s^*$ if $\eta = b$, and $s_0^* \in [s^*, t^*)$ if $\eta < b$.*

Proof. As in Theorem 3.2 in [342] but used (6.3.12) instead of (6.3.14). ■

By Proposition 6.3.2 we see that our results in Theorems 6.3.3 and 6.3.4 improve the corresponding ones in [342] as claimed in the introduction in case $\ell_0 \leq \ell$.

At this point we wonder if hypothesis $\eta \leq b$ can be weakened by finding a larger "universal constant" b. It turns out that this is possible.

We first need the following lemma on majorizing sequences:

Lemma 6.3.5 *Assume there exist nonnegative constants ℓ_0, ℓ, L, η and a parameter $\delta \in [0, 2)$ such that for all $n \geq 0$*

$$\begin{aligned} &\frac{L}{3}\left(\frac{\delta}{2}\right)^{2n}\eta + \frac{2L\left[1 - \left(\frac{\delta}{2}\right)^n\right]}{2 - \delta}\left(\frac{\delta}{2}\right)^n\eta^2 + \ell\left(\frac{\delta}{2}\right)^n\eta \\ &\qquad + \frac{2\ell_0\delta}{2 - \delta}\left[1 - \left(\frac{\delta}{2}\right)^{n+1}\right]\eta \leq \delta. \end{aligned} \tag{6.3.22}$$

Then, iteration $\{s_n\}$ given by (6.3.7) is monotonically increasing, bounded above by $s^{**} = \frac{2\eta}{2-\delta}$ and converges to some s^* such that

$$0 \leq s^* \leq s^{**}. \tag{6.3.23}$$

Moreover the following estimates hold for all $n \geq 0$:

$$0 \leq s_{n+2} - s_{n+1} \leq \frac{\delta}{2}(s_{n+1} - s_n) \leq \left(\frac{\delta}{2}\right)^{n+1}\eta. \tag{6.3.24}$$

Proof. The proof follows along the lines of Lemma 5.1.1. However there are some differences. We must show using induction on the integer $k \geq 0$

$$\frac{L}{3}(s_{k+1} - s_k)^2 + Ls_k(s_{k+1} - s_k) + \ell(s_{k+1} - s_k) + \ell_0\delta s_{k+1} \leq \delta, \tag{6.3.25}$$

$$0 \leq s_{k+1} - s_k, \tag{6.3.26}$$

and

$$0 < 1 - \ell_0 s_{k+1}. \tag{6.3.27}$$

Estimates (6.3.25)–(6.3.27) hold for $k = 0$ by (6.3.22). But then (6.3.6) gives

$$0 \leq s_2 - s_1 \leq \frac{\delta}{2}(s_1 - s_0). \tag{6.3.28}$$

Let us assume (6.3.24)–(6.3.27) hold for all $k \leq n$. We can have in turn:

$$\begin{aligned} &\frac{L}{3}(s_{k+1} - s_k)^2 + L s_k (s_{k+1} - s_k) + \ell (s_{k+1} - s_k) + \ell_0 \delta s_{k+1} \\ &\leq \frac{L}{3}\left[\left(\frac{\delta}{2}\right)^k \eta\right]^2 + L \cdot \frac{1 - \left(\frac{\delta}{2}\right)^k}{1 - \frac{\delta}{2}} \eta \left(\frac{\delta}{2}\right)^k \eta + \ell \left(\frac{\delta}{2}\right)^k \eta \\ &\quad + \ell_0 \delta \frac{1 - \left(\frac{\delta}{2}\right)^{k+1}}{1 - \frac{\delta}{2}} \eta \leq \delta \quad \text{by (6.3.22).} \end{aligned} \tag{6.3.29}$$

We must also show

$$s_k \leq s^{**}. \tag{6.3.30}$$

For $k = 0, 1, 2$ we have

$$s_0 = 0 \leq s^{**}, \ s_1 = \eta \leq s^{**}, \text{ and } s_2 = \eta + \frac{\delta}{2}\eta \leq s^{**}.$$

Assume (6.3.30) holds for all $k \leq n$. It follows from the induction hypothesis

$$\begin{aligned} s_{k+1} &\leq s_k + \frac{\delta}{2}(s_k - s_{k-1}) \leq \cdots \leq \eta + \frac{\delta}{2}\eta + \cdots + \left(\frac{\delta}{2}\right)^k \eta \\ &= \frac{1 - \left(\frac{\delta}{2}\right)^{k+1}}{1 - \frac{\delta}{2}} \eta \leq \frac{2\eta}{2 - \delta} = s^{**}. \end{aligned} \tag{6.3.31}$$

Moreover (6.3.27) holds from (6.3.22) and (6.3.31). Inequality (6.3.26) follows from (6.3.6), (6.3.27) and (6.3.25). Hence, sequence $\{s_n\}$ is monotonically increasing, bounded above by s^{**}, and as such it converges to some s^* satisfying (6.3.23). ∎

Remark 6.3.1 *Condition (6.3.22) can be replaced by a stronger but easier to verify condition*

$$\eta \leq b_0, \tag{6.3.32}$$

where,

$$b_0 = \frac{2}{\beta + \sqrt{\beta^2 + 4\alpha}}, \tag{6.3.33}$$

$$\alpha = \frac{L}{12}(5 - \sqrt{5}), \tag{6.3.34}$$

$$\beta = \frac{\ell}{2} + \frac{2\ell_0}{3 - \sqrt{5}} \tag{6.3.35}$$

for

$$\delta = \sqrt{5} - 1. \tag{6.3.36}$$

Indeed (6.3.22) certainly holds if

$$\frac{L}{3}\left(\frac{\delta}{2}\right)^2 \eta^2 + \frac{L\delta}{2}\eta^2 + \frac{\gamma\delta}{2}\eta + \frac{2\gamma_0\delta}{2-\delta}\eta \leq \delta \tag{6.3.37}$$

since,

$$\left[1 - \left(\frac{\delta}{2}\right)^{n+1}\right]\left(\frac{\delta}{2}\right)^{n+1} \leq \left[1 - \left(\frac{\delta}{2}\right)^{n}\right]\left(\frac{\delta}{2}\right)^{n} \tag{6.3.38}$$

holds for all $n \geq 0$ if $\delta \in [0, \sqrt{5} - 1]$. However (6.3.37) holds by (6.3.32).

Hence, we showed:

Lemma 6.3.6 *If (6.3.22) replaces condition (6.3.22) in Lemma 6.3.5 then the conclusions for sequence $\{s_n\}$ hold.*

Remark 6.3.2 *In Lemma 6.3.5 we wanted to leave (3.4) as uncluttered as possible, since it can be shown to hold in some interesting cases. For example consider Newton's method, i.e. set $L = 0$. See Application 5.1.4 in Section 5.1.*

Note also that b_0 can be larger than b, which means that (3.14) may hold but not $n \leq b$. Indeed, here are two such cases: choose

(a) $L = 1$, $\gamma_0 = \gamma = .5$ and $\eta = .59$

then $b = .58\overline{3}$ *and* $b_0 = .590001591$

(b) $L = .8$, $\gamma = .4$ and $\eta = .7$

then $b = .682762423$ *and* $b_0 = .724927596.$

Below is the main semilocal convergence theorem for Newton's method involving Fréchet differentiable operators and using center-Lipschitz conditions.

Theorem 6.3.7 *Assume: (a) $F \in K^{(1)}(x_0, \ell_0, \ell, L)$, and (b) hypotheses of Lemmas 6.3.5 or 6.3.6 hold. Then, sequence $\{x_n\}$ $(n \geq 0)$ generated by method (4.1.3) is well defined, remains in $\overline{U}(x_0, s^*)$ for all $n \geq 0$, and converges to a solution $x^* \in \overline{U}(x_0, s^*)$ of equation $F(x) = 0$. Moreover the following estimates hold for all $n \geq 0$:*

$$\begin{aligned}\|x_{n+2} - x_{n+1}\| &\leq \frac{L\left[\frac{1}{6}\|x_{n+1} - x_n\| + \frac{1}{2}\|x_n - x_0\|\right] + \frac{\gamma}{2}}{1 - \gamma_0\|x_{n+1} - x_0\|}\|x_{n+1} - x_n\|^2 \\ &\leq \frac{L\left[\frac{1}{6}(s_{n+1} - s_n) + \frac{1}{2}(s_n - s_0)\right] + \frac{\gamma}{2}}{1 - \gamma_0(s_{n+1} - s_0)}(s_{n+1} - s_n)^2 \\ &= s_{n+2} - s_{n+1},\end{aligned} \tag{6.3.39}$$

and

$$\|x_n - x^*\| \leq s^* - s_n, \tag{6.3.40}$$

where, sequence $\{s_n\}$ $(n \geq 0)$ is given by (6.3.6). The solution x^* is unique in $\overline{U}(x_0, s^*)$ if

$$\gamma s^* < 1. \tag{6.3.41}$$

Furthermore, if there exists $R > s^*$ such that $U(x_0, R) \subseteq D$, and

$$\gamma_0(s^* + R) \leq 2, \tag{6.3.42}$$

the solution x^* is unique in $U(x_0, R)$.

Proof. Let us prove that

$$\|x_{k+1} - x_k\| \leq s_{k+1} - s_k \tag{6.3.43}$$

and

$$\overline{U}(x_{k+1}, s^* - s_{k+1}) \subseteq \overline{U}(x_k, s^* - s_k) \tag{6.3.44}$$

hold for all $k \geq 0$. For every $z \in \overline{U}(x_1, s^* - s_1)$

$$\|z - x_0\| \leq \|z - x_1\| + \|x_1 - x_0\| \leq s^* - s_1 + s_1 = s^* - s_0$$

implies $z \in \overline{U}_1(x_1, s^* - s_1)$. Since also

$$\|x_1 - x_0\| = \|F'(x_0)^{-1}F(x_0)\| \leq \eta = s_0,$$

(6.3.43) and (6.3.44) hold for $k = 0$. Given they hold for $n = 0, 1, \ldots, k$, then

$$\|x_{k+1} - x_0\| \leq \sum_{i=1}^{k+1}\|x_i - x_{i-1}\| \leq \sum_{i=1}^{k+1}(s_i - s_{i-1}) = s_{k+1} - s_0 = s_{k+1}$$

and

$$\|x_k + \theta(x_{k+1} - x_k) - x_0\| \le s_k + \theta(s_{k+1} - s_k) < s^*, \quad \theta \in [0,1].$$

Using (4.1.3) we obtain the approximation

$$\begin{aligned} F(x_{k+1}) &= F(x_{k+1}) - F(x_k) - F'(x_k)(x_{k+1} - x_k) \\ &= \int_0^1 [F'(x_k + \theta(x_{k+1} - x_k)) - F'(x_k)](x_{k+1} - x_k) d\theta. \end{aligned} \tag{6.3.45}$$

Then, we get by (6.3.12), (6.3.13) and (6.3.45)

$$\begin{aligned} &\|F'(x_0)^{-1} F(x_{k+1})\| \\ &\quad \le \left\{ L \int_0^1 [\|x_k - x_0\| + \theta\|x_{k+1} - x_k\|](1-\theta) d\theta + \frac{\gamma}{2} \right\} \|x_{k+1} - x_k\|^2 \\ &\le \frac{L}{6}\|x_{k+1} - x_k\|^3 + \frac{L}{2}\|x_k - x_0\| \, \|x_{k+1} - x_k\|^2 + \frac{\gamma}{2}\|x_{k+1} - x_k\|^2 \\ &\quad \le \left\{ L\left[\frac{1}{6}(s_{k+1} - s_k) + \frac{1}{2}(s_k - s_0) \right] + \frac{\gamma}{2} \right\} (s_{k+1} - s_k)^2. \end{aligned} \tag{6.3.46}$$

Using (6.3.12) we get

$$\begin{aligned} \|F'(x_0)^{-1}[F'(x_{k+1}) - F'(x_0)]\| &\le \gamma_0 \|x_{k+1} - x_0\| \\ &\le \gamma_0 s_{k+1} \le \gamma s^* < 1 \quad \text{(by (6.3.22))}. \end{aligned} \tag{6.3.47}$$

It follows from the Banach Lemma on invertible operators 1.3.2 and (6.3.22) that $F'(x_{n+1})^{-1}$ exists so that:

$$\|F'(x_{k+1})^{-1} F'(x_0)\| \le [1 - \gamma_0\|x_{k+1} - x_0\|]^{-1} \le (1 - \gamma_0 t_{k+1})^{-1}. \tag{6.3.48}$$

Therefore by (6.3.2), (6.3.46), and (6.3.48) we obtain in turn

$$\begin{aligned} \|x_{k+2} - x_{k+1}\| &= \|(F'(x_{k+1})^{-1} F'(x_0))(F'(x_0)^{-1} F(x_{k+1}))\| \\ &\le \|F'(x_{k+1})^{-1} F'(x_0)\| \cdot \|F'(x_0)^{-1} F(x_{k+1})\| \\ &\le s_{k+2} - s_{k+1}. \end{aligned} \tag{6.3.49}$$

Thus for every $z \in \overline{U}(x_{k+2}, s^* - s_{k+2})$ we have

$$\begin{aligned} \|z - x_{k+1}\| &\le \|z - x_{k+2}\| + \|x_{k+2} - x_{k+1}\| \\ &\le s^* - s_{k+2} - s_{k+2} - s_{k+1} = s^* - s_{k+1}. \end{aligned} \tag{6.3.50}$$

That is

$$z \in \overline{U}(x_{k+1}, s^* - s_{k+1}). \tag{6.3.51}$$

Estimates (6.3.49), (6.3.50) imply that (6.3.43), (6.3.44) hold for $n = k+1$. The proof of (6.3.43) and (6.3.44) is now complete by induction. Lemma 6.3.5 implies $\{t_n\}$ $(n \geq 0)$ is a Cauchy sequence. From (6.3.43), (6.3.44) $\{x_n\}$ $(n \geq 0)$ becomes a Cauchy sequence too, and as such it converges to some $\overline{U}(x_0, s)$ such that (6.3.40) holds. The combination of (6.3.40) and (6.3.46) yields $F(x^*) = 0$. Finally to show uniqueness let y^* be a solution of equation $F(x) = 0$ in $U(x_0, R)$. It follows from (6.3.12), the estimate

$$\begin{aligned}
&\left\| F'(x_0)^{-1} \int_0^1 [F'(y^* + \theta(x^* - y^*)) - F'(x_0)] \right\| d\theta \\
&\quad \leq \gamma_0 \int_0^1 \|y^* + \theta(x^* - y^*) - x_0\| d\theta \\
&\quad \leq \gamma_0 \int_0^1 [\theta\|x^* - x_0\| + (1-\theta)\|y^* - x_0\|] d\theta \\
&\quad < \frac{\gamma_0}{2}(s^* + R) \leq 1 \quad \text{(by (6.3.42))},
\end{aligned}$$

and the Banach Lemma on invertible operators that linear operator

$$L = \int_0^1 F'(y^* + \theta(x^* - y^*)) d\theta$$

is invertible. Using the identity

$$0 = F(x^*) - F(y^*) = L(x^* - y^*),$$

we deduce $x^* = y^*$. Similarly we show uniqueness in $\overline{U}(x_0, s^*)$ using (6.3.41). ∎

As in [342] we also study the semilocal convergence of King's iteration [208]

$$x_{n+1} = x_n - F'(x_{m[n/m]})^{-1} F(x_n) \quad (n \geq 0) \quad (x_0 \in D) \tag{6.3.52}$$

where $\left[\frac{n}{m}\right]$ denotes the integer part of $\frac{n}{m}$, and Werner's method [345]

$$x_{n+1} = x_n - F'(y_n)^{-1} F(x_n) \tag{6.3.53}$$

$$y_{n+1} = x_{n+1} - \frac{1}{2} F'(y_n)^{-1} F(x_{n+1}) \tag{6.3.54}$$

$$y_0 = x_0. \tag{6.3.55}$$

Both methods improve the computational efficiency with (6.3.53)–(6.3.55) being more preferable because it raises the convergence order to $1 + \sqrt{2}$, although the number of function evaluations doubles when compared with Newton's method.

Using (6.3.12) instead of (6.3.14) or (6.3.19) to compute upper error bounds on $\|F'(x)^{-1} F'(x_0)\|$ (see e.g. (6.3.48)) we can show as in Theorems 4.1 and 4.2 in [342] respectively:

Theorem 6.3.8 *Let $F \in K^{(1)}(x_0, \gamma_0, \gamma, L)$ and assume hypotheses of Lemmas 6.3.5 or 6.3.6 hold. Then sequence $\{x_n\}$ generated by King's method (6.3.52) converges to a unique solution x^* of equation $F(x) = 0$ in $\overline{U}(x_0, s^*)$.*

Theorem 6.3.9 *Let $F \in K^{(2)}(x_0, \ell_0, \ell, L)$ and assume hypotheses of Lemmas 6.3.5 or 6.3.6 hold. Then sequence $\{x_n\}$ generated by Werner's method (6.3.53)–(6.3.55) converges to a unique solution x^* of equation $F(x) = 0$ in $\overline{U}(x_0, s^*)$.*

Note that finer error bounds and a more precise information on the location of the solution x^* are obtained again in the cases of Theorems 6.3.8 and 6.3.9 since more precise majorizing sequences are used.

6.4 A Deformed Method

We assume that F is a twice continuously Fréchet-differentiable operator in this section.

We use the following modification of King's iteration [208] proposed by Werner [345]:

$$x_{n+1} = x_n - F'\left(\frac{x_n + z_n}{2}\right)^{-1} F(x_n), \quad (x_0 \in D) \ (z_0 = x_0) \ (n \geq 0) \tag{6.4.1}$$

$$z_{n+1} = x_{n+1} - F'\left(\frac{x_n + z_n}{2}\right)^{-1} F(x_{n+1}) \quad (n \geq 0). \tag{6.4.2}$$

Although the number of function evaluations increases by one when compared to Newton's method, the order of convergence is raised from 2 to $1 + \sqrt{2}$ [345].

Here in particular using more precise majorizing sequences than before we provide finer error bounds on the distances $\|x_{n+1} - x_n\|$, $\|x_n - x^*\|$ and a more precise information on the location of the solution x^*. Finally, we suggest an approach for even weakening the sufficient convergent condition (6.4.8) for this method along the lines of the works in Section 5.1.

Let $\beta \geq 0$, $\gamma > 0$, and $\gamma_0 > 0$ be given parameters. It is convenient to define real function f by

$$f(t) = \beta - t + \frac{\gamma t^2}{1 - \gamma t}, \quad t \in \left[0, \frac{1}{\gamma}\right], \tag{6.4.3}$$

and scalar sequences $\{t_n\}$, $\{\bar{t}_n\}$ $(n \geq 0)$ by

$$t_{n+1} = t_n - f'(s_n)^{-1} f(t_n) \quad t_0 = r_0 = 0, \tag{6.4.4}$$

$$r_{n+1} = t_{n+1} - f'(s_n)^{-1} f(t_{n+1}) \quad s_n = \frac{t_n + r_n}{2}, \tag{6.4.5}$$

$$\bar{t}_{n+2} = \bar{r}_{n+1} - \frac{f_{n+1}}{\gamma_0 \bar{s}_n - 1} \quad \bar{t}_0 = \bar{r}_0 = 0, \quad \bar{t}_1 = \beta, \tag{6.4.6}$$

$$\overline{r}_{n+1} = \overline{t}_{n+1} - \frac{g_{n+1}}{\gamma_0 \overline{s}_n - 1} \quad \overline{s}_n = \frac{\overline{t}_n + \overline{r}_n}{2}, \tag{6.4.7}$$

where,

$$g_{n+1} = \int_0^1 \frac{2\gamma(\overline{t}_{n+1} - \overline{r}_n)^2(1-\theta)}{[1 - \gamma\overline{r}_n - \gamma(\overline{t}_{n+1} - \overline{r}_n)\theta]^3} d\theta + \frac{1}{4}\int_0^1\int_0^1 \frac{6\gamma^2(\overline{r}_n - \overline{t}_n)^3\theta p(1-\theta)}{\left[1 - \gamma\overline{t}_n - \gamma(\overline{r}_n - \overline{t}_n)\left(\frac{\theta}{2} + p(1-\theta)\right]^4\right.} dpd\theta,$$

and

$$f_{n+1} = \big(f'(\overline{s}_{n+1}) - f'(\overline{s}_n)\big)(\overline{r}_{n+1} - \overline{t}_{n+1}).$$

We need the following lemma on majorizing sequences:

Lemma 6.4.1 *Assume:*

$$\alpha = \beta\gamma \leq 3 - 2\sqrt{2}, \tag{6.4.8}$$

and

$$\gamma_0 = 2\gamma. \tag{6.4.9}$$

Then iterations $\{t_n\}$, $\{\overline{t}_n\}$ are well defined and the following estimates hold:

$$0 = t_0 = r_0 < t_n < r_n < t_{n+1} < r_{n+1} < t^* \quad (n \geq 0), \tag{6.4.10}$$

$$0 = \overline{t}_0 = \overline{r}_0 < \overline{t}_n < \overline{r}_n < \overline{t}_{n+1} < \overline{r}_{n+1} < \overline{t}^* \quad (n \geq 0), \tag{6.4.11}$$

$$\overline{t}_n < t_n \quad (n \geq 2) \tag{6.4.12}$$

$$\overline{r}_n < r_n \quad (n \geq 2) \tag{6.4.13}$$

$$\overline{t}_{n+1} - \overline{t}_n < t_{n+1} - t_n \quad (n \geq 1), \tag{6.4.14}$$

$$\overline{r}_{n+1} - \overline{t}_{n+1} < r_{n+1} - t_{n+1} \quad (n \geq 1), \tag{6.4.15}$$

$$\overline{t}^* - \overline{t}_n \leq t^* - t_n \quad (n \geq 0), \tag{6.4.16}$$

$$\overline{t}^* \leq t^*, \tag{6.4.17}$$

where,

$$\overline{t}^* = \lim_{n\to\infty} \overline{t}_n, \tag{6.4.18}$$

and

$$t^* = \lim_{n\to\infty} t_n = \frac{1 + \alpha - \sqrt{(1+\alpha)^2 - 8\alpha}}{4\gamma} \leq \left(1 + \frac{1}{\sqrt{2}}\right)\beta \leq \left(1 - \frac{1}{\sqrt{2}}\right)\frac{1}{\gamma}. \tag{6.4.19}$$

Proof. Estimates (6.4.10) and (6.4.19) follow using only (6.4.8) (see Lemma 2 in [177, p. 69]. The rest of the proof follows immediately by using induction on n and noticing that

$$-\frac{1}{2\gamma t-1} < -\frac{1}{f'(t)} \quad \text{for all } t \in \left[0, \left(1-\frac{1}{\sqrt{2}}\right)\frac{1}{\gamma}\right]. \tag{6.4.20}$$

Indeed (6.4.20) reduces to showing:

$$0 \leq t < \frac{3}{2\gamma},$$

which is true by the choice of t. ■

Remark 6.4.1 *It follows from Lemma 6.4.1 that if $\{\bar{t}_n\}$ is shown to be a majorizing sequence for $\{x_n\}$ instead of $\{t_n\}$ (shown to be so see Theorem 1 [179]) then we can obtain finer estimates on the distances*

$$\|z_n - x_n\|, \ \|x_{n+1} - z_n\|, \ \|x_n - x^*\|, \ \|z_n - x^*\|$$

and an at least as precise information on the location of the solution x^*. We also note that as it can be seen in the main semilocal convergence theorem that follows all that is achieved under the same hypotheses/computational cost:

Theorem 6.4.2 *Let $F\colon D \subseteq X \to Y$ be a thrice continuously Fréchet differentiable operator. Assume there exist $x_0 \in D$, $b \geq 0$, $\gamma > 0$, $\gamma_0 > 0$ such that $F'(x_0)^{-1} \in L(Y, X)$,*

$$\|F'(x_0)^{-1}F(x_0)\| \leq \beta, \tag{6.4.21}$$

$$\left\|\frac{1}{2}F'(x_0)^{-1}F''(x_0)\right\| \leq \gamma, \tag{6.4.22}$$

$$\|F'(x_0)^{-1}[F'(x) - F'(x_0)]\| \leq \gamma_0\|x - x_0\| \tag{6.4.23}$$

$$\left\|\frac{1}{6}F'(x_0)^{-1}F'''(x)\right\| \leq \frac{\gamma^2}{(1-\gamma\|x-x_0\|)^4} \tag{6.4.24}$$

for all

$$x \in \overline{U}\left(x_0, \left(1-\frac{1}{\sqrt{2}}\right)\frac{1}{\gamma}\right), \tag{6.4.25}$$

$$\overline{U}\left(x_0, \left(1-\frac{1}{\sqrt{2}}\right)\frac{1}{\gamma}\right) \subseteq D \tag{6.4.26}$$

and hypotheses of Lemma 6.4.1 hold. Then iterations $\{x_n\}$, $\{z_n\}$ $(n \geq 0)$ generated by method (6.4.2) are well defined, remain in $\overline{U}(x_0, t^*)$ for all $n \geq 0$, where t^* is *given by (6.4.19) and converge to a unique solution x^* of equation $F(x) = 0$ in $\overline{U}(x_0, (1 - \frac{1}{\sqrt{2}})\frac{1}{\gamma})$. Moreover the following estimates hold:*

$$\|z_n - x_n\| \leq \overline{r}_n - \overline{t}_n < r_n - t_n \tag{6.4.27}$$

$$\|x_{n+1} - z_n\| \leq \overline{t}_{n+1} - \overline{r}_n < t_{n+1} - r_n \tag{6.4.28}$$

$$\|x_n - x^*\| \leq \overline{t}^* - \overline{t}_n \leq t^* - t_n \tag{6.4.29}$$

and

$$\|z_n - x^*\| \leq \overline{t}^* - \overline{r}_n \leq t^* - r_n. \tag{6.4.30}$$

Proof. We only need to show that the left-hand side inequalities in (6.4.27) and (6.4.28) hold. The rest follow from Lemma 2.1, standard majorization techniques and Theorem 1 in [179]. Let $x \in \overline{U}(x_0, t^*)$. It follows from (6.4.9), (6.4.19) and (6.4.23) that

$$\|F'(x_0)^{-1}[F'(x) - F'(x_0)]\| \leq 2\gamma\|x - x_0\| \leq 2\gamma t^* < 1. \tag{6.4.31}$$

Using the Banach Lemma on invertible operators [6] and (6.4.31) we conclude that $F'(x)^{-1}$ exists and

$$\|F'(x)^{-1}F'(x_0)\| \leq \frac{1}{1 - 2\gamma\|x - x_0\|}. \tag{6.4.32}$$

By (6.4.2), (6.4.3)–(6.4.7), (6.4.20) above and (21)–(23) in Theorem 1 in [179, p. 70] we arrive at

$$\|z_{n+1} - x_{n+1}\| \leq -\frac{g_{n+1}}{2\gamma\overline{s}_n - 1} = \overline{r}_{n+1} - \overline{t}_{n+1} < -\frac{f(t_{n+1})}{f'(s_n)} = r_{n+1} - t_{n+1}$$

(with $\overline{r}_n$, $\overline{t}_n$ replacing r_n, t_n respectively in the estimate above (23) in [179]). Note also that (6.4.27) holds for $n = 0$, and by the induction hypothesis

$$\|x_{n+1} - x_0\| \leq \overline{t}_{n+1} < \overline{t}^* \leq t^* \leq \left(1 - \frac{1}{\sqrt{2}}\right)\frac{1}{\gamma}.$$

Therefore the left-hand side of (6.4.27) holds for $n + 1$. We also have

$$\|z_{n+1} - x_0\| \leq \overline{r}_{n+1} \leq \overline{t}^* \leq t^* \leq \left(1 - \frac{1}{\sqrt{2}}\right)\frac{1}{\gamma}.$$

Hence, the same is true for y_{n+1}. Moreover by (24), (25) in [179], and (6.4.20) and (6.4.32) above we have

$$\begin{aligned}\|x_{n+2} - z_{n+1}\| &\leq -\frac{1}{2\gamma\overline{s}_n - 1}(h'(\overline{s}_{n+1}) - h'(\overline{s}_n))(\overline{r}_{n+1} - \overline{t}_{n+1}) \\ &= \overline{t}_{n+2} - \overline{r}_{n+1}.\end{aligned}$$

Hence, the left-hand side of (6.4.28) holds for $n + 1$. The rest of the proof follows exactly as in Lemma 6.4.1 with $\overline{r}_n$, $\overline{t}_n$, $\overline{t}^*$ replacing r_n, t_n and t^* respectively. ■

Remark 6.4.2 *(a) Let us denote by* $\{\bar{\bar{t}}_n\}$ $(n \geq 0)$ *scalar sequence generated as (6.4.6), (6.4.7) by with* $\bar{t}_n$, $\bar{s}_n$, $\bar{r}_n$, g_{n+1} *replaced by* $\bar{\bar{t}}_n$, $\bar{\bar{s}}_n$, $\bar{\bar{r}}_n$, $f(t_{n+1})$, *respectively. It follows from the proof of Theorem 6.4.2 that* $\{\bar{\bar{t}}_n\}$ *is also a majorizing sequence for* $\{x_n\}$ *so that*

$$0 = \bar{\bar{t}}_0 = \bar{\bar{r}}_0 < \bar{\bar{t}}_n < \bar{\bar{r}}_n < \bar{\bar{t}}_{n+1} < \bar{\bar{r}}_{n+1} < \bar{\bar{t}}^* \quad (n \geq 0)$$

$$\bar{\bar{t}}_n < t_n, \quad \bar{\bar{r}}_n < r_n \quad \text{and} \quad \bar{\bar{t}}^* \leq t^*$$

where,

$$\bar{\bar{t}}^* = \lim_{n \to \infty} \bar{\bar{t}}_n.$$

(b) It is expected that since sequence $\{\bar{t}_n\}$ *is finer than* $\{t_n\}$ *condition (6.4.8) can be dropped as a sufficient convergence condition for* $\{\bar{t}_n\}$*, and be replaced by a weaker one. We do not pursue this idea here. However we suggest that such a condition can be found if we work along the lines of the more general majorizing sequences studied in Section 5.1.*

6.5 The Inverse Function Theorem

We provide a semilocal convergence analysis for Newton's method in a Banach space setting using the inverse function theorem. Using a combination of a Lipschitz as well as a center Lipschitz type condition we provide: weaker sufficient convergence conditions; finer error bounds on the distances involved, and a more precise information on the location of the solution than before [340]. Moreover, our results are obtained under the same or less computational cost, as already indicated in Section 5.1.

Let $x_0 \in D$ such that $U(x_0, r) \subseteq D$ for some $r > 0$, and $F'(x_0)^{-1} \in L(Y, X)$.

The inverse function theorem under the above stated conditions guarantees the existence of $\varepsilon > 0$, operator $F'(x_0)^{-1}$: $U(F(x_0), \varepsilon) \subseteq Y \to X$ such that $F'(x_0)^{-1}$ is differentiable,

$$F'(x_0)^{-1}(F(x_0)) = x_0, \tag{6.5.1}$$

and

$$F(F'(x_0)^{-1}(z)) = z \quad \text{for all } z \in U(F(x_0), \varepsilon). \tag{6.5.2}$$

In particular in [341] the center Lipschitz condition

$$\|F'(x_0)^{-1}[F'(x) - F'(x_0)]\| \leq \int_0^{s(x)} L(u)du \quad \text{for all } x \in U(x_0, r) \tag{6.5.3}$$

was used, where $s(x) = \|x - x_0\|$, and L is a positive integrable function on $(0, r]$ to first study the semilocal convergence of the modified Newton method. Then using the same function L and the Lipschitz condition

$$\|F'(x_0)^{-1}[F'(x) - F'(y)]\| \leq \int_{s(x)}^{s(x,y)} L(u)du \tag{6.5.4}$$

for all $x \in U(x_0, r)$, $y \in \overline{U}(x, r - s(x))$, where $s(x, y) = s(x) + \|y - x\| \leq r$, the semilocal convergence of method (6.5.2) was studied in a unified way with the corresponding modified Newton's method. We assume instead of (6.5.3):

$$\|F'(x_0)^{-1}[F'(x) - F'(x_0)]\| \leq \int_0^{s(x)} L_0(u)du \quad \text{for all } x \in U(x_0, r). \tag{6.5.5}$$

By Banach's Lemma on invertible operators 1.3.2 and

$$\|F'(x_0)^{-1}[F'(x) - F'(x_0)]\| \leq \int_0^{s(x)} L_0(u)du < 1, \tag{6.5.6}$$

for all $r_0^0 \in (0, r]$, $x \in U(x_0, r_0^0)$, $F'(x)^{-1}$ exists, and

$$\|F'(x)^{-1}F'(x_0)\| \leq \left[1 - \int_0^{s(x)} L_0(u)du\right]^{-1}, \tag{6.5.7}$$

with r_0^0 satisfying

$$\int_0^{r_0^0} L_0(u)du = 1. \tag{6.5.8}$$

By simply using (6.5.5) instead of (6.5.3) we can show the following improvements of the corresponding results in [340]:

Theorem 6.5.1 *Set $b_0 = \int_0^{r_0^0} L_0(u)$. Assume condition (6.5.5) holds for all $r \geq r_0^0$. Then the following hold:*

$$U\left(F(x_0), \frac{b_0}{\|F'(x_0)^{-1}\|}\right) \subseteq F(U(x_0, r_0^0)); \tag{6.5.9}$$

$F'(x_0)^{-1}$ exists and is differentiable on

$$U\left(F(x_0), \frac{b_0}{\|F'(x_0)^{-1}\|}\right),$$

and the radius of the ball is the best possible (under condition (6.5.5)).

Lemma 6.5.2 *Let*

$$f_0(t) = \beta - t + \int_0^t L_0(u)(t-u)du \quad t \in [0, R], \tag{6.5.10}$$

where R_0 satisfies

$$\frac{1}{R_0}\int_0^{R_0} L_0(u)(R_0 - u)du = 1. \tag{6.5.11}$$

If $\beta \in (0, b_0)$, then f_0 is monotonically decreasing on $[0, r_0^0]$, while it is increasing monotonically on $[r_0^0, R]$ so that

$$f_0(\beta) > 0, \quad f_0(r_0^0) = \beta - b_0 < 0, \quad f_0(R_0) = \beta > 0. \tag{6.5.12}$$

Moreover, f_0 has a unique root in each interval, denoted by r_1^0 and r_2^0, satisfying:

$$\beta < r_1^0 < \frac{r_0}{b_0}\beta < r_0^0 < r_2^0 < R_0. \tag{6.5.13}$$

If (6.5.5) is extended to the boundary, i.e.

$$\|F'(x_0)^{-1}[F'(x) - F'(x_0)]\| \leq \int_0^{s(x)} L_0(u)du \quad x \in \overline{U}(x_0, r), \tag{6.5.14}$$

according to Lemma 6.5.2, Theorem 6.5.1 implies a more precise result:

Proposition 6.5.3 *If*

$$0 < \beta < b_0 = \int_0^{r_0^0} L_0(u)u\, du, \tag{6.5.15}$$

and

$$r \geq r_1^0 \tag{6.5.16}$$

then

$$\overline{U}\left(F(x_0), \frac{\beta}{\|F'(x_0)^{-1}\|}\right) \subseteq F(\overline{U}(x_0, r_1^0)); \tag{6.5.17}$$

$F'(x_0)^{-1}$ exists and is differentiable on $\overline{U}\big(F(x_0), \frac{\beta}{\|F'(x_0)^{-1}\|}\big)$;

$$[F'(x_0)^{-1}]'(y) = F'(x)^{-1}, \quad y = F(x)$$

satisfies (6.5.7), and radius r_1^0 is as small as possible.

Proposition 6.5.4 *Assume:*

$$r \in [r_1^0, r_2^0],$$

$$\beta \in (0, b_0)$$

and condition (6.5.14) hold. Then

$$\overline{U}\left(F(x_0), \frac{\beta}{\|F'(x_0)^{-1}\|}\right) \cap F(U(x_0, r) \setminus \overline{U}(x_0, r_1^0)) = \emptyset. \tag{6.5.18}$$

In particular if $y = 0$ and $\beta = \|F'(x_0)^{-1}F(x_0)\|$, we get:

Theorem 6.5.5 *Assume:*

$$\beta \leq b_0,$$

$$r \in [r_1^0, r_2^0], \quad \text{if } \beta < b_0 \text{ or } r = r_1^0 \text{ if } \beta = b_0.$$

Then, under condition (6.5.14), then equation $F(x) = 0$ has a unique solution

$$x^* \in \overline{U}(x_0 - F'(x_0)^{-1}F(x_0), r_1^0 - \beta) \subseteq \overline{U}(x_0, r_1^0)$$

in the ball $\overline{U}(x_0, r)$.

Remark 6.5.1 *Denote by r_0, b, f, R, r_1, r_2 (the quantities used in [340]) corresponding to r_0^0, b_0, f_0, R_0, r_1^0 and r_2^0. Then in case strict inequality holds in (6.1.14) we have*

$$r_0 < r_0^0, \quad b < b_0, \quad f_0(u) < f(u) \quad u \in [0, r], \ R < R_0, \tag{6.5.19}$$

$$r_1^0 < r_1 \quad \text{and} \quad r_2 < r_2^0, \tag{6.5.20}$$

which improve the corresponding results in [340], and under less computational cost since in practice the computation of L requires the computation of L_0. In case L_0, L are constants we get

$$r_0 = \frac{1}{L}, \ r_0^0 = \frac{1}{L_0}, \ b = \frac{1}{2L}, \ b_0 = \frac{1}{2L_0}, \ R = \frac{2}{L}, \ R_0 = \frac{2}{L_0} \tag{6.5.21}$$

$$f_0(u) = \beta - u + \frac{1}{2}L_0 u^2, \quad f(u) = \beta - u + \frac{1}{2}Lu^2, \tag{6.5.22}$$

$$r_1 = \frac{2\beta}{1 + \sqrt{1 - 2L\beta}}, \quad r_2 = \frac{2\beta}{1 - \sqrt{1 - 2L\beta}}, \tag{6.5.23}$$

$$r_1^0 = \frac{2\beta}{1 + \sqrt{1 - 2L_0\beta}}, \quad \text{and} \quad r_2^0 = \frac{2\beta}{1 - \sqrt{1 - 2L_0\beta}}. \tag{6.5.24}$$

Let us give a numerical example where strict inequality holds in (6.1.14) and therefore favorable estimates (6.5.19) hold:

Example 6.5.1 *Using Example 5.1.3 we again conclude that (5.1.63) (which is the condition used in (4.2.51) in this case) does not hold However our corresponding condition reduces to*

$$h_A = 2L_0\beta \le 1 \tag{6.5.25}$$

or

$$h_A = \frac{2}{3}(1-x)(3-x) \le 1, \tag{6.5.26}$$

which holds on $D_0 = \left[0, \frac{4-\sqrt{10}}{2}\right] \subseteq D$.

In order for us to study the convergence of method (4.1.3) we define scalar iterations $\{t_n\}$, $\{s_n\}$ by

$$t_{n+1} = t_n - \frac{f(t_n)}{f'(t_n)}, \quad t_0 = 0 \tag{6.5.27}$$

and

$$s_{n+2} = s_{n+1} - \frac{\int_{s_n}^{s_{n+1}} L(u)(s_{n+1}-u)du}{f_0'(s_{n+1})}, \quad s_0 = 0, \quad s_1 = \beta. \tag{6.5.28}$$

We can prove the main semilocal convergence theorem for method (4.1.3):

Theorem 6.5.6 *Assume: conditions (6.5.4), (6.5.5);*

$$\beta \le b;$$

and

$$r \ge r_1$$

hold, where b, r_1 *are defined in Remark 6.5.1. Then, sequence* $\{x_n\}$ $(n \ge 0)$ *generated by method (4.1.3) is well defined, remains in* $\overline{U}(x_1, r_1-\beta)$ *for all* $n \ge 0$, *and converges to a solution* $x^* \in \overline{U}(x_1, r_1-\beta)$ *of equation* $F(x) = 0$. *Moreover the following estimates hold for all* $n \ge 0$:

$$\|x_{n+1} - x_n\| \le s_{n+1} - s_n \le t_{n+1} - t_n \tag{6.5.29}$$

$$\|x_n - x^*\| \le r_1 - s_n \le r_1 - t_n. \tag{6.5.30}$$

Furthermore if strict inequality holds in (6.1.14) then

$$s_{n+1} - s_n < t_{n+1} - t_n, \tag{6.5.31}$$

and

$$s_n < t_n \tag{6.5.32}$$

hold for all $n > 1$.

Proof. It can easily be verified by using induction that (6.5.31), (6.5.32) and the right hand sides of double inequalities (6.5.29), (6.5.30) hold. Let us prove:

$$\|x_{k+1} - x_k\| \le s_{k+1} - s_k, \tag{6.5.33}$$

and

$$\overline{U}(x_{k+1}, r_1 - s_{k+1}) \subseteq \overline{U}(x_k, r_1 - s_k), \tag{6.5.34}$$

hold for all $k \ge 0$. For every $z \in \overline{U}(x_1, r_1 - s_1)$,

$$\|z - x_0\| \le \|z - x_1\| + \|x_1 - x_0\| \le r_1 - s_1 + s_1 = r_1 - s_0$$

implies $z \in \overline{U}(x_0, r_1 - s_0)$. Since

$$\|x_1 - x_0\| = \|F'(x_0)^{-1}F(x_0)\| \le \beta = s_1,$$

(6.5.33) and (6.5.34) hold for $k = 0$. Given they hold for $n = 0, 1, \ldots, k$, then

$$\|x_{k+1} - x_0\| \le \sum_{i=1}^{k+1} \|x_i - x_{i-1}\| \le \sum_{i=1}^{k+1} (s_i - s_{i-1}) = s_{k+1} - s_0 = s_{k+1},$$

$$\|x_k + \theta(x_{k+1} - x_k) - x_0\| \le s_k + \theta(s_{k+1} - s_k) \le r_1 \quad \theta \in [0, 1].$$

Using (6.5.2) we obtain the approximation

$$\begin{aligned} F(x_{k+1}) &= F(x_{k+1}) - F(x_k) - F'(x_k)(x_{k+1} - x_k) \\ &= \int_0^1 [F'(x_k + \theta(x_{k+1} - x_k)) - F'(x_k)](x_{k+1} - x_k) d\theta. \end{aligned} \tag{6.5.35}$$

By (6.5.4), (6.5.27), (6.5.28) and the induction hypotheses we obtain in turn:

$$\begin{aligned} &\|F'(x_0)^{-1}F(x_{k+1})\| \le \\ &\le \int_0^1 \int_{s(x_k)}^{s(x_k, x_k + \theta(x_{k+1} - x_k))} L(u) du \|x_{k+1} - x_k\| d\theta \\ &= \int_0^{\|x_{k+1} - x_k\|} L(\|x_k - x_0\| + u)(\|x_{k+1} - x_k\| - u) du \\ &\le \tfrac{1}{\|x_{k+1} - x_k\|^2} \int_0^{\|x_{k+1} - x_k\|} L(\|x_k - x_0\| + u)(\|x_{k+1} - x_k\| - u) du \|x_{k+1} - x_k\|^2 \\ &\le \frac{1}{(s_{k+1} - s_k)^2} \int_0^{s_{k+1} - s_k} L(s_k + u)(s_{k+1} - s_k - u) du \|x_{k+1} - x_k\|^2 \\ &= \int_{s_k}^{s_{k+1}} L(u)(s_{k+1} - u) du \left(\frac{\|x_{k+1} - x_k\|}{s_{k+1} - s_k} \right)^2 \le \int_{s_k}^{s_{k+1}} L(u)(s_{k+1} - u) du \\ &\le \int_{t_k}^{t_{k+1}} L(u)(t_{k+1} - u) du = f(t_{k+1}), \end{aligned} \tag{6.5.36}$$

where in (6.5.36) we used the fact that function

$$g(t) = \frac{1}{t^2} \int_0^t L(s+u)(t-u)du \text{ is nondecreasing}$$

for $t \in [0, r-s]$ [340, p. 175]. By (6.5.7) for $x = x_{k+1}$ we get $F'(x_{k+1})^{-1}$ exists and

$$\begin{aligned} \|F'(x_{k+1})^{-1} F'(x_0)\| &\le \left[1 - \int_0^{\|x_{k+1}-x_0\|} L_0(u)du\right]^{-1} \\ &\le \left(1 - \int_0^{s_{k+1}} L_0(u)du\right)^{-1} \\ &= -f_0'(s_{k+1}) \le -f'(t_{k+1}). \end{aligned} \tag{6.5.37}$$

Using (6.5.2), (6.5.27), (6.5.28), (6.5.36) and (6.5.37) we get

$$\begin{aligned} \|x_{k+2} - x_{k+1}\| &\le \|F'(x_{k+1})^{-1} F'(x_0)\| \cdot \|F'(x_0)^{-1} F(x_{k+1})\| \\ &\le s_{k+2} - s_{k+1} \le t_{k+2} - t_{k+1}. \end{aligned} \tag{6.5.38}$$

Thus for every $z \in \overline{U}(x_{k+2}, r_1 - s_{k+2})$, we have

$$\|z - x_{k+1}\| \le \|z - x_{k+2}\| + \|x_{k+2} - x_{k+1}\| \le r_1 - s_{k+2} + s_{k+1} - s_{k+1} = r_1 - s_{k+1}.$$

That is,

$$z \in \overline{U}(x_{k+1}, r_1 - s_{k+1}). \tag{6.5.39}$$

Estimates (6.5.38) and (6.5.39) imply (6.5.33) and (6.5.34) hold for $n = k+1$. By induction the proof of (6.5.33) and (6.5.34) is completed. Estimate (6.5.30) follows from (6.5.29) by using standard majorization techniques. It also follows that $\{s_n\}$ (and $\{t_n\}$) is a Cauchy sequence. From (6.5.33) and (6.5.34) $\{x_n\}$ becomes a Cauchy sequence too, and as such it converges to some $x^* \in \overline{U}(x_1, r_1 - \beta)$ (since $\overline{U}(x_1, r_1 - \beta)$ is a closed set). By (6.5.5), $\|F'(x_n)\|$ is uniformly bounded. Hence, by

$$F(x_n) + F'(x_n)(x_{n+1} - x_n) = 0,$$

we get $\lim_{n\to\infty} x_n = x^*$. ∎

Remark 6.5.2 *If equality holds in (6.1.14) then $s_n = t_n$ $(n \ge 0)$ and Theorem 6.5.6 reduces to Theorem 3.1 in [340]. Otherwise our error bounds on the distances $\|x_{n+1} - x_n\|$, $\|x_n - x^*\|$ are finer (see (6.5.29)–(6.5.32)) for all $n > 1$ than the ones in [340] using majorizing sequence $\{t_n\}$ instead of the more precise $\{s_n\}$. We also note that our results are obtained under the same computational cost with the ones in [340] since in practice finding L requires finding L_0. The rest of the error bounds in [340] can also be improved by simply using L_0 in (6.5.7) instead of L see Section 5.1. Instead we wonder if the convergence of scalar sequence $\{s_n\}$ can be established under weaker conditions than the ones given in Theorem 6.5.6.*

We showed in Section 5.1 that $\{s_n\}$ is monotonically convergent to s^* if:
(A) If L_0, L are constants then sequence $\{s_n\}$ given by (6.5.28) becomes

$$s_{n+2} = s_{n+1} + \frac{L(s_{n+1}-s_n)^2}{2(1-L_0 s_{n+1})}, \quad s_0 = 0, \quad s_1 = \beta.$$

The advantages in this case have been explained in Section 5.1.

(B) There exists a minimal positive number $R_1 \in [\beta, R^0]$ satisfying equation

$$\beta - r + \frac{\int_0^r L(u)du}{1-\int_0^r L_0(u)du} r = 0. \tag{6.5.40}$$

Moreover R_1 satisfies

$$\int_0^{R_1} L_0(u)du < 1. \tag{6.5.41}$$

Indeed if $s_{n+1} \leq R_1$, then

$$\begin{aligned} s_{n+2} &\leq s_{n+1} + \frac{\int_{s_n}^{s_{n+1}} L(u)du}{1-\int_0^{R_1} L_0(u)du}(s_{n+1}-s_n) \\ &\leq \cdots \leq \beta + \frac{\int_0^{s_1} L(u)du + \int_{s_1}^{s_2} L(u)du + \cdots + \int_{s_n}^{s_{n+1}} L(u)du}{1-\int_0^{R_1} L_0(u)du} s_{n+1} \\ &\leq \beta + \frac{\int_0^{s_{n+1}} L(u)du}{1-\int_0^{R_1} L_0(u)du} s_{n+1} \\ &\leq \beta + \frac{\int_0^{R_1} L(u)du}{1-\int_0^{R_1} L_0(u)du} R_1 = R_1. \end{aligned}$$

Hence, $\{s_n\}$ is monotonically increasing (by (6.5.41)) and bounded above by R_1, and as such it converges to some $s^* \subset (\beta, R_1]$.

(C) Other sufficient convergence conditions for sequence $\{s_n\}$ can be found in Section 5.1 but for more general functions L_0 and L.

Following the proof of Theorem 6.5.6 but using hypotheses (A) or (B) or (C) above we immediately obtain:

Theorem 6.5.7 *Under hypotheses (A) or (B) or (C) above the conclusions of Theorem 6.5.6 for method (4.1.3) hold in the ball* $\overline{U}(x_1, s^*-\beta)$ *and* $s^* \leq r_1$. *Note that* s^* *replaces* r_1 *in the left-hand side of estimate (6.5.30).*

Next we examine the uniqueness of solution x^* in the ball $\overline{U}(x_0, r_1)$ or $\overline{U}(x_0, s^*)$.

Remark 6.5.3 *Assume:*

$$r_1 L_0(r_1) < 1, \tag{6.5.42}$$

$$\int_0^1 L_0[\theta r_1 + (1-\theta)R][\theta r_1 + (1-\theta)R]d\theta < 1. \tag{6.5.43}$$

Then under the hypotheses of Theorem 6.5.6 the solution x^* is unique in $\overline{U}(x_0, r_1)$ if (6.5.42) holds and in $\overline{U}(x_0, R)$ for $R \geq r_1$ if (6.5.43) holds. Indeed, if y^* is a solution of equation $F(x) = 0$ in $\overline{U}(x_0, R)$, then by (6.5.5) we get for $L = \int_0^1 F'(y^* + \theta(x^* - y^*))d\theta$

$$\begin{aligned} &\|F'(x_0)^{-1}(L - F'(x_0))\| \\ &\quad \leq \int_0^1 \int_0^{\|y^*+\theta(x^*-y^*)-x_0\|} L_0(u)du d\theta \\ &\quad \leq \int_0^1 L_0[\theta\|x^* - x_0\| + (1-\theta)\|y^* - x_0\|] \\ &\qquad \cdot [\theta\|x^* - x_0\| + (1-\theta)\|y^* - x_0\|]d\theta < 1 \quad \text{by (6.5.42)}, \end{aligned} \tag{6.5.44}$$

if $y^* \in \overline{U}(x_0, r_1)$ or (6.5.43) if $y^* \in \overline{U}(x_0, R)$. It follows by the Banach Lemma on invertible operators and (6.5.44) that linear operator L is invertible. Using the identity

$$0 = F(x^*) - F(y^*) = L(x^* - y^*)$$

we obtain in both cases

$$x^* = y^*,$$

which establishes the uniqueness of the solution x^* in the corresponding balls $\overline{U}(x_0, r_1)$ and $\overline{U}(x_0, R)$. Replace r_1 by s^* in (6.5.42) and (6.5.43) to obtain uniqueness of x^* in $\overline{U}(x_0, s^*)$ or $\overline{U}(x_0, R)$ under the hypotheses of Theorem 6.5.7. The corresponding problem of uniqueness of the solution x^* was not examined in Theorem 1.5 in [340].

Remark 6.5.4 *Assume operator F is analytic and such that*

$$\|F'(x_0)^{-1}F^{(n)}(x_0)\| \leq n!\gamma^{n-1} \quad (n \geq 2). \tag{6.5.45}$$

Smale in [312] provided a semilocal convergence analysis for method (4.1.3) using (6.5.45). Later in [177] Smale's results were improved by introducing a majorizing function of the form

$$f(t) = \beta - t + \frac{\gamma t^2}{1 - \gamma t} \quad t \in [0, R], \tag{6.5.46}$$

(see also Section 6.4). Define function L by

$$L(u) = \frac{2\gamma}{(1-\gamma u)^3}. \tag{6.5.47}$$

Then conditions (6.5.3) and (6.5.4) are satisfied and function f defined in Remark 6.5.1 coincides with (6.5.46).

Concrete forms of Theorems 6.5.1 and 6.5.5 were given by Theorems 5.1 and 5.2 respectively in [340]. Here we show how to improve on Theorems 5.1 and 5.2 along the lines of our Theorems 6.5.1 and 6.5.5 above.

Indeed simply define functions f_0, L_0 corresponding to f, L by

$$f_0(t) = \beta - t + \frac{\gamma_0 t^2}{1 - \gamma_0 t} \quad t \in [0, R] \tag{6.5.48}$$

and

$$L_0(u) = \frac{2\gamma_0}{(1 - \gamma_0 u)^3}. \tag{6.5.49}$$

Again for this choice of L_0, function f_0 given by (6.5.10) coincides with (6.5.48). Moreover condition (6.5.5) is satisfied. Clearly

$$\gamma_0 \leq \gamma \tag{6.5.50}$$

holds in general. If strict inequality holds in (6.5.50) then our results enjoy the ones corresponding to (6.1.14) benefits (see Remark 6.5.1). In particular as

$$\alpha = \gamma_0 \beta \leq 3 - 2\sqrt{2} \tag{6.5.51}$$

the zeros of f_0 are

$$r_1^0 = \frac{1 + \alpha - \sqrt{(1+\alpha)^2 - 8\alpha}}{4\gamma_0}$$

$$r_2^0 = \frac{1 + \alpha + \sqrt{(1+\alpha)^2 - 8\alpha}}{4\gamma_0}$$

$$r_0^0 = \left(1 - \frac{1}{\sqrt{2}}\right)\frac{1}{\gamma_0}, \quad R_0 = \frac{1}{2\gamma_0} \quad \text{and } b_0 = (3 - 2\sqrt{2})\frac{1}{\gamma_0}.$$

Then the concretizations of e.g., Theorems 6.5.1 and 6.5.5 are given respectively by:

Theorem 6.5.8 *Let $\gamma_0 > 0$ be a given constant. Assume operator F satisfies*

$$\|F'(x_0)^{-1}[F'(x) - F'(x_0)]\| \leq \frac{1}{(1 - \gamma_0\|x - x_0\|)^2} - 1 \tag{6.5.52}$$

for all

$$x \in U\left[x_0, \left(1 - \frac{1}{\sqrt{2}}\right)\frac{1}{\gamma_0}\right] = U.$$

Then $F'(x_0)^{-1} \in L(Y, X)$, and is differentiable in the ball

$$U_0 = U\left[F(x_0), (3 - 2\sqrt{2})\frac{1}{\gamma_0\|F'(x_0)^{-1}\|}\right). \tag{6.5.53}$$

Moreover

$$U_0 \subset U \tag{6.5.54}$$

and the radius in U_0 is the best possible.

Theorem 6.5.9 *Let $\gamma_0 > 0$ be a given constant, and*

$$\alpha = \beta\gamma_0 \le 3 - 2\sqrt{2}.$$

Assume operator F satisfies

$$\|F'(x_0)^{-1}[F'(x) - F'(x_0)]\| \le \frac{1}{(1 - \gamma_0\|x - x_0\|)^2} - 1 \tag{6.5.55}$$

for all $x \in \overline{U}(x_0, r)$, where $r_1^0 \le r < r_2^0$ if $\alpha < 3 - 2\sqrt{2}$, or $r = r_1^0$ if $\alpha = 3 - 2\sqrt{2}$. Then equation $F(x) = 0$ has a unique solution x^* satisfying (6.5.19) in the ball $\overline{U}(x_0, r)$.

The rest of the results in [340] can be improved along the same lines. However we leave the details as exercises to the motivated reader.

Iterative methods of higher order using hypotheses on the second (or higher) Fréchet-derivative as well as results involving outer or generalized inverses can be found in the Exercises that follow.

6.6 Exercises

6.1 Let $F : S \subseteq X \to Y$ be a three times Fréchet-differentiable operator defined on an open convex domain S of Banach space X with values in a Banach space Y. Assume $F'(x_0)^{-1}$ exists for some $x_0 \in S$, $\|F'(x_0)^{-1}\| \le \beta$, $\|F'(x_0)^{-1}F(x_0)\| \le \eta$, $\|F''(x)\| \le M$, $\|F'''(x)\| \le N$, $\|F'''(x) - F'''(y)\| \le L\|x - y\|$ for all $x, y \in S$, and $\bar{U}(x_0, r\eta) \subseteq S$, where

$$A = M\beta\eta, \quad B = N\beta\eta^2, \quad C = L\beta\eta^3,$$

$$a_0 = 1 = c_0, \quad b_0 = \tfrac{2A}{3}, \quad d_0 = \tfrac{A}{2}(1 + A),$$

$$a_{n+1} = \tfrac{a_n}{1 - Aa_n(c_n + d_n)}, \quad b_{n+1} = \tfrac{2A}{3} a_{n+1} c_{n+1},$$

$$c_{n+1} = \tfrac{32}{2187} \frac{27\left[4 + \left(1 + \frac{3}{2}b_n\right)^2\right] A^3 a_n^2 + 18ABa_n + 17c}{b_n^4\left(1 + \frac{3}{2}b_n\right)^4} a_{n+1} d_n^4,$$

$$d_{n+1} = \tfrac{3}{4} b_{n+1}\left(1 + \tfrac{3}{2}b_{n+1}\right) c_{n+1} \quad (n \ge 0),$$

and $r = \lim_{n\to\infty} \sum_{i=0}^{n} (c_i + d_i)$. If $A \in \left[0, \frac{1}{2}\right]$, $B = \left[0, \frac{1}{18A}(P(A) - 17c)\right]$ and $c \in \left[0, \frac{P(A)}{17}\right]$, where $P(A) = 27(A - 1)(2A - 1)\left(A^2 + A + 2\right)\left(A^2 + 2A + 4\right)$.

Then show [186]: Chebyshev-Halley method given by

$$\begin{aligned} y_n &= x_n - F'(x_n)^{-1} F(x_n) \\ H_n &= F'(x_n)^{-1} \left[F'\left(x_n + \tfrac{2}{3}(y_n - x_n)\right) - F'(x_n) \right] \\ x_{n+1} &= y_n - \tfrac{3}{4} H_n \left[I - \tfrac{3}{2} H_n \right] (y_n - x_n) \end{aligned}$$

is well defined, remains in $U(x_0, r\eta)$ and converges to a solution $x^* \in \bar{U}(x_0, r\eta)$ of equation $F(x) = 0$. Moreover, the solution x^* is unique in $U\left(x_0, \frac{2}{M\beta} - r\eta\right)$. Furthermore, the following estimates hold for all $n \geq 0$

$$\|x_n - x^*\| \leq \sum_{i=n}^{\infty} (c_i + d_i)\,\eta \leq \tfrac{3}{2A}\left[1 + \tfrac{A}{2}(1 + A)\right] \tfrac{b_1}{\gamma^{1/3}} \sum_{i=1}^{\infty} \gamma^{4^{i-1}/3},$$

where $\gamma = \frac{b_2}{b_1}$.

6.2. Consider the scalar equation [185]

$$f(x) = 0.$$

Using the degree of logarithmic convexity of f

$$Lf(x) = \frac{f(x)\, f''(x)}{f'(x)^2},$$

the convex acceleration of Newton's method is given by

$$x_{n+1} = F(x_n) = x_n - \frac{f(x_n)}{f'(x_n)} \left[1 + \frac{Lf(x_n)}{2(1 - Lf(x_n))} \right] \quad (n \geq 0)$$

for some $x_0 \in \mathbb{R}$. Let $k \geq 1754877$, the interval $[a, b]$ satisfying $a + \frac{2k-1}{2(k-1)} \frac{f(b)}{f'(b)} \leq b$ and $x_0 \in [a, b]$ with $f(x_0) > 0$, and $x_0 \geq a + \frac{2k-1}{2(k-1)} \frac{f(b)}{f'(b)}$. If $|Lf(x)| \leq \frac{1}{k}$ and $Lf'(x) \in \left[\frac{1}{k}, 2(k-1)^2 - \frac{1}{k}\right)$ in $[a, b]$, then show: Newton's method converges to a solution x^* of equation $f(x) = 0$ and $x_{2n} \geq x^*$, $x_{2n+1} \leq x^*$ for all $n \geq 0$.

6.3 Consider the midpoint method [68], [158]:

$$\begin{aligned} y_n &= x_n = \Gamma_n F(x_n), \quad \Gamma_n = F'(x_n)^{-1}, \\ z_n &= x_n + \tfrac{1}{2}(y_n - x_n), \\ x_{n+1} &= x_n - \bar{\Gamma}_n F(x_n), \quad \bar{\Gamma}_n = F'(z_n)^{-1} \quad (n \geq 0), \end{aligned}$$

for approximating a solution x^* of equation $F(x) = 0$. Let $F : \Omega \subseteq X \to Y$ be a twice Fréchet-differentiable operator defined on an open convex subset of a Banach space X with values in a Banach space Y. Assume:

(1) $\Gamma_0 \in L(Y, X)$ for some $x_0 \in \Omega$ and $\|\Gamma_0\| \le \beta$;
(2) $\|\Gamma_0 F(x_0)\| \le \eta$;
(3) $\|F''(x)\| \le M \quad (x \in \Omega)$;
(4) $\|F''(x) - F''(y)\| \le K\|x - y\| \quad (x, y \in \Omega)$.

Denote $a_0 = M\beta\eta$, $b_0 = K\beta\eta^2$. Define sequence $a_{n+1} = a_n f(a_n)^2 g(a_n, b_n)$, $b_{n+1} = b_n f(a_n)^3 g(a_n, b_n)^2$, $f(x) = \frac{2-x}{2-3x}$, and $g(x, y) = \frac{x^2}{(2-x)^2} + \frac{7y}{24}$. If $0 < a_0 < \frac{1}{2}$, $b_0 < h(a_0)$, where

$$h(x) = \tfrac{96(1-x)(1-2x)}{7(2-x)^2}, \quad \bar{U}(x_0, R\eta) \subseteq \Omega,$$
$$R = \tfrac{2}{2-a_0}\tfrac{1}{1-\Delta}, \quad \Delta = f(a_0)^{-1},$$

then show: midpoint method $\{x_n\}$ $(n \ge 0)$ is well defined, remains in $\bar{U}(x_0, R\eta)$ and converges at least R-cubically to a solution x^* of equation $F(x) = 0$. The solution x^* is unique in $U(x_0, \frac{2}{M\beta} - R\eta) \cap \Omega$ and

$$\|x_{n+1} - x^*\| \le \tfrac{2}{2-a_0}\gamma^{\frac{3^n-1}{2}}\tfrac{\Delta^n}{1-\Delta}\eta.$$

6.4 Consider the multipoint method in the form [68], [158]:

$$\begin{aligned} y_n &= x_n - \Gamma_n F(x_n), \quad \Gamma_n = F'(x_n)^{-1}, \\ z_n &= x_n + \theta(y_n - x_n), \\ H_n &= \tfrac{1}{\theta}\Gamma_n[F'(x_n) - F'(z_n)], \quad \theta \in (0, 1], \\ x_{n+1} &= y_n + \tfrac{1}{2}H_n(y_n - x_n) \quad (n \ge 0), \end{aligned}$$

for approximating a solution x^* of equation $F(x) = 0$. Let F be a twice-Fréchet-differentiable operator defined on some open convex subset Ω of a Banach space X with values in a Banach space Y. Assume:

(1) $\Gamma_0 \in L(Y, X)$, for some $x_0 \in X$ and $\|\Gamma_0\| \le \beta$;
(2) $\|\Gamma_0 F(x_0)\| \le \eta$;
(3) $\|F''(x)\| \le M, \quad (x \in \Omega)$;
(4) $\|F''(x) - F''(y)\| \le K\|x - y\|^p, \quad (x, y) \in \Omega,\ K \ge 0,\ p \in [0, 1]$.

Denote $a_0 = M\beta\eta$, $b_0 = K\beta\eta^{1+p}$ and define sequence

$$\begin{aligned} a_{n+1} &= a_n f(a_n)^2 g_\theta(a_n, b_n), \\ b_{n+1} &= b_n f(a_n)^{2+p} g_\theta(a_n, b_n)^{1+p}, \\ f(x) &= \tfrac{2}{2-2x-x^2} \quad \text{and} \\ g_\theta(x, y) &= \tfrac{x^3+4x^2}{8} + \tfrac{[2+(p+2)\theta^p]y}{2(p+1)(p+2)}. \end{aligned}$$

Suppose $a_0 \in \left(0, \frac{1}{2}\right)$ and $b_0 < h_p(a_0, \theta)$, where

$$h_p(x, \theta) = \tfrac{(p+1)(p+2)}{4[2+(p+2)\theta^p]} (1 - 2x)\left(8 - 4x^2 - x^3\right).$$

Then, if $\bar{U}(x_0, R\eta) \subseteq \Omega$, where $R = \left(1 + \frac{a_0}{2}\right) \frac{1}{1-\gamma\Delta}$, $\Delta = f(a_0)^{-1}$ show: iteration $\{x_n\}$ $(n \geq 0)$ is well defined, remains in $\bar{U}(x_0, R\eta)$ for all $n \geq 0$ and converges with R-order at least $2 + p$ to a solution x^* of equation $F(x) = 0$. The solution x^* is unique in $U(x_0, \frac{2}{M\beta} - R\eta) \cap \Omega$. Moreover, the following estimates hold for all $n \geq 0$

$$\|x_n - x^*\| \leq \left[1 + \tfrac{a_0}{2} \gamma^{\left(\frac{(2+p)^n - 1}{1+p}\right)}\right] \gamma^{\left(\frac{(2+p)^n - 1}{1+p}\right)} \frac{\Delta^n}{1 - \gamma^{(2+p)^n} \Delta} \eta,$$

where $\gamma = \frac{a_1}{a_0}$.

6.5. Consider the multipoint iteration [191]:

$$\begin{aligned}
y_n &= x_n - \Gamma_n F(x_n), \quad \Gamma_n = F'(x_n)^{-1}, \\
z_n &= x_n - \tfrac{2}{3} \Gamma_n F(x_n), \\
H_n &= \Gamma_n \left[F'(z_n) - F'(x_n)\right], \\
x_{n+1} &= y_n - \tfrac{3}{4} \left[I + \tfrac{3}{2} H_n\right]^{-1} H_n (y_n - x_n) \quad (n \geq 0),
\end{aligned}$$

for approximation equation $F(x) = 0$. Let $F : \Omega \subseteq X \to Y$ be a three times Fréchet-differentiable operator defined on some convex subset Ω of a Banach space X with values in a Banach space Y. Assume $F'(x_0)^{-1} \in L(Y, X)$ $(x_0 \in \Omega)$, $\|\Gamma_0\| \leq \alpha$, $\|\Gamma_0 F(x_0)\| \leq \beta$, $\|F''(x)\| \leq M$, $\|F'''(x)\| \leq N$, and $\|F'''(x) - F'''(y)\| \leq k \|x - y\|$ for all $x, y \in \Omega$. Denote $\theta = M\alpha\beta$, $w = N\alpha\beta^2$ and $\delta = K\alpha\beta^3$. Define sequences

$$\begin{aligned}
a_0 &= c_0 = 1, \quad b_0 = \tfrac{2}{3}\theta, \quad d_0 = \tfrac{2-\theta}{2(1-\theta)}, \\
a_{n+1} &= \tfrac{a_n}{1 - \theta a_n d_n}, \quad b_{n+1} = \tfrac{2}{3} \theta a_{n+1} c_{n+1}, \\
c_{n+1} &= \tfrac{8(2-3b_n)^4}{(4-3b_n)^4} \left[\tfrac{a_n^2}{(2-3b_n)^2} \theta^3 + \tfrac{17}{108} \delta + \tfrac{w a_n}{3(2-3b_n)} \theta\right] a_{n+1} d_n^4
\end{aligned}$$

and

$$d_{n+1} = \tfrac{4-3b_{n+1}}{4-6b_{n+1}} c_{n+1} \quad (n \geq 0).$$

Moreover, assume: $\bar{U}(x_0, R\beta) \subseteq \Omega$, where

$$\begin{aligned}
R &= \lim_{n \to \infty} \sum_{i=0}^{n} d_i, \quad \theta \in \left(0, \tfrac{1}{2}\right), \\
0 &\leq \delta < \frac{27(2\theta - 1)\left(\theta^3 - 8\theta^2 + 16\theta - 8\right)}{17(1-\theta)^2}, \\
0 &\leq \omega < \frac{3(2\theta - 1)\left(\theta^3 - 8\theta^2 + 16\theta - 8\right)}{4\theta(1-\theta)^2} - \frac{17\delta}{36\theta}.
\end{aligned}$$

Then show: iteration $\{x_n\}$ $(n \geq 0)$ is well defined, remains in $U(x_0, R\beta)$ for all $n \geq 0$ and converges to a solution x^* of equation $F(x) = 0$. Furthermore, the solution x^* is unique in $U\left(x_0, \frac{2}{\alpha M} - R\beta\right)$ and for all $n \geq 0$

$$\|x_n - x^*\| \leq \sum_{i \geq n} d_i \beta \leq \beta \tfrac{(2-\theta)}{2(1-\theta)} \tfrac{1}{\gamma^{1/3}} \sum_{j \geq n} \gamma^{4^{i+1}/3},$$

where $\gamma = \frac{b_1}{b_0}$.

6.6. Consider the multipoint iteration [157]:

$$\begin{aligned} y_n &= x_n - F'(x_n)^{-1} F(x_n) \\ G_n &= \left[F'(x_n + p(y_n - x_n)) - F'(x_n)\right](y_n - x_n), \quad p \in (0,1], \\ x_{n+1} = y_n &= \tfrac{1}{2p} F'(y_n)^{-1} G_n \quad (n \geq 0) \end{aligned}$$

for approximating a solution x^* of equation $F(x) = 0$. Let $F : \Omega \subseteq X \to Y$ be a continuously Fréchet-differentiable operator in an open convex domain Ω which is a subset of a Banach space X with values in a Banach space Y. Let $x_0 \in \Omega$ such that $\Gamma_0 = F'(x_0)^{-1} \in L(Y, X)$; $\|\Gamma_0\| \leq \beta$, $\|y_0 - x_0\| \leq \eta$, $p = \frac{2}{3}$, and $\|F'(x) - F'(y)\| \leq K\|x - y\|$ for all $x, y \in \Omega$. For $b_0 = K\beta\eta$, define $b_n = b_{n-1} f(b_{n-1})^2 g(b_{n-1})$, where

$$f(x) = \tfrac{2(1-x)}{x^2 - 4x + 2} \quad \text{and} \quad g(x) = \tfrac{x\left(x^2 - 8x + 8\right)}{8(1-x)^2}.$$

If $b_0 < r = .2922...$, where r is the smallest positive root of the polynomial $q(x) = 2x^4 - 17x^3 + 48x^2 - 40x + 8$, and $\bar{U}(x_0, \frac{1}{K\beta}) \subseteq \Omega$, then show: iteration $\{x_n\}$ $(n \geq 0)$ is well defined, remains in $\bar{U}(x_0, \frac{1}{K\beta})$ and converges to a solution x^* of equation $F(x) = 0$, which is unique in $U(x_0, \frac{1}{K\beta})$.

6.7. Consider the biparametric family of multipoint iterations [159]:

$$\begin{aligned} y_n &= x_n - \Gamma_n F(x_n), \quad z_n = x_n + p(y_n - x_n), \quad p \in [0,1], \\ H_n &= \tfrac{1}{p} \Gamma_n \left[F'(z_n) - F'(x_n)\right], \\ x_{n+1} &= y_n - \tfrac{1}{2} H_n (I + \alpha H_n)(y_n - x_n), \quad (n \geq 0) \end{aligned}$$

where $\Gamma_n = F'(x_n)^{-1}$ $(n \geq 0)$ and $\alpha = -2\beta \in \mathbb{R}$. Assume $\Gamma_0 = F'(x_0)^{-1} \in L(Y, X)$ exists at some $x_0 \in \Omega_0 \subseteq X$, $F : \Omega_0 \subseteq X \to Y$ twice Fréchet-differentiable, X, Y Banach spaces, $\|\Gamma_0\| \leq \beta$, $\|\Gamma_0 F(x_0)\| \leq \eta$, $\|F''(x)\| \leq M$, $x \in \Omega_0$ and $\|F''(x) - F''(y)\| \leq K\|x - y\|$ for all $x, y \in \Omega_0$. Denote $a_0 = M\beta\eta$, $b_0 = k\beta\eta^2$. Define sequences

$$a_{n+1} = a_n f(a_n)^2 g(a_n, b_n), \quad b_{n+1} = b_n f(a_n)^3 g(a_n, b_n)^2,$$

where

$$f(x) = 2\left[2 - 2x - x^2 - |\alpha|\, x^3\right]^{-1},$$

and

$$g(x, y) = \tfrac{|\alpha|^2}{8} x^5 + \tfrac{|\alpha|}{4} x^4 + \tfrac{1+4|\alpha|}{8} x^3 + \tfrac{|1+\alpha|}{2} x^2 + \tfrac{1-p}{4} xy + \tfrac{2+3p}{12} y$$

for some real parameters α and p. Assume:

$$a_0 \in \left(0, \tfrac{1}{2}\right), \quad b_0 < P = \frac{3\left(8 - 16a_0 - 4a_0^2 + 7a_0^3 + 2a_0^4\right)}{3a_0 + 2},$$

$|\alpha| < \min\{6, r\}$, $p \in (0, 1]$ and $p < h(|\alpha|)$, where r is a positive root of

$$h(x) = \tfrac{1}{6b_0(1-a_0)}\Big[\left(24 - 48a_0 - 12a_0^2 + 21a_0^3 + 6a_0^4 - 2b_0(3a_0 + 2)\right) + 6a_0^2\left(2a_0^3 + 3a_0^2 - 6a_0 - 2\right)x + 3a_0^5(2a_0 - 1)x^2\Big],$$

$$\bar{U}(x_0, R\eta) \subseteq \Omega_0, \; R = \left[1 + \tfrac{a_0}{2}(1 + |\alpha|\, a_0)\right]\tfrac{1}{1-\gamma\Delta}, \; \gamma = \tfrac{a_1}{a_0}, \; \Delta = f(a_0)^{-1}.$$

Then show: iteration $\{x_n\}$ $(n \geq 0)$ is well defined, remains in $\bar{U}(x_0, R\eta)$ for all $n \geq 0$ and converges to a unique solution x^* of equation $F(x) = 0$ in $U(x_0, \frac{2}{M\beta} - R\eta) \cap \Omega_0$. The following estimates hold for all $n \geq 0$:

$$\|x_n - x^*\| \leq \left[1 + \tfrac{1}{2}\gamma^{\frac{3^n-1}{2}} a_0\left(1 + |\alpha|\,\gamma^{\frac{3^n-1}{2}} a_0\right)\right]\gamma^{\frac{3^n-1}{2}} \tfrac{\Delta^n}{1-\gamma^{3^n}\Delta}\eta.$$

6.8. Let f be a real function, x^* a simple root of f and G a function satisfying $G(0) = 1$, $G'(0) = \frac{1}{2}$ and $|G''(0)| < +\infty$. Then show: iteration

$$x_{n+1} = x_n - G(Lf(x_n))\frac{f(x_n)}{f'(x_n)}, \quad (n \geq 0)$$

where

$$Lf(x) = \frac{f(x)\, f''(x)}{f'(x)^2}$$

is of third order for an appropriate choice of x_0. This result is due to Gander [167]. Note that function G can be chosen

$G(x) = 1 + \frac{x}{2}$ (Chebyshev method);
$G(x) = 1 + \frac{x}{2-x}$ ();
$G(x) = 1 + \frac{x}{2(1-x)}$ (Super-Halley method).

6.9. Consider the super-Halley method [68] for all $n \geq 0$ in the form:

$$F(x_n) + F'(x_n)(y_n - x_n) = 0,$$
$$3F(x_n) + 3F'(x_n)\left[x_n + \tfrac{2}{3}(y_n - x_n)\right](y_n - x_n) + 4F'(y_n)(x_{n+1} - y_n) = 0,$$

for approximating a solution x^* of equation $F(x) = 0$. Let $F : \Omega \subseteq X \to Y$ be a three-times Fréchet-differentiable operator defined on an open convex subset Ω of a Banach space X with values in a Banach space Y. Assume:

(1) $\Gamma_0 = F'(x_0)^{-1} \in L(Y, X)$ for some $x_0 \in \Omega$ with $\|\Gamma_0\| \leq \beta$;
(2) $\|\Gamma_0 F(x_0)\| \leq \eta$;
(3) $\|F''(x)\| \leq M \; (x \in \Omega)$;
(4) $\|F'''(x) - F'''(y)\| \leq L\|x - y\| \; (x, y \in \Omega), \; (L \geq 0)$.

Denote by $a_0 = M\beta\eta$, $c_0 = L\beta\eta^3$, and define sequences

$$a_{n+1} = a_n f(a_n)^2 g(a_n, c_n),$$
$$c_{n+1} = c_n f(a_n)^4 g(a_n, c_n)^3,$$

where

$$f(x) = \tfrac{(1-x)}{x^2-4x+2} \quad \text{and} \quad g(x, y) = \tfrac{1}{8}\left[\tfrac{x^3}{(1-x)^2} + \tfrac{17}{27}y\right].$$

Suppose: $a_0 \in \left(o, \frac{1}{2}\right)$, $c_0 < h(a_0)$, where

$$h(x) = \frac{27(2x-1)(x-1)\left(x-3+\sqrt{5}\right)\left(x-3-\sqrt{5}\right)}{17(1-x)^2},$$
$$\bar{U}(x_0, R\eta) \subseteq \Omega, \quad R = \left[1 + \tfrac{a_0}{2(1-a_0)}\right]\tfrac{1}{1-\Delta},$$

and $\Delta = f(a_0)^{-1}$. then show: iteration $\{x_n\}$ $(n \geq 0)$ is well defined, remains in $\bar{U}(x_0, R\eta)$ for all $n \geq 0$ and converges to a solution x^* of equation $F(x) = 0$. the solution x^* is unique in $U(x_0, \frac{2}{M\beta} - R\eta) \cap \Omega$ and

$$\|x_n - x^*\| \leq \left[1 + \frac{a_0\gamma^{\frac{4^n-1}{3}}}{2(1-a_0)}\right]\gamma^{\frac{4^n-1}{3}}\frac{\Delta^n}{1-\gamma^{4^n}\Delta}\eta \quad (n \geq 0),$$

where $\gamma = \frac{a_1}{a_0}$.

6.10. Consider the multipoint iteration method [161]:

$$y_n = x_n - F'(x_n)^{-1} F(x_n)$$
$$G_n = F'(x_n)^{-1}\left[F'\left(x_n + \tfrac{2}{3}(y_n - x_n)\right) - F'(x_n)\right],$$
$$x_{n+1} = y_n - \tfrac{3}{4}G_n\left[I - \tfrac{3}{2}G_n\right](y_n - x_n) \quad (n \geq 0),$$

for approximating a solution x^* of equation $F(x) = 0$. Let $F : \Omega \subseteq X \to Y$ be a three times Fréchet-differentiable operator defined on an open convex subset Ω of a Banach space X with values in a Banach space Y. Assume:

(1) $\Gamma_0 = F'(x_0)^{-1} \in L(Y, X)$ exists for some $x_0 \in \Omega$ and $\|\Gamma_0\| \leq \beta$;
(2) $\|\Gamma_0 F(x_0)\| \leq \eta$;
(3) $\|F''(x)\| \leq M \, (x \in \Omega)$;
(4) $\|F'''(x)\| \leq N \, (x \in \Omega)$;
(5) $\|F'''(x) - F'''(y)\| \leq L \|x - y\| \;\; (x, y \in \Omega), \; (L \geq 0)$.

Denote by $a_0 = M\beta\eta$, $b_0 = N\beta\eta^2$ and $c_0 = L\beta\eta^3$. Define the sequence

$$\begin{aligned} a_{n+1} &= a_n f(a_n)^2 g(a_n, b_n, c_n), \\ b_{n+1} &= b_n f(a_n)^3 g(a_n, b_n, c_n)^2, \\ c_{n+1} &= c_n f(a_n)^4 g(a_n, b_n, c_n)^3, \end{aligned}$$

where

$$f(x) = \frac{2}{2-2x-x^2-x^3},$$

and

$$g(x, y, z) = \frac{1}{216}\left[27x^3\left(x^2+2x+5\right)+18xy+17z\right].$$

If $a_0 \in \left(0, \frac{1}{2}\right)$, $17c_0 + 18a_0b_0 < p(a_0)$, where

$$p(x) = 27(1-x)(1-2x)\left(x^2+x+2\right)\left(x^2+2x+4\right),$$
$$\bar{U}(x_0, R\eta) \subseteq \Omega, \quad R = \left[1 + \tfrac{a_0}{2}(1+a_0)\right]\frac{1}{1-\gamma\Delta}, \quad \gamma = \frac{a_1}{a_0}, \quad \Delta = f(a_0)^{-1},$$

then show: iteration $\{x_n\}$ $(n \geq 0)$ is well defined, remains in $\bar{U}(x_0, R\eta)$ for all $n \geq 0$ and converges to a solution x^* of equation $F(x) = 0$, which is unique in $U(x_0, \frac{2}{M\beta} - R\eta) \cap \Omega$. Moreover the following estimates hold for all $n \geq 0$:

$$\|x_n - x^*\| \leq \left[1 + \tfrac{a_0}{2}\gamma^{\frac{4^n-1}{3}}\left(1 + a_0\gamma^{\frac{4^n-1}{3}}\right)\right]\gamma^{\frac{4^n-1}{3}}\frac{\Delta^n}{1-\gamma^{4^n}\Delta}\eta.$$

6.11. Consider the Halley method [68], [131]

$$x_{n+1} = x_n - \left[I - L_F(x_n)\right]^{-1} F'(x_n)^{-1} F(x_n) \quad (n \geq 0)$$

where

$$L_F(x) = F'(x)^{-1} F''(x) F'(x)^{-1} F(x),$$

for approximating a solution x^* of equation $F(x) = 0$. Let $F : \Omega \subseteq X \rightarrow$ be a twice Fréchet-differentiable operator defined on an open convex subset of a Banach space X with values in a Banach space Y. Assume:

(1) $F'(x_0)^{-1} \in L(Y, X)$ exists for some $x_0 \in \Omega$;

(2) $\|F'(x_0)^{-1} F(x_0)\| \leq \beta$;

(3) $\|F'(x_0)^{-1} F''(x_0)\| \leq \gamma$;

(4) $\|F'(x_0)^{-1} [F''(x) - F''(y)]\| \leq M \|x - y\| \quad (x, y \in \Omega)$.

If

$$\beta \leq \frac{2[2\sqrt{\gamma^2+2M}+\gamma]}{3[\sqrt{\gamma^2+2M}+\gamma]^2}, \quad \bar{U}(x_0, r_1) \subseteq \Omega,$$

$(r_1 \leq r_2)$ where r_1, r_2 are the positive roots of $h(t) = \beta - t + \frac{\gamma}{2}t^2 + \frac{M}{6}t^3$, then show: iteration $\{x_n\}$ $(n \geq 0)$ is well defined, remains in $\bar{U}(x_0, r_1)$ for all $n \geq 0$ and converges to a unique solution x^* of equation $F(x) = 0$ in $\bar{U}(x_0, r_1)$. Moreover, the following estimates hold for all $n \geq 0$:

$$\|x^* - x_{n+1}\| \leq (r_1 - t_{n+1}) \left(\frac{\|x^*-x_n\|}{r_1-t_n}\right)^3,$$

$$\frac{(\lambda_2\theta)^{3^n}}{\lambda_2-(\lambda_2\theta)^{3^n}} (r_2 - r_1) \leq r_1 - t_n \leq \frac{\lambda_1\theta}{\lambda_1-(\lambda_1\theta)^{3^n}} (r_2 - r_1),$$

$$\theta = \frac{r_1}{r_2}, \quad \lambda_1 = \sqrt{\frac{(r_0-r_2)^2+r_0r_2}{(r_0-r_1)^2+r_0r_1}} \leq 1, \quad \lambda_2 = \frac{\sqrt{3}}{2},$$

$-r_0$ is the negative root of h, and $t_{n+1} = H(t_n)$, where

$$H(t) = t - \frac{h(t)/h'(t)}{1 - \frac{1}{2}L_h(t)}, \qquad L_h(t) = \frac{h(t)/h''(t)}{h'(t)^2}.$$

6.12. Consider the iteration [68], [158]

$$x_{n+1} = x_n - [I + T(x_n)]\,\Gamma_n F(x_n) \quad (n \geq 0),$$

where $\Gamma_n = F'(x_n)^{-1}$ and $T(x_n) = \frac{1}{2}\Gamma_n A \Gamma_n F(x_n)$ $(n \geq 0)$, for approximating a solution x^* of equation $F(x) = 0$. Here $A : X \times X \to Y$ is a bilinear operator with $\|A\| = \alpha$, and $F : \Omega \subseteq X \to Y$ is a Fréchet-differentiable operator defined on an open convex subset Ω of a Banach space X with values in a Banach space Y. Assume:

(1) $F'(x_0)^{-1} = \Gamma_0 \in L(Y, X)$ exists for some $x_0 \in \Omega$ with $\|\Gamma_0\| \leq \beta$;

(2) $\|\Gamma_0 F(x_0)\| \leq \eta$;

(3) $\|F'(x) - F'(y)\| \leq k \|x - y\| \, (x, y \in \Omega)$.

Let a, b be real numbers satisfying $a \in \left[0, \frac{1}{2}\right), b \in (0, \sigma)$, where

$$\sigma = \frac{2\left[2a^2-3a-1+\sqrt{1+8a-4a^2}\right]}{a(1-2a)}.$$

Set $a_0 = 1$, $c_0 = 1$, $b_0 = \frac{b}{2}$ and $d_0 = 1 + \frac{b}{2}$. Define sequence

$$a_{n+1} = \frac{a_n}{1 - a a_n d_n}, \qquad c_{n+1} = \frac{a_{n+1}}{2}\left[a + \frac{b}{(1+b_n)^2}\right] d_n^2,$$
$$b_{n+1} = \frac{b}{2} a_{n+1} c_{n+1}, \qquad d_{n+1} = (1 + b_{n+1})\, c_{n+1}$$

and $r_{n+1} = \sum_{k=0}^{n+1} d_k$ $(n \geq 0)$. If $a = \kappa\beta\eta \in \left[0, \frac{1}{2}\right)$, $\bar{U}(x_0, r\eta) \subseteq \Omega$, $r = \lim_{n\to\infty} r_n$, $\alpha \in \left[0, \frac{\sigma}{\beta\eta}\right)$, then show: iteration $\{x_n\}$ $(n \geq 0)$ is well defined, remains in $\bar{U}(x_0, r\eta)$ and converges to a solution x^* of equation $F(x) = 0$, which is unique in $U(x_0, \frac{2}{\kappa\beta} - r\eta)$. Moreover the following estimates hold for all $n \geq 0$

$$\|x_{n+1} - x_n\| \leq d_n \eta$$

and

$$\|x^* - x_{n+1}\| \leq (r - r_n)\,\eta = \sum_{k=n+1}^{\infty} d_k \eta.$$

6.13. Consider the Halley-method [68], [175] in the form:

$$y_n = x_n - \Gamma_n F(x_n), \qquad \Gamma_n = F'(x_n)^{-1},$$
$$x_{n+1} = y_n + \tfrac{1}{2} L_F(x_n)\, H_n (y_n - x_n) \qquad (n \geq 0),$$
$$L_F(x_n) = \Gamma_n F''(x_n)\, \Gamma_n F(x_n), \qquad H_n = [I - L_F(x_n)]^{-1} \qquad (n \geq 0),$$

for approximating a solution x^* of equation $F(x) = 0$. Let $F : \Omega \subseteq X \to Y$ be a twice Fréchet-differentiable operator defined on an open convex subset Ω of a Banach space X with values in a Banach space Y. Assume:

(1) $\Gamma_0 \in L(Y, X)$ exists for some $x_0 \in \Omega$ with $\|\Gamma_0\| \leq \beta$;
(2) $\|F''(x)\| \leq M$ $(x \in \Omega)$;
(3) $\|F''(x) - F''(y)\| \leq N \|x - y\|$ $(x, y \in \Omega)$;
(4) $\|\Gamma_0 F(x_0)\| \leq \eta$;
(5) the polynomial $p(t) = \frac{k}{2}t^2 - \frac{1}{\beta}t + \frac{\eta}{\beta}$, where $M^2 + \frac{N}{2\beta} \leq k^2$ has two positive roots r_1 and r_2 with $(r_1 \leq r_2)$.

Let sequences $\{s_n\}$, $\{t_n\}$, $(n \geq 0)$ be defined by

$$s_n = t_n - \frac{p(t_n)}{p'(t_n)}, \qquad t_{n+1} = s_n + \frac{1}{2}\frac{L_p(t_n)}{1 - L_p(t_n)}(s_n - t_n) \quad (n \geq 0).$$

If, $\bar{U}(x_0, r_1) \subseteq \Omega$, then show: iteration $\{x_n\}$ $(n \geq 0)$ is well defined, remains in $U(x_0, r_1)$ for all $n \geq 0$ and converges to a solution x^* of equation $F(x) = 0$.

Moreover if $r_1 < r_2$ the solution x^* is unique in $\bar{U}(x_0, r_2)$. Furthermore, the following estimates hold for all $(n \geq 0)$

$$\|x^* - x_n\| \leq r_1 - t_n = \frac{(r_2 - r_1)\,\theta^{4^n}}{1 - \theta^{4^n}}, \qquad \theta = \tfrac{r_1}{r_2}.$$

6.14. Consider the two-point method [68], [153]:

$$\begin{aligned} y_n &= x_n - F'(x_n)^{-1} F(x_n), \\ H_n &= \tfrac{1}{p} F'(x_n)^{-1} \left[F'(x_n + p(y_n - x_n)) - F'(x_n)\right], \qquad p \in (0,1], \\ x_{n+1} &= y_n - \tfrac{1}{2} H_n \left[I + H_n\right]^{-1} (y_n - x_n) \qquad (n \geq 0), \end{aligned}$$

for approximating a solution x^* of equation $F(x) = 0$. Let $F : \Omega \subseteq X \to Y$ be a twice-Fréchet-differentiable operator defined on an open convex subset Ω of a Banach space X with values in a Banach space Y. Assume:

(1) $\Gamma_0 = F'(x_0)^{-1} \in L(Y, X)$ exists for some $x_0 \in \Omega$ with $\|\Gamma_0\| \leq \beta$;

(2) $\|\Gamma_0 F(x_0)\| \leq \eta$;

(3) $\|F''(x)\| \leq M \;\; (x \in \Omega)$;

(4) $\|F''(x) - F''(y)\| \leq K \|x - y\|, \qquad x, y \in \Omega$.

Denote by $a_0 = M\beta\eta$, $b_0 = K\beta\eta^2$. Define sequences

$$\begin{aligned} a_{n+1} &= a_n f(a_n)^2 g_p(a_n, b_n), \\ b_{n+1} &= b_n f(a_n)^3 g_p(a_n, b_n)^2, \end{aligned}$$

where $f(x) = \frac{2(1-x)}{x^2-4x+2}$ and $g_p(x,y) = \frac{3x^3+2y(1-x)[(1-6p)x+(2+3p)]}{24(1-x)^2}$. If $a_0 \in \left(0, \frac{1}{2}\right)$, $b_0 < h_p(h_0)$, where

$$h_p(x) = \frac{3(2x-1)(x-2)\left(x-3+\sqrt{5}\right)\left(x-3-\sqrt{5}\right)}{2(1-x)[(1-6p)x+2+3p]}, \qquad \bar{U}(x_0, \tfrac{\eta}{a_0}) \subseteq \Omega$$

then show: iteration $\{x_n\}$ $(n \geq 0)$ is well defined, remains in $\bar{U}(x_0, \frac{\eta}{a_0})$ for all $n \geq 0$ and converges to a solution x^* of equation $F(x) = 0$, which is unique in $U(x_0, \frac{\eta}{a_0})$. Moreover, the following estimates hold for all $n \geq 0$:

$$\|x^* - x_n\| \leq \left[1 + \frac{a_0 \gamma^{\frac{3^n-1}{2}}}{2(1-a_0)}\right] \gamma^{\frac{3^n-1}{2}} \frac{\Delta^n}{1-\Delta}\eta,$$

where $\gamma = \frac{a_1}{a_0}$ and $\Delta = f(a_0)^{-1}$.

6.15. Consider the two-step method:

$$y_n = x_n - F'(x_n)^{-1} F(x_n)$$
$$x_{n+1} = y_n - F'(x_n)^{-1} F(y_n) \quad (n \geq 0);$$

for approximating a solution x^* of equation $F(x) = 0$. Let $F : \Omega \subseteq X \subseteq Y$ be a Fréchet-differentiable operator defined on an open convex subset Ω of a Banach space X with values in a Banach space Y. Assume:

(1) $\Gamma_0 = f'(X_0)^{-1} \in L(Y, X)$ for some $x_0 \in \Omega$, $\|\Gamma_0\| \leq \beta$;

(2) $\|\Gamma_0 F(x_0)\| \leq \eta$;

(3) $\|F'(x) - F'(y)\| \leq K \|x - y\| \ (x, y \in \Omega)$.

Denote $a_0 = k\beta\eta$ and define the sequence $a_{n+1} = f(a_n)^2 g(a_n) a_n \ (n \geq 0)$, where $f(x) = \frac{2}{2-2x-x^2}$ and $g(x) = x^2(x+4)/8$. If $a_0 \in \left(0, \frac{1}{2}\right)$, $\bar{U}(x_0, R\eta) \subseteq \Omega$, $R = \frac{1+\frac{a_0}{2}}{1-\gamma\Delta}$, $\gamma = \frac{a_1}{a_0}$ and $\Delta = f(a_0)^{-1}$, then show: iteration $\{x_n\} \ (n \geq 0)$ is well defined, remains in $\bar{U}(x_0, R\eta)$ for all $n \geq 0$ and converges to a solution x^* of equation $F(x) = 0$, which is unique in $U(x_0, \frac{2}{K\beta} - R\eta) \cap \Omega$. Moreover, the following estimates hold for all $n \geq 0$

$$\|x_n - x^*\| \leq \left[1 + \tfrac{a_0}{2}\gamma^{\frac{3^n-1}{2}}\right] \gamma^{\frac{3^n-1}{2}} \frac{\Delta^n}{1 - \gamma^{3^n}\Delta} \eta \ (n \geq 0).$$

6.16 (a) Assume there exist non-negative parameters K, M, L, ℓ, μ, η, $\delta \in [0, 1]$ such that:

$$L \leq K, \tag{1}$$
$$\ell + 2\mu < 1, \tag{2}$$

and

$$h_\delta \equiv \left(K + L\delta + \tfrac{4M}{2-\delta}\right)\eta + \delta\ell + 2\mu \leq \delta. \tag{3}$$

Show: iteration $\{t_n\} \ (n \geq 0)$ given by

$$t_0 = 0, \quad t_1 = \eta, \quad t_{n+2} = t_{n+1} + \tfrac{K(t_{n+1}-t_n)+2(Mt_n+\mu)}{2(1-\ell-Lt_{n+1})}(t_{n+1} - t_n) \quad (n \geq 0)$$

is nondecreasing, bounded above by t^{**}, and converges to some t^* such that

$$0 \leq t^* \leq \tfrac{2\eta}{2-\delta} \equiv t^{**}.$$

Moreover, the following estimates hold for all $n \geq 0$

$$0 \leq t_{n+2} - t_{n+1} \leq \tfrac{\delta}{2}(t_{n+1} - t_n) \leq \left(\tfrac{\delta}{2}\right)^{n+1} \eta.$$

(b) Let $F\colon D \subseteq X \to Y$ be a Fréchet-differentiable operator. Assume: there exist an approximation $A(x) \in L(X,Y)$ of $F'(x)$, an open convex subset D_0 of D, $x_0 \in D_0$, a bounded outer inverse $A^{\#}$ of $A(x_0)$, and parameters $\eta > 0$, $K > 0$, $M \geq 0$, $L \geq 0$, $\mu \geq 0$, $\ell \geq 0$ such that (1)–(3) hold

$$\|A^{\#}F(x_0)\| \leq \eta,$$
$$\|A^{\#}[F'(x) - F'(y)]\| \leq K\|x - y\|,$$
$$\|A^{\#}[F'(x) - A(x)]\| \leq M\|x - x_0\| + \mu,$$

and

$$\|A^{\#}[A(x) - A(x_0)]\| \leq L\|x - x_0\| + \ell$$

for all $x, y \in D_0$, *and*

$$\bar{U}(x_0, t^*) \subseteq D_0$$

Show: sequence $\{x_n\}$ $(n \geq 0)$ generated by Newton-like method with

$$A\,(x_n)^{\#} = [I + A^{\#}(A(x_n) - A(x_0))]^{-1}A^{\#}$$

is well defined, remains in $U(x_0, s^*)$ for all $n \geq 0$ and converges to a unique solution x^* of equation $A^{\#}F(x) = 0$, $\bar{U}\,(x_0, t^*) \cap D_0$

Moreover, the following estimates hold for all $n \geq 0$

$$\|x_{n+1} - x_n\| \leq t_{n+1} - t_n,$$

and

$$\|x_n - x^*\| \leq t^* - t_n.$$

(c) Assume:

- there exist an approximation $A(x) \in L(X,Y)$ of $F'(x)$, a simple solution $x^* \in D$ of equation (1), a bounded outer inverse $A_1^{\#}$ of $A(x^*)$ and non-negative parameters $\bar{K}$, $\bar{L}$, $\bar{M}$, $\bar{\mu}$, $\bar{\ell}$, such that:

$$\|A_1^{\#}[F'(x) - F'(y)]\| \leq \bar{K}\|x - y\|,$$
$$\|A_1^{\#}[F'(x) - A(x)]\| \leq \bar{M}\|x - x^*\| + \bar{\mu},$$

and

$$\|A_1^{\#}[A(x) - A(x^*)]\| \leq \bar{L}\|x - x^*\| + \bar{\ell}$$

for all $x, y \in D$;

- equation

$$\left(\frac{\bar{K}}{2}+\bar{M}+\bar{L}\right)r+\bar{\mu}+\bar{\ell}-1=0$$

has a minimal non-negative zero r^* satisfying

$$\bar{L}r+\bar{\ell}<1,$$

and

$$U(x^*,r^*)\subseteq D.$$

Show: sequence $\{x_n\}$ $(n\geq 0)$ generated by Newton-like method is well defined, remains in $U(x^*,r^*)$ for all $n\geq 0$ and converges to x^* provided that $x_0\in U(x^*,r^*)$. Moreover, the following estimates hold for all $n\geq 0$:

$$\begin{aligned}&\|x^*-x_{n+1}\|\leq\\&\leq\frac{1}{1-\bar{L}\|x^*-x_n\|-\bar{\ell}}\left[\frac{\bar{K}}{2}\|x^*-x_n\|+(\bar{M}\|x^*-x_n\|+\bar{\mu})\right]\|x^*-x_n\|\\&<\frac{(\frac{\bar{K}}{2}+\bar{M})r^*+\bar{\mu}}{1-\bar{L}r^*-\bar{\ell}}\|x^*-x_n\|.\end{aligned}$$

6.17 (a) Let $F:D\subseteq X\to Y$ be an m-times Fréchet-differentiable operator ($m\geq 2$ integer).
Assume:

(a_1) there exist an open convex subset D_0 of D, $x_0\in D_0$, a bounded outer inverse $F'(x_0)^{\#}$ of $F'(x_0)$, and constants $\eta>0$, $\alpha_i\geq 0$, $i=2,...,m+1$ such that for all $x,y\in D_0$ the following conditions hold:

$$\|F'(x_0)^{\#}(F^{(m)}(x)-F^{(m)}(x_0))\|\leq\varepsilon,\quad \varepsilon>0,\tag{6.6.1}$$
$$\forall x\in U(x_0,\delta_0)\text{ and some }\delta_0>0.$$

$$\|F'(x_0)^{\#}F(x_0)\|\leq\eta,$$
$$\|F'(x_0)^{\#}F^{(i)}(x_0)\|\leq\alpha_i$$

the positive zeros s of p' is such that

$$p(s)\leq 0,$$

where,

$$p(t)=\eta-t+\frac{\alpha_2 t^2}{2!}+\cdots+\frac{\alpha_m+\varepsilon}{m!}t^m.$$

Show: polynomial p has only two positive zeros denoted by t^*, t^{**} $(t^*\leq t^{**})$.

(a$_2$)

$$\bar{U}(x_0, \delta) \subseteq D_0, \quad \delta = \max\{\delta_0, t^*, t^{**}\}.$$

(a$_3$) $\delta_0 \in [t^*, t^{**}]$ or $\delta_0 > t^{**}$.

Moreover show: sequence $\{x_n\}$ $(n \geq 0)$ generated by Newton's method with $F'(x_n)^{\#} = [I + F'(x_0)^{\#}(F'(x_n) - F'(x_0))]^{-1} F'(x_0)^{\#}$ $(n \geq 0)$ is well defined, remains in $U(x_0, t^*)$ and converges to a solution $x^* \in \bar{U}(x_0, t^*)$ of equation $F'(x_0)^{\#} F(x) = 0$;

- the following estimates hold for all $n \geq 0$

$$\|x_{n+1} - x_n\| \leq t_{n+1} - t_n$$

and

$$\|x_n - x^*\| \leq t^* - t_n,$$

where $\{t_n\}$ $(n \geq 0)$ is a monotonically increasing sequence converging to t^* and generated by

$$t_0 = 0, \quad t_{n+1} = t_n - \tfrac{p(t_n)}{p'(t_n)}.$$

(b) Let $F \colon D \subseteq X \to Y$ be an m-times Fréchet-differentiable operator ($m \geq 2$ an integer). Assume:

(b$_1$) condition (1) holds;

(b$_2$) there exists an open convex subset D_0 of D, $x_0 \in D_0$, and constants $\alpha, \beta, \eta \geq 0$ such that for any $x \in D_0$ there exists an outer inverse $F'(x)^{\#}$ of $F'(x)$ satisfying $N(F'(x)^{\#}) = N(F'(x_0)^{\#})$ and

$$\|F'(x_0)^{\#} F(x_0)\| \leq \eta,$$

$$\left\| F'(y)^{\#} \int_0^1 F''[x + t(y - x)](1 - t)dt(y - x)^2 \right\| \leq$$
$$\leq \left[\tfrac{\alpha_m + \varepsilon}{m!} \|y - x\|^{m-2} + \cdots + \tfrac{\alpha_2}{2!} \right] \|y - x\|^2,$$

for all $x, y \in D_0$,

$$\left[\tfrac{\alpha_m + \varepsilon}{m!} \eta^{m-2} + \cdots + \tfrac{\alpha_2}{2!} \right] \eta < 1,$$

and

$$\bar{U}(x_0, r) \subseteq D_0 \quad \text{with } r = \min\left\{ \tfrac{\eta}{1 - r_0}, \delta_0 \right\},$$

where,

$$r_0 = \left[\tfrac{\alpha_m + \varepsilon}{m!} \eta^{m-2} + \cdots + \tfrac{\alpha_2}{2!} \right] \eta.$$

Show: sequence $\{x_n\}$ $(n \geq 0)$ generated by Newton's method is well defined, remains in $\bar{U}(x_0, r)$ for all $n \geq 0$ and converges to a solution x^* of $F'(x_0)^{\#}F(x) = 0$ with the iterates satisfying $N(F'(x_n)^{\#}) = N(F'(x_0)^{\#})$ $(n \geq 0)$. Moreover, the following estimates hold for all $n \geq 0$

$$\|x_{n+1} - x_n\| \leq r_0^n \|x_1 - x_0\|,$$
$$\|x^* - x_n\| \leq \tfrac{r_0^n}{1-r_0} \|x_1 - x_0\|,$$

and

$$\|x_n - x_0\| \leq \tfrac{1-r_0^n}{1-r_0} \|x_1 - x_0\| \leq \tfrac{1-r_0^n}{1-r_0} \eta \leq r.$$

6.18 Let X and Y be Banach spaces, and let L be a bounded linear operator on X into Y. A linear operator $M : Y \to X$ is said to be an inner inverse of L if $LMA = L$. A linear operator $M : Y \to X$ is an outer inverse of L if $MLM = M$. Let L be an $m \times n$ matrix, with $m > n$. Any outer inverse M of L will be an $n \times m$ matrix. Show:

(a) If $\text{rank}(L) = n$, then L can be written as

$$L = A \begin{bmatrix} I \\ 0 \end{bmatrix},$$

where I is the $n \times n$ identity matrix, and A is an $m \times m$ invertible matrix.The $n \times m$ matrix

$$M = \begin{bmatrix} I & B \end{bmatrix} A^{-1}$$

is an outer inverse of L for any $n \times (m - n)$ matrix B.

(b) If rank $(L) = r < n$, then L can be written as

$$L = A \begin{bmatrix} I & 0 \\ 0 & 0 \end{bmatrix} C,$$

where A is an $m \times m$ invertible matrix, I is the $r \times r$ identity matrix, and C is an $n \times n$ invertible matrix. If E is an outer (inner) inverse of the matrix $\begin{bmatrix} I & 0 \\ 0 & 0 \end{bmatrix}$, then the $n \times m$ matrix

$$M = C^{-1}EA^{-1}$$

is an outer (inner) inverse of L.

(c) E is both an inner and an outer inverse of $\begin{bmatrix} I & 0 \\ 0 & 0 \end{bmatrix}$ if and only if E can be written in the form

$$E = \begin{bmatrix} I & M \\ C & CM \end{bmatrix} = \begin{bmatrix} I \\ C \end{bmatrix} \begin{bmatrix} I & M \end{bmatrix}.$$

(d) For any $(n-r)\times r$ matrix T, the matrix $E=\begin{bmatrix} I & 0 \\ T & 0 \end{bmatrix}$ is an outer inverse of $\begin{bmatrix} I & 0 \\ 0 & 0 \end{bmatrix}$.

6.19 Let $F: D \subseteq X \to Y$ be a Fréchet-differentiable operator between two Banach space X and Y, $A(x) \in L(X,Y)$ $(x \in D)$ be an approximation to $F'(x)$. Assume that there exist an open convex subset D_0 of D, $x_0 \in D_0$, a bounded outer inverse $A^{\#}$ of $A (= A(x_0))$ and constants $\eta, k > 0, M, L, \mu, l \geq 0$ such that for all $x, y \in D_0$ the following conditions hold:

$$\|A^{\#}F(x_0)\| \leq \eta, \quad \|A^{\#}(F'(x) - F'(y))\| \leq k\|x-y\|,$$
$$\|A^{\#}(F'(x) - A(x))\| \leq M\|x - x_0\| + \mu,$$
$$\|A^{\#}(A(x) - A)\| \leq L\|x - x_0\| + l, \quad b := \mu + l < 1.$$

Assume $h = \sigma\eta \leq \frac{1}{2}(1-b)^2$, $\sigma := \max(k, M+L)$, and $\bar{U} = \bar{U}(x_0, t^*) \subseteq D_0, t^* = \frac{1-b-\sqrt{(1-b)^2-2h}}{\sigma}$. Then show

(i) sequence $\{x_n\}$ $(n \geq 0)$ generated by $x_{n+1} = x_n - A(x_n)^{\#} F(x_n)$ $(n \geq 0)$ with $A(x_n)^{\#} = \left[I + A^{\#}(A(x_n) - A)\right]^{-1} A^{\#}$ remains in U and converges to a solution $x^* \in \bar{U}$ of equation $A^{\#}F(x) = 0$.

(ii) equation $A^{\#}F(x) = 0$ has a unique solution in $\bar{U} \cap \left\{R\left(A^{\#}\right) + x_0\right\}$, where

$$\widetilde{U} = \begin{cases} (x_0, t^*) \cap D_0 & \text{if } h = \frac{1}{2}(1-b)^2 \\ U(x_0, t^{**}) \cap D_0 & \text{if } h < \frac{1}{2}(1-b)^2, \end{cases}$$

$$R(A^{\#}) + x_0 := \left\{x + x_0 : x \in R(A^{\#})\right\},$$

and

$$t^* = \frac{1-b+\sqrt{(1-b)^2-2h}}{\sigma}$$

(iii) $\|x_{n+1} - x_n\| \leq t_{n+1} - t_n$, $\|x^* - x_n\| \leq t^* - t_n$, where $t_0 = 0$, $t_{n+1} = t_n + \frac{f(t_n)}{g(t_n)}$, $f(t) = \frac{\sigma}{2}t^2 - (1-b)t + \eta$, and $g(t) = 1 - Lt - \ell$.

6.20 Let $F: D \subseteq X \to Y$ be a Fréchet-differentiable operator between two Banach spaces X and Y and let $A(x) \in L(X,Y)$ be an approximation of $F'(x)$. Assume that there exist an open convex subset D_0 of D, a point $x_0 \in D_0$ and constants $\eta, k > 0$ such that for any $x \in D_0$ there exists an outer inverse $A(x)^{\#}$ of $A(x)$ satisfying $N(A(x)^{\#}) = N(A^{\#})$, where $A = A(x_0)$ and $A^{\#}$ is a bounded outer inverse of A, and for this outer inverse the following conditions hold:

$$\|A^{\#} F(x_0)\| \leq \eta,$$
$$\|A(y)^{\#} (F'(x + t(y - x)) - F'(x))\| \leq kt \|x - y\|$$

for all $x, y \in D_0$ and $t \in [0,1]$, $h = \frac{1}{2}k\eta < 1$ and $\bar{U}(x_0, r) \subseteq D_0$ with $r = \frac{\eta}{1-h}$. Then show sequence $\{x_n\}$ $(n \geq 0)$ generated by $x_{n+1} = x_n - A(x_n)^{\#} F(x_n)$ $(n \geq 0)$ with $A(x_n)^{\#}$ satisfying $N(A(x_n)^{\#}) = N(A^{\#})$ remains in $\bar{U}(x_0, r)$ and converges to a solution x^* of equation $A^{\#} F(x) = 0$.

6.21 Show that Newton's method with outer inverses $x_{n+1} = x_n - F'(x_n)^{\#}$ $(n \geq 0)$ converges quadratically to a solution $x^* \in \widetilde{U} \cap \{R(F'(x_0)^{\#}) + x_0\}$ of equation $F'(x_0)^{\#} F(x) = 0$ under the conditions of Exercise 6.4 with $A(x) = F'(x)$ $(x \in D_0)$.

6.22 Let $F : D \subseteq X \to Y$ be Fréchet-differentiable and assume that $F'(x)$ satisfies a Lipschitz condition

$$\|F'(x) - F'(y)\| \leq L \|x - y\|, \quad x, y \in D.$$

Assume $x^* \in D$ exists with $F(x^*) = 0$. Let $a > 0$ such that $U\left(x^*, \frac{1}{a}\right) \subseteq D$. Suppose there is an

$$F'(x^*)^{\#} \in \Omega(F'(x^*)) = \{B \in L(Y, X) : BF'(x^*)B = B, \ B \neq 0\}$$

such that $\|F'(x^*)^{\#}\| \leq a$ and for any $x \in U\left(x^*, \frac{1}{3La}\right)$, the set $\Omega(F'(x))$ contains an element of minimum norm. Then show there exists a ball $U(x^*, r) \subseteq D$ with $cr < \frac{1}{3aL}$ such that for any $x_0 \in U(x^*, r)$ the sequence $\{x_n\}$ $(n \geq 0)$ $x_{n+1} = x_n - F'(x_n)^{\#} F(x_n)$ $(n \geq 0)$ with

$$F'(x_0)^{\#} \in \operatorname{argmin}\{\|B\| \,|B \in \Omega(F'(x_0))\}$$

and with $F'(x_n)^{\#} = (I + F'(x_0)^{\#}(F'(x_n) - F'(x_0)))^{-1} F'(x_0)^{\#}$ converges quadratically to $\bar{x}^* \in U\left(x_0, \frac{1}{La}\right) \cap \{R(F'(x_0)^{\#}) + x_0\}$, which is a solution of equation $F'(x_0)^{\#} F(x) = 0$. Here, $R(F'(x_0)^{\#}) + x_0 = \{x + x_0 : x \in R(F'(x_0)^{\#})\}$.

Chapter 7

Equations with Nonsmooth Operators

In recent years, Newton-type methods have been generalized to solve many problems in mathematical programming. An excellent treatment of these generalizations in the form of a genearlized equation was inaugurated by S. M. Robinson [297], [298]. Relevant work has been conducted by Josephy [200]–[203] and Argyros [76], [81]–[89]. When applied to an optimization problem Newton-type methods constitute a type of sequential quadratic programming (SQP) algorithms. These algorithms are very efficient and among the most promising for solving large-scale problems. See Gill, Murray and Wright [168] for more discussions on the numerical aspects of these algorithms.

There have been several previous papers dealing with Newton's method for solving non-Fréchet differentiable operators in normed spaces Kojima and Shindo [209] analized Newton and quasi-Newton methods for systems of piecewise continuously differentiable operators, and Kummer [215] extended the Kojima-Shindo result to real Banach spaces and studied Newton's method defined by a large family of linear continuous operators. Catinas [117], Chen and Yamamoto [121] Nguen and Zabrejko [363], Argyros [18], [34], [68] have provided results on equations involving a Fréchet differentiable term and another term whose differentiability is not assumed.

Robinson [302] discusses a Newton method for nonsmooth operators which admits a certain kind of a point-based approximating operator. He then provides a semilocal convergence analysis using the Newton-Kantorovich hypothesis and the nondiscrete mathematical induction (see Potra and Ptak [284], [286].

In this chapter we further exploit the weaker version of the Newton-Kantorovich theorem. The benefits of this approach have already been explained in Section 5.1.

In particular in the first section we show that under weaker hypotheses and same computational cost we can provide a finer convergence analysis and a better information on the location of the solution than the corresponding one in [302].

Moreover, in sections 2–4 we examine the semilocal convergence of Newton's method under even weaker hypotheses. Furthermore in Sections 5 and 6 we do the same for the secant method under different hypotheses whose advantages over each other have already been explained in Section 5.1.

Finally as a future research we suggest to the motivated reader to further exploit our work here to improve the results obtained by Robinson [302], Pang et. all [267], Qi [287], Han [178], Shapiro [309], Ip et all [197] concerning Bouligand and other types of derivatives in connection with Newton-type methods.

7.1 Convergence of Newton-Kantorovich Method. Case 1: Lipschitzian Assumptions

The classical Newton-Kantorovich method (4.1.3) replaces the operator whose solution is sought by an approximate operator that we hope can be solved easier. A solution of the approximation (linearization) can be used to restart the process. Under certain conditions a sequence is being generated that converges to x^*.

In particular in the case of the Newton–Kantorovich theorem (see Section 4.1) the approximating operators are linearizations of the form $F(x) + F'(x)(x - z)$ where given x we solve for z.

Robinson [302] extended this theorem to hold for nonsmooth operators, where the linearization is no longer available. This class includes smooth operators as well as nonsmooth reformulations of variational inequalities. He used as an alternative a point-based-approximation which imitates the properties of linearization of Newton's method.

Here we extend Robinson's work, and in particular we show that it is possible under weaker hypotheses, and the same computational cost to provide finer error bounds on the distances involved, and a more precise information on the location of the solution x^*.

We need the definitions.

Definition 7.1.1 *Let F be an operator from a closed subset D of a metric space (X, m) to a normed linear space Y. Operator F has a point-based approximation PBA on D if there exist an operator $A\colon D \times D \to Y$ and a number $\eta \geq 0$ such that for each u and v in D,*

(a) $\|F(v) - A(u, v)\| \leq \frac{1}{2}\eta m(u, v)^2$, and

(b) operator $A(u, \cdot) - A(v, \cdot)$ is Lipschitzian on D with modulus $\eta m(u, v)$.

We then say operator A is a PBA for F with modulus η.

Moreover if for a given point $x_0 \in D$

(c) operator $A(u, \cdot) - A(x_0, \cdot)$ is Lipschitzian on D with modulus $\eta_0 m(u, x_0)$, then we say operator A is a PBA for F with moduli (η, η_0) at the point x_0.

Note that if F has a Fréchet-derivative that is Lipschitzian on a closed, convex set D with modulus η, and X is also a normed linear space then by choosing

$$A(u, v) = F(u) + F'(u)(v - u) \tag{7.1.1}$$

(used in method (4.1.3)), we get by (a)

$$\|F(v) - F(u) - F'(u)(v-u)\| \leq \frac{1}{2}\eta\|v-u\|^2,$$

whereas (b) and (c) are obtained by the Lipschitzian property of F'.

As already noted in [300] the existence of a PBA does not necessarily imply differentiability.

Method (4.1.3) is applicable to any operator having a PBA A, if solving the equation

$$A(x,z) = 0$$

for z provided that x is given is possible.

We also need the definition concerning bounds for the inverses of operators:

Definition 7.1.2 *Let X, D, and Y be as in Definition 7.1.1, and $F\colon D \to Y$. Then we define the number:*

$$\delta(F,D) = \inf\left\{\frac{\|F(u)-F(v)\|}{m(u,v)},\ u \neq v,\ u,v \in D\right\}.$$

Clearly, if $\delta(F,D) \neq 0$ then F is 1–1 on D.

We also define the center-Lipschitzian number

$$\delta_0(F,D) = \inf\left\{\frac{\|F(u)-F(x_0)\|}{m(u,x_0)},\ u \neq x_0,\ u \in D\right\}.$$

Set $d = \delta(F,D)$ and $d_1 = \delta_0(F,D)$.

At this point we need two lemmas. The first one is the analog of the Banach Lemma on invertible operators 1.3.2. The second one gives conditions under which operator F is 1–1.

Lemma 7.1.3 *Let (X,d) be a Banach space, D a closed subset of X, and Y a normed linear space. Let F and G be functions from D into Y, G being Lipschitzian with modulus η and center-Lipschitzian with modulus η_0. Let $x_0 \in D$ with $F(x_0) = y_0$. Assume:*

(a) $U(y_0,\alpha) \subseteq F(D)$,

(b) $0 \leq \eta < d$,

(c) $U(x_0, d_1^{-1}\alpha) \subseteq D$, and

(d) $\theta_0 = (1-\eta_0 d_1^{-1})\alpha - \|G(x_0)\| \geq 0$.

Then the following hold:

$$U(y_0,\theta_0) \subseteq (F+G)\big(U(x_0,d_1^{-1}\alpha)\big)$$

and

$$0 < d - \eta \leq \delta(F+G,D).$$

Proof. Let $y \in U(y_0, \theta_0)$, $x \in U(x_0, d_1^{-1}\alpha)$ and define $Ty(x) = F^{-1}(y - G(x))$. Then we can have in turn

$$\begin{aligned}\|y - G(x) - y_0\| &\leq \|y - y_0\| + \|G(x) - G(x_0)\| + \|G(x_0)\| \\ &\leq \theta_0 + \eta_0 d_1^{-1}\alpha + \|G(x_0)\| = \alpha.\end{aligned}$$

That is the set $Ty(x) \neq \emptyset$ and in particular it contains a single point because $d > 0$. The rest follows exactly as in Lemma 3.1 in [300, p. 298] (see also the more detailed proof of Lemma 7.2.2). ∎

Remark 7.1.1 *In general*

$$\eta_0 \leq \eta, \quad \text{and} \quad d \leq d_1 \tag{7.1.2}$$

hold and $\frac{\eta}{\eta_0}$, $\frac{d_1}{d}$ *can be arbitrarily large (see Section 5.1). If* η_0 *is replaced by* η *and* d_1 *by* d *in Lemma 1 then our result reduces to the corresponding one in [Lemma 3.1, [300]]. Denote* θ_0 *by* θ *if* η_0 *is replaced by* η *and* d_1 *by* d*. Then in case strict inequality holds in any of the inequalities in (7.1.2), we get*

$$\theta < \theta_0, \tag{7.1.3}$$

which improves the corresponding Lemma 3.1 in [302], and under the same computational cost, since in practice the computation of η *requires the computation of* η_0 *or* d_1 *respectively.*

Lemma 7.1.4 *Let* X *and* Y *be normed linear spaces, and let* D *be a closed subset of* X*. For* $x_0 \in D$ *and* $F\colon D \to Y$ *assume: function* $A\colon D \times D \to Y$ *is a PBA for* F *on* D *with moduli* (η, η_0) *at the point* x_0*; there exists* $x_0 \in D$ *such that*

$$U(x_0, \rho) \subseteq D.$$

Then the following hold:

$$\delta(F, U(x_0, \rho)) \geq d - \eta_1 \rho \tag{7.1.4}$$

where,

$$d = \delta(A(x_0, \cdot), D), \quad \text{and} \quad \eta_1 = \eta_0 + \frac{1}{2}\eta.$$

Moreover, if $d - \eta_1\rho > 0$ *then* F *is 1–1 on* $U(x_0, \rho)$*.*

Proof. It follows as in the proof of Lemma 2.4 in [302, p. 295] but there are some differences. Set $x = \frac{x_1 + x_2}{2}$ for x_1 and x_2 in $U(x_0, \rho)$. Then we can write:

$$\begin{aligned}F(x_1) - F(x_2) = &[F(x_1) - A(x, x_1)] + [A(x, x_1) - A(x, x_2)] \\ &+ [A(x, x_2) - F(x_2)].\end{aligned} \tag{7.1.5}$$

We will find bounds on each one of the quantities inside the brackets above. First by (a) of Definition 7.1.1, we obtain

$$\|F(x_i) - A(x, x_i)\| \leq \frac{1}{2}\eta\|x - x_i\|^2 = \frac{1}{8}\eta\|x_1 - x_2\|^2. \tag{7.1.6}$$

Using the triangle inequality we get

$$\begin{aligned}\|A(x,u) - A(x,v)\| &\geq \|A(x_0,u) - A(x_0,v)\| \\ &- \|[A(x,u) - A(x_0,u)] - [A(x,v) - A(x_0,v)]\|.\end{aligned} \tag{7.1.7}$$

By the definition of d, and (7.1.7) we get in turn

$$\begin{aligned}\delta(A(x,\cdot), D) \geq \delta(A(x_0,\cdot), D) - \sup\{\|[A(x,u) - A(x_0,u)] \\ - [A(x,v) - A(x_0,v)]\|/\|u - v\|, \ u \neq v, \ u, v \in D\} \\ \geq d - \eta_0\|x - x_0\| \geq d - \eta_0\rho.\end{aligned} \tag{7.1.8}$$

Hence by (7.1.5), (7.1.8) and (c) of Definition 7.1.1 we have

$$\|F(x_1) - F(x_2)\| \geq \left[(d - \eta_0\rho) - \frac{1}{4}\eta\|x_1 - x_2\|\right]\|x_1 - x_2\|$$

and for $x_1 \neq x_2$,

$$\begin{aligned}\|F(x_1) - F(x_2)\|/\|x_1 - x_2\| &\geq d - \eta_0\rho - \frac{1}{4}\eta\|x_1 - x_2\| \\ &\geq d - \eta_0\rho - \frac{1}{4}\eta\|(x_1 - x_0) + (x_0 - x_2)\| \geq d - \eta_0\rho - \frac{1}{2}\eta\rho.\end{aligned}$$

■

Remark 7.1.2 *If equality holds in (7.1.2), then Lemma 7.1.4 reduces to Lemma 2.4 in [302]. Otherwise our result provides a wider range for* $\delta(F, U(x_0, \rho))$.

Let r_0 and d_0 be positive numbers.

It is convenient to define scalar sequences $\{t_n\}$, $\{t_n\}$ by

$$t_{n+2} - t_{n+1} + \frac{d_0^{-1}\eta(t_{n+1} - t_n)^2}{2(1 - d_0^{-1}\eta_0 t_{n+1})}, \ t_0 = 0, \ t_1 = r_0 \tag{7.1.9}$$

and

$$s_{n+2} = s_{n+1} + \frac{d_0^{-1}\eta(s_{n+1} - s_n)^2}{2(1 - d_0^{-1}\eta s_{n+1})}, \ s_0 = 0, \ s_1 = r_0. \tag{7.1.10}$$

We set

$$h_A = d_0^{-1}\overline{\eta}r_0, \quad \overline{\eta} = \frac{\eta_0 + \eta}{2} \tag{7.1.11}$$

and

$$h_K = d_0^{-1}\eta r_0. \tag{7.1.12}$$

Note that

$$\frac{1}{2}h_K \leq h_A \leq h_K. \tag{7.1.13}$$

If strict inequality holds in (7.1.2) then so does in (7.1.13). It was shown in Section 5.1 that sequence $\{t_n\}$ converges monotonically to some $t^* \in (0, 2r_0]$ provided that

$$h_A \leq 1. \tag{7.1.14}$$

Moreover by the Newton–Kantorovich theorem for nonsmooth equations 4.2.4 sequence $\{s_n\}$ converges monotonically to some s^* with

$$0 < s^* = \frac{r_0}{h_K}\left[1 - (1 - 2h_K)^{1/2}\right] \leq 2r_0 \tag{7.1.15}$$

provided that the famous Newton–Kantorovich hypotheses

$$h_K \leq 1. \tag{7.1.16}$$

holds.

Note also that

$$h_K \leq 1 \Rightarrow h_A \leq 1 \tag{7.1.17}$$

but not vice versa unless equality holds in (7.1.2).

Under hypotheses (7.1.14) and (7.1.16) we showed in Section 5.1 (for $\eta_0 < \eta$):

$$t_n < s_n \quad (n \geq 2) \tag{7.1.18}$$

$$t_{n+1} - t_n < s_{n+1} - s_n \quad (n \geq 1) \tag{7.1.19}$$

$$t^* - t_n \leq s^* - s_n \tag{7.1.20}$$

and

$$t^* \leq s^* \leq 2r_0. \tag{7.1.21}$$

Note that $\{s_n\}$ was essentially used as a majorizing sequence for $\{x_n\}$ in Theorem 3.2 in [302] for Newton's method.

We can state the following semilocal convergence theorem for Newton's method involving nonsmooth operators:

Theorem 7.1.5 *Let $F\colon D \subseteq X \to Y$ be a continuous operators $x_0 \in D$, and $r_0 > 0$. Suppose that $A\colon D \times D \to Y$ is a PBA for F with moduli (η, η_0) at the point $x = x_0$. Moreover assume:*

(a) $\delta(A(x_0, \cdot), D) \geq d_0 > 0$;
(b) $0 < h_A \leq \frac{1}{2}$;
(c) for each $y \in U(0, d_0(t^* - r_0))$ *equation* $A(x_0, x) = y$ *has a solution* x;
(d) the solution $S(x_0)$ *of* $A(x_0, S(x_0)) = 0$ *satisfies* $\|S(x_0) - x_0\| \leq r_0$; *and (e)* $U(x_0, t^*) \subseteq D$.

Then the Newton iteration defining x_{n+1} *by*

$$A(x_n, x_{n+1}) = 0 \quad (n \geq 0)$$

is well defined, remains in $U(x_0, t^*)$ *for all* $n \geq 0$ *and converges to a solution* $x^* \in U(x_0, t^*)$ *of equation* $F(x) = 0$. *Moreover the following estimates hold for all* $n \geq 0$:

$$\|x_{n+1} - x_n\| \leq t_{n+1} - t_n \tag{7.1.22}$$

and

$$\|x_n - x^*\| \leq t^* - t_n. \tag{7.1.23}$$

Furthermore if ρ *given in Lemma 7.1.4 satisfies*

$$t^* < \rho < d_0 \eta_1^{-1} \tag{7.1.24}$$

then x^* *is the unique solution of equation* $F(x) = 0$ *in* $U(x_0, \rho)$.

Proof. We use Lemma 7.1.3 with quantities F, G, α, x_0 and y_0 replaced by $A(x_0, \cdot)$, $A(x_1, \cdot) - A(x_0, \cdot)$, $d_1(t^* - r_0)$, $x_1 = S(x_0)$, and 0 respectively. Hypothesis (a) of the Lemma follows from the fact that $A(x_0, x) = y$ has a unique solution x_1 (since $\delta(A(x_0, \cdot), D)) \geq d_0 > 0$). For hypothesis (b) we have

$$\delta(A(x_1, \cdot), D) \geq \delta(A(x_0, \cdot), D) - \eta_0 \|x_1 - x_0\| \geq d_0 - \eta_0 r_0 > 0. \tag{7.1.25}$$

Condition (c) of Lemma 7.1.3 follows immediately from the above choices. The condition $h_A \leq \frac{1}{2}$ is equivalent to $2r_0(\eta + \eta_0) \leq d_0$. We have $d_1 = \delta_0(F, D) \geq \delta(F, D) \geq d_0$, and $\|G(x_1)\| \leq \frac{1}{2}\eta_0 r_0^2 \leq \frac{1}{2}\eta r_0^2$. Then condition (d) of Lemma 7.1.3 follows from the estimate

$$\begin{aligned}
\theta &\geq \left(1 - \frac{\eta_0 r_0}{d_1}\right) d_1(t^* - r_0) - \frac{\eta}{2} r_0^2 = (d_1 - \eta_0 r_0)(t^* - r_0) - \frac{\eta}{2} r_0^2 \\
&\geq (d_0 - \eta_0 r_0)(t^* - r_0) - \frac{\eta_0}{2} r_0^2 = (d_0 - \eta_0 t_1)(t^* - t_1) - \frac{\eta}{2} t_1^2 \\
&= (d_0 - \eta_0 r_0)(t^* - t_2) + (d_0 - \eta_0 t_1)(t_2 - t_1) - \frac{\eta}{2}(t_1 - t_0)^2 \\
&= (d_0 - \eta_0 r_0)(t^* - t_2) \geq \eta_0 r_0 (t^* - t_2) \geq 0.
\end{aligned}$$

It follows from Lemma 7.1.3 that for each $y \in U(0, t^* - \|x_1 - x_0\|)$, the equation $A(x_1, z) = y$ has a unique solution $z = x_2$ in D since $\delta(A(x_1, \cdot), D) > 0$.

We also have $A(x_0, x_1) = A(x_1, x_2) = 0$ and $A(x_1, x_1) = F(x_1)$. By Definition 7.1.2 and

$$\begin{aligned}\|x_2 - x_1\| &\leq \delta(A(x_1, \cdot), D)^{-1} \|A(x_1, x_2) - A(x_1, x_1)\| \\ &\leq d_0^{-1}(1 - d_0^{-1}\eta_0\|x_1 - x_0\|)^{-1}\|A(x_0, x_1) - F(x_1)\| \\ &\leq d_0^{-1}(1 - d_0^{-1}\eta_0\|x_1 - x_0\|)^{-1}\frac{1}{2}\eta\|x_1 - x_0\|^2 \leq t_2 - t_1.\end{aligned}$$

Hence, we showed:

$$\|x_{n+1} - x_n\| \leq t_{n+1} - t_n, \tag{7.1.26}$$

and

$$U(x_{n+1}, t^* - t_{n+1}) \subseteq U(x_n, t^* - t_n) \tag{7.1.27}$$

hold for $n = 0, 1$.

Moreover for every $z \in \overline{U}(x_1, t^* - t_1)$,

$$\|z - x_0\| \leq \|z - x_1\| + \|x_1 - x_0\| \leq t^* - t_1 + t_1 = t^* - t_0,$$

implies $z \in U(x_0, t^* - t_0)$. Given they hold for $n = 0, 1, \ldots, j$, then

$$\|x_{j+1} - x_0\| \leq \sum_{i=1}^{j+1} \|x_i - x_{i-1}\| \leq \sum_{i=1}^{j+1}(t_i - t_{i-1}) = t_{j+1} - t_0 = t_{j+1}.$$

The induction can be completed by simply replacing above x_0, x_1 by x_n, x_{n+1} respectively, and noticing that $\|x_{n+1} - x_n\| \leq r_0$. Indeed for the crucial condition (d) of Lemma 7.1.3 assigning F as $A(x_n, \cdot)$, G as $A(x_{n+1}, \cdot) - A(x_n, \cdot)$, α as $d_1(t^* - t_{n+1})$ we have, since $d_1 = \delta_0(F, D) \geq \delta(F, D) \geq d_0 - \eta_0 t_n$:

$$\begin{aligned}\theta_0 &\geq \left[1 - \frac{\eta(t_{n+1} - t_n)}{d_1}\right] d_1(t^* - t_{n+1}) - \frac{\eta}{2}(t_{n+1} - t_n)^2 \\ &= [d_1 - \eta(t_{n+1} - t_n)](t^* - t_{n+1}) - \frac{\eta}{2}(t_{n+1} - t_n)^2 \\ &\geq (d_0 - \eta t_{n+1})(t^* - t_{n+1}) - \frac{\eta}{2}(t_{n+1} - t_n)^2 \\ &= (d_0 - \eta t_{n+1})(t^* - t_{n+2}) + (d_0 - \eta t_{n+1})(t_{n+2} - t_{n+1}) - \frac{\eta}{2}(t_{n+1} - t_n)^2 \\ &= (d_0 - \eta t_{n+1})(t^* - t_{n+2}) \geq 0.\end{aligned}$$

It follows by hypothesis (b) that scalar sequence $\{t_n\}$ is Cauchy. From (7.1.26) and (7.1.27) it follows $\{x_n\}$ is Cauchy too in a Banach space X and as such it converges to some $x^* \in U(x_0, t^*)$. Furthermore we have

$$\|F(x_{n+1})\| = \|F(x_{n+1}) - A(x_n, x_{n+1})\| \leq \frac{1}{2}\eta\|x_{n+1} - x_n\|^2 \leq \frac{1}{2}\eta(t_{n+1} - t_n)^2.$$

That is $\|F(x_{n+1})\|$ converges to zero as $n \to \infty$. Therefore by the continuity of F we deduce $F(x^*) = 0$. Estimate (7.1.23) follows from (7.1.22) by using standard majorization techniques. Moreover for the uniqueness part using Lemma 7.1.4 and hypothesis (7.1.24) we deduce that x^* is an isolated zero of F, since F is 1–1 on $U(x_0, t^*)$. ■

Remark 7.1.3 *(a) If equality hold in (7.1.2), then our theorem reduces to the corresponding Theorem 3.2 in [302]. Otherwise according to (7.1.17)–(7.1.21) our theorem provides under weaker conditions, and the same computational cost finer error bounds on the distances involved and a more precise information on the location of the solution x^*. Moreover the uniqueness of the solution x^* is shown in a larger ball (see (7.1.24) and compare it with 3.7 in [302] (i.e. compare (7.1.24) with $s^* < \rho < d_0\eta^{-1}$).*

(b) Condition (c) of Theorem 7.1.5 can be replaced by the stronger but more practical:

(b)′ For each fixed x_0 and y in $U(0, d_0(t^* - r_0)) = U$ (or in $U(0, d_0 r_0)$) operator Q given by

$$Q(x) = x - y + A(x_0, x)$$

is a contraction on U and maps U into itself. It then follows by the contraction mapping principle (see Chapter 3) that equation $A(x_0, x) = y$ has a solution x in U.

We can also provide a local convergence result for Newton's method:

Theorem 7.1.6 *Let $F\colon D \subseteq X \to Y$ be as in Theorem 7.1.5. Assume: $x^* \in D$ is a simple solution of equation $F(x) = 0$; operator F has a point-based approximation A on D with moduli (ℓ, ℓ_0) at the point x^*. Moreover assume that the following hold:*

(a) $\delta(A(x^*, \cdot), D) \geq d^* > 0$;

(b) $d^* > (\ell_0 + \frac{1}{2}\ell) r^*$;

(c) for each $y \in U(0, d^* r^*)$ and provided $x_0 \in U^0(x^*, r^*)$ the equation $A(x_0, x) = y$ has a solution x satisfying $\|x - x^*\| \leq r^*$; and

(d) $U(x^*, r^*) \subseteq D$.

Then Newton's method is well defined, remains in $U(x^*, r^*)$ for all $n \geq 0$ and converges to x^*, so that

$$\|x_{n+1} - x^*\| \leq \frac{\ell \|x_n - x^*\|^2}{2(d^* - \ell_0 \|x_n - x^*\|)} \quad (n \geq 0). \tag{7.1.28}$$

Proof. It follows immediately by replacing x_0 by x^* in Lemma 7.1.4 and in appropriate places in the proof of Theorem 7.1.5. ■

Remark 7.1.4 *(a) The local convergence of Newton's method was not studied in [302]. However if it was it would have been as in Theorem 7.1.6 with $\ell_0 = \ell$. Note however that as in (7.1.2) in general*

$$\ell_0 \leq \ell \tag{7.1.29}$$

holds, and $\frac{\ell}{\ell_0}$ can be arbitrarily large. Denote by r_R^* the convergence radius to have been obtained by Robinson. It then follows by (b) of Theorem 7.1.6 that

$$r_1^* = \frac{2d^*}{3\ell} \leq r^* = \frac{2d^*}{2\ell_0 + \ell}. \tag{7.1.30}$$

Moreover in case strict inequality holds in (7.1.29), then so does in (7.1.30). This observation is very important in computational mathematics since our approach allows a wider choice for initial guesses x_0. Moreover the corresponding error bounds are obtained if we set $\ell = \ell_0$ in (7.1.29). Therefore our error bounds are also finer (for $\ell_0 < \ell$) and under the same computational cost. (b) If operator A is given by the linearization (7.1.1), then $A(x_n, x_{n+1}) = 0$ is a linear equation. Thus we have the Newton's method for differentiable operators. Otherwise, equation $A(x_n, x_{n+1}) = 0$ applies to a wide class of important real life problems for which it is not a linear equation. Such problems are discussed below.

Application 7.1.7 ***(Application to Normal Maps.)*** *An important class of nonsmooth functions that have tractable point-based approximations. This class was given in [302] and is a reformulation of the class of variational inequalities on a Hilbert space, so in particular it includes the finite-dimensional generalized equations investigated by Josephy [200]–[203], as well as many other problems of practical importance. For many members of this class, computational methods are available to solve the Newton subproblems, and therefore the method we suggest here can be applied not only as a theoretical device, but also as a practical computational method for solving problems.*

The functions we shall treat are the so-called normal maps. We define these, and discuss briefly their significance, before relating them to point-based approximations. Let C be a closed convex subset of a Hilbert space H, and denote by Π the metric projector from H onto C. Let Ω be an open subset of H meeting G, and let F be a function from Ω to H. The normal map F_C is defined from the set $\Pi^{-1}(\Omega)$ to H by

$$F_C(z) = F(\Pi(z)) + (z - \Pi(z)).$$

A more extensive discussion of these normal maps is in [301], where we show how the first order necessary optimality conditions for nonlinear optimization, as well as linear and nonlinear complementarity problems and more general variational inequalities, can all be expressed compactly and conveniently in the form of $F_C(z) = 0$ involving normal maps. This is true because, as is easily verified, the variational problem of finding an $x_0 \in C$ such that for each $c \in C$, $\langle F(x_0), c - x_0 \rangle \geq 0$ is completely equivalent to the normal-map equation $F_C(z_0) = 0$ through the transformation $z_0 = F(x_0)$. However, the use of normal maps sometimes enable one to gain insight into special properties of problem classes that might have remained obscure in the formalism of variational inequalities. A particular example of this is the characterization of the local and global homeomorphism properties of linear normal maps given in [301] and improved in [291].

As the functions F and the sets C used in the constructioin of these normal maps are those naturally appearing in the optimizatioin or variational problems, the normal-map technique provides a simple and convenient device for formulating a wide variety of important problems from applications in single, unified manner. Results established for such normal maps will then apply to the entire range of these applicatioins, so that there is no need to derive them separately for each different type of problem.

Results of this kind already established include an implicit-funciton theorem and related machinery in [300], characterizations in [301] of those normal maps derived from linear thransformations that are also global or local homeomorphisms. As we shall now show, the results on Newton's method that we establish here will apply to normal maps, therefore encompassing in particular the problems treated in the work already cited on application of Newton's method to certain optimization problems and variational inequalities.

Our results on local convergence of Newton's method require, in essence, the existence of a PBA that is a local homeomorhism, together with some closeness conditions. To show that these apply to normal maps, we must indicate how one can construct PBAs for such maps. Recalling our earlier demonstration that the linearization operation yields a PBA for the function being linearized, one might ask if one could ninearize the function appearing in a normal map. This does not seem easy, as such functions are in general not differentiable. However, we shall show that a simple modification of the linearization procedure will work very well. The key observation is is that although the function F_C will be differentiable (indeed, often smooth), and therefore will have an easily constructible PBA. We now show that if such a PBA exists for F, then one can easily be constructed for F_C.

Proposition 7.1.8 *Let H be a Hilbert space and C a nonempty closed convex subset of H, and denote by Π the metric projector on C. Let F be a function from C to H, and let A be a function from $C \times C$ to H. If A is a PBA for F on C, then the function $G : H \times H \to H$ defined by $G(w, z) = A(\Pi(w), \cdot)_C(z)$ is a PBA for F_C on H. Further, the constant in the definition of PBA can be taken to be the same for G as for A.*

Before giving the proof we note that by the definition of normal map, $G(w, z)$ is $A(\Pi(w), \Pi(z)) + (z - \Pi(z))$. Therefore in determininng values of this function one only evaluates A on the set $C \times C$, where it is defined.

Proof. By hypothesis there is a constant ϕ for which A has the two properties given in Definition 7.1.1. We shall show that the same two properties hold for G, again with the constant ϕ. First, let w and z be any elements of H. Then

$$\begin{aligned}
&\|F_C(z) - G(w,z)\| = \\
&= \|F(\Pi(z)) + (z - \Pi(z)) - [A(\Pi(w), \Pi(z)) + (z - \Pi(z))]\| \\
&= \|F(\Pi(z)) - A(\Pi(w), \Pi(z))\| \\
&\leq \phi/2 \|\Pi(w) - \Pi(z)\|^2 \\
&\leq \phi/2 \|w - z\|^2,
\end{aligned}$$

where we used the facts that A was a PBA for F on C and that the metric projector is nonexpansive. Therefore G satisfies the first property of a PBA. For the second, suppose that z and z' are two elements of H; we show that $G(z,\cdot) - G(z',\cdot)$ is Lipschitzian on H with modulus $\phi \|z - z'\|$. Indeed, let x and y be elements of H. Then

$$\begin{aligned}
&\|[G(z,x) - G(z',x)] - [G(z,y) - G(z',y)]\| = \\
&= \|[A(\Pi(z), \Pi(x)) + (x - \Pi(x)) - A(\Pi(z'), \Pi(x)) - (x - \Pi(x))] - \\
&- [A(\Pi(z), \Pi(y)) + (y - \Pi(y)) - A(\Pi(z'), \Pi(y)) - (y - \Pi(y))]\| \\
&= \|[A(\Pi(z), \Pi(x)) - A(\Pi(z'), \Pi(x))] - \\
&[A(\Pi(z), \Pi(y)) - A(\Pi(z'), \Pi(y))]\| \\
&\leq \phi \|\Pi(z) - \Pi(z')\| \|\Pi(x) - \Pi(y)\| \\
&\leq \phi \|z - z'\| \|x - y\|,
\end{aligned}$$

where we used the second property of a PBA (for A) along with the nonexpansivity of Π. ∎

7.2 Convergence of Newton-Kantorovich Method. Case 2: Hölderian assumptions

In this section we weaken further assumptions made in Section 7.1 by considering Hölderian instead of Lipschitzian assumptions in order to cover a wider range of problems than the ones discussed in Section 7.1 or in [302].

Numerical examples are also provided to show that our results apply where others fail.

We need the definitions:

Definition 7.2.1 *Let F be an operator from a closed subset D of a metric space (X, m) to a normed linear space Y. We say F has a point-based approximation p-(PBA) at some point $x_0 \in D$ if there exist an operator $A\colon D \times D \to Y$ and scalars k, k_0 such that for each u and v in D and some $p \in (0,1]$:*

$$\|F(v) - A(u,v)\| \leq \frac{1}{1+p} k m(u,v)^{1+p}, \tag{7.2.1}$$

$$\|[A(u,x) - A(v,x)] - [A(u,y) - A(v,y)]\| \leq \frac{2}{1+p} km(u,v)^{1+p}, \tag{7.2.2}$$

and

$$\|[A(u,x) - A(x_0,x)] - [A(u,y) - A(x_0,y)]\| \leq \frac{2}{1+p} k_0 m(u,v)^{1+p} \tag{7.2.3}$$

for all $x, y \in D$.

Let us assume that D is a convex set and X is also a normed linear space for both cases that follow.

Here are some cases where the above hold:

(a) Let $p = 1$. Then see Section 7.1.

(b) If F has a Fréchet derivative that is p-Hölderian on D with modulus k and center-p-Hölderian (at x_0) with modulus k_0 then with the choice of A given by (7.1.1) we get

$$\|F(v) - F(u) - F'(u)(v-u)\| \leq \frac{k}{1+p} \|u-v\|^{1+p}$$

whereas again (7.2.2) and (7.2.3) are equivalent to the Hölderian and center Hölderian property of operator F' (see also Section 5.1).

We need the following generalization of the classical Banach Lemma on invertible operators 1.3.2:

Lemma 7.2.2 *Let (X,d) be a Banach space, D a closed subset of X, and Y a normed linear space. Let F, G be operators from D into Y, with G being Lipschitzian with modulus k, and G being center Lipschitzian with modulus k_0. Let $x_0 \in D$, and set $F(x_0) = y_0$. Assume that:*

(a) $U(y_0,\alpha) \subseteq F(D)$;

(b) $0 \leq k < d$;

(c) $U(x_0, d_1^{-1}\alpha) \subseteq D$, and

(d) $\theta = (1 - k_0 d_1^{-1})\alpha - \|G(x_0)\| \geq 0$.

Then the following hold:

$$U(y_0,\theta) \leq (F+G)\big(U(x_0, d_1^{-1}\alpha)\big), \tag{7.2.4}$$

and

$$\delta(F+G, D) \geq d - k > 0. \tag{7.2.5}$$

Proof. Define operator $Ty(x) = F^{-1}(y - G(x))$ for $y \in U(y_0,\theta)$, and $x \in U(x_0, d_1^{-1}\alpha)$. We can get:

$$\begin{aligned}\|y - G(x) - y_0\| &\leq \|y - y_0\| + \|G(x) - G(x_0)\| + \|G(x_0)\| \\ &\leq \theta + k_0(d_1^{-1}\alpha) + \|G(x_0)\| = \alpha.\end{aligned}$$

Therefore $Ty(x)$ is a singleton set since $\delta > 0$. That is Ty is an operator on $U(x_0, (d_1^{-1}\alpha)^{1/p})$. This operator maps $U(x_0, d_1^{-1}\alpha)$ into itself. Indeed for $x \in U(x_0, d_1^{-1}\alpha)$

$$d(Ty(x), x_0) = d\big(F^{-1}(y - G(x)), F^{-1}(y_0)\big) \le d_1^{-1}\alpha,$$

so we get:

$$d(Ty(x), x_0) \le d_1^{-1}\alpha.$$

Moreover let u, v be in $U(x_0, d_1^{-1}\alpha)$, then

$$d(Ty(u), Ty(v)) \le d^{-1} d(F^{-1}(y - G(u)), F^{-1}(y - G(v))) \le d^{-1} k d(u, v)$$

so

$$d(Ty(u), Ty(v)) \le d^{-1} k d(u, v). \tag{7.2.6}$$

It follows by hypothesis (b) and (7.2.6) that Ty is a contraction operator (see Chapter 3) mapping closed ball $U(x_0, d_1^{-1}\alpha)$ into itself and as such it has a unique fixed point $x(y)$ in this ball, since

$$\begin{aligned}\delta(F+G, D) &= \inf\left\{ \frac{\|[F(u) - F(v)] + [G(u) - G(v)]\|}{m(u,v)}, \right.\\ &\qquad \left. u \ne v,\ u, v \in D \right\}\\ &\ge \delta - \sup\left\{ \frac{\|G(u) - G(v)\|}{m(u,v)},\ u \ne v,\ u, v \in D \right\}\\ &\ge \delta - k > 0. \end{aligned} \tag{7.2.7}$$

Furthermore (7.2.7) shows $F + G$ is one-to-one on D. ∎

The following lemma can be used to show uniqueness of the solution on the semilocal case and convergence of Newton's method in the local case:

Lemma 7.2.3 *Let X and Y be normed linear spaces, and let D be a closed subset of X. Let $F\colon D \to Y$, and let A be a p-(PBA) for F on D at the point $x_0 \in D$. Denote by d_0 the quantity $\delta(A(x_0, \cdot), D)$. If $x_0 \in D$ and $U(x_0, \rho) \subseteq D$, then*

$$\delta(F, U(x_0, \rho)) \ge d_0 - (2k_0 + k)\frac{\rho^p}{1+p}.$$

In particular, if $d_0 > (2k_0 + k)\frac{\rho^p}{1+p}$ then F is one-to-one on $U(x_0, \rho)$.

Proof. Let y, z be points in $U(x_0, \rho)$. Set $x = \frac{y+z}{z}$. We can write

$$F(y) - F(z) = [F(y) - A(x, y)] + [A(x, y) - A(x, z)] + [A(x, z) - F(z)].$$

Then we get

$$\|F(y) - A(x,y)\| \leq \frac{k}{1+p}\|x-y\|^{1+p} = \frac{k}{2^{1+p}(1+p)}\|y-x\|^{1+p},$$

and similarly

$$\|F(z) - A(x,z)\| \leq \frac{k}{2^{1+p}(1+p)}\|x-z\|^{1+p}.$$

Using the estimate

$$\begin{aligned}\|A(x,u) - A(x,v)\| \geq \|A(x_0,u) - A(x_0,v)\| \\ - \|[A(x,u) - A(x_0,u)] - [A(x,v) - A(x_0,v)]\|,\end{aligned}$$

we get in turn:

$$\begin{aligned}
&\delta(A(x,\cdot),D) = \\
&= \inf\left\{\tfrac{\|A(x,u)-A(x,v)\|}{\|u-v\|} \mid u \neq v,\ u,v \in D\right\} \\
&\geq \delta(A(x_0,\cdot),D) - \sup\left\{\tfrac{\|[A(x,u)-A(x_0,u)]-[A(x,v)-A(x_0,v)]\|}{\|u-v\|},\right. \\
&\qquad \left. u \neq v,\ u,v \in D\right\} \\
&\geq d_0 - \frac{2}{1+p}k_0\|x-x_0\|^p \geq d_0 - \frac{2k_0}{1+p}\rho^p.
\end{aligned}$$

Hence, we get

$$\|F(y) - F(z)\| \geq \left(d_0 - \frac{2k_0\rho^p}{1+p}\right)\|y-z\| - \frac{k}{2^p(1+p)}\|y-z\|^{1+p},$$

and for $y \neq z$

$$\begin{aligned}\frac{\|F(y)-F(z)\|}{\|y-z\|} &\geq d_0 - \frac{2k_0\rho^p}{1+p} - \frac{k}{2^p(1+p)}\|y-z\|^p \\ &\geq d_0 - (2k_0+k)\frac{\rho^p}{1+p}.\end{aligned}$$

■

Remark 7.2.1 *Lemmas 7.2.2 and 7.2.3 can be reduced to the corresponding ones in [302] if $p = 1$, $d_1 = d = d_0$, and $k_0 = k$. In the case $p = 1$ and $k_0 < k$ or $d_1 > d$ our Lemma 7.2.2 improves the range for θ, whereas our Lemma 7.2.3 enlarges ρ given in [300]. Note that in general*

$$k_0 \leq k \quad \text{and} \quad d \leq d_1$$

hold and $\frac{k}{k_0}$, $\frac{d_1}{d}$ can be arbitrarily large (see Section 5.1).

It is convenient for us to define scalar iteration $\{t_n\}$ $(n \geq 0)$ by

$$t_{n+2} = t_{n+1} + \frac{k(t_{n+1} - t_n)^{1+p}}{d_0(1+p)\left[1 - \frac{2k_0}{1+p} d_0^{-1} t_{n+1}^p\right]}, \quad t_0 = 0, \ t_1 = \eta \tag{7.2.8}$$

for some $\eta \geq 0$.

The convergence of sequence $\{t_n\}$ was studied in Section 5.1 and in [90]. In particular, we showed that if

$$h = \left[k + \frac{2k_0}{1+p}\left(\frac{p+1}{p}\right)^p\right] d_0^{-1} \eta^p \leq 1$$
$$\text{for } p \neq 1 \text{ or } h = (k_0 + k)\eta d_0^{-1} \leq 1 \text{ for } p = 1, \tag{7.2.9}$$

then sequence $\{t_n\}$ is monotonically convergent to its upper bound t^* with

$$\eta \leq t^* \leq \frac{p+1}{p}\eta = r. \tag{7.2.10}$$

Note also that these conditions are obtained as special cases from Lemma 5.1.1.

Conditions (7.2.9) were compared favourably in [90] with corresponding ones already in the literature [303]. That is why we decided to use them here over the earlier ones.

It will be shown that $\{t_n\}$ is a majorizing sequence for Newton's iteration $\{x_n\}$ $(n \geq 0)$ to be defined in the main semilocal convergence theorem that follows:

We can state and prove the main semilocal convergence result for Newton's method involving p-(PBA).

Theorem 7.2.4 *Let X and Y be Banach spaces, D a closed convex subset of X, $x_0 \in D$, and F a continuous operator from D into Y. Suppose that F has a p-(PBA) approximation at x_0. Moreover assume:*

(a) $\delta(A(x_0, \cdot), D) \geq d_0 > 0$;

(b) condition (7.2.9) holds;

(c) for each $y \in U(0, d_0(t^* - \eta))$ the equation $A(x_0, x) = y$ has a solution x;

(d) the solution $S(x_0)$ of $A(x_0, S(x_0)) = 0$ satisfies $\|S(x_0) - x_0\| \leq \eta$; and

(e) $U(x_0, t^*) \subseteq D$.

Then the Newton iteration defining x_{n+1} by

$$A(x_n, x_{n+1}) = 0 \quad (n \geq 0) \tag{7.2.11}$$

remains in $U(x_0, r)$, and converges to a solution $x^* \in U(x_0, r)$ of equation $F(x) = 0$. Moreover the following estimates hold for all $n \geq 0$:

$$\|x_{n+1} - x_n\| \leq t_{n+1} - t_n \tag{7.2.12}$$

and

$$\|x_n - x^*\| \leq t^* - t_n \tag{7.2.13}$$

where sequence $\{t_n\}$ is defined by (7.2.8) and $t^* = \lim_{n\to\infty} t_n$.

Proof. We use Lemma 7.2.2 with quantities F, G, x_0, and y_0 replaced by $A(x,\cdot)$, $A(x_1,\cdot)-A(x,\cdot)$, $x_1 = S(x_0)$, and 0 respectively. Hypothesis (a) of Lemma 1 follows from the fact that $A(x_0,x) = y$ has a unique solution x_1 (since $\delta(A(x_0,\cdot),D) > 0$). For hypothesis (b) we have

$$\delta(A(x_1,\cdot),D) \geq \delta(A(x_0,\cdot),D) - \frac{2}{1+p}k_0\eta^p \geq d_0 - \frac{2}{1+p}k_0\eta^p$$
$$> 0 \qquad \text{(by (7.2.9))}.$$

To show (c) we must have $U(x_1, t^* - \eta) \subseteq D$. Instead by hypothesis (e) it suffices to show:

$$U(x_1, t^* - \eta) \subset U(x_0, t^*),$$

which is true since

$$\|x_1 - x_0\| + \eta \leq t^* - \eta + \eta = t^*.$$

We also have by (7.2.2)

$$\|A(x_0,x_1) - A(x_1,x_1)\| \leq \frac{1}{1+p}k\|x_0 - x_1\|^{1+p}.$$

The quantity

$$\left[1 - \frac{2}{1+p}k_0\eta^p d_0^{-1}\right]\frac{d_0\eta}{p} - \frac{1}{1+p}k\eta^{1+p} \tag{7.2.14}$$

is nonnegative by (7.2.9), and is also an upper bound on θ given by (d) in Lemma 7.2.2, since θ is bounded above by

$$\left[1 - \frac{2}{1+p}k_0\eta^p d_1^{-1}\right]\frac{d_0\eta}{p} - \|A(x_0,x_1) - A(x_1,x_1)\|.$$

It follows from Lemma 7.2.2 that for each $y \in U(0, t^* - \|x_1 - x_0\|)$, the equation $A(x_1, z) = y$ has a unique solution $z = x_2$ in D, since $\delta(A(x_1,\cdot),D) > 0$. We also have $A(x_0,x_1) = A(x_1,x_2) = 0$ and $A(x_1,x_1) = F(x_1)$. By Definition 7.2.1 we get in turn:

$$\begin{aligned}\|x_2 - x_1\| &\leq \delta(A(x_1,\cdot),D)^{-1}\|A(x_1,x_2) - F(x_1)\| \\ &\leq d_0^{-1}\left[1 - d_0^{-1}\frac{2k_0}{1+p}\|x_1 - x_0\|^p\right]^{-1}\frac{k}{1+p}\|x_1 - x_0\|^{1+p} \\ &\leq d_0^{-1}\left[1 - d_0^{-1}\frac{2k_0}{1+p}(t_1 - t_0)^p\right]^{-1}\frac{k}{1+p}(t_1 - t_2)^{1+p} \leq t_2 - t_1.\end{aligned}$$

Hence we showed:

$$\|x_{n+1} - x_n\| \leq t_{n+1} - t_n \tag{7.2.15}$$

and

$$U(x_{n+1}, t^* - t_{n+1}) \subset U(x_n, t^* - t_n) \tag{7.2.16}$$

hold for $n = 0, 1$. Moreover for every $v \in U(x_1, t^* - t_1)$

$$\|v - x_0\| \leq \|v - x_1\| + \|x_1 - x_0\| \leq t^* - t_1 + t_1 = t^* - t_0,$$

implies $v \in U(x_0, t^* - t_0)$. Given they hold for $n = 0, 1, \ldots, j$, then

$$\|x_{j+1} - x_0\| \leq \sum_{i=1}^{j+1} \|x_i - x_{i-1}\| \leq \sum_{i=1}^{j+1} (t_i - t_{i-1}) = t_{j+1} - t_0 = t_{j+1}.$$

The induction for (7.2.15), (7.2.16) can easily be completed by simply replacing above x_0, x_1 by x_n, x_{n+1} respectively. Indeed the crucial quantity corresponding to (7.2.14) will be

$$\left[1 - \frac{2k_0}{1+p}\left(\frac{p+1}{p}\right)^p \eta^p d_0^{-1}\right] \frac{d_0 \eta}{p} - \frac{k}{1+p}\eta^{1+p}$$

or

$$\left[\frac{k}{1+p} + \frac{2k_0}{1+p}\left(\frac{p+1}{p}\right)^p\right] d_0^{-1} \eta^p \leq 1,$$

which is nonnegative by (7.2.9). Furthermore by (7.2.9) scalar sequence $\{t_n\}$ is Cauchy. From (7.2.15) and (7.2.16) it follows $\{x_n\}$ is Cauchy too in a Banach space X, and as such it converges to some $x^* \in U(x_0, t^*)$. Moreover we have by (7.2.2) and (7.2.15)

$$\begin{aligned} \|F(x_{n+1})\| &= \|F(x_{n+1}) - A(x_n, x_{n+1})\| \leq \frac{k}{1+p}\|x_{n+1} - x_n\|^{1+p} \\ &\leq \frac{k}{1+p}(t_{n+1} - t_n)^{1+p}. \end{aligned}$$

Therefore $\|F(x_{n+1})\|$ converges to zero as $n \to \infty$. By the continuity of F we deduce $F(x^*) = 0$. Estimate (7.2.13) now follows from (7.2.12) by using standard majorization techniques. ∎

Remark 7.2.2 *The uniqueness of the solution x^* was not considered in Theorem 7.2.4. Indeed, we do not know if under the conditions stated above the solution x^* is unique, say in $U(x_0, t^*)$. However using Lemma 7.2.2 we can obtain a uniqueness result, so that if ρ satisfies*

$$t^* < \rho < \left[\frac{(1+p)^{d_0}}{2k_0 + k}\right]^{1/p}$$

then operator F is one-to-one in a neighborhood of x^*, since $x^* \in U(x_0, t^*)$. Note that r can replace t^* in Theorem 7.2.4 as well as in the above double inequality. That is x^* is an isolated zero of F in this case.

We can also provide a local convergence result for Newton's method:

Theorem 7.2.5 *Let $F\colon D \subseteq X \to Y$ be as in Theorem 7.2.4. Assume: $x^* \in D$ is an isolated zero of F; operator F has a p-(PBA) on D with modulus (ℓ, ℓ_0) at the point x^*. Moreover assume that the following hold:*

(a) $\delta(A(x^*, \cdot), D) \geq d^* > 0$;

(b) $d^* \geq \frac{2\ell_0 + \ell}{1+p}(r^*)^p$;

(c) for each $y \in U(0, d^* r^*)$ and provided $x_0 \in U^0(x^*, r^*)$ the equation $A(x_0, x) = y$ has a solution x satisfying

$$\|x - x^*\| \leq r^*$$

and

$$U(x^*, r^*) \subseteq D.$$

Then Newton's method generated by (7.2.11) is well defined, remains in $U(x^*, r^*)$ for all $n \geq 0$, and converges to x^*.

Proof. Simply replace x_0 by x^* in Lemma 7.2.3 and in the appropriate places in the proof of Theorem 7.2.4. ■

Example 7.2.1 *Let $X = Y = \mathbf{R}$, $D = [0, b]$ for some $b > 1$. Define function F on D by*

$$F(x) = \frac{x^{1+\frac{1}{i}}}{1+\frac{1}{i}} - b_0 x + b_1$$

for some real parameters b_0 and b_1, and an integer $i > 1$. Then no matter how operator A is chosen corresponding to (7.2.2), (7.2.3), conditions in [302] won't hold (i.e. set $p = 1$ in both (7.2.2) and (7.2.3)). However if for example A is given by (7.1.1) our conditions (7.2.2)–(7.2.3) hold for say $i = 2$, $p = \frac{1}{2}$, $k_0 = k = 1$ and $x_0 = 1$.

A more interesting application for Theorem 7.2.4 is given by Example 3.3.1 (alternative approach).

7.3 Convergence of Newton-Kantorovich Method. Case 3: Local Lipschitzian Assumptions

We need the definitions suitabloe for further weakening earlier hypotheses:

Definition 7.3.1 *Let F be an operator from a closed subset D of a metric space (X, m) to a normed linear space Y. Let $x_0 \in D$, $\varepsilon \geq 0$, and $\varepsilon_0 \geq 0$. We say F has an $(x_0, \varepsilon_0, \varepsilon)$ point-based approximation (PBA) on D if there exist an operator $A\colon D \times D \to Y$ and $R > 0$ such that for all $u, v, x, y \in U(x_0, R)$:*

(a)

$$U(x_0, R) \subseteq D, \tag{7.3.1}$$

(b)

$$\|F(v) - A(u,v)\| \le \frac{\varepsilon}{2} m(u,v), \tag{7.3.2}$$

(c)

$$\|[A(u,x) - A(v,x)] - [A(u,y) - A(v,y)]\| \le \varepsilon m(u,v), \tag{7.3.3}$$

and (d)

$$\|[A(u,x) - A(x_0,x)] - [A(u,y) - A(x_0,y)]\| \le \varepsilon_0 m(u,v). \tag{7.3.4}$$

Let X be also a normed linear space. Then, if F has a Fréchet-derivative F' which is $\frac{\varepsilon}{2}$-continuous on D, and we set $A(u,v) = F(u) + F'(u)(v-u)$, (b) gives

$$\|F(v) - F(u) - F'(u)(v-u)\| \le \frac{\varepsilon}{2}\|v-u\| \tag{7.3.5}$$

provided that D is also a convex set, where as Parts (c) and (d) are equivalent to the continuity property of F'.

Conditions (7.3.1)–(7.3.4) are weaker than the ones considered in [302, p. 293] or in Section 7.1. The advantages of such an approach have already been explained in Section 5.1 for the classical case. Similar advantages can be found in the nonsmooth case.

Let $\delta(F,D)$, $\delta_0(F,D)$, δ, δ_1 be as $\delta(F,D)$, $\delta_0(F,D)$, d, d_1 in Definition 7.1.2 respectively.

We need the following generalization of the classical Banach Lemma 1.3.2 on invertible operators in order to examine both the semilocal as well as the local convergence of Newton's method:

Lemma 7.3.2 *Let X be a Banach space, D a closed subset of X, and Y a normed linear space. Let F, G be operators from D into Y. Let $x_0 \in D$ and set $F(x_0) = y_0$. Let also G be Lipschitzian with modulus ε and center Lipschitzian at x_0 with modulus ε_0. Assume that:*

(a) $\overline{U}(y_0, \alpha) \subseteq F(D)$;

(b) $0 \le \varepsilon < \delta$;

(c) $\overline{U}(x_0, \delta_1^{-1}\alpha) \subseteq D$, and

(d) $\theta = (1 - \varepsilon_0\delta_1^{-1})\alpha - \|G(x_0)\| \ge 0$.

Then the following hold:

$$\overline{U}(y_0, \theta) \le (F+G)\big(\overline{U}(x_0, \delta_1^{-1}\alpha)\big)$$

and

$$\delta(F+G, D) \ge \delta - \eta > 0.$$

Proof. Define operator $Ty(x) = F^{-1}(y - G(x))$ for each fixed $y \in \overline{U}(y_0, \theta)$, and $x \in \overline{U}(x_0, \delta_1^{-1}\alpha)$. We can get

$$\begin{aligned}\|y - G(x) - y_0\| &\leq \|y - y_0\| + \|G(x) - G(x_0)\| + \|G(x_0)\| \\ &\leq \theta + \varepsilon_0(\delta_1^{-1}\alpha) + \|G(x_0)\| = \alpha.\end{aligned}$$

Therefore $Ty(x)$ is a singleton set since $\delta > 0$. That is Ty is an operator on $\overline{U}(x_0, \delta_1^{-1}\alpha)$. This operator maps $\overline{U}(x_0, \delta_1^{-1}\alpha)$ into itself. Indeed for $x \in \overline{U}(x_0, \delta_1^{-1}\alpha)$, we get:

$$m(Ty(x), x_0) = m(F^{-1}(y - G(x)), F^{-1}(y_0)) \leq \delta_1^{-1}\alpha.$$

Moreover let u, v be in $\overline{U}(x_0, \delta_1^{-1}\alpha)$, then

$$m(Ty(u), Ty(v)) \leq \delta^{-1}e[G(u), G(v)] \leq \delta^{-1}\varepsilon m(u, v).$$

It follows by hypothesis (7.3.5) that Ty is a contraction operator mapping closed ball $\overline{U}(x_0, \delta_1^{-1}\alpha)$ into itself and as such it has a unique fixed point $x(y)$ in this ball, since

$$\begin{aligned}\delta(F + G, D) &= \inf\{\|[F(u) - F(v)] + [G(u) - G(v)]\|,\ u \neq v,\ u, v, \in D\} \\ &\geq \delta - \sup\left\{\frac{\|G(u) - G(v)\|}{m(u, v)},\ u \neq v,\ u, v \in D\right\} \geq \delta - \eta > 0.\end{aligned}$$

Furthermore the above estimate shows $F + G$ is one-to-one on D. ∎

Remark 7.3.1 *In general*

$$\varepsilon_0 \leq \varepsilon \quad \text{and} \quad \delta \leq \delta_1 \tag{7.3.6}$$

hold. If strict inequality holds in any of the inequalities above, then Lemma 7.3.2 improves the corresponding Lemma 3.1 in [300] since the range for θ is enlarged and under the same computational cost, since the computation of ε (or δ) in practice requires the evaluation of ε_0 (or δ_1). Note also that $\frac{\varepsilon}{\varepsilon_0}$ and $\frac{\delta_1}{\delta}$ can be arbitrarily large (see Section 5.1).

The following lemma helps to establish the uniqueness of the solution x^* in the semilocal case.

Lemma 7.3.3 *Let X and Y be normed linear spaces, let D be a closed subset of X, and $x_0 \in D$, $\varepsilon_0 \geq 0$ and $\varepsilon \geq 0$. Let F has an $(x_0, \varepsilon_0, \varepsilon)$ (PBA) on D. Denote by d the quantity $\delta(A(x_0, \cdot), D)$. If $\overline{U}(x_0, \rho) \subseteq D$, then*

$$\delta(F, \overline{U}(x_0, \rho)) \geq d - \varepsilon_0\rho - \frac{1}{2}\varepsilon.$$

In particular, if $d > \varepsilon_0\rho + \frac{1}{2}\varepsilon$ then F is one-to-one on $\overline{U}(x_0, \rho)$.

Proof. Let y, z be points in $\overline{U}(x_0, \rho)$. Set $x = \frac{y+z}{z}$. We can write

$$F(y) - F(z) = [F(y) - A(x,y)] + [A(x,y) - A(x,z)] + [A(x,z) - F(z)].$$

Then we get

$$\|F(y) - A(x,y)\| \leq \frac{\varepsilon}{2}\|x - y\| = \frac{\varepsilon}{4}\|x - y\|,$$

and similarly,

$$\|F(z) - A(x,z)\| \leq \frac{\varepsilon}{4}\|x - z\|.$$

Using the estimate

$$\begin{aligned}\|A(x,u) - A(x,v)\| \geq \|A(x_0,u) - A(x_0,v)\| \\ - \|[A(x,u) - A(x_0,u)] - [A(x,v) - A(x_0,v)]\|,\end{aligned}$$

we get in turn

$$\begin{aligned}&\delta(A(x,\cdot), D) = \\ &= \inf\left\{\tfrac{\|A(x,u)-A(x,v)\|}{\|u-v\|},\ u \neq v,\ u,v \in D\right\} \geq \delta(A(x_0,\cdot), D) \\ &- \sup\left\{\tfrac{\|[A(x,u)-A(x_0,u)]-[A(x,v)-A(x_0,v)]\|}{\|u-v\|},\ u \neq v,\ u,v \in D\right\} \\ &\geq d - \varepsilon_0\|x - x_0\| \geq d - \varepsilon_0\rho.\end{aligned}$$

Hence, we get

$$\|F(y) - F(z)\| \geq (d - \varepsilon_0\rho)\|y - z\| - \frac{1}{2}\varepsilon\|y - z\|$$

and for $y \neq z$

$$\frac{\|F(y) - F(z)\|}{\|y - z\|} \geq d - \varepsilon_0\rho - \frac{1}{2}\varepsilon\,.$$

■

From now on we set $D = \overline{U}(x_0, R)$.

We can now state and prove the main semilocal convergence theorem for Newton's method:

Theorem 7.3.4 *Let X and Y be Banach spaces, D a closed convex subset of X, $x_0 \in D$, $\varepsilon > 0$, $\varepsilon_0 \geq 0$, and F a continuous operator from D into Y. Suppose that F has an $(x_0, \varepsilon_0, \varepsilon)$ (PBA) approximation on D. Moreover assume:*

(a) $\delta(A(x_0,\cdot), D) \geq d_0 \geq \varepsilon > 0$;

(b) $2\varepsilon_0\eta \leq \min\{2(d_0 - \varepsilon), d_0 + \frac{1}{2}\varepsilon\}$;

(c) for each $y \in \overline{U}(0, d_0\eta)$ the equation $A(x_0, x) = y$ has a solution x;
(d) the solution $S(x_0)$ of $A(x_0, S(x_0)) = 0$ satisfies $\|S(x_0) - x_0\| \leq \eta$, and
(e) $\overline{U}(x_0, 2\eta) \subseteq D$.
Then the Newton iteration defining x_{n+1} by

$$A(x_n, x_{n+1}) = 0 \quad (n \geq 0) \tag{7.3.7}$$

remains in $\overline{U}(x_0, 2\eta)$, and converges to a solution $x^* \in \overline{U}(x_0, 2\eta)$ of equation $F(x) = 0$. Moreover the following estimates hold for all $n \geq 0$:

$$\|x_{n+1} - x_n\| \leq t_{n+1} - t_n, \tag{7.3.8}$$

and

$$\|x_n - x^*\| \leq 2\eta - t_n \tag{7.3.9}$$

where $\lim_{n\to\infty} t_n = 2\eta$

$$t_0 = 0, \quad t_1 = \eta, \quad t_{n+2} = t_{n+1} + \frac{1}{2}(t_{n+1} - t_n).$$

Proof. We use Lemma 7.3.2 with quantities F, G, x_0, and y_0 replaced by $A(x, \cdot)$, $A(x_1, \cdot) - A(x, \cdot)$, $x_1 = S(x_0)$, and 0 respectively. Hypothesis (a) of Lemma 7.3.2 follows from the fact that $A(x_0, x) = y$ has a unique solution x_1 (since $\delta(A(x_0, \cdot), D) > 0$). For hypothesis (b) we have

$$\delta(A(x_1, \cdot), U) \geq \delta(A(x_0, \cdot), U) - \varepsilon_0\|x_1 - x_0\| \geq d_0 - \varepsilon_0\eta > 0.$$

To show (c) we must have $\overline{U}(x_1, \eta) \subseteq D$. Instead by hypothesis (e) of the theorem it suffices to show:

$$\overline{U}(x_1, \eta) \subset \overline{U}(x_0, 2\eta),$$

which is true since

$$\|x_1 - x_0\| + \eta \leq 2\eta.$$

Note that

$$\|A(x_0, x_1) - A(x_1, x_1)\| \leq \frac{\varepsilon}{2}\|x_0 - x_1\| \leq \frac{\varepsilon}{2}\eta$$

and since $d_0 \leq d_1 = \delta_0(A(x_0, \cdot), D)$, we get

$$\begin{aligned}
\theta &\geq (1 - 2\eta\varepsilon_0 d_1^{-1})d_0\eta - \|A(x_0, x_1) - A(x_1, x_1)\| \\
&\geq (1 - 2\eta\varepsilon_0 d_0^{-1})d_0\eta - \frac{1}{2}\varepsilon\eta = \left[d_0 - 2\varepsilon_0\eta - \frac{1}{2}\varepsilon\right]\eta \geq 0,
\end{aligned}$$

by hypothesis (b). That is condition (d) of the Lemma is also satisfied. It now follows from Lemma 7.3.2 that for each $y \in \overline{U}(0, 2\eta - \|x_1 - x_0\|)$, the equation $A(x_1, z) = y$ has a unique solution $z = x_2$ in U since $\delta(A(x_1, \cdot), D) > 0$. We also have $A(x_0, x_1) = A(x_1, x_2) = 0$, and $A(x_1, x_1) = F(x_1)$. By Definition 7.1.2 we get in turn:

$$\begin{aligned}\|x_2 - x_1\| &\leq \delta(A(x_1, \cdot), U)^{-1}\|A(x_1, x_2) - A(x_1, x_1)\| \\ &\leq d_0^{-1}(1 - d_0^{-1}\varepsilon\|x_1 - x_0\|)^{-1}\|A(x_0, x_1) - F(x_1)\| \\ &\leq d_0^{-1}(1 - d_0^{-1}\varepsilon_0\|x_1 - x_0\|)^{-1}\frac{\varepsilon}{2}\|x_1 - x_0\| \leq \frac{\eta}{2} = t_2 - t_1,\end{aligned}$$

by hypothesis (b). Hence, we showed:

$$\|x_{n+1} - x_n\| \leq t_{n+1} - t_n, \tag{7.3.10}$$

and

$$\overline{U}(x_{n+1}, 2\eta - t_{n+1}) \subseteq \overline{U}(x_n, 2\eta - t_n) \tag{7.3.11}$$

hold for $n = 0, 1$. Moreover for every $z \in \overline{U}(x_1, 2\eta - t_1)$,

$$\|z - x_0\| \leq \|z - x_1\| + \|x_1 - x_0\| \leq 2\eta - t_1 + t_1 = 2\eta - t_0,$$

implies $z \in \overline{U}(x_0, 2\eta - t_0)$. Given they hold for $n = 0, 1, \ldots, k$, then

$$\|x_{k+1} - x_0\| \leq \sum_{i=1}^{k+1}\|x_i - x_{i-1}\| \leq \sum_{i=1}^{k+1}(t_i - t_{i-1}) = t_{k+1} - t_0 = t_{k+1}.$$

The induction can easily be completed by simply replacing above x_0, x_1 by x_n, x_{n+1} respectively. Furthermore scalar sequence $\{t_n\}$ is Cauchy. From (7.3.10) and (7.3.11) it follows $\{x_n\}$ is Cauchy too in a Banach space X and as such it converges to some $x^* \in \overline{U}(x_0, 2\eta)$. Then we have

$$\begin{aligned}\|F(x_{n+1})\| = \|F(x_{n+1} - A(x_n, x_{n+1})\| &\leq \frac{\varepsilon}{2}\|x_{n+1} - x_n\| \\ &\leq \frac{\varepsilon}{2}(t_{n+1} - t_n).\end{aligned}$$

That is $\|F(x_{n+1})\|$ converges to 0 as $n \to \infty$. By the continuity of F we deduce $F(x^*) = 0$. Estimate (7.3.9) follows from (7.3.8) by using standard majorization techniques. Finally note

$$t_{n+2} = t_{n+1} + \frac{1}{2}(t_{n+1} - t_n) = \cdots = \frac{1 - \left(\frac{1}{2}\right)^{n+1}}{1 - \frac{1}{2}}\eta$$

and consequently

$$\lim_{n\to\infty} t_n = 2\eta.$$

■

Remark 7.3.2 *The uniqueness of the solution x^* was not studied in Theorem 7.3.4. Indeed, we do not know if under the conditions stated above the solution x^* is unique, say in $\overline{U}(x_0, 2\eta)$. However, we can obtain a uniqueness result by using Lemma 7.3.3 at the point x_0, so that if ρ satisfies*

$$2\eta < \rho < \left(d_0 - \tfrac{1}{2}\varepsilon\right)\varepsilon_0^{-1} = \varepsilon_1 \quad \textit{for } \varepsilon \in (0, 2d_0),\ \varepsilon > 0,\ \varepsilon_0 > 0,\ \textit{and } \overline{U}(x_0, \varepsilon_1) \subseteq D,$$

then operator F is one-to-one in a neighborhood of x^*, since $x^* \in \overline{U}(x_0, 2\eta)$. Therefore x^* is an isolated zero of F in this case.

Example 7.3.1 *Let us show how to choose operator A in some interesting cases. Let $B(u) \in L(X, Y)$ the space of bounded linear operators from X into Y, and define A by*

$$A(u, v) = F(u) + B(u)(v - u).$$

This is the class of the so-called Newton-like methods (see Section 5.1). Moreover assume there exist nonnegative, nondecreasing functions $\beta_t(p, q)$ and $\gamma(p)$ on $[0, +\infty) \times [0, +\infty)$ and $[0, +\infty)$ respectively, and $F'(u) \in L(X, Y)$ such that:

$$\|F'(x + t(y - x)) - B(x_0)\| \le \beta_t(\|x - x_0\|, \|y - x_0\|)$$

and

$$\|B(x) - B(x_0)\| \le \gamma(\|x - x_0\|)$$

for all $x, y \in \overline{U}(x_0, R)$, $t \in [0, 1]$. Then we can have

$$F(v) - F(u) - B(u)(v - u) = \int_0^1 [F'(u + t(v - u)) - B(x_0)](v - u)dt + (B(x_0) - B(u))(v - u).$$

Therefore we can define

$$\beta_t(p, q) = \ell[(1 - t)p + tq] \quad (\ell > 0)$$

and

$$\gamma(p) = \ell p.$$

In the special case of Newton's method, i.e. when $B(u) = F'(u)$ we can set

$$\varepsilon_0 = \varepsilon = 4\ell[\beta_t(R, R) + \gamma(R)] = 4\ell R.$$

In this case crucial condition (b) of Theorem 7.3.4 becomes

$$R \le \frac{d_0}{4\ell(1 + 2\eta)}.$$

We must also have $\overline{U}(x_0, 2\eta) \subseteq \overline{U}(x_0, R)$, *i.e.*

$$2\eta \leq R.$$

By combining all the above we deduce:

$$R \in \left[2\eta, \frac{d_0}{4\ell(1+2\eta)}\right]$$

provided that:

$$\eta \in \left[0, \frac{-\ell + \sqrt{\ell(\ell + d_0)}}{4\ell}\right].$$

Consider function F *defined on* $[0, \infty)$ *and given by*

$$F(x) = \frac{x^{1+\frac{1}{n}}}{1+\frac{1}{n}} \quad (n > 1).$$

Then F' *is center-Lipschitz with modulus* 1 *since for* $x_0 = 1$ *(see Section 5.2). However* F' *is not Lipschitz on* $[0, \infty)$. *Therefore the Newton–Kantorovich Theorem in the form given in [302] cannot apply here, whereas our Theorem 7.3.4 can.*

7.4 Convergence of the Secant Method Using Majorizing Sequences

We noticed (see the numerical example at the end of this section) that such an approximations of the kind discussed in the previous sections may not exist. Therefore in order to solve a wider range of problems we introduce a more flexible and precise point-based-approximation which is more suitable for Newton-like methods and in particular for Secant-type iterative procedures (see Section 5.1).

A local as well as a semilocal convergence analysis for the Secant method is provided, and our approach is justified through numerical examples.

We need a definition of a point-based-approximation (PBA) for operator F which is suitable for the Secant method.

Definition 7.4.1 *Let* F *be an operator from a closed subset* D *of a metric space* (X, m) *into a normed linear space* Y. *Operator* F *has a (PBA) on* D *at the point* $x_0 \in D$ *if there exists an operator* $A \colon D \times D \times D \to Y$ *and scalars* ℓ_0, ℓ *such that* u, v, w, x, y *and* z *in* D,

$$\|F(w) - A(u, v, w)\| \leq \ell m(u, w) m(v, w), \tag{7.4.1}$$

$$\|[A(x, y, z) - A(x_0, x_0, z)] - [A(x, y, w) - A(x_0, x_0, w)]\| \\ \leq \ell_0 [m(x, x_0) + m(y, x_0)] m(z, w), \tag{7.4.2}$$

and

$$\begin{aligned}\|[A(x,y,z) - A(u,v,z)] - [A(x,y,w) - A(u,v,w)]\| \\ \leq \ell[m(x,u) + m(u,v)]m(z,w).\end{aligned} \tag{7.4.3}$$

We then say A is a (PBA) for F.

This definition is suitable for the application of the Secant method. Indeed let X be also a normed linear space, D a convex set and F having a divided difference of order one on $D \times D$ denoted by $[x,y;F]$ and satisfying the standard condition [68], [99], [213], [217] (see also Chapter 2):

$$\|[u,v;F] - [w,x;F]\| \leq \ell(\|u-w\| + \|v-x\|) \tag{7.4.4}$$

for all u, v, w and x in D. If we set

$$A(u,v,w) = F(v) + [u,v;F](w-v) \tag{7.4.5}$$

then (7.4.1) becomes

$$\|F(w) - F(v) - [u,v;F](w-v)\| \leq \ell\|u-w\|\,\|v-w\|, \tag{7.4.6}$$

whereas (7.4.2) and (7.4.3) are equivalent to property (7.4.4) of linear operator $[\cdot,\cdot;F]$. Note that a (PBA) does not imply differentiability.

It follows by (7.4.1) that one way of finding a solution x^* of equation $F(x) = 0$ is to solve for w the equation

$$A(x,y,w) = 0 \tag{7.4.7}$$

provided that x and y are given.

We state the following generalization of the classical Banach Lemma 1.3.2 on invertible operators. The proof can be found in Lemma 7.1.3.

Lemma 7.4.2 *Let X, D and Y be as in Definition 7.4.1. Assume further X is a Banach space. Let F and G be operators from D into Y with G being Lipschitzian with modulus ℓ and center-Lipschitzian with modulus ℓ_0. Let $x_0 \in D$ with $F(x_0) = y_0$. Assume that:*

$$U(y_0,\alpha) \subseteq F(D); \tag{7.4.8}$$

$$0 \leq \ell < d; \tag{7.4.9}$$

$$U(x_0, d_1^{-1}\alpha) \subseteq D, \tag{7.4.10}$$

and

$$\theta_0 = (1 - \ell_0 d_1^{-1})\alpha - \|G(x_0)\| \geq 0. \tag{7.4.11}$$

Then the following hold:

$$U(y_0, \theta_0) \subseteq (F+G)(U(x_0, d_1^{-1}\alpha)) \tag{7.4.12}$$

and

$$\delta(F+G, D) \geq d - \ell > 0. \tag{7.4.13}$$

The advantages of this approach have been explained in Remark 7.1.1.

The following lemma is used to show uniqueness of the solution in the semilocal case and convergence of Secant method in the local case.

Lemma 7.4.3 *Let X and Y be normed linear spaces, and let D be a closed subset of X. Let $F\colon D \to Y$, and let A be a (PBA) for operator F on D at the point $x_0 \in D$. Denote by d the quantity $\delta(A(x_0, x_0, \cdot), D)$. If $U(x_0, \rho) \subseteq D$, then*

$$\delta(F, U(x_0, \rho)) \geq d - (2\ell_0 + \ell)\rho. \tag{7.4.14}$$

In particular, if

$$d - (2\ell_0 + \ell)\rho > 0,$$

then F is one-to-one on $U(x_0, \rho)$.

Proof. Let w, z be points in $U(x_0, \rho)$. We can write

$$\begin{aligned} F(w) - F(z) = {} & [F(w) - A(x, y, w)] + [A(x, y, w) - A(x, y, z)] \\ & + \ [A(x, y, z) - F(z)] \end{aligned} \tag{7.4.15}$$

By (7.4.1) we can have

$$\|F(w) - A(x, y, w)\| \leq \ell \|x - w\| \, \|y - w\|$$

and

$$\|F(z) - A(x, y, z)\| \leq \ell \|x - z\| \, \|y - z\|.$$

Moreover we can find

$$\begin{aligned} \|A(x, y, u) - A(x, y, v)\| \geq {} & \|A(x_0, x_0, u) - A(x_0, x_0, v)\| \\ & - \ \|[A(x, y, u) - A(x_0, x_0, u)] - [A(x, y, v) - A(x_0, x_0, v)]\| \end{aligned}$$

and therefore

$$\begin{aligned} & \delta(A(x, y, \cdot), D) \geq \\ & \geq \delta(A(x_0, x_0, \cdot), D) \\ & - \sup \left\{ \frac{\|[A(x, y, u) - A(x_0, x_0, u)] - [A(x, y, v) - A(x_0, x_0, v)]\|}{\|u - v\|}, \right. \\ & \qquad \qquad \left. u \neq v, \ u, v \in D \right\} \\ & \geq d - \ell_0(\|x - x_0\| + \|y - x_0\|) \geq d - 2\ell_0\rho. \end{aligned}$$

Furthermore, we can now have

$$\begin{aligned}\|F(w) - F(z)\| &\geq (d - 2\ell_0\rho)\|w - z\| - \ell[\|x - w\| \, \|y - w\| + \|x - z\| \, \|y - z\|] \\ &\geq (d - 2\ell_0\rho)\|w - z\| - \frac{\ell}{2}\|w - z\|^2\end{aligned}$$

and for $w \neq z$,

$$\frac{\|F(w) - F(z)\|}{\|w - z\|} \geq d - (2\ell_0 + \ell)\rho. \tag{7.4.16}$$

■

Remark 7.4.1 *In order for us to compare our result with the corresponding Lemma 2.4 in [302, p. 294], first note that if: (a) equality holds in both inequalities in (7.3.6), $u = v$ and $x = y$ in (7.4.1)–(7.4.3), then our result reduces to Lemma 2.4 by setting $\frac{k}{2} = \ell = \ell_0$. (b) Strict inequality holds in any of the inequalities in (7.3.6), $u = v$ and $x = y$ then our Lemma 7.4.3 improves (enlarges) the range for ρ, and under the same computational cost. The implications of that are twofold (see Theorems 7.4.5 and 7.4.6 that follows): in the semilocal case the uniqueness ball is more precise, and in the local case the radius of convergence is enlarged.*

We will need our result on majorizing sequences for the Secant method. The proof using conditions (C_1)–(C_3) can be found in [94], whereas for well-known condition (C_4) see e.g. [306]. Detailed comparisons between conditions (C_1)–(C_4) were given in [93]. In particular if strict inequality holds in (7.3.6) (first inequality) the error bounds under (C_1)–(C_3) are more precise and the limit of majorizing sequence more accurate than under condition (C_4).

Note also that these conditions are obtained as special cases of Lemma 5.1.1.

Lemma 7.4.4 *Assume for $\eta \geq 0$, $c \geq 0$, $d_0 > 0$: (C_1) for all $n \geq 0$ there exists $\delta \in [0, 1)$ such that:*

$$(C_1) \quad \ell\delta^n(1 + \delta)\eta + \frac{\delta\ell_0}{1 - \delta}(2 - \delta^{n+2} - \delta^{n+1})\eta + \delta\ell_0 c \leq \delta d_0,$$

and

$$\frac{\delta\ell_0}{1 - \delta}(2 - \delta^{n+2} - \delta^{n+1})\eta + \delta\ell_0 c < d_0,$$

or

(C_2) there exists $\delta \in [0, 1)$ such that:

$$\ell(1 + \delta)\eta + \frac{2\delta\ell_0\eta}{1 - \delta} + \delta\ell_0 c \leq \delta d_0,$$

and

$$\frac{2\delta\ell_0\eta}{1 - \delta} + \delta\ell_0 c < d_0,$$

or

(C_3) *there exists* $a \in [0,1]$ *and*

$$\delta \in \begin{cases} \left[0, \frac{-1+\sqrt{1+4a}}{2a}\right], & a \neq 0 \\ [0,1), & a = 0 \end{cases}$$

such that

$$(\ell + \delta \ell_0)(c+\eta) \le d_0 \delta, \quad \eta \le \delta c \quad \textit{and} \quad \ell_0 \le a\ell,$$

or

(C_4) $d_0^{-1}\ell c + 2\sqrt{d_0^{-1}\ell\eta} \le 1$ *for* $\ell_0 = \ell$.

Then,

(a) iteration $\{t_n\}$ $(n \ge -1)$ *given by*

$$t_{-1} = 0, \quad t_0 = c, \quad t_1 = c + \eta,$$
$$t_{n+2} = t_{n+1} + \frac{d_0^{-1}\ell(t_{n+1} - t_{n-1})}{1 - d_0^{-1}\ell_0[t_{n+1} - t_0 + t_n]}(t_{n+1} - t_n), \tag{7.4.17}$$

is non-decreasing, bounded above by r

$$r = \frac{\eta}{1-\delta} + c,$$

and converges to some t^* *such that*

$$0 \le t^* \le r.$$

Moreover, the following estimates hold for all $n \ge 0$*:*

$$0 \le t_{n+2} - t_{n+1} \le \delta(t_{n+1} - t_n) \le \delta^{n+1}\eta.$$

(b) Iteration $\{s_n\}$ $(n \ge -1)$ *given by*

$$s_{-1} - s_0 = c, \quad s_0 - s_1 = \eta,$$
$$s_{n+1} = s_{n+2} + \frac{d_0^{-1}\ell(s_{n-1} - s_{n+1})}{1 - d_0^{-1}\ell_0[(s_0 + s_{-1}) - (s_n + s_{n+1})]}(s_n - s_{n+1}), \tag{7.4.18}$$

provided that $s_{-1} \ge 0$, $s_0 \ge 0$, $s_1 \ge 0$ *is non-increasing, bounded below by* s *given by*

$$s = s_0 - \frac{\eta}{1-\delta},$$

and converges to some s^* *such that*

$$0 \le s^* \le s.$$

Moreover, the following estimates hold for all $n \geq 0$*:*

$$0 \leq s_{n+1} - s_n \leq \delta(s_n - s_{n+1}) \leq \delta^{n+1}\eta.$$

We denote by (C)*, Conditions* (C_1) *or* (C_2) *or* (C_3) *or* (C_4)*.*

We can state and prove the main semilocal convergence result for the Secant method involving a (PBA).

Theorem 7.4.5 *Let* X *and* Y *be Banach spaces,* D *a closed convex subset of* X*,* x_{-1}*, and* $x_0 \in D$ *with* $\|x_{-1} - x_0\| \leq c$*, and* F *a continuous operator from* D *into* Y*. Suppose operator* F *has a (PBA) on* D *at the point* x_0*. Moreover assume:*

$$\delta(A(x_{-1}, x_0, \cdot), D) \geq d_0 > 0;$$

Condition (C) *holds; for each* $y \in U(x_0, d_0(t^* - t_1)$ *the equation* $A(x_{-1}, x_0, x) = y$ *has a solution* x*; the solution* $S(x_{-1}, x_0)$ *of* $A(x_{-1}, x_0, S(x_{-1}, x_0)) = 0$ *satisfies*

$$\|S(x_{-1}, x_0) - x_0\| \leq \eta;$$

and

$$U(x_0, t^*) \subseteq D.$$

Then the Secant iteration defining x_{n+1} *by*

$$A(x_{n-1}, x_n, x_{n+1}) = 0 \tag{7.4.19}$$

remains in $U(x_0, t^*)$*, and converges to a solution* $x^* \in U(x_0, t^*)$ *of equation* $F(x) = 0$*. Moreover the following estimates hold for all* $n \geq 0$*:*

$$\|x_{n+1} - x_n\| \leq t_{n+1} - t_n \tag{7.4.20}$$

and

$$\|x_n - x^*\| \leq t^* - t_n, \tag{7.4.21}$$

where sequence $\{t_n\}$ *is defined by (7.4.17) and* $t^* = \lim_{n\to\infty} t_n$*.*

Proof. We use Lemma 7.4.2 with quantities F, G, x_0 and y_0 replaced by $A(w, x, \cdot)$, $A(x_0, x_1, \cdot) - A(v, x, \cdot)$, $x_1 = S(x_{-1}, x_0)$, and 0 respectively. Hypothesis (7.4.8) of the Lemma follows from the fact that $A(x_{-1}, x_0, x) = y$ has a unique solution x_1 (since $d_0 = \delta(A(x_{-1}, x_0, \cdot), D) > 0$).

For hypothesis (7.4.9) we have using (7.4.2)

$$\begin{aligned}\delta(A(x_0, x_1, \cdot), D) &\geq \delta(A(x_{-1}, x_0, \cdot), D) - \ell_0(\|x_0 - x_{-1}\| + \|x_1 - x_0\|)\\ &\geq d_0 - \ell_0 t_1 > 0 \quad \text{by } (C)).\end{aligned} \tag{7.4.22}$$

To show (7.4.10) we must have $U(x_1, t^* - t_1) \subseteq D$. Instead by hypothesis $U(x_0, t^*) \subseteq D$, it suffices to show:

$$U(x_1, t^* - t_1) \subset U(x_0, t^*)$$

which is true since

$$\|x_1 - x_0\| + t^* - t_1 \le t^*.$$

We also have by (7.4.1)

$$\|A(x_{-1}, x_0, x_1) - A(x_1, x_1, x_1)\| \le \ell(t_1 - t_{-1})(t_1 - t_0). \tag{7.4.23}$$

It follows by (7.4.11) and (7.4.23) that $\theta_0 \ge 0$ if

$$\theta_0 \ge [1 - \ell_0 d_1^{-1}(c + \eta)]d_0\eta - \ell(t_1 - t_{-1})(t_1 - t_0) \ge 0, \tag{7.4.24}$$

which is true by (7.4.17) and since $d_0 \le d_1 = \delta_0(A(x_{-1}, x_0, \cdot), D)$. It follows from Lemma 7.4.2 that for each $y \in U(0, r - \|x_1 - x_0\|)$, the equation $A(x_0, x_1, z) = y$ has a unique solution since $\delta(A(x_0, x_1, \cdot), D) > 0$. We also have

$$A(x_0, x_1, x_2) = A(x_{-1}, x_0, x_1) = 0.$$

By Definition 7.1.2, the induction hypothesis, (7.4.1) and (7.4.17) we get in turn:

$$\begin{aligned} \|x_2 - x_1\| &\le \delta(A(x_0, x_1, \cdot), D)^{-1}\|A(x_{-1}, x_0, x_1) - F(x_1)\| \\ &\le \frac{\ell\|x_{-1} - x_0\|\,\|x_0 - x_1\|}{d_0 - \ell_0(\|x_0 - x_{-1}\| + \|x_1 - x_0\|)} \\ &\le \frac{d_0^{-1}\ell(t_1 - t_{-1})(t_1 - t_0)}{1 - d_0^{-1}\ell_0 t_1} = t_2 - t_1. \end{aligned} \tag{7.4.25}$$

Hence we showed:

$$\|x_{n+1} - x_n\| \le t_{n+1} - t_n, \tag{7.4.26}$$

and

$$U(x_{n+1}, t^* - t_{n+1}) \subset U(x_n, t^* - t_n) \tag{7.4.27}$$

hold for $n = 0, 1$. Moreover for every $v \in U(x_1, t^* - t_1)$

$$\|v - x_0\| \le \|v - x_1\| + \|x_1 - x_0\| \le t^* - t_1 + t_1 - t_0,$$

implies $v \in U(x_0, t^* - t_0)$. Given they hold for $n = 0, 1, \ldots, j$, then

$$\|x_{j+1} - x_0\| \le \sum_{i=1}^{j+1} \|x_i - x_{i-1}\| \le \sum_{i=1}^{j+1}(t_i - t_{i-1}) = t_{j+1} - t_0.$$

The induction for (7.4.26) and (7.4.27) can easily be completed by simply replacing x_{-1}, x_0, x_1, by x_{n-1}, x_n, x_{n+1} respectively. Indeed corresponding to (7.4.24) we have:

$$\theta_0 \geq [1-\ell_0 d_1^{-1}(t_{n+1}-t_0+t_n)]d_0(t_{n+2}-t_{n+1})-\ell(t_{n+1}-t_{n-1})(t_{n+1}-t_n) = 0, \quad (7.4.28)$$

which is true by (7.4.17). Scalar sequence $\{t_n\}$ is Cauchy. From (7.4.26) and (7.4.27) it follows $\{x_n\}$ is Cauchy too in a Banach space X, and as such it converges to some $x^* \in U(x_0, t^*)$. Moreover we have by (7.4.1) and (7.4.26)

$$\begin{aligned}\|F(x_{n+1})\| &= \|F(x_{n+1}) - A(x_{n-1}, x_n, x_{n+1})\| \\ &\leq \ell\|x_n - x_{n-1}\| \, \|x_n - x_{n+1}\| \\ &\leq \ell(t_n - t_{n-1})(t_{n+1} - t_n) \longrightarrow 0 \quad \text{as } n \to \infty.\end{aligned}$$

By the continuity of F we deduce $F(x^*) = 0$. Finally estimate (7.4.21) follows from (7.4.20) by using standard majorization techniques. ■

Remark 7.4.2 *The uniqueness of the solution x^* was not considered in Theorem 7.4.5. Indeed, we do not know if under the conditions stated above the solution x^* is unique, say in $U(x_0, t^*)$. However using Lemma 7.4.3 we can obtain a uniqueness result, so that if ρ satisfies*

$$t^* < \rho < \frac{d_0}{2\ell_0 + \ell}, \quad (7.4.29)$$

then operator F is one-to-one in a neighborhood of x^*, since $x^* \in U(x_0, t^*)$. That is x^* is an isolated zero of F in this case.

The corresponding local convergence result for the Secant method is given by:

Theorem 7.4.6 *Assume: $x^* \in D$ is an isolated zero if F on D; operator F has a (PBA) on D at the point x^* of modulus (L, L_0). Moreover assume that the following hold:*

$$\delta(A(x^*, x^*, \cdot), D) \geq d^* > 0;$$

$$0 \leq r^* < \frac{d^*}{2L_0 + 3L};$$

for each $y \in U(0, r^* d^*)$ and provided $x_{-1}, x_0 \in U^0(x^*, r^*)$ the equation $A(x_{-1}, x_0, x) = y$ has a solution x satisfying

$$\|x - x^*\| \leq r^*;$$

and

$$U(x^*, r^*) \subseteq D.$$

Then Secant method $\{x_n\}$ *generated by (7.4.19) is well defined, remains in* $U(x^*, r^*)$ *for all* $n \geq 0$, *and converges to* x^*. *Moreover the following estimates hold for all* $n \geq 0$:

$$\|x_{n+1} - x^*\| \leq \frac{(d^*)^{-1} L(\|x_{n-1} - x_n\| + \|x_n - x^*\|)}{1 - (d^*)^{-1} L_0(\|x_{n-1} - x^*\| + \|x_n - x^*\|)} \|x_n - x^*\|. \quad (7.4.30)$$

Proof. The proof is omitted as it is similar to Theorem 7.4.5. Note that local results were not given in [302]. ∎

Example 7.4.1 *We show how to choose operator A in cases not covered in [302]. Let* $X = Y = (\mathbf{R}^2, \|\cdot\|_\infty)$. *Consider again the system given in Example 5.1.1. Set for* $v = (v_1, v_2)$, $\|v\|_\infty = \|(v_1, v_2)\|_\infty = \max\{|v_1|, |v_2|\}$, $F(v) = P(v) + Q(v)$, $P = (P_1, P_2)$, $Q = (Q_1, Q_2)$. *Define*

$$P_1(v) = 3v_1^2 v_2 + v_2^2 - 1, \quad P_2(v) = v_1^4 + v_1 v_2^3 - 1,$$

$$Q_1(v) = |v_1 - 1|, \quad Q_2(v) = |v_2|.$$

We shall take divided differences of order one $[x, y; P]$, $[x, y; Q] \in M_{2\times 2}(\mathbf{R})$ *to be for* $w = (w_1, w_2)$:

$$[v, w, P]_{i,1} = \frac{P_i(w_1, w_2) - P_i(v_1, w_2)}{w_1 - v_1},$$

$$[v, w, P]_{i,2} = \frac{P_i(v_1, w_2) - P_i(v_1, v_2)}{w_2 - v_2}$$

provided that $w_1 \neq v_1$ *and* $w_2 \neq v_2$. *If* $w_1 = v_1$ *or* $w_2 = v_2$ *replace* $[x, y; P]$ *by* P'. *Similarly we define*

$$[v, w; Q]_{i,1} = \frac{Q_i(w_1, w_2) - Q_i(v_1, w_2)}{w_1 - v_1},$$

$$[v, w; Q]_{i,2} = \frac{Q_i(v_1, w_2) - Q_i(v_1, v_2)}{w_2 - v_2}$$

for $w_1 \neq v_1$ *and* $w_2 \neq v_2$. *If* $w_1 = v_1$ *or* $w_2 = v_2$ *replace* $[x, y; Q]$ *by the zero* 2×2 *matrix in* $M_{2\times 2}(\mathbf{R})$. *We consider three interesting choices for operator A:*

$$A(v, v, w) = P(v) + Q(v) + P'(v)(w - v), \quad (7.4.31)$$

$$A(u, v, w) = P(v) + Q(v) + ([u, v; P] + [u, v; Q])(w - v) \quad (7.4.32)$$

and

$$A(u, v, w) = P(v) + Q(v) + (P'(v) + [u, v; Q])(w - v). \quad (7.4.33)$$

Using method (7.4.31) for $x_0 = (1,0)$, and both methods (7.4.32) and (7.4.33) for $x_{-1} = (5,5)$, $x_0 = (1,0)$ we obtain the following three tables respectively:

We did not verify the hypotheses of Theorem 7.4.5 for the above starting points. However, it is clear that the hypotheses of Theorem 7.4.5 are satisfied for all three methods for starting points closer to the solution

$$x^* = (.894655373334687, .327826521746298)$$

chosen from the lists of the same tables given in Section 5.1.

Note that the results in [303] cannot apply here because operator A no matter how it is chosen cannot satisfy the Lipschitz conditions (a) or (b) in Definition 2.1 in [302, p. 293] needed for the application of Theorem 3.2 in the same paper.

Another possible application is also given in the example at the end in Exercise 7.5.1.

7.5 Exercises

In the following two exercises we examine convergence of Newton's method using Hölderian operators for $p \in [0,1)$ and $p > 1$ respectively.

7.5.1 An Exercise

We need the following definition of a point-based approximation for p-Hölderian operators:

Definition 7.5.1 *Let f be an operator from a closed subset D of a metric space (X, m) to a normed linear space Y, and let $x_0 \in D$, $p \in (0,1]$. We say f has a point-based approximation (PBA) on D at $x_0 \in D$ if there exist an operator $A\colon D \times D \to Y$ and scalars ℓ, ℓ_0 such that for each u, and v in D,*

$$\|f(v) - A(u,v)\| \leq \ell m(u,v)^p \tag{7.5.1}$$

$$\|[A(u,x) - A(v,x)] - [A(u,y) - A(v,y)]\| \leq 2\ell m(u,v)^p \tag{7.5.2}$$

and

$$\|[A(u,v) - A(x_0,x)] - [A(u,y) - A(x_0,y)]\| \leq 2\ell_0 m(u,v)^p \tag{7.5.3}$$

for all $x, y \in D$.

Assume X is also a normed linear space and D is a convex set. Then this definition is suitable for operators f where Fréchet derivative f' is $(p-1)$-Hölderian with modulus $p\ell$ and $(p-1)$-center Hölderian with modulus $p\ell_0$. Indeed if we set

$$A(u,v) = f(u) + f'(u)(v-u),$$

then equation $F(x) = 0$ says

$$\|F(v) - f(u) - f'(u)(v-u)\| \leq \ell \|v-u\|^p,$$

where as parts (7.5.2) and (7.5.3) are equivalent to the Hölderian and center Hölderian property of f'.

Definition 7.5.2 *Let f be an operator from a metric space (X, m) into a normed linear space Y. We let*

$$\delta(f, X) = \inf\left\{\frac{\|F(u) - F(v)\|}{m(u,v)^p},\ u \neq v,\ u, v \in X\right\}. \tag{7.5.4}$$

Note that by (7.5.4) it follows F^{-1} (if it exists) is $\frac{1}{p}$-Hölderian with modulus $\delta(f,X)^{-1/p}$. We also define

$$\delta_0(f, X) = \inf\left\{\frac{\|F(u) - F(x_0)\|}{m(u,x_0)^p},\ u \neq x_0,\ u \in D\right\}.$$

Set $\delta = \delta(f, D)$ and $\delta_1 = \delta_0(f, D)$.

Lemma 7.5.3 *Let X be a Banach space, D a closed subset of X, and Y a normed linear space. Let f, y be operators from D into Y, g being p-Hölderian with modulus ℓ and center p-Hölderian with modulus ℓ_0. Let $x_0 \in D$ with $f(x_0) = y_0$. Assume that:*

$$U(y_0, \alpha) \subseteq F(D); \tag{7.5.5}$$

$$0 \leq \ell < \delta; \tag{7.5.6}$$

$$U\big(x_0, (\delta_1^{-1}\alpha)^{1/p}\big) \subseteq D; \tag{7.5.7}$$

and

$$\theta = (1 - \ell_0\delta_1^{-1})\alpha - \|g(x_0)\| \geq 0. \tag{7.5.8}$$

Then show the following hold:

$$(f+g)\big(U(x_0, (\delta_1^{-1}\alpha)^{1/p}\big) \supseteq U(y_0, \theta) \tag{7.5.9}$$

and

$$\delta(f+g, D) \geq \delta - \ell > 0. \tag{7.5.10}$$

Lemma 7.5.4 *Let X and Y be normed linear spaces, and let D be a closed subset of X. Let $f\colon D \to Y$, and let A be a PBA for f on D at $x_0 \in D$. Denote by d the quantity $\delta(A(x_0,\cdot), D)$. If $U(x_0, \rho) \subseteq D$, then show*

$$\delta(f, U(x_0, \rho)) \geq \delta_0 - 2\ell_0 - \frac{\ell}{2^{p-1}}.$$

In particular, if $\delta_0 > 2\ell_0 + \frac{\ell}{2^{p-1}}$, then show f is one-to-one on $U(x_0, \rho)$.

We need the following result on fixed points (see also Section 3.3):

Theorem 7.5.5 *Let* $Q\colon X \subset X \to Y$ *be a continuous operator, let* $p \in [0,1)$, $q \geq 0$ *and* $x_0 \in D$ *be such that*

$$\|Q(x) - Q(y)\| \leq q\|x-y\|^p \quad \textit{for all } x, y \in D \tag{7.5.11}$$

$$d \in [0,1), \tag{7.5.12}$$

and

$$U(x_0, r) \subseteq D \tag{7.5.13}$$

where,

$$b = q^{\frac{1}{1-p}}, \tag{7.5.14}$$

$$d = b^{-1}\|x_0 - Q(x_0)\|, \quad \textit{and} \quad r = \frac{b}{1-d^p}. \tag{7.5.15}$$

Then show: sequence $\{x_n\}$ $(n \geq 0)$ *generated by successive substitutions:*

$$x_{n+1} = Q(x_n) \quad (n \geq 0) \tag{7.5.16}$$

converges to a fixed point $x^* \in U(x_0, r)$ *of operator* Q*, so that for all* $n \geq 0$*:*

$$\|x_{n+1} - x_n\| \leq d^{p^n} b, \tag{7.5.17}$$

and

$$\|x_n - x^*\| \leq \frac{d^{np}}{1-d^p} b. \tag{7.5.18}$$

Moreover if

$$d_1 = b^{-1}(\|x_0 - Q(x_0)\| + qr) < 1, \tag{7.5.19}$$

or

$$q^{1/p} r < 1 \tag{7.5.20}$$

then show x^* *is the unique fixed point of* Q *in* $U(x_0, r)$.

If instead of (7.5.12) and (7.5.19) we assume

$$h = 2d^p < 1, \tag{7.5.21}$$

and

$$2b \leq q^{-1/p} \tag{7.5.22}$$

then the conclusions of Theorem 7.5.5 hold in the ball $U(x_0, r_0)$ where

$$r_0 = 2b. \tag{7.5.23}$$

We set

$$\|x_0 - Q(x_0)\| \leq \eta. \tag{7.5.24}$$

We can state the main semilocal convergence result for Newton's method involving a p-(PBA) approximation for f.

Theorem 7.5.6 *Let X and Y be Banach spaces, D a closed convex subset of X, $x_0 \in D$, and F a continuous operator from D into Y. Suppose that F has a p-(PBA) approximation at x_0. Moreover assume:*

$$\delta(A(x_0, \cdot), D) \geq \delta_0 > 0, \tag{7.5.25}$$

$$\eta \leq b, \quad 2\ell_0 < \delta_0, \quad (1 - 2\ell_0\delta_1^{-1}r_0)\delta_0\eta - \ell\eta^p \geq 0 \tag{7.5.26}$$

and conditions (7.5.8), (7.5.21), (7.5.22) are satisfied for $\alpha = \delta_0\eta$ and $q = \ell(\delta_0 - 2\ell_0)^{-1}$, respectively where,

$$\delta_1 = \delta_0(A(x_0, \cdot), D).$$

for each $y \in U(0, \delta_0\eta)$ the equation $A(x_0, x) = y$ has a solution x; the solution $T(x_0)$ of $A(x_0, T(x_0)) = 0$ satisfies $\|x_0 - T(x_0)\| \leq \eta$; and $U(x_0, r_0) \subseteq D$, where r_0 is given by (7.5.23).

Then show: the Newton iteration defining x_{n+1} by

$$A(x_n, x_{n+1}) = 0 \tag{7.5.27}$$

remains in $U(x_0, r_0)$, and converges to a solution $x^* \in U(x_0, r_0)$ of equation $F(x) = 0$, so that estimates (7.5.17) and (7.5.18) hold.

The uniqueness of the solution x^* was not considered in Theorem 7.5.6. Using Lemma 7.5.4 show: we can obtain a uniqueness result so that if

$$r_0 < \rho \quad \text{and} \quad \delta_0 > 2\ell_0 + \frac{\ell}{2^{p-1}},$$

then operator F is one-to-one in a neighborhood of x^*, since $x^* \in U(x_0, r_0)$. That is show: x^* is an isolated zero of F in this case.

7.5.2 Another Exercise

With the notations introduced in the previous subsection, assume $p > 1$.

We need the following result on fixed points (see also Section 3.3):

Theorem 7.5.7 *Let $Q\colon D \subset X \to X$ be an operator, p, q scalars with $p > 1$, $q \geq 0$, and x_0 a point in D such that*

$$\|Q(x) - Q(y)\| \leq q\|x - y\|^p \quad \text{for all } x, y \in D; \tag{7.5.28}$$

equation

$$2^{p-1}qr^p - r + \|x_0 - Q(x_0)\| = 0 \tag{7.5.29}$$

has a unique positive solution r in $I = \left[\|x_0 - Q(x_0)\|, \frac{1}{2}q^{\frac{1}{1-p}}\right)$; and

$$U(x_0, r) \subseteq D.$$

Then show: sequence $\{x_n\}$ $(n \geq 0)$ generated by successive substitutions

$$x_{n+1} = Q(x_n) \quad (n \geq 0) \tag{7.5.30}$$

converges to a unique fixed point $x^* \in U(x_0, r)$ of operator Q, so that for all $n \geq 1$

$$\|x_{n+1} - x_n\| \leq d\|x_n - x_{n-1}\| \leq d^n\|x_0 - Q(x_0)\| \tag{7.5.31}$$

and

$$\|x_n - x^*\| \leq \frac{d^n}{1-d}\|x_0 - Q(x_0)\| \tag{7.5.32}$$

where,

$$d = (2r)^{p-1}q. \tag{7.5.33}$$

Remark 7.5.1 *Conditions can be given to guarantee the existence and uniqueness of r. Indeed define scalar function h by*

$$h(t) = 2^{p-1}qt^p - t + \eta, \quad \|x_0 - Q(x_0)\| \leq \eta. \tag{7.5.34}$$

By the intermediate value theorem show (7.5.29) has a solution r in I if

$$h\left(\frac{1}{2}q^{\frac{1}{1-p}}\right) \leq 0 \tag{7.5.35}$$

or if

$$\eta \leq \frac{1}{2}q^{\frac{1}{1-p}}(1 - q^{-1}) = \eta_0 \tag{7.5.36}$$

and

$$q \geq 1. \tag{7.5.37}$$

This solution is unique if

$$\eta \leq \eta_1 \tag{7.5.38}$$

where,

$$\eta_1 = \min\left\{\eta_0, \frac{1}{2}(pq)^{\frac{1}{1-p}}\right\}. \tag{7.5.39}$$

Indeed if (7.5.37) and (7.5.38) hold, it follows that

$$h'(t) \leq 0 \quad on \quad I_1 = [\eta, \eta_1] \subset I.$$

Therefore h crosses the t-axis only once (since h is nonincreasing on I_1). Set

$$\overline{\eta} = \min\{\eta_1, \eta_2\} \tag{7.5.40}$$

where,

$$\eta_2 = (2^p q)^{\frac{1}{1-p}}. \tag{7.5.41}$$

It follows from (7.5.34) and (7.5.40) that if

$$\eta \leq \overline{\eta}, \tag{7.5.42}$$

then

$$r \leq 2\eta = r_0. \tag{7.5.43}$$

We can state the main semilocal convergence theorem for Newton's method involving a p-(PBA) ($p > 1$) approximation for f.

Theorem 7.5.8 *Let X and Y be Banach spaces, D a closed convex subset of X, $x_0 \in D$, and F a continuous operator from D into Y. Suppose that F has a p-(PBA) approximation at x_0. Moreover assume: $\delta(A(x_0, \cdot), D) \geq \delta_0 > 0$;*

$$2\ell_0 < \delta_0, \quad (1 - 2\ell_0\delta_1^{-1}r_0)\delta_0\eta - \ell\eta^p \geq 0 \tag{7.5.44}$$

where,

$$\delta_1 = \delta_0(A(x_0, \cdot), D);$$

(7.5.32) in Lemma 7.1.3 and conditions (7.5.37), (7.5.42) in Remark 7.5.1 hold for $\alpha = \delta_0\eta$ and $q = \ell(\delta_0 - 2\ell_0)^{-1}$; for each $y \in U(0, \delta_0\eta)$ the equation $A(x_0, x) = y$ has a solution x; the solution $T(x_0)$ of $A(x_0, T(x_0)) = 0$ satisfies $\|x_0 - T(x_0)\| \leq \eta$, and

$$U(x_0, r_0) \subseteq D,$$

where r_0 is given by (7.5.43). Then show: Newton iteration remains in $U(x_0, r_0)$, and converges to a solution $x^* \in U(x_0, r_0)$ of equation $F(x) = 0$, so that estimates (7.5.31) and (7.5.32) hold.

Remark 7.5.2 *Our Theorem 7.5.8 compares favorably with Theorem 3.2 in [302, p. 298]. First of all the latter theorem cannot be used when e.g.* $p \in [1,2)$ *(see the example that follows). In the case* $p = 2$ *our condition (7.5.44) becomes for* $\delta_0 = \delta_1$

$$h_0 = \delta_0^{-1}(\ell + 4\ell_0)\eta \leq 1 \tag{7.5.45}$$

where as the corresponding one in [302] becomes

$$h = 4\delta_0^{-1}\ell\eta \leq 1. \tag{7.5.46}$$

Clearly (7.5.45) is weaker than (7.5.46) if

$$\frac{\ell}{\ell_0} > \frac{4}{3}. \tag{7.5.47}$$

But $\frac{\ell}{\ell_0}$ *can be arbitrarily large. Therefore our Theorem 7.5.8 can be used in cases when Theorem 3.2 in [302] cannot when* $p = 2$.

Example 7.5.1 *Let* $X = \mathbf{R}^{m-1}$, $m \geq 2$ *an integer and define matrix operator* Q *on* X *by*

$$Q(z) = M + M_1(z), \quad z = (z_1, z_2, \ldots, z_{m-1}), \tag{7.5.48}$$

where M *is a real* $(m-1) \times (m-1)$ *matrix,*

$$M_1(z) = m_{ij} = \begin{cases} 0 & i \neq j, \\ z_i^p, & i = j \end{cases}$$

and e.g. $p \in [1,2)$.

Operators Q of the form (7.5.48) appear in many discretization studies in connection with the solution of two boundary value problems [68]. Clearly no matter how operator A is chosen the conditions in Definition 2.1 in [302, p. 293] cannot hold, whereas our result can apply.

Chapter 8

Applications of the weaker version of the Newton-Kantorovich theorem

Diverse further applications of the results in Chapter 5 are provided here.

8.1 Comparison with Moore Theorem

In this section we are concerned with the problem of approximating a locally unique solution x^* of equation

$$F(x) = 0 \tag{8.1.1}$$

where $F \colon D \subseteq \mathbb{R}^k \to \mathbb{R}^k$ is continuously differentiable on an open convex set D, and k is a positive integer.

Rall in [290] compared the theorems of Kantorovich and Moore. This comparison showed that the Kantorovich theorem has only a slight advantage over the Moore theorem with regard to sensitivity and precision, while the latter requires less computational cost. Later Neumaier and Shen [255] showed that when the derivative in the Krawczyk operator is replaced with a suitable slope then the corresponding existence theorem is at least as effective as the Kantorovich theorem with respect to sensitivity and precision. At the same time Zuhe and Wolfe [369] showed that the hypotheses in the affine invariant form of the Moore theorem are always implied by the Newton-Kantorovich theorem 4.2.4 but not necessarily vice versa.

Here we show that this implication is not true in general for a weaker version of theorem 4.2.4.

We will need the following semilocal convergence theorem for Newton's method due to Deuflhard and Heindl [143]:

Theorem 8.1.1 *Let F be a Fréchet-differentiable operator defined on an open convex subset D of a Banach space X with values in a Banach space Y. Suppose that $x_0 \in D$ is such that the Newton-Kantorovich hypotheses (4.2.51) holds and:*

$$F'(x_0)^{-1} \in L(Y, X), \tag{8.1.2}$$

$$\|F'(x_0)^{-1}F(x_0)\| \leq \eta, \tag{8.1.3}$$

$$\|F'(x_0)^{-1}[F'(x) - F'(y)]\| \leq \ell\|x - y\| \quad \text{for all } x, y \in D, \tag{8.1.4}$$

and

$$\bar{U}(x_0, r_1) \subseteq D \tag{8.1.5}$$

where,

$$r_1 = \frac{1 - \sqrt{1-h}}{\ell} \quad (\ell \neq 0). \tag{8.1.6}$$

Then sequence $\{x_n\}$ generated by method (4.1.3) is well defined, remains in $U(x_0, r_1)$ for all $n \geq 0$, and converges to a solution x^* of equation $F(x) = 0$ which is unique in $\overline{U}(x_0, r_1) \cup (D \cap U(x_0, r_2))$, where,

$$r_2 = \frac{1 + \sqrt{1-h}}{\ell}. \tag{8.1.7}$$

Moreover the following estimates hold for all $n \geq 0$:

$$\|x_{n+1} - x_n\| \leq s_{n+1} - s_n \tag{8.1.8}$$

and

$$\|x_n - x^*\| \leq r_1 - s_n, \tag{8.1.9}$$

where

$$s_0 = 0, \quad s_1 = \eta, \quad s_{n+2} = s_{n+1} + \frac{\ell(s_{n+1} - s_n)^2}{2(1 - \ell s_{n+1})} \quad (n \geq 0). \tag{8.1.10}$$

Remark 8.1.1 *In the case of Moore's theorem [240] suppose that hypotheses of Theorem 8.1.1 are valid with $X = Y = \mathbb{R}^k$, $x_0 = z = mid(\underline{x}_\gamma)$, where $\underline{x}_\gamma \in I(D)$ is given by*

$$\underline{x}_\gamma = U_\infty[z, \gamma] = [z - \gamma e, z + \gamma e], \; e = (1, \ldots, 1)^T \in \mathbb{R}^n, \; \gamma > 0, \tag{8.1.11}$$

and define the Krawczyk operator

$$\underline{K}(\underline{x}_\gamma) = w - \gamma\{F'(z)^{-1}\underline{F}[z, \underline{x}_\gamma] - I\}[-e, e], \tag{8.1.12}$$

where

$$w = z - F'(z)^{-1}F(z), \tag{8.1.13}$$

and

$$\underline{F}[z, \underline{x}_\gamma] = \int_0^1 \underline{F}'(z + t(\underline{x}_\gamma - z))dt \tag{8.1.14}$$

in which integration is defined as in [240].

The following result relates a weaker version of Theorem 8.1.1 with Moore's theorem.

Theorem 8.1.2 *Assume for* $x_0 = z = mid(\underline{x}_\gamma)$*: hypotheses* (8.1.2) *and* (8.1.3) *of Theorem 8.1.1 hold;*

$$\|F'(x_0)^{-1}[F'(x) - F'(x_0)]\| \leq \ell_0\|x - x_0\| \quad \textit{for all } x \in D; \tag{8.1.15}$$

$$\bar{h} = 2\ell_0\eta \leq 1; \tag{8.1.16}$$

and

$$\bar{U}(x_0, r_3) \subseteq D, \tag{8.1.17}$$

where,

$$r_3 = \frac{1 - \sqrt{1 - \bar{h}}}{\ell_0} \quad (\ell_0 \neq 0). \tag{8.1.18}$$

Then $\underline{K}(\underline{x}_\gamma) \subseteq \underline{x}_\gamma$,

$$\underline{K}\left(\underline{x}_\gamma\right) \subseteq \underline{x}_\gamma, \tag{8.1.19}$$

where $\underline{x}_\gamma$, $\underline{K}$ *are given by* (8.1.11) *and* (8.1.12)–(8.1.14) *respectively.*

Proof. If $u \in \underline{K}(\underline{x}_\gamma)$, then $u = w + v$

$$v = \gamma\{[F'(z)^{-1}\underline{F}[z, \underline{x}_\gamma] - I\}[-e, e] \tag{8.1.20}$$

(by (8.1.13) and (8.1.14)). We have in turn

$$\begin{aligned} \|z - u\|_\infty &\leq \|z - w\|_\infty + \|v\|_\infty \\ &\leq \eta + \gamma\|F'(z)^{-1}\{\underline{F}[z, \underline{x}_\gamma] - F'(z)\}\|_\infty \\ &\leq \eta + \tfrac{1}{2}\ell_0\gamma^2. \end{aligned} \tag{8.1.21}$$

Hence (8.1.19) holds if

$$\eta + \tfrac{1}{2}\ell_0\gamma^2 \leq \gamma, \tag{8.1.22}$$

which is true by (8.1.16)–(8.1.18). ■

Remark 8.1.2 *(a) Note that Moore's theorem [240] guarantees the existence of a solution x^* of equation* (8.1.1) *if* (8.1.19) *holds. (b) Theorem 8.1.2 reduces to Theorem 8.1.2 in [369] if $\ell_0 = \ell$. However in general*

$$\ell_0 \leq \ell \tag{8.1.23}$$

holds. Note also that

$$h \leq 1 \Rightarrow \bar{h} \leq 1 \tag{8.1.24}$$

but not vice versa unless if $\ell_0 = \ell$. Hence our Theorem 8.1.2 weakens Theorem 2 in [369], and can be used in cases not covered by the latter.

An example was given in [255] and [369] to show that the hypotheses of the affine invariant form of the Moore theorem may be satisfied even when those of Theorem 8.1.1 are not.

Example 8.1.1 *(used also in Section 5.1) Let $k = 1$, $a \in \left[0, \frac{1}{2}\right)$, $x \in \underline{x}_\gamma = [a, 2-a]$ and define function $F\colon \underline{x}_\gamma \to \mathbb{R}$ by*

$$F(x) = x^3 - a. \tag{8.1.25}$$

Choose $z = mid(\underline{x}_\gamma)$ then (4.2.51) becomes

$$h = \tfrac{4}{3}(1-a)(2-a) > 1 \quad \textit{for all } a \in \left[0, \tfrac{1}{2}\right). \tag{8.1.26}$$

That is, there is no guarantee that method generated by (4.1.3) converges to a solution of $F(x) = 0$, since the Newton–Kantorovich hypothesis (4.2.51) is violated. However using (8.1.11)–(8.1.14) and (8.1.25) we get

$$\underline{K}(\underline{x}_\gamma) = \tfrac{1}{3}[a^3 - 6a^2 + 10a + 2, -a^3 + 3a^2 - 8a + 6] \tag{8.1.27}$$

and if

$$a \in \left[.44, \tfrac{1}{2}\right) \tag{8.1.28}$$

then (8.1.19) holds. That is Moore's theorem guarantees convergence of method (4.1.3) to a solution x^* of equation $F(x) = 0$ provided that (8.1.28) holds.

However we can do better. Indeed, since $\ell_0 = 3 - a$

$$\bar{h} = \tfrac{2}{3}(1-a)(3-a) \leq 1 \tag{8.1.29}$$

if

$$a \in \left[\tfrac{4-\sqrt{10}}{2}, \tfrac{1}{2}\right), \tag{8.1.30}$$

which improves (8.1.28).

Remark 8.1.3 *If* (4.2.51) *holds as a strict inequality method* (4.1.3) *converges quadratically to* x^*. *However* (8.1.17) *guarantees only the linear convergence to* x^* *of the modified method*

$$y_{n+1} = y_n - F'(y_0)^{-1}F(y_n) \quad (y_0 = x_0), \ (n \geq 0). \tag{8.1.31}$$

*In practice the quadratic convergence of method (4.1.3) is desirable. In Section 5.1. we found (see condition (*5.1.56*)) a condition weaker than* (4.2.51) *but stronger than* (8.1.16) *so that the quadratic convergence of method (4.1.3) is guaranteed (see condition (5.1.56)).*

8.2 Miranda Theorem

In this section we are concerned with the problem of approximating a locally unique solution x^* of an equation

$$F(x) = 0, \tag{8.2.1}$$

where F is defined on an open convex subset S of $\mathbb{R}^n$ (n a positive integer) with values in $\mathbb{R}^n$.

We showed in Section 5.1 that the Newton–Kantorovich hypothesis (4.2.51) can always be replaced by the weaker (5.1.56). Here we first weaken the generalization of Miranda's theorem (Theorem 4.3 in [227]). Then we show that operators satisfying the weakened Newton–Kantorovich conditions satisfy those of the weakened Miranda's theorem.

In order for us to compare our results with earlier ones we need to list the following theorems guaranteeing the existence of solution x^* of equation (8.2.1)(see Chapter 4).

Theorem 8.2.1 *Let* $F: S \to \mathbf{R}^n$ *be a Fréchet-differentiable operator. Assume: there exists* $x_0 \in S$ *such that* $F'(x_0)^{-1} \in L(Y, X)$, *and set*

$$G(x) = F'(x_0)^{-1}F(x) \quad (x \in S); \tag{8.2.2}$$

there exists an $\eta \geq 0$ *such that*

$$\|G(x_0)\| \leq \eta; \tag{8.2.3}$$

there exists an $\ell > 0$ *such that*

$$\|G'(x) - G'(y)\| \leq \ell\|x - y\| \quad \textit{for all } x, y \in S; \tag{8.2.4}$$

$$h = 2\ell\eta \leq 1, \tag{8.2.5}$$

and

$$\bar{U}(x_0, r^*) \subseteq S, \tag{8.2.6}$$

where,

$$r^* = \frac{1 - \sqrt{1-h}}{\ell}. \tag{8.2.7}$$

Then there exists a unique solution $x^* \in \bar{U}(x_0, r^*)$ of equation $F(x) = 0$.

Remark 8.2.1 *Theorem 8.2.1 is the portion of the Newton–Kantorovich theorem (4.2.4). The following theorem is due to Miranda [236], which is a generalization of the intermediate value theorem:*

Theorem 8.2.2 *Let $b > 0$, $x_0 \in R^n$, and define*

$$Q = \{x \in \mathbb{R}^n \mid \|x - x_0\|_\infty \leq b\}, \tag{8.2.8}$$

$$Q_k^+ = \{x \in Q \colon x_k = x_{0,k} + b\}, \tag{8.2.9}$$

and

$$Q_k^- = \{x \in Q \colon x_k = x_{0,k} - d\} \quad \text{for } k = 1, \ldots, n. \tag{8.2.10}$$

Let $F = (F_k)$ $(1 \leq k \leq n) \colon Q \to R^n$ be a continuous operator satisfying for all $k = 1, 2, \ldots, n$,

$$F_k(x) F_k(y) \leq 0 \quad \text{for all } x \in Q_k^+ \text{ and } y \in Q_k^-. \tag{8.2.11}$$

Then there exists $x^* \in Q$ such that $F(x^*) = 0$. The following result connects Theorems 8.2.1 and 8.2.2.

Theorem 8.2.3 *Suppose $F \colon S \to R^n$ satisfies all hypotheses of Theorem 8.2.1 in the maximum norm, then G satisfies the conditions of Theorem 8.2.3 on $\bar{U}(x_0, r^*)$.*

In the elegant study [227] a generalization of Theorem 8.2.3 was given (Theorem 4.3). We first weaken this generalization.

Let $\mathbb{R}^n$ be equipped with a norm denoted by $\|\cdot\|$ and $\mathbb{R}^{n\times n}$ with a norm $\|\cdot\|$ such that $\|M \cdot x\| \leq \|M\| \cdot \|x\|$ for all $M \in \mathbb{R}^{n\times n}$ and $x \in \mathbb{R}^n$. Choose constants $c_0, c_1 > 0$ such that for all $x \in \mathbb{R}^n$

$$c_0 \|x\|_\infty \leq \|x\| \leq c_1 \|x\|_\infty, \tag{8.2.12}$$

since all norms on finite dimensional spaces are equivalent.

Set

$$c = \frac{c_0}{c_1} \leq 1. \tag{8.2.13}$$

Definition 8.2.4 *Let $S \subseteq \mathbb{R}^n$ be an open convex set, and let $G\colon S \to \mathbb{R}^n$ be a differentiable operator on S. Let $x_0 \in S$, and assume:*

$$G'(x_0) = I \quad \text{(the identity matrix)} \tag{8.2.14}$$

there exists $\eta \geq 0$ such that

$$\|G(x_0)\| \leq \eta; \tag{8.2.15}$$

there exists an $\ell_0 \geq 0$ such that

$$\|G'(x) - G'(x_0)\| \leq \ell_0 \|x - x_0\| \quad \text{for all } x \in S. \tag{8.2.16}$$

Define:

$$h_0 = 2\ell_0\eta. \tag{8.2.17}$$

We say that G satisfies the weak center-Kantorovich conditions in x_0 if

$$h_0 \leq 1. \tag{8.2.18}$$

We also say that G satisfies the strong center-Kantorovich conditions in x_0 if

$$h_0 \leq \frac{c^2}{2}. \tag{8.2.19}$$

Moreover define

$$r_1 = \frac{c - \sqrt{c^2 - h_0}}{\ell_0}, \tag{8.2.20}$$

$$r_2 = \frac{c + \sqrt{c^2 - h_0}}{\ell_0} \tag{8.2.21}$$

and

$$R = [r_1, r_2] \quad \text{for } \ell_0 \neq 0. \tag{8.2.22}$$

Furthermore if $\ell_0 = 0$, define

$$r_1 = \frac{\eta}{c} \tag{8.2.23}$$

and

$$R = [r_1, \infty). \tag{8.2.24}$$

Remark 8.2.2 *The weak and strong center-Kantorovich conditions are equivalent only for the maximum norm.*

As in [196] we need to define certain concepts. Let $r > 0$, $x_0 \in \mathbb{R}^n$, and define

$$\begin{aligned} U(r) &= \{z \in \mathbb{R}^n \mid \|z\| \le r\}, && (8.2.25)\\ U(x_0, r) &= \{x = x_0 + z \in \mathbb{R}^n \mid z \in U(r)\}, && (8.2.26)\\ U_k^+(r) &= \{z \in \mathbb{R}^n \mid \|z\| = r, z_k = \|z\|_\infty\}, && (8.2.27)\\ U_k^-(r) &= \{z \in \mathbb{R}^n \mid \|z\| = r, z_k = -\|z\|_\infty\}, && (8.2.28)\\ U_k^+(x_0, r) &= \{x = x_0 + z \in \mathbb{R}^n \mid z \in U_k^+(r)\}, && (8.2.29)\\ U_k^-(x_0, r) &= \{x = x_0 + z \in \mathbb{R}^n \mid z \in U_k^-(r)\} \quad \text{for all } k = 1, 2, \ldots, n. && (8.2.30) \end{aligned}$$

We show the main result:

Theorem 8.2.5 *Let $G\colon S \to R^n$ be a differentiable operator defined on an open convex subset of R^n. Assume G satisfies the strong center-Kantorovich conditions. Then, for any $r \in R$ with $U(x_0, r) \subseteq S$ the following hold:*

$$\text{(a)} \qquad U = U(r) = U(x_0, r) \quad \textit{is a Miranda domain [227]}, \tag{8.2.31}$$

and

$$U_1 = U_1(r) = \{U_1^+(x_0, r),\ U_1^-(x_0, r), \ldots, U_n^+(x_0, r),\ U_n^-(x_0, r)\} \tag{8.2.32}$$

is a Miranda partition [227] of the boundary ∂U. It is a canonical Miranda partition [227] for $r > 0$ and a trivial Miranda domain and partition for $r = 0$;

$$\text{(b)} \qquad G_k(x) \ge 0 \quad \textit{for all} \quad x \in U_k^+(x_0, r), \quad k = 1, \ldots, n \tag{8.2.33}$$

and

$$G_k(x) \le 0 \quad \textit{for all} \quad x \in U_k^-(x_0, r), \quad k = 1, \ldots, n; \tag{8.2.34}$$

(c) G satisfies the Miranda conditions on (U, U_1); (d) if $G(x_0) = 0$ and $\ell_0 > 0$, then G satisfies the Miranda conditions on (U, U_1) for any $r \in \left[0, \frac{2c}{\ell_0}\right]$ such that $U(x_0, r) \subseteq S$.

Proof. The first point of the theorem follows exactly as in Theorem 4.3 in [227]. For the rest we follow the proof of Theorem 4.3 in [227] (which is essentially the reasoning of Theorem 8.2.3) but with some differences stretching the use of center-Lipschitz condition (8.2.16) instead of the stronger Lipschitz condition (8.2.4) which is not really needed in the proof. However it was used in both proofs mentioned above. Using the intermediate value theorem for integration we first obtain the identity

$$G(x_0 + rz) - G(x_0) = \tag{8.2.35}$$

$$\begin{aligned}
&= \int_0^1 G'(x_0 + rtz)rz\,dt \\
&= \int_0^1 [G'(x_0 + rtz) - G'(x_0)]rz\,dt + \int_0^1 G'(x_0)rz\,dt \\
&= \int_0^1 [G'(x_0 + rtz) - G'(x_0)]rz\,dt + rz\int_0^1 dt \quad (G'(x_0) = I).
\end{aligned} \tag{8.2.36}$$

Let e_k denote the k-th unit vector. Then we can have:

$$G_k(x_0 + rz) = G_k(x_0) + \int_0^1 e_k^T[G'(x_0 + rtz) - G'(x_0)]rz\,dt + rz_k, \tag{8.2.37}$$

and by (8.2.16)

$$\begin{aligned}
&\left|\int_0^1 e_k^T[G'(x_0 + rtz) - G'(x_0)]rz\,dt\right| \leq \\
&\leq \int_0^1 \left|e_k^T[G'(x_0 + rtz) - G'(x_0)]rz\,dt\right| \\
&\leq \int_0^1 \frac{1}{c_1}\|(G'(x_0 + rtz) - G'(x_0))rz\|\,dt \\
&\leq \int_0^1 \frac{1}{c_1}\|G'(x_0 + rtz) - G'(x_0)\|\,\|rz\|\,dt \\
&\leq \frac{1}{c_1}\int_0^1 \ell_0\|rtz\|\,\|rz\|\,dt \\
&= \frac{\ell_0 r^2}{c_1}\int_0^1 t\,dt = \frac{\ell_0 r^2}{2c_1}.
\end{aligned} \tag{8.2.38}$$

Let $z \in U_k^+(1)$. Using (8.2.38), and

$$|G_k(x_0)| \leq \frac{1}{c_1}\|G(x_0)\| \leq \frac{\eta}{c_1}, \tag{8.2.39}$$

$$z_k \geq \frac{1}{c_2} \quad \text{for } u \in U_k^+(1), \tag{8.2.40}$$

$$z_k \leq -\frac{1}{c_2} \quad \text{for } u \in U_k^-(1), \tag{8.2.41}$$

we get from (8.2.37)

$$G_k(x_0 + rz) \geq -|G(x_0)| - \frac{\ell_0 r^2}{2c_1} + rz_k \geq -\frac{\eta}{c_1} - \frac{\ell_0 r^2}{2c_1} + \frac{r}{c_2} \geq 0 \tag{8.2.42}$$

for $r \in R$ (by (8.2.19) and (8.2.22)). Similarly,

$$G_k(x_0 + rz) \leq |G(x_0)| + \frac{\ell_0 r^2}{2c_1} + rz_k \leq \frac{\eta}{c_1} + \frac{\ell_0 r^2}{2c_1} - \frac{r}{c_2} \leq 0 \tag{8.2.43}$$

for $r \in R$. If $G(x_0) = 0$, let $\eta = 0$ which implies $h_0 = 0$. ■

Remark 8.2.3 *If $\ell = \ell_0$, then our Theorem 8.2.5 becomes Theorem 4.3 in [227]. Moreover if $\|\cdot\|$ is the maximum norm then Theorem 8.2.5 becomes Theorem 8.2.3. However in general*

$$\ell_0 \leq \ell. \tag{8.2.44}$$

Hence the strong Kantorovich condition is such that

$$h_1 = 2\ell\eta \leq \frac{c^2}{2}, \tag{8.2.45}$$

$$h_1 \leq \frac{c^2}{2} \Rightarrow h_0 \leq \frac{c^2}{2}. \tag{8.2.46}$$

Similarly the Kantorovich condition (8.2.5) is such that

$$h \leq 1 \Rightarrow h_2 = 2\ell_0\eta \leq 1, \tag{8.2.47}$$

but not vice versa unless if $\ell_0 = \ell$. If strict inequality holds in (8.2.44) and conditions (8.2.45) or (8.2.5) *are not satisfied then the conclusions of Theorem 4.3 in [227] respectively do es not necessarily hold. However if* (8.2.8) *holds the conclusions of our Theorem 8.2.5 hold.*

Remark 8.2.4 *Condition* (8.2.5) *guarantees the quadratic convergence of Newton-Kantorovich method to x^*. However this is not the case for condition* (8.2.18). *To rectify this and still use a condition weaker than* (8.2.5) *(or* (8.2.45)*) define*

$$p(\delta) = p_\delta = (\ell + \delta\ell_0)\eta, \quad \delta \in [0, 2). \tag{8.2.48}$$

We showed in Section 5.1 that if

$$p_\delta \leq \delta, \tag{8.2.49}$$

$$\frac{2\ell_0\eta}{2-\delta} \leq 1, \tag{8.2.50}$$

and

$$\frac{\ell_0\delta^2}{2-\delta} \leq \ell \tag{8.2.51}$$

replace (8.2.5) *then Newton-Kantorovich method converges to $x^* \in \bar{U}(x_0, r_3)$ with $r_3 \leq r^*$. Moreover finer error bounds on the distances between iterates or between* *iterates and x^* are obtained. If we restrict $\delta \in [0,1]$, then (8.2.50) and (8.2.51) hold if only (8.2.49) is satisfied. Choose $\delta = 1$ for simplicity in (8.2.49). Then again*

$$h \leq 1 \Rightarrow p_1 \leq 1, \tag{8.2.52}$$

$$h_1 \leq \frac{c^2}{2} \Rightarrow p_1 \leq \frac{c^2}{2} \tag{8.2.53}$$

but not vice versa unless if $\ell = \ell_0$. Hence if (8.2.18) is replaced by

$$p_1 \leq \frac{c^2}{2} \tag{8.2.54}$$

then all conclusions of Theorem 8.2.5 hold.

8.3 Computation of Orbits

In this section we are concerned with the problem of approximating shadowing orbits for dynamical systems. It is well known in the theory of dynamical systems that actual computations of complicated orbits rarely produce good approximations to the trajectory. However under certain conditions, the computed section of an orbit lies in the shadow of a true orbit. Hence using product spaces and the results in Section 5.1, we show that the sufficient conditions for the convergence of Newton's method to a true orbit can be weakened under the same computational cost as in the elegant work by Hadeller in [177]. Moreover the information on the location of the solutions is more precise and the corresponding error bounds are finer.

Let f be a Fréchet-differentiable operator defined on an open convex subset D of a Banach space X with values in X.

The operator f defines a local dynamical system as follows:

$$x_{n+1} = f(x_k) \quad (x_0 \in D) \tag{8.3.1}$$

as long as $x_k \in D$.

A sequence $\{x_m\}_{i=0}^{N}$ in D with $x_{m+1} = f(x_m)$, $i = 0, ..., N-1$ is called an orbit. Any sequence $\{x_m\}_{i=0}^{N}$, $x_m \in D$, $m = 0, ..., N$ is called a pseudo-orbit of length N.

We can now pass to product spaces. Let $y = X^{N+1}$ equipped with maximum norm. The norm $\mathbf{x} = (x_0, ..., x_N) \in Y$ is given by

$$\|\mathbf{x}\| = \max_{0 \leq m \leq N} \|x_m\|.$$

Set $S = D^{N+1}$. Let $F : S \to Y$ be an operator associated with $\dot{f}$:

$$F(\mathbf{x}) = F \begin{bmatrix} x_o \\ \vdots \\ x_N \end{bmatrix} = \begin{bmatrix} f(x_0) - x_1 \\ \vdots \\ f(x_N) \end{bmatrix}.$$

Assume there exist constants l_0, l, L_0, L such that:

$$\begin{aligned}\left\|f'(u)-f'(x_0)\right\| &\leq l_0\left\|u-x_0\right\|\\ \left\|f'(u)-f'(v)\right\| &\leq l\left\|u-v\right\|\\ \left\|F'(\mathbf{u})-F'(\mathbf{x}_0)\right\| &\leq L_0\left\|\mathbf{u}-\mathbf{x}_0\right\|\\ \left\|F'(\mathbf{u})-F'(\mathbf{v})\right\| &\leq L\left\|\mathbf{u}-\mathbf{v}\right\|\end{aligned}$$

for all $u, v \in D$, $\mathbf{u}, \mathbf{v} \in S$.

From now on we assume: $l_0 = L_0$ and $l = L$.

For $y \in X$ define an operator

$$F_y : S \rightarrow Y$$

by

$$F_y(\mathbf{x}) = \left\| \begin{array}{c} f(x_0) - x_1 \\ \vdots \\ f(x_N) - y \end{array} \right\|.$$

It follows that $F_y'(\mathbf{x}) = F'(\mathbf{x})$.

As in [166] define the quantities

$$a(\mathbf{x}) = \max_{0 \leq i \leq N} \sum_{j=i}^{N} \left\|f'(x_i) \ldots f'(x_j)\right\|^{-1},$$

$$b(\mathbf{x}) = \max_{0 \leq i \leq N-1} \left\| \sum_{j=i}^{N-1} f'(x_i) \ldots f'(x_j)\left(f(x_j) - x_{j+1}\right) \right\|,$$

and

$$b_y(\mathbf{x}) = b(\mathbf{x}) + \left\|f'(x_N)\left(f(x_N) - y\right)\right\|$$

for $x \in Y$.

That is $a(x)$ is the operator norm of $F'(x)^{-1}$ and by (x) is the norm of the Newton convection $F'(\mathbf{x})^{-1} F_y(\mathbf{x})$.

Remark 8.3.1 *The interpretation to the measures is given by: (a) The dilation measures $a(\mathbf{x})$ and by $(\mathbf{x})$ are the norm of $F'(x)^{-1}$ and the norm of the Newton-correction $F'(x)^{-1} F_y(x)$, respectively; (b) the solution of equation*

$$F_y(\mathbf{x}) = 0$$

yields a section $(x_0, \ldots, x_N)$ of length $N+1$ of a true orbit which meets the prescribed point y at the N-th iteration step [177].

We obtain the following existence and uniqueness result for a true orbit:

Theorem 8.3.1 *Let* $\mathbf{x} \in Y$, $y \in X$ $a(\mathbf{x})$, *by* $(\mathbf{x})$ *be as above and* $\{x_i\}_{i=0}^{N}$ *be a pseudo-orbit. Assume:*

$$h_0 = \bar{L}, \quad a(\mathbf{x})\, by(\mathbf{x}) \leq \tfrac{1}{2}, \;\; \bar{L} = \tfrac{L_0 + L}{2} \tag{8.3.2}$$

and

$$\overline{U}(\mathbf{x}_0 = \mathbf{x}, \quad r^* = 2by(\mathbf{x}) = \{\mathbf{z} \in Y \mid \|\mathbf{x} - \mathbf{z}\| \leq r^*\}) \subseteq S.$$

Then there is a unique true orbit $\mathbf{x}^* = (x_0^*, ..., x_N^*)$ *inside* $U(\mathbf{x}, r^*)$ *satisfying* $f(x_N^*) = y$.

We also have a more neutral form of Theorem 8.3.1:

Theorem 8.3.2 *Let* $\{x_i\}_{i=0}^{N}$ *be a pseudo-orbit of length* $N+1$. *assume:*

$$h_0^1 = \bar{L}\, a(\mathbf{x})\, b(\mathbf{x}) \leq \tfrac{1}{2} \tag{8.3.3}$$

and

$$\overline{U}(\mathbf{x}, r_1^* = 2b(x)) \subseteq S,$$

where $a(\mathbf{x})$, $b(\mathbf{x})$, r^*, $\bar{L}$ *are as defined above. Then there is a unique true orbit*

$$\mathbf{x}^* = (x_0^*, ..., x_N^*) \in U(\mathbf{x}, r_1^*)$$

satisfying $f(x_N^*) = f(x_N)$.

Remark 8.3.2 *If*

$$L_0 = L, \tag{8.3.4}$$

then Theorems 8.3.1 and 8.3.2 reduce to Theorems 1 and 2 in Hadeller [177] respectively. However in general

$$L_0 \leq L \tag{8.3.5}$$

The conditions corresponding to (8.3.2) and (8.3.3) respectively in Theorems 1 and 2 in [177] are given by

$$h = La(\mathbf{x})\, by(\mathbf{x}) \leq \tfrac{1}{2} \tag{8.3.6}$$

and

$$h^1 = L\, a(\mathbf{x})\;\, b(\mathbf{x}) \leq \tfrac{1}{2}. \tag{8.3.7}$$

It follows from (8.3.2), (8.3.3), (8.3.5), (8.3.6) and (8.3.7) that:

$$h \leq \tfrac{1}{2} \Longrightarrow h_0 \leq \tfrac{1}{2} \tag{8.3.8}$$

$$h^1 \leq \tfrac{1}{2} \Longrightarrow h_0^1 \leq \tfrac{1}{2} \tag{8.3.9}$$

but not viceversa unless if (8.3.4) holds. Hence we managed to weaken the sufficient convergence conditions given in [177], and under the same computational cost, since the evaluation of L requires in precise the evaluation of L_0.

Moreover the information on the location of the true orbit is more precise and the corresponding error bounds are finer.

8.4 Continuation Methods

In this section we are concerned with the problem of approximating a solution x^* of equation

$$F(x) = 0, \tag{8.4.1}$$

where F is an operator mapping a subset D of a normed linear space X into X.

An efficient way of approximating x^* is to imbed (8.4.1) into an equation of the form

$$H(t, x) = 0 \tag{8.4.2}$$

where $t \in [0, 1] = J$ is a parameter. The imbedding is chosen so that at $t = 0$, the solution of (8.4.2) is a known point x_0, while at $t = 1$, the solution x_1 of (8.4.2) also solves (8.4.1).

A popular choice of mapping H is

$$H(t, x) = F(x) + (t - 1)F(x_0). \tag{8.4.3}$$

We then suppose that there exist a continuous solution curve $x^* : J \to D$ of (8.4.2) starting at x_0. The continuation method continues in an unspecified way along or near the curve $x = x(t)$ from the initial point $x_0 = x(0)$ to the terminal point $x_1 = x(1)$.

Here we revisit the elegant work by Avila in [103] and show that it is possible by providing a more precise local convergence analysis to provide under the same or weaker hypotheses and computational cost a larger convergence radius and finer error bounds on the distances involved. That is we can enlarge the set of initial guesses and reduce the number of steps required to achieve a certain error tolerance between x^* and its k-th approximation. We use the popular Newton's method and assume for simplicity that $X = R^n$ although our results can be immediately extended to Banach spaces along the lines of our work in Section 5.1.

To avoid repetition we refer the reader to Avila [103] for the definitions of the concepts introduced here. Despite the fact that all results obtained in [103] can be rewritten to justify the benefits claimed at the introduction we only prove the main ones out of those and only suggest how the rest can be modified.

We need the following result concerning an extension of the strongness of the derivative.

Lemma 8.4.1 *Let D be an open subset of R^n. Assume $H\colon J \times D \subset J \times R^n \to R^n$ has a partial Fréchet-derivative with respect to x at a point $(t^*, x^*) \in J \times D$ for which $H(t^*, x^*) = 0$. Moreover let $U_0 \subset D$ be an open neighborhood of x^* and $I_0 \subset J$ a relatively open neighborhood of t^*, and assume that the operator $A\colon I_0 \times U_0 \to L(R^n, R^n)$ is continuous at (t^*, x^*) and that $A(t^*, x^*)$ has a bounded inverse. Then [103] there exists a relatively open interval $I \subset I_0$ and a ball $U = \overline{U}(x^*, \delta) \subset U_0$, $\delta > 0$, on which the operator*

$$G\colon I \times U \to R^n, \ G(t, x) = x - A(t, x)^{-1} H(t, x) \tag{8.4.4}$$

for all $(t, x) \in I \times U$ is well defined and F has a partial Fréchet derivative with respect to x at (t^*, x^*) given by

$$G_x(t^*, x^*) = I - A(t^*, x^*)^{-1} H_x(t^*, x^*). \tag{8.4.5}$$

Moreover if there exist positive constants ℓ_0 and ℓ such that

$$\|A(t^*, x^*)^{-1}[A(t, x^*) - A(t^*, y)]\| \leq \ell_0 \|x^* - y\| \tag{8.4.6}$$

and

$$\|A(t^*, x^*)^{-1}[A(t, x) - A(t, y)]\| \leq \ell \|x - y\| \tag{8.4.7}$$

for all (t, x), $(t, y) \in I_0 \times U_0$, then the continuity of H at (t^*, x^*) and strongness of $H_x(t^*, x^*)$ implies the strongness of $G_x(t^*, x^*)$.

Proof. Let us fix $\varepsilon \in (0, 1)$ and set $\delta = \frac{\varepsilon}{\ell_0}$ so that

$$U = \overline{U}(x^*, \delta) \subset S_0, \quad I = \{t \in J \mid |t - t^*| < \delta\} \subset I_0,$$

and

$$\|A(t^*, x^*)^{-1}[A(t, x) - A(t^*, x^*)]\| \leq \ell_0 \delta = \varepsilon < 1 \ \text{ for all } (t, x) \in I \times S. \tag{8.4.8}$$

It follows by (8.4.8) and the Banach Lemma on invertible operators 1.3.2 that $A(t, x)^{-1}$ exists, and

$$\|A(t, x)^{-1} A(t^*, x^*)\| \leq (1 - \varepsilon)^{-1} \quad \text{for all } (t, x) \in I \times S. \tag{8.4.9}$$

By restricting δ if necessary we can have

$$\begin{aligned}
&\|A(t^*, x^*)^{-1}[H(t, x) - H(t, y)]\| \\
&\qquad \leq \|A(t^*, x^*)^{-1}[H(t, x) - H(t, y) - H_x(t^*, x^*)(x - y)]\| \\
&\qquad\qquad + \|A(t^*, x^*)^{-1} H_x(t^*, x^*)(x - y)\| \\
&\qquad \leq (\varepsilon + \|A(t^*, x^*)^{-1} H_x(t^*, x^*)\|)\|x - y\|
\end{aligned}$$

and

$$\|A(t^*,x^*)^{-1}H(t,y)\| \le \varepsilon \quad \text{for all } (t,y) \in I \times S. \tag{8.4.10}$$

Using (8.4.4)–(8.4.7), (8.4.9)–(8.4.10) and the strongness of $H_x(t^*,x^*)$ we obtain in turn:

$$\begin{aligned}
&\|G(t,x) - G(t,y) - G_x(t^*,x^*)(x-y)\| \\
&\quad \le \|A(t^*,x^*)^{-1}[H(t,x) - H(t,y) - H_x(t^*,x^*)(x-y)]\| \\
&\qquad + \|A(t^*,x^*)^{-1}[A(t,x) - A(t^*,x^*)] \\
&\qquad\qquad \cdot A(t,x)^{-1}A(t^*,x^*)A(t^*,x^*)^{-1}(H(t,x) - H(t,y))\| \\
&\qquad + \|A(t,y)^{-1}A(t^*,x^*)A(t^*,x^*)^{-1}(A(t,x) \\
&\qquad - A(t,y))A(t,x)^{-1}A(t^*,x^*)A(t^*,x^*)^{-1}H(t,y)\| \le \varepsilon a\|x-y\|,
\end{aligned}$$

where,

$$a = 1 + \frac{1}{1-\varepsilon}(\varepsilon + \|A(t^*,x^*)^{-1}H_x(t^*,x^*)\|) + \left(\frac{1}{1-\varepsilon}\right)^2 \ell, \tag{8.4.11}$$

which shows that $G_x(t^*,x^*)$ is strong. The rest of the result follows by a classical lemma due to Ortega and Rheinboldt [263]. ∎

Remark 8.4.1 *In general*

$$\ell_0 \le \ell \tag{8.4.12}$$

holds and $\frac{\ell}{\ell_0}$ *can be arbitrarily large (for* t *sufficiently close to* t^**). If* $\ell = \ell_0$ *Lemma 8.4.1 can be reduced to Lemma 3.1 in [103]. However in that Lemma* δ *is chosen is an unspecified (in terms of* ε*) way, whereas in our proof it is specified* $\left(\delta = \frac{\ell_0}{\varepsilon}\right)$*. Moreover if strict inequality holds* I*,* U *are enlarged,* a *is smaller than the corresponding* a *call it* a_1 *given in [103, p. 109] by*

$$a_1 = \beta + 4\beta^2(\varepsilon + \|H_x(t^*,x^*)\|) + 4\beta^2\ell \tag{8.4.13}$$

where,

$$\beta = \|A(t^*,x^*)^{-1}\|, \tag{8.4.14}$$

and the strongness of $G_x(t^*,x^*)$ *is established on a larger cross product* $I \times S$*. The above benefits are achieved under the same hypotheses and computational cost since in practice the computation of* ℓ *requires that of* ℓ_0*. Furthermore a wider choice of initial guesses to start Newton's method becomes available since* U *is enlarged. We also note that our result is given in an affine invariant form. The advantages of such an approach have been explained in Section 5.1. Finally we can see from the from of the Lemma that condition (8.4.7) can be replaced by the stronger*

$$\|A(t^*,x^*)^{-1}[A(t^1,x) - A(t^2,y)]\| \le \ell_1\|x-y\| \tag{8.4.15}$$

for all (t^1, x), $(t^2, y) \in I_0 \times U_0$. *Clearly* $\ell \leq \ell_1$, $\ell_0 \leq \ell_1$, *and again* $\frac{\ell_1}{\ell_0}$ *can be arbitrarily large. Under this modification* t *does not have to necessarily be close enough to* t^* *for the conclusions of the Lemma to hold (with* ℓ_1 *replacing* ℓ *throughout the statment and the proof of the Lemma).*

In the result that follows we provide a local convergence analysis for method (4.1.3):

Theorem 8.4.2 *Let* $H\colon J \times D \subset J \times R^n \to R^n$, *where* D *is an open subset of* R^n *and* H *has partial Fréchet derivatives with respect to both* t *and* x. *Assume* $x^*\colon J \to D$ *is a continuous solution of the equation* $H(t, x) = 0$, $t \in J$. *It follows by the openness of* D *that there exists* $r^* > 0$ *such that*

$$D_0 = \{v \in X \ \|x^* - v(t^*)\| < r_0 \ \ \textit{for some } t^* \in J\} \subseteq D. \tag{8.4.16}$$

Moreover assume:

Theorem 8.4.3 H_t *is continuous on* $J \times D_0$, $L = H_{x^*}(t^*, x^*)^{-1}$ *exists, and*

$$\|LH_t(t, v(t))\| \leq \alpha \ \ \textit{for all } t \in J; \tag{8.4.17}$$

$$\|L[H_x(t, x) - H_x(t, y)]\| \leq \ell\|x - y\| \tag{8.4.18}$$

and

$$\|L(L^{-1} - H_v(t, v))\| \leq \ell_0\|x^* - v\| \tag{8.4.19}$$

for all $t \in J$ *and for all* $x, y, v \in D_0$;

Theorem 8.4.4 *for* $\gamma = \sqrt{2\alpha h\ell}$, *and*

$$q = \frac{1 - \gamma}{\ell_0} \tag{8.4.20}$$

r_0 *satisfies*

$$r_0 \leq q. \tag{8.4.21}$$

Moreover let h_1 *be the largest value in* $\left[0, \frac{1}{2\alpha\beta_1^2\ell}\right]$ *such that*

$$r_1^1(h_1) \leq r_0, \tag{8.4.22}$$

where,

$$r_1^1(h) = \frac{1}{\beta_1\ell}\left(1 - \sqrt{1 - 2\alpha\beta_1^2\ell h}\right). \tag{8.4.23}$$

Then, if the integer N *is chosen so that*

$$0 < \Delta t = \frac{1}{N_1} \leq h_1 \tag{8.4.24}$$

Newton's method

$$x^k = x^{k-1} - H_x(k\Delta t, x^{k-1})^{-1}H(k\Delta t, x^{k-1}) \tag{8.4.25}$$

$$k = 1, \ldots, N-1, \quad x^0 = x(0);$$

$$x^k = x^{k-1} - H_x(1, x^{k-1})^{-1}H(1, x^{k-1}), \quad k = N, N+1, \ldots \tag{8.4.26}$$

is feasible.

Proof. By hypotheses and the Banach Lemma on invertible operators $H_x(t,x)^{-1}$ exists and

$$\|H_x(t,x)^{-1}L\| \leq \frac{1}{1-\ell_0 r_0} = \beta_1. \tag{8.4.27}$$

We then set $e_k = \|x^{k-1} - x(k\Delta t)\|$ and exactly as in Theorem 4.3 in [103, p. 113] we get

$$e_{k+1} \leq a e_k^2 + c, \tag{8.4.28}$$

where,

$$a = \frac{\beta_1 \ell}{2} \quad \text{and} \quad c = \beta_1 \alpha. \tag{8.4.29}$$

By hypotheses $1 - 4ac \geq 0$. Therefore it follows from Lemma 4.2 in [103, p. 112] that

$$e_k \leq \frac{1}{2a}\left(1 - \sqrt{1-4ac}\right) = r_1^1(h_1). \tag{8.4.30}$$

■

Remark 8.4.2 *According to the hypotheses of Theorem 4.3 in [103] the parameter β in practice will be given by*

$$\beta = \frac{\|H_{x^*}(t^*, x^*)^{-1}\|}{1 - \|H_{x^*}(t^*, x^*)^{-1}\|\ell_2 r_0}, \tag{8.4.31}$$

where ℓ_2 is the constant appearing in hypothesis (iii) in that theorem given by

$$\|H_x(t,x) - H_x(t,y)\| \leq \ell_2\|x-y\| \text{ for all } t \in J, \text{ and all } x, y \in D_0. \tag{8.4.32}$$

Clearly $\frac{\ell_2}{\ell_0}$ can be arbitrarily large. In particular if

$$\ell_0 < \ell_2\|H_x(t^*, x^*)^{-1}\|, \tag{8.4.33}$$

then

$$\beta_1 < \beta, \tag{8.4.34}$$

which implies that

$$h_1^{-1} = N_1 < N = h_0^{-1} \qquad [3, p.\ 113], \tag{8.4.35}$$

and is the largest value in $\left[0, \frac{1}{2\alpha\beta_1^2\ell_2}\right]$ *such that*

$$r_1(h_0) \le r_0, \qquad [3, p.113], \tag{8.4.36}$$

where,

$$r_1(h) = \frac{1}{\beta\ell_2}\left(1 - \sqrt{1 - 2\alpha\beta^2\ell_2 h}\right). \tag{8.4.37}$$

Note also that in this case

$$r_1(h_0) < r_1^1(h_1). \tag{8.4.38}$$

That is our approach and under the same hypotheses and computational cost has the following advantages over the one given in Theorem 4.3 in [103]: a smaller number of steps; a larger convergence radius; finer upper bounds on the distances involved and the results are given in affine invariant form. As a simple example where strict inequality holds in (8.4.12), consider

$$n = 1, \quad D_0 = \overline{U}(0, 1), \quad J = [1, +\infty), \quad (t^*, x^*) = (1, 0)$$

and define operator H *on* $J \times D_0$ *by*

$$H(t, x) = e^x - t. \tag{8.4.39}$$

Then it can easily be seen using (8.4.18) and (8.4.19) that

$$\ell_0 = e - 1 < e = \ell. \tag{8.4.40}$$

The rest of the results in [103] can now be improved if rewritten with (β_1, h_1) *replacing* (β, h)*, see, e.g., Theorem 4.7 and Corollary 4.8.*

8.5 Trust Regions

Approximate solution of equation involves a complex problem: choice of an initial approximation x_0 sufficiently close to the true solution. The method of random choice is often successful. Another frequently used method is to replace

$$F(x) = 0 \tag{8.5.1}$$

by a "similar" equation and to regard the exact solution of the latter as the initial approximation x_0. Of course, there are no general "prescriptions" for admissible initial approximations. Nevertheless, one can describe various devices suitable for extensive classes of equations.

As usual, let F map X into Y. To simplify the exposition, we shall assume that F is defined throughout X. Assume that $G(x.\lambda)$ $(x \in X; 0 \leqq \lambda \leqq 1)$ is an operator with values in Y such that

$$G(x;1) \equiv Fx \quad (x \in X), \tag{8.5.2}$$

and the equation

$$G(x;0) = 0 \tag{8.5.3}$$

has an obvious solution x^0. for example, the operator $G(x;\lambda)$ might be defined by

$$G(x;\lambda) = Fx - (1-\lambda)Fx^0. \tag{8.5.4}$$

Consider the equation

$$G(x;\lambda) = 0. \tag{8.5.5}$$

Suppose that equation (8.5.5) has a continuous solution $x = x(\lambda)$, defined for $0 \leqq \lambda \leqq 1$ and satisfying the condition

$$x(0) = x^0. \tag{8.5.6}$$

Were the solution $x(\lambda)$ known,

$$x^* = x(1) \tag{8.5.7}$$

would be a solution of equation (8.5.1). One can thus find a point x_0 close to x^* by approximating $x(\lambda)$.

Our problem is thus to approximate the implicit function defined by (8.5.5) and the initial conditions (8.5.6). Global propositions are relevant here theorems on implicit functions defined on the entire interval $[0,1]$. the theory of implicit functions of this type is at present insufficiently developed.

The idea of extending solutions with respect to a parameter is due to S.N. Bernstein [322]; it has found extensive application in various theoretical and applied problems.

Assume that the operator $G(x;\lambda)$ is differentiable with respect to both x and λ, in the sense that there exist linear operators $G'_x(x;\lambda)$ mapping X and Y and elements $G'_x(x;\lambda) \in Y$ such that

$$\lim_{\|h\|+|\Delta\lambda|\to 0} \frac{\left\|G(x+h;\lambda+\Delta\lambda)-G(x;\lambda)-G'_x(x;\lambda)h-G'_x(x;\lambda)\Delta\lambda\right\|}{\|h\|+|\Delta\lambda|} = 0.$$

The implicit function $x(\lambda)$ is then a solution of the differential equation

$$G'_x(x;\lambda)\frac{dx}{d\lambda} + G'_x(x;\lambda) = 0, \tag{8.5.8}$$

satisfying the initial condition (8.5.6). conditions for existence of a solution of this Cauchy problem defined on $[0, 1]$ are precisely conditions for existence of the implicit function. Assuming the existence of a continuous operator

$$\Gamma(x; \lambda) = [G'_x(x; \lambda)]^{-1}, \tag{8.5.9}$$

we can rewrite equation (8.5.8) as

$$\tfrac{dx}{d\lambda} = -\Gamma(x; \lambda) G'_x(x; \lambda). \tag{8.5.10}$$

One must bear in mind that Peano's Theorem is false for ordinary differential equations in Banach spaces. Therefore, even in the local existence theorem for equation (8.5.10) with condition (8.5.6) one must assume that the right-hand side of the equation satisfies certain smoothness conditions. However, there are no sufficient smoothness conditions for the existence of a global extension of the solution to the entire interval $0 \leqq \lambda \leqq 1$. We shall only mention a trivial fact: if the equation

$$\tfrac{dx}{d\lambda} = f(x; \lambda) \tag{8.5.11}$$

in a Banach space satisfies the local existence theorem for some initial condition, and

$$\|f(x; \lambda)\| \leqq a + b\|x\| \quad (0 \leqq \lambda \leqq 1; x \in X), \tag{8.5.12}$$

then every solution of equation (8.5.11) can be extended to the entire interval $0 \leqq \lambda \leqq 1$. Thus, if

$$\left\|\Gamma(x; \lambda) G'_y(x; \lambda)\right\| \leqq a + b\|x\|, \tag{8.5.13}$$

then equation (8.5.5) defines an implicit function which satisfies (8.5.6) and is defined for $0 \leqq \lambda \leqq a$. consequently, condition (8.5.13) guarantees that equation (8.5.1) is solvable and its solution can be constructed by integrating the differential equation (8.5.10).

To approximate a solution $x(\lambda)$ of equation (8.5.10), one can use, for example, Euler's method. To this end, divide the interval $[0, 1]$ into m subintervals by points

$$\lambda_0 = 0 < \lambda_1 < \cdots < \lambda_m = 1. \tag{8.5.14}$$

The approximate values $x(\lambda_i)$ of the implicit function $x(\lambda)$ are then determined by the equalities $x(\lambda_0) = x^0$ and

$$x(\lambda_{i+1}) = x(\lambda_i) - \Gamma[x(\lambda_i); \lambda_i] G'_\lambda[x(\lambda_i); \lambda_i](\lambda_{i+1} - \lambda_i). \tag{8.5.15}$$

The element $x(\lambda_m)$ is in general close to the solution x^* of equation (8.5.1) and one may therefore expect it to fulfill the demands imposed on initial approximations for iterative solution of equation (8.5.1). We emphasize that (8.5.15) does not describe an iterative process; it only yields a finite sequence of operations, whose result is an element which may be a suitable initial approximation for iterative solution of equation (8.5.1).

Other constructions may be used to approximate the implicit function $x(\lambda)$. Partition the interval $[0,1]$ by the points (8.5.14). The point $x(\lambda_1)$ is a solution of the equation $G(x,\lambda_1)=0$. Now $x(\lambda_0)=x^0$ is a suitable initial approximation to $x(\lambda_1)$. Approximate $x(\lambda_1)$ by performing a fixed number of steps of some iterative process. The result is an element x_1 which should be fairly close to $x(\lambda_1)$. x_1 is obtained from x^0 by a certain operator

$$x_1 = W\left[x^0; G(x;\lambda_1)\right].$$

Now regard x_1 as an initial approximation to the solution $x(\lambda_2)$ of the equation $G(x;\lambda_2)$, and proceed as before. The result is an element x_2:

$$x_2 = W\left[x_1; G(x,\lambda_2)\right].$$

Continuing in this way, we obtain a finite set of points

$$x_{i+1} = W\left[x_i; G(x,\lambda_i)\right] \quad (i=0,1,...,m-1), \tag{8.5.16}$$

the last of which x_m, may be regarded as an initial approximation for iterative solution of equation (8.5.1)

If the operator W represents one iteration of the method, formula (8.5.16) is

$$x_{i+1} = x_i - \left[G'_x(x_i;\lambda_{i+1})\right]^{-1} G(x_i;\lambda_{i+1}) \quad (i=1,...,m-1). \tag{8.5.17}$$

Using the algorithm proposed by H.T. Kung [216] (see also [322]) we show that the number of steps required to compute a good starting point x_0 (to be precised later) can be significantly reduced.

This observation is very important in computational mathematics.

In Chapter 5 we showed the following semilocal convergence theorem for Newton's method (4.1.3) which essentially states the following:

If

$$F'(x_0)^{-1} \text{ exists, } \left\|F'(x_0)^{-1}\right\| \le \beta_0, \tag{8.5.18}$$

$$\left\|F'(x_0)^{-1} F(x_0)\right\| \le \xi_0, \tag{8.5.19}$$

$$\left\|F'(x) - F(x_0)\right\| \le K_0 \left\|x - x_0\right\|, \tag{8.5.20}$$

$$\left\|F'(x) - F'(y)\right\| \le K \left\|x - y\right\|, \tag{8.5.21}$$

for all $x,y \in \overline{U}(x_0,r)$

$$h_0 = \beta_0 L \xi_0 < \tfrac{1}{2} \tag{8.5.22}$$

where,

$$L = \tfrac{K_0+K}{2}, \tag{8.5.23}$$

$$2\xi_0 \le r, \tag{8.5.24}$$

and

$$\overline{U}\left(x_0, r\right) \subseteq D, \tag{8.5.25}$$

then sequence $\{x_n\}$ $(n \geq 0)$ generated by Newton's method (4.1.3) is well defined, remains in $\overline{U}\left(x_0, r_0\right)$ for all $n \geq 0$ and converges quadratically to a unique solution $x^* \in U\left(x_0, r\right)$ of equation $F\left(x\right) = 0$. Moreover we have

$$\left\|x_0 - x^*\right\| \leq 2\xi_0. \tag{8.5.26}$$

Remark 8.5.1 *In general*

$$K_0 \leq K \tag{8.5.27}$$

holds. If equality holds in (8.5.27) then the result stated above reduces to the famous Newton-Kantorovich theorem and (8.5.22) to the Newton-Kantorovich hypothesis (4.2.51). If strict inequality holds in (8.5.27) then (8.5.22) is weaker than the Newton-Kantorovich hypothesis

$$\overline{h}_0 = \beta_0 K \xi_0 < \tfrac{1}{2}. \tag{8.5.28}$$

Moreover the error bounds on the distances $\left\|x_{n+1} - x_n\right\|, \left\|x_n - x^*\right\|$ $(n \geq 0)$ *are finer and the information on the location of the solution more precise.*

Note also that the computational cost of obtaining (K_0, K) is the same as the one for K since in practice evaluating K requires finding K_0.

Hence all results using (8.5.28) instead of (8.5.22) can now be challenged to obtain more information. That is exactly what we are doing here. In particular motivated by the elegant work of H.T. Kung [216] on good starting points for Newton's method we show how to improve on these results of we use our theorem stated above instead of the Newton-Kantorovich theorem.

Definition 8.5.1 *We say x_0 is a good starting point for approximating x^* by Newton's method or a good starting point for short if conditions (8.5.18)-(8.5.25) hold.*

Note that the existence of a good starting point implies the existence of a solution x^* of equation $F\left(x\right) = 0$ in $U\left(x_0, 2\xi_0\right)$.

We provide the following theorem / Algorithm which improves the corresponding ones given in [216, Thm. 4.1] to obtain good starting points.

Theorem 8.5.2 *Let $F : D \subseteq X \rightarrow Y$ be a Fréchet-differentiable operator. If F' satisfies center-Lipschitz, Lipschitz conditions (8.5.20), (8.5.21) respectively on* $U\left(x_0, 2r\right)$

$$\left\|F\left(x_0\right)\right\| \leq \eta_0$$

$$\left\|F'\left(x\right)^{-1}\right\| \leq \beta \text{ for all } x \in U\left(x_0, 2r\right), \tag{8.5.29}$$

$$U\left(x_0, 2r\right) \subseteq D, \tag{8.5.30}$$

and

$$\beta\eta_0 < \tfrac{r}{2}, \tag{8.5.31}$$

then there exists a solution x^* of equation $F(x) = 0$ in $U(x_0, 2r)$.

Proof. Simply use L instead of K in the proof of Theorem 4.1 in [216, p. 11] including the algorithm there which is essentially repeated here with some modifications: **Algorithm A:** The goal of this algorithm is to produce starting point (under the conditions of Theorem 1) for approximating x^*.

1. Set $h_0 \longleftarrow \beta^2 L\eta_0$ and $i \longleftarrow 0$. Choose any number δ in $\left(0, \frac{1}{2}\right)$.

2. If $h_1 < \frac{1}{2}, x_i$ is a good starting point for approximating x^* and algorithn A terminates.

3. Set $\lambda_i \longleftarrow \left(\frac{1}{2} - \delta\right)/h_i$, and

$$F_i(x) \longleftarrow [F(x) - F(x_i)] + \lambda_i F(x_i). \tag{8.5.32}$$

4. (It is shown in the proof that x_i is a good starting point for approximating a zero, denoted by x_{i+1}, of F_i, Apply Newton's method to F_i, starting from x_i, to find x_{i+1}.

5. (Assume that the exact x_{i+1} is found.) Set $\eta_{i+1} \longleftarrow \|F(x_{i+1})\|$ and

$$h_{i+1} \longleftarrow \beta^2 K\eta_{i+1}.$$

6. Set $i \longleftarrow i+1$, and return back to Step 2.

In the following we prove algorithm works. First we note that $\lambda_i \in (0,1)$ and by (8.5.32)

$$\eta_{i+1} = (1 - \lambda_i)\eta_i. \tag{8.5.33}$$

We shall prove by induction that

$$\|x_i - x_{i-1}\| \le 2\beta\lambda_{i-1}\eta_{i-1}, \tag{8.5.34}$$

and

$$\|x_i - x_0\| \le r. \tag{8.5.35}$$

They trivially hold for $i = 0$. Suppose that (8.5.34) and (8.5.35) hold and $h_i \ge \frac{1}{2}$. By (8.5.32)

$$\beta^2 L\|f_i(x_i)\| \le \beta^2 L\lambda_i\eta_i = \lambda_i h_i = \tfrac{1}{2} - \delta, \tag{8.5.36}$$

and by (8.5.33)

$$2\beta \|f_i(x_i)\| \leq 2\beta\lambda_i\eta_i < 2\beta\eta_i \leq 2\beta\eta_0 < r.$$

Further, by (8.5.35), we have $U(x_i, r) \subseteq U(x_0, 2r)$. Hence x_i is a good starting point for approximating the zero x_{i+1} of f_i . From (8.5.30), we know

$$\|x_{i+1} - x_i\| \leq 2\beta\lambda_i\eta_i. \tag{8.5.37}$$

Hence (8.5.34) holds with im replaced by $i+1$. By (8.5.37), (8.5.33) and (8.5.31), we have

$$\begin{aligned} \|x_{i+1} - x_0\| &\leq \|x_{i+1} - x_i\| + \|x_i - x_{i-1}\| + \cdots + \|x_1 - x_0\| \\ &\leq 2\beta(\lambda_i\eta_i + \lambda_{i-1}\eta_{i-1} + \cdots + \lambda_0\eta_0) \\ &\leq 2\beta((1-\lambda_{i-1})\eta_{i-1} + \lambda_{i-1}\eta_{i-1} + \cdots + \lambda_0\eta_0) \\ &= 2\beta(\eta_{i-1} + \lambda_{i-2}\eta_{i-2} + \cdots + \lambda_0\eta_0) \\ &\leq \cdots \\ &\leq 2\beta\eta_0 < r, \end{aligned} \tag{8.5.38}$$

i.e., (8.5.35) holds with i replaced by $i+1$. We now assume that (8.5.34) and (8.5.35) hold and $h_1 < \frac{1}{2}$. By (8.5.31) and (8.5.33), $2\beta\|f(x_i)\| = 2\beta\eta_i < 2\beta\eta_0 < r$. Further by (8.5.35), $s_r(x_i) \subseteq s_{2r}(x_0)$. Hence x_i is a good starting point for approximating α. It remains to show that the loop starting from step 2 is finite. Suppose that $h_0 \geq \frac{1}{2}$. since $\lambda_i \in (0,1)$ for all i, we have

$$\lambda_i = \frac{\frac{1}{2}-\delta}{\beta^2 L\eta_i} = \frac{\frac{1}{2}-\delta}{\beta^2 L(1-\lambda_{i-1})\eta_{i-1}} > \frac{\frac{1}{2}-\delta}{\beta^2 L\eta_{i-1}} = \lambda_{i-1}, \text{ for all } i.$$

Hence by (8.5.33)

$$\begin{aligned} \eta_{i+1} &= (1-\lambda_i)\eta_i < (1-\lambda_0)\eta_i \\ &< \cdots \\ &< (1-\lambda_0)^{i+1}\eta_0. \end{aligned}$$

This implies that $h_i < \frac{1}{2}$ when $\beta^2 L(1-\lambda_0)^i\eta_0 < \frac{1}{2}$, i.e., when

$$(1-\lambda_0)^i < \frac{1}{2h_0}. \tag{8.5.39}$$

Since $1 - \lambda_0 < 1$, (8.5.39) is satisfied for large i. therefore when i is large enough, $h_i < \frac{1}{2}$ and hence Algorithm A terminates. ■

Remark 8.5.2 *As already noted in [216] Theorem 8.5.2 is trivial for the scalar case ($f : R \to R$), since the mean value theorem can be used. Some of the assumptions of Theorem 8.5.2 can be weakened. Avila for example in [103, Theorem 4.3] instead of (8.5.31) used a more complicated condition involving β, K, and η_0. However the* *idea algorithm is basically different from Algorithm A. Note also that if $K_0 = K$ then out Theorem 8.5.2 reduces to Theorem 4.1 in [216]. We now modify Algorithm A to make it work in Banach spaces without necessarily assuming that the exact zero of x_{i+1} of F_i can be found using Newton's method (4.1.3).*

Theorem 8.5.3 *Under the hypotheses of Theorem 8.5.2 a good starting point for approximating solution x^* of equation $F(x) = 0$ can be obtained in $N(\delta, K_0, K)$ Newton steps, δ is any number in $\left(0, \frac{1}{2}\right)$,*

$$N(\delta, K_0, K) = \begin{cases} 0, & \text{if } h_0 = \beta^2 L\eta_0 \leq \frac{1}{2} - \delta \\ I(\delta, K_0, K) \cdot J(\delta, K_0, K), & \text{otherwise,} \end{cases}$$

where, $I(\delta, K_0, K)$ is the smallest integer i such that:

$$\left[1 - \frac{\frac{1}{2} - \delta}{h_0}\right]^i \leq \left[\tfrac{1}{2} - \delta\right] / h_0, \tag{8.5.40}$$

and $J(\delta, K_0, K)$ is the smallest integer j such that:

$$2^{\frac{1}{j-1}} (1 - 2\delta)^{2^j - 1} (a + \beta\eta_0) \leq r - 2\beta\eta_0, \tag{8.5.41}$$

$$\tfrac{1}{2^j} (1 - 2\delta)^{2^j - 1} (a + \beta\eta_0) \leq a, \tag{8.5.42}$$

where,

$$a = \min\left(\tfrac{r}{2} - \beta\eta_0, \tfrac{\delta}{2\beta L}\right). \tag{8.5.43}$$

Proof. Simply use L instead of K in the proof of Theorem 4.2 in [216, p. 16], and the following algorithm: **Algorithm B.**

1. Set $h_0 \longleftarrow \beta^2 L\eta_0, \overline{x}_0 \longleftarrow x_0$ and $i \longleftarrow 0$. Choose any number δ in $\left(0, \frac{1}{2}\right)$.

2. If $h_i \leq \frac{1}{2} - \delta, \overline{x}_i$ is a good starting point for approximating x^* and Algogitm B terminates.

3. Set $\lambda_i \longleftarrow \left(\frac{1}{2} - \delta\right) / h_i$,

$$F_i(x) \longleftarrow [F(x) - \eta_i F(x_0)/\eta_0] + \lambda_i \eta_i F(x_0)/\eta_0,$$

and

$$\eta_{i+1} \longleftarrow (1 - \lambda_i)\eta_i. \tag{8.5.44}$$

4. Apply Newton's method to F_i, starting from $\overline{x}_i$, to find an approxiation $\overline{x}_{i+1}$ to a zero x_{i+1} of F_i such that

$$\|\overline{x}_{i+1} - x_{i+1}\| \leq r - 2\beta\eta_0, \tag{8.5.45}$$

and

$$\left\|F_i'(\overline{x}_{i+1})^{-1} F_i(\overline{x}_{i+1})\right\| \leq \min\left(\tfrac{r}{2} - \beta\eta_{i+1}, \tfrac{\delta}{2\beta L}\right). \tag{8.5.46}$$

5. Set $h_{i+1} \longleftarrow \beta^2 L \eta_{i+1}$.

6. Set $i \longleftarrow i+1$ and return to Step 2.

Note that the h_i, λ_i, η_i, f_i, x_i in Algorithm A are the same h_i, λ_i, η_i, f_i, x_i in algorithm B. Note also that by (8.5.45) and (8.5.38) we have

$$\begin{aligned}\|\overline{x}_i - x_0\| &\le \|\overline{x}_i - x_i\| + \|x_i - x_0\| \\ &\le (r - 2\beta\eta_0) + 2\beta\eta_0 = r, \quad \forall i.\end{aligned} \tag{8.5.47}$$

It is clear that if $h_0 < \frac{1}{2} - \frac{\delta}{2}$, $\overline{x}_0$ is a good starting point for approximating α. Now suppose $h_0 > \frac{1}{2} - \frac{\delta}{2}$. since $\overline{x}_0 = x_0$, in the proof of Theorem 8.5.2 we have shown that $\overline{x}_0$ is a good starting point for approximating x_1 a zero of f_0. Let z_j denote the jth Newton iterate starting from $\overline{x}_0$ for approximating x_1. Since

$$\beta^2 L \|f_0(\overline{x}_0)\| = \beta^2 L \lambda_1 \eta_0 = \tfrac{1}{2} - \delta,$$

it is known (see e.g. Rall [242, Section 22]) that

$$\|z_j - x_1\| \le \tfrac{1}{2^{j-1}} (1 - 2\delta)^{2^j - 1} \left\| [f'(x_0)]^{-1} f(x_0) \right\|$$

and

$$\left\| [f_0'(z_j)]^{-1} f_0(z_j) \right\| \le \tfrac{1}{2^j} (1 - 2\delta)^{2^j - 1} \left\| [f'(x_0)]^{-1} f(x_0) \right\|.$$

Hence we may let $\overline{x}_1$ be z_j for j large enough, say, $j = j(\delta)$, then

$$\|\overline{x}_1 - x_1\| \le r - 2\beta\eta_0,$$

and

$$\left\| [f_0'(\overline{x}_1)]^{-1} f_0(\overline{x}_1) \right\| \le \min\left(\tfrac{r}{2} - \beta\eta_1, \tfrac{\delta}{2\beta L} \right),$$

i.e., (8.5.45) and (8.5.46) hold for $i = 0$. Suppose that (8.5.45) and (8.5.46) hold. Then

$$\begin{aligned}&\left\| [f'(\overline{x}_{i+1})]^{-1} f(\overline{x}_{i+1}) \right\| \le \\ &\le \left\| [f_i'(\overline{x}_{i+1})]^{-1} f_i(\overline{x}_{i+1}) \right\| + \left\| [f_i'(\overline{x}_{i+1})]^{-1} [f_i'(\overline{x}_{i+1}) - f_i(\overline{x}_{i+1})] \right\|. \\ &\le \min\left(\tfrac{r}{2} - \beta\eta_{i+1}, \tfrac{\delta}{2\beta L} \right) + \beta\eta_{i+1},\end{aligned} \tag{8.5.48}$$

and

$$\begin{aligned}&\left\| \left[f_{i+1}'(\overline{x}_{i+1}) \right]^{-1} f_{i+1}(\overline{x}_{i+1}) \right\| \le \\ &\le \left\| [f'(\overline{x}_{i+1})]^{-1} [f(\overline{x}_{i+1}) - \eta_{i+1} f(x_0)/\eta_0] \right\| \\ &\le \left\| \left[[f'(\overline{x}_{i+1})]^{-1} \lambda_{i+1}\eta_{i+1} f(x_0)/\eta_0 \right] \right\| \\ &\le \left\| [f'(\overline{x}_{i+1})]^{-1} f_i(\overline{x}_{i+1}) \right\| + \lambda_{i+1}\beta\eta_{i+1}.\end{aligned} \tag{8.5.49}$$

Suppose that $h_{i+1} < \frac{1}{2} - \delta$. We want to prove that $\overline{x}_{i+1}$ is a good starting point for approximating α. By (8.5.48)

$$\begin{aligned}
&\beta L \left\| \left[f'\left(\overline{x}_{i+1}\right)\right]^{-1} f\left(\overline{x}_{i+1}\right) \right\| \leq \\
&\leq \beta L \cdot \tfrac{\delta}{2\beta L} + h_{i+1} < \tfrac{\delta}{2} + \tfrac{1}{2} - \delta = \tfrac{1}{2} - \tfrac{\delta}{2}.
\end{aligned}$$

Let $a = \left\| \left[f'\left(\overline{x}_{i+1}\right)\right]^{-1} f\left(\overline{x}_{i+1}\right) \right\|$. If $x \in U\left(\overline{x}_{i+1}, 2a\right)$, then

$$\begin{aligned}
\left\|x - x_0\right\| &\leq \left\|x - \overline{x}_{i+1}\right\| + \left\|\overline{x}_{i+1} - x_0\right\| \\
&\leq 2a + r \\
&\leq 2\left(\tfrac{r}{2} - \beta\eta_{i+1} + \beta\eta_{i+1}\right) + r = 2r,
\end{aligned} \tag{8.5.50}$$

i.e., $x \in U\left(x_0, 2r\right)$. hence $\overline{x}_{i+1}$ is a good starting point for approximating α. we now assume that $h_{i+1} > \frac{1}{2} - \delta$, and want to prove that $\overline{x}_{i+1}$ is a good starting point for approximating x_{i+2}, a zero of f_{i+1}. We have by (8.5.49) and (8.5.46),

$$\begin{aligned}
&\beta L \left\| \left[f_{i+1}'\left(\overline{x}_{i+1}\right)\right]^{-1} f_{i+1}\left(\overline{x}_{i+1}\right) \right\| \\
&\leq \tfrac{\delta}{2} + \lambda_{i+1}\beta^2 L \eta_{i+1} \\
&= \tfrac{\delta}{2} + \tfrac{1}{2} - \delta = \tfrac{1}{2} - \tfrac{\delta}{2}.
\end{aligned}$$

Let $b = \left\| \left[f_{i+1}'\left(\overline{x}_{i+1}\right)\right]^{-1} f_{i+1}\left(\overline{x}_{i+1}\right) \right\|$. If $x \in U\left(\overline{x}_{i+1}, 2b\right)$, as in (8.5.50) we can prove that $x \in U\left(x_0, 2r\right)$. hence $\overline{x}_{i+1}$ is a good starting point for approximating x_{i+2}. By the same argument as used for obtaining $\overline{x}_0$ and by (8.5.49), one can prove that if $\overline{x}_{i+2}$ is set to be the $J\left(\delta\right)$ -th Newton iterate starting from $\overline{x}_{i+1}$, then

$$\left\|\overline{x}_{i+2} - x_{i+1}\right\| \leq r - 2\beta\eta_0,$$

and

$$\left\| \left[f_{i+1}'\left(\overline{x}_{i+2}\right)\right]^{-1} f_{i+1}\left(\overline{x}_{i+2}\right) \right\| \leq \min\left(\tfrac{r}{2} - \beta\eta_{i+2}, \tfrac{\delta}{2\beta L}\right),$$

i.e. (8.5.45) and (8.5.46) hold with i replaced by $i+1$. This shows that we need to perform at most $J\left(\delta\right)$ Newton steps at step 4 of Algorithm B to obtain each $\overline{x}_{i+1}$. Therefore, for any $\delta \in \left(0, \frac{1}{2}\right)$, to obtain a good starting point we need to perform at most $N\left(\delta\right) = I\left(\delta\right) \cdot J\left(\delta\right)$ Newton steps. ■

Remark 8.5.3 *As noted in [216] δ should not be chosen to minimize the complexity of Algorithm B. Instead, δ should be chosen to minimize the complexity of algorithm:*

1. *Search Phase: Perform algorithm B.*
2. *Iteration Phase: Perform Newton's method starting from the point obtained by Algorithm B.*

An upper bound on the complexity of the iteration phase is the time needed to carry out $T(\delta, K_0, K, \varepsilon)$ is the smallest integer K such that

$$\frac{1}{2^{K-1}}(1-2\delta)^{2^K-1}(a+\beta\eta_0) \leq \varepsilon. \tag{8.5.51}$$

Note also that if $K_0 = K$ our Theorem 8.5.3 reduces to Theorem 4.2 in [216, p. 15].

Hence we showed the following result:

Theorem 8.5.4 *Under the hypotheses of Theorem 8.5.3 the time needed to find a solution x^* of equation $F(x) = 0$ inside a ball of radius ε is bounded above by the time needed to carry out $R(\delta, K_0, K, \varepsilon)$ Newton steps, where*

$$R(\varepsilon, K_0, K) = \min_{0<\delta<\frac{1}{2}} [N(\delta, K_0, K, \varepsilon) + T(\delta, K_0, K, \varepsilon)], \tag{8.5.52}$$

where $N(\delta, K_0, K, \varepsilon)$ and $T(\delta, K_0, K, \varepsilon)$ are givey by (8.5.32) and (8.5.51) respectively.

Remark 8.5.4 *If $K_0 = K$ Theorem 8.5.4 reduces to Theorem 4.3 in [216, p. 20]. In order for us to compare our results with the corresponding ones in [216] we computed the values of $R(\varepsilon, K_0, K)$ for F satisfying the conditions of Theorem 4.3 in [216] and Theorem 8.5.4 above with*

$$\beta\eta_0 \leq .4r, \tag{8.5.53}$$

and

$$1 \leq h_0 = \beta^2 L\eta_0 \leq 10, \tag{8.5.54}$$

and for ε equal to $10^{-i}r$, $1 \leq i \leq 10$.

The following table gives the results for $\varepsilon = 10^{-6}r$. Note that by I we mean $I(\delta_0, K, K)$, $I_{\alpha K}$ we mean $I(\delta_0, K_0, K)$ with $K_0 = \alpha K$, $\alpha \in [0,1]$. similarly for J, N and T.

Remark 8.5.5 *It follows from the table that our results significantly improve the corresponding ones in [216] and under the same computational cost. Suppose for example that $h_0 = 9$, $\delta = .159$. Kung found that the search phase can be done in 255 Newton steps and the iteration phase in 5 Newton steps. That is a root can be located inside a ball of radius $10^{-6}r$ using 260 Newton steps. However for $K_0 = .9K$, $K_0 = .5K$ and $K_0 = 0$ the corresponding Newton steps are 245,179 and 104 respectively which constitute a significant improvement. At the end of his paper Kung asked whether the number of Newton steps used by this procedure is close to the minimum. It is now clear from our approach that the answer is no (in general). Finally Kung proposed the open question: Suppose that the conditions of the Newton-kantorovich theorem hold: Is Newton's method optimal or close to optimal, in terms of the numbers of function and derivative equations required to approximate the solution x^* of equation $F(x) = 0$ to within a given tolerance ε? Clearly according to our approach the answer is no (see also Remark 8.5.1).*

h_0	δ_0	I	J	N	T	R	$I_{.9K}$	$N_{.9K}$	$R_{.9K}$	$I_{.3K}$	$N_{.3K}$	$R_{.3K}$	I_{0K}	N_{0K}	R_{0K}
1	.165	3	2	6	5	11	3	6	11	2	4	9	1	2	8
2	.103	8	3	24	6	30	7	21	27	5	15	21	2	6	12
3	.118	16	3	48	6	54	14	42	48	10	30	36	5	15	21
4	.129	25	3	75	6	81	23	69	75	16	48	54	9	27	33
5	.137	35	3	105	6	111	33	99	105	30	90	96	13	39	45
6	.144	47	3	141	5	146	44	132	137	31	93	98	17	51	56
7	.149	59	3	177	5	182	55	165	170	40	120	125	22	66	71
8	.154	72	3	216	5	221	67	201	206	41	123	128	28	84	89
9	.159	85	3	255	5	260	80	240	245	58	174	179	33	99	104
10	.163	99	3	295	5	302	93	279	284	68	204	209	39	117	122

8.6 Convergence on a Cone

In this section we are concerned with the problem of approximating a solution x^* of equation (5.1.1).

We propose the Newton–Kantorovich method (5.1.47) to generate a sequence approximating x^*.

This study is motivated by the elegant work in [116], where X is a real Banach space ordered by a closed convex cone K. We note that passing from scalar majorants to vector majorants enlarges the range of applications, since the latter uses the spectral radius which is usually smaller than its norm used by the former.

Here using finer vector majorants than before (see [116]) we show under the same hyotheses:

(a) sufficient convergence conditions can be obtained which are always weaker than before.

(b) finer error bounds on the distances involved and an at least as precise information on the location of the solution x^* are provided.

Several applications are provided. In particular we show as a special case that the Newton–Kantorovich hypothesis (4.2.51) is weakened. Finally we study the local convergence of method (5.1.47).

In order to make the study as self-contained as possible we need to introduce some concepts involving K-normed spaces.

Let X be a real Banach space ordered by a closed convex cone K. We say that cone K is regular if every increasing sequence $\gamma_1 \le \gamma_2 \le \cdots \le \gamma_n \le \cdots$ which is bounded above, converges in norm. If $\gamma_n^0 \le \gamma_n \le \gamma_n^1$ and $\lim_{n\to\infty} \gamma_n^0 = \lim_{n\to\infty} \gamma_n^1 = \gamma^*$, the regularity of K implies $\lim_{n\to\infty} \gamma_n = \gamma^*$.

Let $\alpha, \beta \in X$ the conic segment $\langle \alpha, \beta \rangle = \{\gamma \mid \alpha \le \gamma \le \beta\}$. An operator Q in X is called positive if $Q(\gamma) \in K$ for all $\gamma \in K$. Denote by $L(X, X)$ the space of all bounded linear operators in X, and $L_{\text{sym}}(X^2, X)$ the space of bilinear, symmetric, bounded operators from X^2 to X. By the standard linear isometry between $L(X^2, X)$, and $L(X, L(X, X))$, we consider the former embedded into the latter.

Let D be a linearly connected subset of K, and φ be a continuous operator from D into $L(X, X)$ or $L(X, L(X, X))$. We say that the line integral of φ is independent of the path if for every polygonal line L in D, the line integral depends only on the initial and final point of L. We define

$$\int_{r_0}^{r} \varphi(t)dt = \int_0^1 \varphi[(1-s)r_0 + sr](r - r_0)ds. \tag{8.6.1}$$

We need the definition of K-normed space [365]:

Definition 8.6.1 *Let X be a real linear space. Then X is said to be K-normed if operator $]\cdot[: X \to Y$ satisfies:*

$$\begin{aligned} &]x[\ \geq 0 \quad (x \in X); \\ &]x[\ = 0 \Leftrightarrow x = 0; \\ &]rx[\ = |r|\,]x[\quad (x \in X, r \in \mathbf{R}); \end{aligned}$$

and

$$]x + y[\ \leq\ ]x[\ +\ ]y[\quad (x, y \in X).$$

Let $x_0 \in X$ and $r \in K$. Then we denote

$$\overline{U}(x_0, r) = \{x \in X \mid\ ]x - x_0[\ \leq r\}. \tag{8.6.2}$$

Using K-norm we can define convergence on X. A sequence $\{y_n\}$ $(n \geq 0)$ in X is said to be

(1) convergent to a limit $y \in X$ if

$$\lim_{n\to\infty}\ ]y_n - y[\ = 0 \quad \text{in } X$$

and we write

$$(X) - \lim_{n\to\infty} y_n = y;$$

(2) a Cauchy sequence if

$$\lim_{m,n\to\infty}\ ]y_m - y_n[\ = 0.$$

The space X is complete if every Cauchy sequence is convergent.

We use the following conditions:

F is differentiable on the K-ball $U(x_0, R)$, and for every $r \in S = \langle 0, R\rangle$ there exist positive operators $w_0(r)$, $\overline{w}(r) \in L_{\text{sym}}(X^2, X)$ such that for all $z \in X$

$$](F'(x) - F'(x_0))(z)[\;\leq w_0(r)(]x - x_0[\,,]z[) \tag{8.6.3}$$

for all $x \in \overline{U}(x_0, r)$,

$$](F'(x) - F'(y))(z)[\;\leq \overline{w}(r)(]x - y[\,,]z[) \tag{8.6.4}$$

for all $x, y \in \overline{U}(x_0, r)$, where operators $w_0, \overline{w}\colon S \to L_{\text{sym}}(X^2, X)$ are increasing. Moreover the line integral of $\overline{w}$ (similarly for w_0) is independent of the path, and the same is true for the operator $w\colon S \to L(X, X)$ given by

$$w(r) = \int_0^r \overline{w}(t)dt. \tag{8.6.5}$$

Note that in general

$$w_0(r) \leq \overline{w}(r) \quad \text{for all} \quad r \in S. \tag{8.6.6}$$

The Newton–Leibniz formula holds for F on $\overline{U}(x_0, R)$:

$$F(x) - F(y) = \int_x^t F'(z)dz, \tag{8.6.7}$$

for all segments $[x, y] \in \overline{U}(x_0, R)$; for every $r \in S$ there exists a positive operator $w_1(r) \in L(X, X)$ such that

$$]G(x) - G(y)[\;\leq w_1(r)\,(]x - y[) \quad \text{for all} \quad x, y \in \overline{U}(x_0, r), \tag{8.6.8}$$

where $w_1 : S \to L(X, X)$ is increasing and the line integral of w_1 is independent of the path;

Operator $F'(x_0)$ is invertible and satisfies:

$$]F'(x_0)(y)[\;\leq b]y[\quad \text{for all} \quad y \in Y \tag{8.6.9}$$

for some positive operator $b \in L(X, X)$.

Let

$$\eta = \left]F'(x_0)^{-1}(F(x_0) + G(x_0))\right[, \tag{8.6.10}$$

and define operator $f\colon S \to X$ by letting

$$f(r) = \eta + b\int_0^r w(t)dt + b\int_0^r w_1(t)dt. \tag{8.6.11}$$

By the monotonicity of operators w, w_1 we see that f is order convex, i.e. for all $r, \overline{r} \in S$ with $r \leq \overline{r}$,

$$f((1 - s)r + s\overline{r}) \leq (1 - s)f(r) + sf(\overline{r}) \quad \text{for all } s \in [0, 1]. \tag{8.6.12}$$

We will use the following results whose proofs can be found in [116]:

Lemma 8.6.2 *(a) If Lipschitz condition (8.6.4) holds then*

$$](F'(x+y) - F'(x))(z)[\ \leq (w(r+]y[) - w(r))(]z[) \tag{8.6.13}$$

for all $r, r+\,]y[\, \in S$, $x \in \overline{U}(x_0, r)$, $z \in X$;

(b) If Lipschitz condition (8.6.8) holds then

$$]G(x+y) - G(x)[\leq \int_r^{r+]y[} w_1(t)dt \quad \text{for all} \quad r, r+\,]y[\, \in S,\ x \in \overline{U}(x_0, r). \tag{8.6.14}$$

Denote by $Fix(f)$ the set of all fixed points of the operator f.

Lemma 8.6.3 *Assume:*

$$\text{Fix}(f) \neq \emptyset. \tag{8.6.15}$$

Then there is a minimal element r^* *in* $Fix(f)$ *which can be found by applying the method of successive approximations*

$$r = f(r) \tag{8.6.16}$$

with 0 *as the starting point.*
The set

$$B(f, r^*) = \{r \in S\colon \lim_{n \to \infty} f^n(r) = r^*\} \tag{8.6.17}$$

is the attracting zone of r^*.

Remark 8.6.1 *[116] Let* $r \in S$. *If*

$$f(r) \leq r, \tag{8.6.18}$$

and

$$\langle 0, r \rangle \cap \text{Fix}(f) = \{r^*\} \tag{8.6.19}$$

then,

$$\langle 0, r \rangle \subseteq B(f, r^*). \tag{8.6.20}$$

Note that the successive approximations

$$\varepsilon_{n+r} = \delta(\varepsilon_n) \quad (\varepsilon_0 = r) \quad (n \in N) \tag{8.6.21}$$

converges to a fixed point ε^* *of* f, *satisfying* $0 \leq s^* \leq r$. *Hence, we conclude* $s^* = r^*$, *which implies* $r \in B(f, r^*)$.

In particular, Remark 8.6.1 implies

$$\langle 0, (1-s)r^* + sr\rangle \subseteq B(f, r^*) \tag{8.6.22}$$

for every $r \in \text{Fix}(f)$ with $\langle 0, r\rangle \cap \text{Fix}(f) = \{r^*, r\}$, and for all $\lambda \in [0, 1)$.

In the scalar case $X = \mathbf{R}$, we have

$$B(f, r^*) = [0, r^*] \cup \{r \in S\colon r^* < r, f(q) < q, (r^* < q \leq r)\}. \tag{8.6.23}$$

We will also use the notation

$$E(r^*) = \bigcup_{r \in B(f,r^*)} \overline{U}(x_0, r). \tag{8.6.24}$$

Returning back to method (5.1.47), we consider the sequences of approximations

$$r_{n+1} = r_n - (bw_0(r_n) - I)^{-1}(f(r_n) - r_n) \quad (r_0 = 0,\ n \geq 0) \tag{8.6.25}$$

and

$$\overline{r}_{n+1} = \overline{r}_n - (bw(\overline{r}_n) - I)^{-1}(f(\overline{r}_n) - \overline{r}_n) \quad (\overline{r}_0 = 0,\ n \geq 0) \tag{8.6.26}$$

for the majorant equation (8.6.16).

Lemma 8.6.4 *If operators*

$$I - bw_0(r) \quad r \in [0, r^*) \tag{8.6.27}$$

are invertible with positive inverses, then sequence $\{r_n\}$ $(n \geq 0)$ *given by (7) is well defined for all* $n \geq 0$*, monotonically increasing and convergent to* r^**.*

Proof. We first show that if $a_1 \leq a_2$, $a_2 \neq r^*$, then

$$(I - bw_0(a_1))^{-1}\theta \leq (I - bw_0(a_2))^{-1}\theta \quad \text{for all} \quad \theta \in K. \tag{8.6.28}$$

We have

$$\begin{aligned}\theta + bw_0(a_2)\theta + \cdots + (bw_0(a_2))^n\theta &= (I - bw_0(a_2))^{-1}\theta \\ &\quad - (bw_0(a_2))^{n+1}(I - bw_0(a_2))^{-1}\theta \leq \theta_2,\end{aligned} \tag{8.6.29}$$

where,

$$\theta_2 = (I - bw_0(a_2))^{-1}\theta. \tag{8.6.30}$$

Using the monotonicity of w we get

$$\theta + bw_0(a_1)\theta + \cdots + (bw_0(a_1))^n\theta \leq \theta + bw_0(a_1)\theta + \cdots + (bw_0(a_1))^n\theta \leq \theta_2, \tag{8.6.31}$$

which implies increasing sequence $\{(bw_0(a_1))^n\theta\}$ $(n \geq 0)$ is bounded in a regular cone K, and as such it converges to some

$$\theta_1 = (I - bw_0(a_1))^{-1}\theta \tag{8.6.32}$$

with

$$\theta_1 \leq \theta_2.$$

We also need to show that the operator

$$g(r) = r - (bw_0(r) - I)^{-1}(f(r) - r) \tag{8.6.33}$$

is increasing on $\langle 0, r^*\rangle - \{r^*\}$. Let $b_1 \leq b_2$ with $b_1, b_2 \in [0, r^*]$, then using (8.6.6) and (8.6.33) we obtain in turn

$$\begin{aligned} & g(b_2) - g(b_1) = \\ & = b_2 - b_1 - (bw_0(b_2) - I)^{-1}(f(b_2) - b_2) + (bw_0(b_1) - I)^{-1}(f(b_1) - b_1) \end{aligned} \tag{8.6.34}$$

$$\begin{aligned} & = b_2 - b_1 - (bw_0(b_1) - I)^{-1}[(f(b_2) - b_2) - (f(b_1) - b_1)] \\ & + \ [(bw_0(b_1) - I)^{-1} - (bw_0(b_2) - I)^{-1}](f(b_2) - b_2) \\ & = (I - bw_0(b_1))^{-1}[f(b_2) - f(b_1) - bw_0(b_1)(b_2 - b_1)] \\ & + \ (I - bw_0(b_2))^{-1}(bw_0(b_2) - bw_0(b_1))(I - bw_0(b_1))^{-1}(f(b_2) - b_2) \geq 0, \end{aligned} \tag{8.6.35}$$

since all terms in the right hand of equality (8.6.35) are in the cone. Moreover, g leaves $\langle 0, r^*\rangle - \{r^*\}$ invariant, since

$$0 = g(0) \leq g(r) \leq g(r^*) = r^*. \tag{8.6.36}$$

Hence sequence

$$r_{n+1} = g(r_n) \quad (r_0 = 0, \ n \geq 0) \tag{8.6.37}$$

is well defined for all $n \geq 0$, lie in the set $\langle 0, r^*\rangle - \{r^*\}$, and is increasing. Therefore the limit of this sequence exists. Let us call it r_1^*. The point r_1^* is a fixed point of f in $\langle 0, r^*\rangle$. But r^* is the unique fixed point of f in $\langle 0, r^*\rangle$. Hence, we deduce

$$r_1^* = r^*. \tag{8.6.38}$$

■

Remark 8.6.2 *If equality holds in (8.6.6) then sequence $\{\overline{r}_n\}$ becomes $\{r_n\}$ $(n \geq 0)$ and Lemma 8.6.4 reduces to the corresponding Lemma in [116, p. 555]. Moreover as it can easily be seen using induction on n*

$$r_{n+1} - r_n \leq \overline{r}_{n+1} - \overline{r}_n, \tag{8.6.39}$$

and

$$r_n \leq \overline{r}_n \tag{8.6.40}$$

for all $n \geq 0$. *Furthermore if strict inequality holds in (8.6.6) so does in (8.6.39) and (8.6.40). If* $\{r_n\}$ $(n \geq 0)$ *is a majorizing sequence for method (5.1.47), then (8.6.39) shows that the error bounds on the distances* $\|x_{n+1} - x_n\|$ *are improved. It turns out that this is indeed the case.*

We can show the semilocal convergence theorem for method (5.1.47).

Theorem 8.6.5 *Assume:*
hypotheses (8.6.4), (8.6.5), (8.6.7), (8.6.8), (8.6.9), (8.6.15) hold, and operators (8.6.27) are invertible with positive inverses.
Then sequence $\{x_n\}$ $(n \geq 0)$ *generated by Newton–Kantorovich method (5.1.48) is well defined, remains in the K-ball* $U(x_0, r^*)$ *for all* $n \geq 0$ *and converges to a solution* x^* *of equation* $F(x)+G(x) = 0$ *in* $E(r^*)$, *where* $E(r^*)$ *is given by (8.6.24). Moreover the following estimates hold for all* $n \geq 0$:

$$]x_{n+1} - x_n[\, \leq r_{n+1} - r_n, \tag{8.6.41}$$

and

$$]x^* - x_n[\, \leq r^* - r_n, \tag{8.6.42}$$

where sequence $\{r_n\}$ *is given by (8.6.25).*

Proof. We first show (8.6.41) using induction on $n \geq 0$ (by(8.6.10)). For $n = 0$;

$$]x_1 - x_0[\, = \,]F'(x_0)^{-1}(F(x_0) + G(x_0))[\, = \eta = r_1 - r_0. \tag{8.6.43}$$

Assume:

$$]x_k - x_{k-1}[\, \leq r_k - r_{k-1} \tag{8.6.44}$$

for $k = 1, 2, \ldots, n$. Using (8.6.44) we get

$$]x_n - x_0[\, \leq \sum_{k=1}^{n}]x_k - x_{k-1}[\, \leq \sum_{k=1}^{n} (r_k - r_{k-1}) = r_n. \tag{8.6.45}$$

Define operators $Q_n \colon X \to X$ by

$$Q_n = -F'(x_0)^{-1}[F'(x_n) - F'(x_0)]. \tag{8.6.46}$$

By (8.6.3) and (8.6.9) we get

$$\begin{aligned}]Q_n(z)[&=]F'(x_0)^{-1}(F'(x_n) - F'(x_0))(z)[\\ &\leq b](F'(x_n) - F'(x_0))(z)[\\ &\leq bw_0(r_n)(]z[), \end{aligned} \tag{8.6.47}$$

and

$$]Q_n^i(z)[\leq (bw_0(r_n))^i(]z[) \quad (i \geq 1). \tag{8.6.48}$$

Hence, we have:

$$\sum_{i=0}^{\infty}]Q_n^i(z)[\leq \sum_{j=0}^{\infty} (bw_0(r_n))^i(]z[). \tag{8.6.49}$$

That is series $\sum_{i=0}^{\infty} Q_n^i(z)$ is convergent in X. Hence operator $I - Q_n$ is invertible, and

$$](I - Q_n)^{-1}(z)[\leq (I - bw_0(r_n))^{-1}(]z[). \tag{8.6.50}$$

Operator $F'(x_n)$ is invertible for all $n \geq 0$, since $F'(x_n) = F'(x_0)(I - Q_n)$, and for all $x \in Y$ we have:

$$\begin{aligned}]F'(x_n)^{-1}(x)[&=](I - Q_n)^{-1} F'(x_0)^{-1}(x)[\\ &\leq (I - bw_0(r_n))^{-1}(]F'(x_0)^{-1}(x)[) \\ &\leq (I - bw_0(r_n))^{-1}(b]x[). \end{aligned} \tag{8.6.51}$$

Using (5.1.47) we obtain the approximation

$$\begin{aligned}]x_{n+1} - x_n[&=]F'(x_n)^{-1}(F(x_n) + G(x_n)) \\ &- F'(x_n)^{-1}(F'(x_{n-1})(x_n - x_{n-1}) + F(x_{n-1}) + G(x_{n-1}))[. \end{aligned} \tag{8.6.52}$$

It now follows from (8.6.3)–(8.6.9), (8.6.11), (8.6.25) and (8.6.52)

$$\begin{aligned} &]x_{n+1} - x_n[\leq \\ &\leq]F'(x_n)^{-1}(F(x_n) - F(x_{n-1}) - F'(x_{n-1})(x_n - x_{n-1})[\\ &\quad +]F'(x_n)^{-1}(G(x_n) - G(x_{n-1})) \\ &\leq (I - bw_0(r_n))^{-1} \\ &\quad \times \left\{ b \right] \int_0^1 (F'((1-\lambda)x_{n-1} + \lambda x_n) - F'(x_{n-1}))(x_n - x_{n-1}) d_\lambda \left[\right\} \\ &\quad + (I - bw_0(r_n))^{-1}(b]G(x_n) - G(x_{n-1})[) \\ &\leq (I - bw_0(r_n))^{-1} \left\{ b \int_0^1 (w((1-\lambda)r_{n-1} + \lambda r_n) - w(r_{n-1}))(r_n - r_{n-1}) d_\lambda \right\} \\ &\quad + (I - bw_0(r_n))^{-1} \left(b \int_{r_{n-1}}^{r_n} w_1(t) dt \right) \\ &= (I - bw_0(r_n))^{-1} \left\{ b \int_{r_{n-1}}^{r_n} w(t) dt - bw(r_{n-1})(r_n - r_{n-1}) + b \int_{r_{n-1}}^{r_n} w_1(t) dt \right\} \end{aligned} \tag{8.6.53}$$

$$
\begin{aligned}
&= (I - bw_0(r_n))^{-1}(f(r_n) - f(r_{n-1}) - bw(r_{n-1})(r_n - r_{n-1}))\\
&= (I - bw_0(r_n))^{-1}((f(r_n) - r_n) - (f(r_{n-1}) - r_{n-1})\\
&- (bw(r_{n-1}) - I)(r_n - r_{n-1}))\\
&= (I - bw_0(r_n))^{-1}((f(r_n) - r_n) - (f(r_{n-1}) - r_{n-1})\\
&- (bw(r_{n-1}) - I)(r_n - r_{n-1}))\\
&\leq (I - bw_0(r_n))^{-1}((f(r_n) - r_n) - (f(r_{n-1}) - r_{n-1})\\
&- (bw_0(r_{n-1}) - I)(r_n - r_{n-1})) \qquad (8.6.54)\\
&= (I - bw_0(r_n))^{-1}(f(r_n) - r_n) = r_{n+1} - r_n. \qquad (8.6.55)
\end{aligned}
$$

By Lemma 8.6.4 sequence $\{r_n\}$ $(n \geq 0)$ converges to r^*. Hence $\{x_n\}$ is a convergent sequence, and its limit is a solution of equation (5.1.1). Therefore x_n converges to x^*. Finally (8.6.42) follows from (8.6.41) by using standard majorization techniques. The uniqueness part is omitted since it follows exactly as in Theorem 2 in [116]. ∎

Remark 8.6.3 *It follows immediately from (8.6.53) and (8.6.54) that sequence*

$$
\begin{aligned}
t_0 &= t_0,\\
t_1 &= \eta,\\
t_{n+1} - t_n &= (I - bw_0(t_n))^{-1}\left\{b\int_{t_{n-1}}^{t_n} w(t)dt - bw(t_{n-1})(t_n - t_{n-1})\right.\\
&\left. + \; b\int_{t_{n-1}}^{t_n} w_1(t)dt\right\} \quad (n \geq 1) \qquad (8.6.56)
\end{aligned}
$$

is also a finer majorizing sequence of $\{x_n\}$ $(n \geq 0)$ *and converges to some* t^* *in* $\langle 0, r^*\rangle$. *Moreover the following estimates hold for all* $n \geq 0$

$$
\begin{aligned}
]x_1 - x_0[\; &\leq \; t_1 - t_0 = r_1 - r_0, \qquad (8.6.57)\\
]x_{n+1} - x_n[\; &\leq \; t_{n+1} - t_n \leq r_{n+1} - r_n, \qquad (8.6.58)\\
]x^* - x_n[\; &\leq \; t^* - t_n \leq r^* - r_n, \qquad (8.6.59)\\
t_n &\leq r_n, \qquad (8.6.60)
\end{aligned}
$$

and

$$t^* \leq r^*. \qquad (8.6.61)$$

That is $\{t_n\}$ *is a finer majorizing sequence than* $\{r_n\}$ *and the information on the location of the solution* x^* *is more precise. Therefore we wonder if studying the convergence of* $\{t_n\}$ *without assuming (8.6.15) can lead to weaker sufficient convergence conditions for method (5.1.47). In Theorem 8.6.7 we answer this question.*

But first we need the following lemma on majorizing sequences for method (5.1.47).

Lemma 8.6.6 *If:*
there exist parameters $\eta \geq 0$, $\delta \in [0,2)$ such that: Operators

$$I - bw_0 \left[2(2I - \delta I)^{-1} \left(I - \left(\tfrac{\delta I}{2}\right)^{n+1}\right) \eta\right] \tag{8.6.62}$$

be positive, invertible and with positive inverses for all $n \geq 0$;

$$2(I - bw_0(\eta))^{-1} \left[bw_1(\eta) + b \int_0^1 w(s\eta)ds - bw(0)\right] \leq \delta I, \tag{8.6.63}$$

and

$$\begin{aligned} 2b \int_0^1 w&\left[2(2I - \delta I)^{-1} \left(I - \left(\tfrac{\delta I}{2}\right)^{n+1}\right) \eta + s \left(\tfrac{\delta I}{2}\right)^{n+1} \eta\right] ds \\ &- 2bw\left[2(2I - \delta I)^{-1} \left(I - \left(\tfrac{\delta I}{2}\right)^{n+1}\right) \eta\right] \\ &+ 2bw_1\left[2(2I - \delta I)^{-1} \left(I - \left(\tfrac{\delta I}{2}\right)^{n+1}\right) \eta\right] \\ &+ \delta bw_0\left[2(2I - \delta I)^{-1} \left(I - \left(\tfrac{\delta I}{2}\right)^{n+1}\right) \eta\right] \\ \leq&\ \delta I, \quad \textit{for all } n \geq 0. \end{aligned}$$

Then iteration $\{t_n\}$ $(n \geq 0)$ given by (8.6.56) is non-decreasing, bounded above by

$$t^{**} = 2(2I - \delta I)^{-1}\eta, \tag{8.6.64}$$

and converges to some t^* such that:

$$0 \leq t^* \leq t^{**}. \tag{8.6.65}$$

Moreover the following estimates hold for all $n \geq 0$:

$$0 \leq t_{n+2} - t_{n+1} \leq \tfrac{\delta I}{2}(t_{n+1} - t_n) \leq \left(\tfrac{\delta I}{2}\right)^{n+1} \eta. \tag{8.6.66}$$

Proof. We must show:

$$2(I - bw_0(t_{k+1}))^{-1}\left[b \int_0^1 w(t_k + s(t_{k+1} - t_k))ds - bw(t_k) + bw_1(t_{k+1})\right] \leq \delta I, \tag{8.6.67}$$

and operators

$$0 \leq t_{k+1} - t_k, \tag{8.6.68}$$
$$I - bw_0(t_{k+1}), \tag{8.6.69}$$

positive, invertible and with positive inverses. Estimate (8.6.67) can then follow immediately from (8.6.68)–(8.6.70). Using induction on the integer k, we get for $k = 0$

$$2(I - bw_0(t_1))^{-1}\left[b\int_0^1 w[t_0 + s(t_1 - t_0)]ds - bw(t_1) + w_1(t_1)\right] \leq \delta I,$$

$$bw_0(t_1) < I, \tag{8.6.70}$$

by the initial conditions. But (8.6.56) then gives

$$0 \leq t_2 - t_1 \leq \tfrac{\delta I}{2}(t_1 - t_0). \tag{8.6.71}$$

Assume (8.6.68)–(8.6.70) hold for all $k \leq n+1$. Using (8.6.62)–(8.6.64) we obtain in turn:

$$\begin{aligned}
&2b\int_0^1 w[t_{k+1} - s(t_{k+2} - t_{k+1})]ds - 2bw(t_{k+1}) + 2bw_1(t_{k+1}) + \delta bw_0(t_{k+1})\\
&\leq 2b\int_0^1 w\left[2(2I - \delta I)^{-1}\left(I - \left(\tfrac{\delta I}{2}\right)^{k+1}\right)\eta + s\left(\tfrac{\delta I}{2}\right)^{k+1}\eta\right]ds\\
&\quad - 2bw\left[2(2I - \delta I)^{-1}\left(I - \left(\tfrac{\delta I}{2}\right)^{k+1}\right)\eta\right]\\
&\quad + 2bw_1\left[2(2I - \delta I)^{-1}\left(I - \left(\tfrac{\delta I}{2}\right)^{k+1}\eta\right)\right]\\
&\quad + \delta bw_0\left[2(2I - \delta I)^{-1}\left(I - \left(\tfrac{\delta I}{2}\right)^{k+1}\right)\eta\right]\\
&\leq \delta I.
\end{aligned} \tag{8.6.72}$$

Moreover, we show:

$$t_k \leq t^{**}. \tag{8.6.73}$$

We have:

$$t_0 = \eta \leq t^{**},\ t_1 = \eta \leq t^{**},\ t_2 \leq \eta + \tfrac{\delta I}{2}\eta = \tfrac{2I+\delta I}{2}\eta \leq t^{**}.$$

Assume (8.6.73) holds for all $k \leq n+1$. It follows from (8.6.56), (8.6.68)–(8.6.70):

$$\begin{aligned}
t_{k+2} &\leq t_{k+1} + \tfrac{\delta I}{2}(t_{k+1} - t_k) \leq t_k + \tfrac{\delta I}{2}(t_k - t_{k-1}) + \tfrac{\delta I}{2}(t_{k+1} - t_k)\\
&\leq \cdots \leq \eta + \tfrac{\delta I}{2}\eta + \left(\tfrac{\delta I}{2}\right)^2\eta + \cdots + \left(\tfrac{\delta I}{2}\right)^{k+1}\eta\\
&= 2(2I - \delta I)^{-1}\left[I - \left(\tfrac{\delta I}{2}\right)^{k+2}\right]\eta \leq 2(2I - \delta I)^{-1}\eta = t^{**}.
\end{aligned} \tag{8.6.74}$$

Hence, sequence $\{t_n\}$ $(n \geq 0)$ converges to some t^* satisfying (8.6.66). ■

We can show the main semilocal convergence theorem for method (5.1.47).

Theorem 8.6.7 *Assume:*
hypotheses (8.6.3)–(8.6.5), (8.6.7)–(8.6.9), (8.6.62)–(8.6.64) hold, and

$$t^{**} \leq R, \tag{8.6.75}$$

where t^{**} is given by (8.6.65).
Then sequence $\{x_n\}$ $(n \geq)$ generated by Newton–Kantorovich method (5.1.47) is well defined, remains in the K-ball $U(x_0, t^*)$ for all $n \geq 0$ and converges to a solution x^* of equation $F(x) + G(x) = 0$, which is unique in $E(t^*)$. Moreover the following error bounds hold for all $n \geq 0$:

$$]x_{n+1} - x_n[\leq t_{n+1} - t_n \tag{8.6.76}$$

and

$$]x^* - x_n[\leq t^* - t_n, \tag{8.6.77}$$

where sequence $\{t_n\}$ $(n \geq 0)$ and t^* are given by (8.6.56) and (8.6.66) respectively.

Proof. The proof is identical to Theorem 8.6.5 with sequence t_n replacing r_n until the derivation of (8.6.53). But then the right hand side of (8.6.53) with these changes becomes $t_{n+1} - t_n$. By Lemma 8.6.6 $\{t_n\}$ converges to t^*. Hence $\{x_n\}$ is a convergent sequence, its limit converges to a solution of equation $F(x) + G(x) = 0$. Therefore $\{x_n\}$ converges to x^*. Estimate (8.6.77) follows from (8.6.76) by using standard majorization techniques. The uniqueness part is omitted since it follows exactly as in Theorem 2 in [116]. ∎

Remark 8.6.4 *Conditions (8.6.62), (8.6.64) can be replaced by the stronger but easier to check*

$$I - bw_0[2(2I - \delta I)^{-1}\eta] \tag{8.6.78}$$

and

$$\begin{aligned} 2b \int_0^1 & w\left[2(2I - \delta I)^{-1}\eta + s\left(\tfrac{\delta I}{2}\right)\eta\right] s - 2bw[2(2I - \delta I)^{-1}\eta] \\ & + 2bw_1[2(2I - \delta I)^{-1}\eta] + \delta b w_0[2(2I - \delta I)^{-1}\eta] \\ & \leq \delta I, \end{aligned} \tag{8.6.79}$$

respectively.

Application 8.6.8 *Assume operator $]\ [$ is given by a norm $\|\cdot\|$ and set $G(x) = 0$ for all $x \in \overline{U}(x_0, R)$. Choose for all $r \in S$, $b = 1$ for simplicity,*

$$\overline{w}(r) = \ell r, \tag{8.6.80}$$

$$w_0(r) = \ell_0 r \tag{8.6.81}$$

and

$$w_1(r) = 0. \tag{8.6.82}$$

With these choices our conditions reduces to the ones in Section 5.1 that have already been compared favourably with the Newton-Kantorovich theorem.

Remark 8.6.5 *The results obtained here hold under even weaker conditions. Indeed, since (8.6.4) is not "directly" used in the proofs above, it can be replaced by the weaker condition (8.6.13) throughout this study. As we showed in Lemma 8.6.2*

$$(6) \Rightarrow (15)$$

but not necessarily vice versa unless if operator w is convex [116, p. 674].

The local convergence for method (5.1.47) was not examined in [111]. Let x^* be a simple solution of equation (5.1.1), and assume $F(x^*)^{-1} \in L(Y, X)$. Moreover assume with x^* replacing x_0 that hypotheses (8.6.3), (8.6.4), (8.6.5), (8.6.7), (8.6.8), (8.6.9) hold. Then exactly as in (8.6.53) but using the local conditions, and the approximation

$$x_{n+1} - x^* = \tag{8.6.83}$$

$$= [F'(x_n)^{-1} F'(x^*)] F'(x^*)^{-1}$$
$$\cdot \left\{ \int_0^1 [F'(x_n + t(x^* - x_n)) - F'(x_n)](x^* - x_n)dt + \; (G(x^*) - G(x_n)) \right\} \tag{8.6.84}$$

we can show the following local result for method (5.1.47).

Theorem 8.6.9 *Assume there exists a minimal solution $r^* \in S$ of equation*

$$p(r) = 0, \tag{8.6.85}$$

where,

$$p(r) = b \int_0^1 [w((1+s)r) - w(r)]ds + bw_0(r) + bw_1(r) - 1. \tag{8.6.86}$$

Then, sequence $\{x_n\}$ $(n \geq 0)$ generated by Newton–Kantorovich method (5.1.47) is well defined, remains in $\overline{U}(x^*, r^*)$ for all $n \geq 0$, and converges to x^* provided that $x_0 \in U(x^*, r^*)$.
Moreover the following estimates hold for all $n \geq 0$

$$]x^* - x_{n+1}[\; \leq \varepsilon_{n+1}, \tag{8.6.87}$$

where,

$$\varepsilon_{n+1} = \frac{b \int_0^1 (w((1+s)]x_n - x^*[) - w(]x_n - x^*[))ds]x_n - x^*[+ \int_0^{]x_n - x^*[} w_1(s)ds}{1 - bw_0(]x_n - x^*[)} \tag{8.6.88}$$

$(n \geq 0)$.

Application 8.6.10 *Returning back to the choices of Application 8.6.8 and using (8.6.85) we get*

$$r^* = \tfrac{1}{2\ell_0+\ell}\,. \tag{8.6.89}$$

See also Section 5.1.

8.7 Convergence on Lie Groups

In this section we are concerned with the problem of approximating a solution x^* of the equation

$$F(y) = 0, \tag{8.7.1}$$

where F is an operator mapping a Lie group to its corresponding Lie algebra.

Most numerical algorithms are formulated on vector spaces with some type of a Banach space structure that allows the study of the convergence properties of the iteration. Recently, there has been an interest in studying numerical algorithms on manifolds (see [265] and the references there). Note that a suitable choice of coordinates can be interpreted locally to be the Euclidean space, which justifies the study of such algorithms on manifolds.

In particular, Newton's method for solving equation (8.7.1) has already been introduced in the elegant work in [265], where the quadratic convergence of the algorithm was established provided that the initial guess is chosen from a ball of a certain convergence radius.

Here by using more precise majorizing sequences we show that under the same hypotheses and computational cost a larger radius of convergence and finer error bounds can be obtained than before [265]. This observation allows a wider choice of initial guesses for starting Newton's method and shows that fewer iterations are required to find the solution than it is suggested in [265].

Finally we provide a numerical example to justify what we stated in the above paragraph.

Our study is divided as follows: first we provide the background in condensed form needed for this paper, then we introduce Newton's method and state some results obtained in [265] that are needed for the main local convergence results appearing later in the study.

A Lie group $(G, \cdot)$ is understood to have the structure of a smooth manifold, i.e. the group product and the inversion are smooth operations in the differentiable structure embedded in the manifold. The dimension of the Lie group equals the one of the corresponding manifold, and we assume that it is finite.

We will need the definitions [265]:

Definition 8.7.1 *A Lie algebra g of a Lie group $(G, \cdot)$ is the tangent space at i (i the identity element of G), $TG/_i$ equipped with the Lie bracket $[-,-]\colon g \times g \to g$*

defined for $u, v \in g$ *by*

$$[u,v] = \left.\frac{\partial^2}{\partial t \partial s}\right|_{t=s=0} y(t)z(s)y(t)^{-1} \tag{8.7.2}$$

where, $y(t)$, $z(s)$ *are two smooth curves in* G *satisfying* $y(0) = z(0) = e$, $y'(0) = u$ *and* $z'(0) = v$.

Definition 8.7.2 *The exponential map is an operator*

$$\exp\colon g \to G : u \to \exp(u).$$

Given $u \in G$, *the left invariant vector field* $X_u\colon y \to L'_y(u)$ *(where* $L'_y(g) = TG/_y$*) determines an integral curve* $\gamma_u(t)$ *solving the initial value problem*

$$\gamma'_u(t) = X_u(\gamma_u(t))$$
$$\gamma_u(0) = e.$$

If

$$N(i) = \exp(\overline{N}(0)) \tag{8.7.3}$$

where $\overline{N}(0)$ *is an open neighborhood of* $o \in g$ *then for any* $y \in N(i)$

$$y = \exp(v) \quad \text{for some } v \in \overline{N}(0)$$

and if

$$\exp(u) = \exp(v) \in N(i)$$

for some $u, v \in \overline{N}(0)$, *then* $u = v$. *If* G *is Abelian, then*

$$\exp(u+v) = \exp(u)\exp(v) = \exp(v)\exp(u) \tag{8.7.4}$$

for all $u, v \in g$. *In the non-Abelian case (8.7.4) is replaced by*

$$\exp(w) = \exp(u)\exp(v), \tag{8.7.5}$$

where w *is given by the (BCH) formula [265, p. 114]*

$$w = u + v + \frac{1}{2}[u,v] + \frac{1}{12}\left([u,[u,v]] + [v,[v,u]]\right) + \cdots \tag{8.7.6}$$

for $u, v \in \overline{N}(0)$.

Definition 8.7.3 *The differential of exp at* $u \in g$ *is a map*

$$\exp'_u : g \to TG/_{\exp(u)}$$
$$v \to \left.\frac{d}{dt}\right|_{t=0} \exp(u+v)$$

expressed by $d\exp_u\colon g \to g$ *by*

$$d\exp_u = L'_{(\exp(u))^{-1}} \circ \exp'_u = L'_{\exp(-u)} \circ \exp'_u . \tag{8.7.7}$$

Definition 8.7.4 *The differential of operator F at a point $y \in G$ is a map*

$$F'_y \colon TG\big|_y \to Tg/_{f(y)} \cong g$$

defined by

$$F'_y(\Delta y) = \frac{d}{dt}\bigg|_{t=0} F\big(y \cdot \exp(tL'_{y^{-1}}(\Delta y))\big) \tag{8.7.8}$$

for any tangent vector $\Delta y \in TG/_y$ to the manifold G at y.

It follows:

$$F'_y(\Delta y) = \lim_{t \to 0} \frac{F(y \cdot z) - F(y)}{t}.$$

As in (8.7.7), the differential F'_y can be expressed in terms of an operator Df_y $g \to g$ given by

$$dF_y = (F \circ L_y)' = F'_y \circ L'_y, \tag{8.7.9}$$

and consequently

$$dF_y(u) = F'_y(L'_y u) = \frac{d}{dt}\bigg|_{t=0} F\big(y \cdot \exp(tu)\big). \tag{8.7.10}$$

We can now define Newton's method on the Lie group $(G, \cdot)$ using (8.7.8) as follows:

Choose $y_0 \in G$ and determine the differential F'_{y_0} by (8.7.8). Then find $\Delta y_0 \in TG/_{y_0}$ satisfying the approximation

$$F'_{y_0}(\Delta y_0) + F(y_0) = 0,$$

and then set $y_1 = y_0 \exp\big(L'_{L_0^{-1}}(\Delta y_0)\big)$. Hence it follows by (8.7.9) that Newton's method can be formally written as:

Given $y_n \in G$, determine dF_{y_n} according to (8.7.10);

Find $u_n \in g$ such that

$$dF_{y_n}(u_n) + F(y_n) = 0; \tag{8.7.11}$$

compute

$$y_{n+1} = y_n \exp(u_n). \tag{8.7.12}$$

It is well known that a Lie group $(G, \cdot)$ is a manifold and as such it is also a topological space. Therefore the notion of convergence is defined with respect to this topology. We find it convenient to introduce a Banach space structure on the Lie algebra g, and use it to analyze convergence.

As already mentioned before we only discuss finite dimensional Lie algebras. Therefore every linear operator $h\colon g \to g$ is bounded. Consequently we can define its norm subordinate to that on g by

$$\|h\| = \sup_{u \neq 0} \frac{\|h(u)\|}{\|u\|} = \sup_{\|u\|=1} \|h(u)\| < \infty.$$

We also need the definitions concerning convergence of sequences:

Definition 8.7.5 *A sequence $\{y_n\}$ $(n \geq 0)$ in $N(i)$ converges to i if and only if corresponding sequence implicitly defined by $y_n = \exp(u_n)$ $(n \geq 0)$ converges to 0, in the sense that*

$$\lim_{n \to \infty} \|u_n\| = 0. \tag{8.7.13}$$

Definition 8.7.6 *If there exists $y \in G$ such that*

$$y^{-1} y_n = \exp(u_n) \in N(i) \tag{8.7.14}$$

and (8.7.13) holds, then we say sequence $\{y_n\}$ converges to y. Moreover the order of convergence is $\lambda \geq 0$ if

$$\|u_{n+1}\| \leq O(\|u_n\|^{\lambda}). \tag{8.7.15}$$

The continuity of the Lie bracket $[-,-]$ guarantees the existence of a constant $\mu \geq 0$ such that

$$\|[u, v]\| \leq \mu \|u\| \cdot \|v\| \quad \textit{for all } u, v \in g.$$

We need the Lemmas whose proofs can be found in [265]:

Lemma 8.7.7 *There exists a constant $\delta > 0$ and a real function $f = f(t)$ analytic in the disk $|t| \leq \delta$, non-decreasing for $t \geq 0$, with $h(0) = 1$, such that for all $u, v \in g$*

$$\|u\| + \|v\| \leq \beta \Rightarrow \|w\| \leq \|u + v\| f(\mu\beta) \tag{8.7.16}$$

and

$$\|u\| \leq \frac{\beta}{2}, \quad \|u\| + \|v\| \leq \beta \Rightarrow \|u - v\| \leq \|v\| f(\mu\beta), \tag{8.7.17}$$

for any $\beta \leq \frac{\delta}{\mu}$, where w is given by (8.7.6).

Lemma 8.7.8 *For $y \in G$, $u \in g$ and $t \in R$ we have*

$$\frac{d}{dt} F(y \cdot \exp(-tu)) = -dF_{y \exp(-tu)}(u). \tag{8.7.18}$$

We also need the radius

$$r = \max_{\|u\|=1} \{t \in R, \ su \in \overline{N}(0), \quad \textit{for all } s \in [0,t]\} > 0 \tag{8.7.19}$$

of the maximal ball in g centered at 0 and contained in $\overline{N}(0)$. Moreover let $r^* \in [0, r]$. We can define the closed ball with center $y \in G$ and of radius r^* by

$$U(y, r^*) = \{z \in G, \ y^{-1} \cdot z = \exp(u) \in N(i), \ \|u\| \leq r^*\}. \tag{8.7.20}$$

With the notation introduced above, we can state and prove the following local convergence result for Newton's method (8.7.11)–(8.7.12):

Theorem 8.7.9 *Assume:*
Operator dF_y is one-to-one, dF_y^{-1} exists and

$$\|dF_y^{-1}\| \leq a \tag{8.7.21}$$

for some positive constant a; there exist constants $r_0 \in (0, r]$, $\ell_0 \geq 0$ and $\ell \geq 0$ such that for all $z \in U(y, r_0)$, $u \in g$ with $\|u\| \leq r_0$ the following hold:

$$\|dF_{y\exp(u)} - dF_y\| \leq \ell_0 \|u\|, \tag{8.7.22}$$

and

$$\|dF_{z\exp(u)} - dF_z\| \leq \ell \|u\|. \tag{8.7.23}$$

Then for each fixed $q \in [0,1]$ there exists $r^* > 0$ such that, if $y_0 \in U^0(y, r^*)$ sequence $\{y_n\}$ $(n \geq 0)$ generated by Newton's method (8.7.11)–(8.7.12) is well defined, remains in $U^0(y, r^*)$ for all $n \geq 0$ and converges quadratically to y. Moreover the following relations hold for all $n \geq 0$:

$$y^{-1} y_n = \exp(v_n), \tag{8.7.24}$$

$$\lim_{n \to \infty} v_n = 0, \tag{8.7.25}$$

and

$$\|v_{n+1}\| \leq c \|v_n\|^2, \tag{8.7.26}$$

where,

$$c = \frac{a \ell f(3\mu r^*)}{2(1 - \ell_0 a r^*)}, \tag{8.7.27}$$

and function f is given in Lemma 8.7.7.

Proof. Let $\delta > 0$ and $f(-)$ be as in Lemma 8.7.7. Set $b = \sup_{t \in [0,\delta]} f(t) = f(\delta) \geq 1$, and define

$$r^* = \min\left(p, \frac{r_0}{1+q}, \frac{\delta}{3\mu}\right), \tag{8.7.28}$$

where,

$$p = \frac{2q}{(b\ell + 2q\ell_0)a}. \tag{8.7.29}$$

Let $y_0 \in U^0(y, r^*)$. It follows $y^{-1}y_0 = \exp(v_0)$ for some $v_0 \in g$ with $\|v_0\| < r^*$. Using (8.7.22) we can get

$$\|dF_y^{-1}(dF_y - dF_{y_0})\| \leq \|dF_y^{-1}\| \, \|dF_y - dF_{y_0}\| \leq a\ell_0\|v_0\| < a\ell_0 r^* \leq 1 \tag{8.7.30}$$

by the choice of r^*. By (8.7.30) and the Banach Lemma on invertible operators 1.3.2 $dF_{y_0}^{-1}$ exists, and

$$\|dF_{y_0}^{-1}\| \leq \frac{\|dF_y^{-1}\|}{1 - \|dF_y^{-1}(dF_y - dF_{y_0})\|} \leq \frac{a}{1 - a\ell_0\|v_0\|} < \frac{a}{1 - a\ell_0 r^*}. \tag{8.7.31}$$

Define u_0 by (8.7.11). Then using Lemma 8.7.8, we can have in turn

$$\begin{aligned} u_0 + v_0 &= dF_{y_0}^{-1}[F(y) - F(y_0) + dF_y(v_0)] \\ &= dF_{y_0}^{-1}\int_0^1 \left[\frac{d}{dt}F(y_0\exp(-tv_0)) + dF_{y_0}(v_0)\right]dt \\ &= dF_{y_0}^{-1}\int_0^1 [dF_{y_0}(v_0) - dF_{y_0\exp(-tv_0)}(v_0)]dt. \end{aligned} \tag{8.7.32}$$

Moreover using (8.7.23), (8.7.31) and (8.7.32), we get since $\|tv_0\| < r^*$ for all $t \in [0,1]$

$$\begin{aligned} \|u_0 + v_0\| &\leq \|dF_{y_0}^{-1}\| \int_0^1 \|dF_{y_0}(v_0) - dF_{y_0\exp(-tv_0)}(v_0)\|dt \\ &\leq \|dF_{y_0}^{-1}\| \int_0^1 \ell\|tv_0\| \, \|v_0\|dt \\ &\leq \frac{\|dF_{y_0}^{-1}\|\ell\|v_0\|^2}{2} < \frac{a\ell}{2(1 - a\ell_0\|v_0\|)}\|v_0\|^2 \leq q\|v_0\| \end{aligned} \tag{8.7.33}$$

Then by the choice of r^* and (8.7.33) we obtain:

$$\|u_0\| \leq \|v_0\| + \|u_0 + v_0\| \leq (1+q)\|v_0\| \leq (1+q)r^* \leq r_0 < r,$$

which implies $u_0 \in N(0)$. Therefore we have

$$y^{-1}y_1 = \exp(v_0)\exp(u_0) = \exp(u_1). \tag{8.7.34}$$

Hence by (8.7.6) and (8.7.34)

$$v_1 = v_0 + u_0 + \frac{1}{2}[v_0, u_0] + \cdots \in g.$$

By inequality (8.7.16) with $\beta = 3r^* \leq \frac{\delta}{\mu}$ we get

$$\|v_1\| \leq \|v_0 + u_0\| f(3\mu r^*) \leq c\|v_0\|^2 < \|v_0\| < r^*,$$

by the choice of r^*. That is $y_1 \in U^0(y, r^*)$. The rest follows by induction and noticing that $v_n \to 0$ as $n \to \infty$ since $\|v_{n+1}\| < \|v_n\|$ for all $n \geq 0$ holds. ■

Remark 8.7.1 *In general*

$$\ell_0 \leq \ell \tag{8.7.35}$$

holds and $\frac{\ell}{\ell_0}$ can be arbitrarily large. If

(a) $\ell_0 = \ell$, and $q = 1$, our results are reduced to the ones obtained in [265, Theorem 4.6, p. 137]. Denote the convergence radius obtained there by r_1^* and the convergence ratio by c_1. In this case we have

$$r^* = r_1^* \quad \text{and} \quad c = c_1. \tag{8.7.36}$$

(b) $\ell_0 = \ell$ and $q \in [0, 1)$, then we get by (8.7.27) and (8.7.28)

$$r_1^* < r^*, \tag{8.7.37}$$

and

$$c < c_1. \tag{8.7.38}$$

(c) $\ell_0 < \ell$ and $q \in [0, 1]$, then using (8.7.27) and (8.7.28) we deduce again (8.7.37) and (8.7.38).

That is in cases (b) and (c) we obtain a larger convergence radius r^* and a smaller convergence ratio c than the corresponding r_1^* and c_1 found in [265, p. 137]. Note that these advantages are obtained under the same computational cost, since in practice the evaluation of Lipschitz constant ℓ requires the evaluation of the center Lipschitz constant ℓ_0.

We complete this study with a numerical example. But first we need some background on some matrix groups. The set $GL(k, R)$ of all invertible $k \times k$ matrices with real coefficients forms a Lie group called the general linear group under standard matrix multiplication. This Lie group has some interesting subgroups one of which is the orthogonal group $O(k, R)$ consisting of orthogonal matrices y such that $y^T y = I_k$ where I_k is the identity $k \times k$ matrix. The Lie algebra of $GL(k, R)$ is denoted by $g\ell(k, R)$ and it is the linear space of all real $k \times k$ matrices such that $[u, v] = uv - vu$. In particular the Lie algebra of $O(k, R)$ is the set of all $k \times k$ skew-symmetric matrices

$$sq(k, R) = \{v \in g\ell(k, R),\ v^T + v = 0\}.$$

Example 8.7.1 *We consider the initial value problem*

$$\begin{aligned} y' &= yg(y) \\ y(0) &= y^0 \end{aligned} \tag{8.7.39}$$

for $y \in G = O(N, R)$, where $g(y) = sq(N, R)$ and $y^{(0)}$ is a random initial guess for Newton's method (8.7.11)–(8.7.12). For simplicity we take $N = 1$. Choose $y(t) = e^t$, $g(t) = e^{-t}$. Then by Lemma 8.7.7 we can get $f(t) = 1$ and $a = b = 1$. Note that (8.7.39) corresponds to the equation

$$F(y) = 0 \tag{8.7.40}$$

where

$$F(y) = e^y - 1. \tag{8.7.41}$$

We now restrict F on $U(y, r^*) = U(0, 1)$. By using (8.7.10), (8.7.22) and (8.7.23) we can get

$$r^* = \frac{2}{(2\ell_0 + \ell)a}, \tag{8.7.42}$$

$$r_1^* = \frac{2}{3\ell a}, \tag{8.7.43}$$

$\ell = e$ and $\ell_0 = e - 1$. That is (8.7.35) holds as strict inequality. In particular we obtain

$$r_1^* = .245252961 < r^* = .254028662.$$

Moreover (8.7.38) holds. Note that the results concerning Version 2 of Newton's method also studied in [265] can be immediately improved along the same lines. However we leave the details to the motivated reader.

8.8 Finite Element Analysis and Discretizations

In this Section we use our weaker version of the Newton-Kantorovich theorem introduced in Section 5.1 to show existence and uniqueness of solutions for strongly nonlinear elliptic boundary value problems [324]. Moreover, a priori estimates are obtained as a direct consequence of this theorem. This study is motivated by the work in [324] where the Newton-Kantorovich theorem 4.2.4 was used. The advantages of our approach have been explained in detail in Section 5.1.

Finally examples of boundary value problems where our results apply are provided.We assume the following:

($\mathbf{A_1}$) There are Banach spaces $Z \subseteq X$ and $U \subseteq Y$ such that the inclusions are continuous, and the restriction of F to Z, denoted again by F, is a Fréchet-differentiable operator from Z to U.

($\mathbf{A_2}$) For any $v \in Z$ the derivative $F'(v) \in L(Z, U)$ can be extended to $F'(v) \in L(X, Y)$ and it is:

- locally Lipschitz continuous on Z, i.e., for any bounded convex set $T \subseteq Z$ there exists a positive constant c_1 depending on T such that:

$$\|F'(v) - F'(w)\| \leq c_1 \|v - w\|, \tag{8.8.1}$$

for all $v, w \in T$

- center locally Lipschitz continuous at a given $u_0 \in Z$, i.e., for any bounded convex set $T \subseteq Z$, with $u_0 \in T$, there exists a positive constant c_0 depending on u_0 and T such that:

$$\|F'(v) - F'(u_0)\| \leq c_0 \|v - u_0\|, \tag{8.8.2}$$

for all $v \in T$.

($\mathbf{A_3}$) There are Banach spaces $V \subseteq Z$ and $W \subseteq U$ such that the inclusions are continuous. We suppose that there exists a subset $S \subseteq V$ for which the following holds: "if $F'(u) \in L(V, W)$ is an isomorphism between V and W at $u \in S$, then $F'(u) \in L(X, Y)$ is an isomorphism between X and Y as well."

To define discretized solutions of $F(u) = 0$, we introduce the finite dimensional subspaces $S_d \subseteq Z$ and $S_d \subseteq U$ parametrized by d, $0 < d < 1$ with the following properties:

($\mathbf{A_4}$) There exists $r \geq 0$ and a positive constant c_2 independent of d such that

$$\|v_h\|_Z \leq \frac{c_2}{d^r} \|v_h\|_X, \quad \text{for all } v_d \in S_d. \tag{8.8.3}$$

($\mathbf{A_5}$) There exists projection $\Pi_d : X \longrightarrow S_d$ for each S_d such that, if $u_0 \in S$ is a solution of $F(u) = 0$, then

$$\lim_{d \to o} d^{-r} \|u_0 - \Pi_d u_0\|_X = 0 \tag{8.8.4}$$

and

$$\lim_{d \to o} d^{-r} \|u_0 - \Pi_d u_0\|_Z = 0 \tag{8.8.5}$$

We can show the following result concerning the existence of locally unique solutions of discretized equations:

Theorem 8.8.1 *Assume that conditions (A_1)-(A_5) hold. Suppose $F'(u_0) \in L(V, W)$ is an isomorphism, and $u_0 \in S$. Moreover, assume $F'(u_0)$ can be decomposed into $F'(u_0) = Q + R$, where $Q \in L(X, Y)$ and $R \in L(X, Y)$ is compact. The discretized nonlinear operator $F_d : Z \to U$ is defined by*

$$F_d(u) = (I - P_d)Q(u) + P_d F(u), \tag{8.8.6}$$

where I is the identity of Y, and $P_d : Y \to S_d$ is a projection such that

$$\lim_{d \to o} \|v - P_d v\|_Y = 0, \quad \text{for all } v \in Y, \tag{8.8.7}$$

and

$$(I - P_d)Q(v_d) = 0, \quad \text{for all } v_d \in S_d. \tag{8.8.8}$$

Then, for sufficiently small $d > 0$, there exists $u_d \in S_d$ such that $F_d(u_d) = 0$, and u_d is locally unique. Moreover, the following estimate holds

$$\|u_d - \Pi_d(u_0)\| \leq l_1 \|u_0 - \Pi_d(u_0)\|, \tag{8.8.9}$$

where l_1 is a positive constant independent of h.

Proof. The proof is similar to the corresponding one in [324, Th. 2.1, p. 126]. However, there are some crucial differences where weaker (8.8.2) is used (needed) instead of stronger condition (8.8.1). Step 1. We claim that there exists a positive constant c_3, independent of d, such that, for sufficiently small $h > 0$,

$$\|F'{}_d(\Pi_d(u_0))v_d\|_Y \geq c_3 \|v_d\|_X, \quad \text{for all } v_d \in S_d. \tag{8.8.10}$$

From (A_3) and $u_0 \in S$, $F'(u_0) \in L(X, Y)$ is an isomorphism. Set $B_0 = \|F'(u_0)^{-1}\|$. We can have in turn

$$\begin{aligned} F'{}_d(\Pi_d(u_0))v_d = {} & F'(u_0)v_d + P_d(F'(\Pi_d(u_0)) - F'(u_0))v_d \\ & - (I - P_d)(-Q + F'(u_0))v_d. \end{aligned} \tag{8.8.11}$$

Since $-Q + F'(u_0) \in L(X, Y)$ is compact we get by (8.8.7) that

$$\lim_{d \to o} \|(I - P_d)(-Q + F'(u_0)\| = 0. \tag{8.8.12}$$

By (8.8.7) there exists a positive constant c_4 such that

$$\sup_{d > 0} \|P_d\| \leq c_4. \tag{8.8.13}$$

That is, using (8.8.2) we get

$$\|P_d(F'(\Pi_d(u_0)) - F'(u_0))\| \leq c_0 c_4 \|\Pi_d(u_0) - u_0\|. \tag{8.8.14}$$

Hence, by (8.8.5) we can have

$$\|F_d'(\Pi_d(u_0))v_d\| \geq \left(\tfrac{1}{B_0} - \delta(d)\right) \|v_d\| , \tag{8.8.15}$$

where $\lim_{d\to o} \delta(d) = 0$, and (8.8.10) holds with $c_3 = \frac{B_0^{-1}}{2}$. Step 2. We shall show:

$$\lim_{d\to o} d^{-r} \left\|F_d'(\Pi_d(u_0))^{-1} F_d(\Pi_d(u_0)\right\| = 0. \tag{8.8.16}$$

Note that

$$\begin{aligned}\|F_d(\Pi_d(u_0))\| &\leq c_4 \|F_d(\Pi_d(u_0)) - F_d(u_0)\| \\ &\leq c_4 \int_0^1 \|G_t\| \, dt \, \|\Pi_d(u_0) - u_0\| \\ &\leq c_4 c_5 \|u_0 - \Pi_d(u_0)\| ,\end{aligned} \tag{8.8.17}$$

where,

$$G_t = F'((1-t)u_0 + t\Pi_d(u_0)), \tag{8.8.18}$$

and we used

$$\begin{aligned}\|G_t\| &\leq \|G_t - F'(u_0)\| + \|F'(u_0)\| \\ &\leq c_0 t \|\Pi_d(u_0)u_0 - \Pi_d(u_0)\| + \|F'(u_0)\| \\ &\leq c_5,\end{aligned} \tag{8.8.19}$$

where c_5 is independent of d. The claim is proved. Step 3. We use our modification of the Newton-Kantorovich theorem with the following choices: $A = S_d \subseteq Z$ with norm $d^{-r} \|w_d\|_X$, $B = S_d \subseteq U$ with norm $d^{-r} \|w_d\|_Y$, $x_0 = \Pi_d(u_0)$, $F = F_d$. Notice that $\|S\|_{L(A,B)} = \|S\|_{L(X,Y)}$, for any linear operator $S \in L(S_d, S_d)$. By step 1, we know that $F_d'(\Pi_d(u_0)) \in L(S_d, S_d)$ is an isomorphism. It follows from (8.8.1) and (A_4) that for any $w_d, v_d \in S_d$,

$$\begin{aligned}\|F_d'(w_d) - F_d'(v_d)\| &\leq c_1 c_4 \|w_d - v_d\|_Z \\ &\leq c_1 c_2 c_4 d^{-r} \|w_d - v_d\|_X .\end{aligned} \tag{8.8.20}$$

Similarly, we get using (8.8.2) and (A_4) that

$$\|F_d'(w_d) - F_d'(\Pi_d(u_0))\| \leq c_0 c_2 c_4 d^{-r} \|w_d - x_0\|_X . \tag{8.8.21}$$

Hence assumptions on the Lipschitz constants are satisfied with

$$l = c_1 c_2 c_3^{-1} c_4 \text{ and } l_0 = c_0 c_2 c_3^{-1} c_4. \tag{8.8.22}$$

From step 2, we may take sufficiently small $d > 0$ such that $(l_0 + l)\eta \leq 1$, where

$$\eta = d^{-r} \left\|F_d'(\Pi_d(u_0))^{-1} F_d(\Pi_d(u_0))\right\|_X . \tag{8.8.23}$$

That is, assumption (5.1.56) is satisfied. Hence, for sufficiently small $d > 0$ there exists a locally unique $u_d \in S_d$ such that $F_d(u_d) = 0$ and

$$\begin{aligned} \|u_d - \Pi_d(u_0)\|_X &\leq 2 \cdot d^r \eta \leq 2c_3^{-1} \|F_d(\Pi_d(u_0))\|_Y \\ &\leq 2c_3^{-1} c_4 c_5 \|u_0 - \Pi_d(u_0)\|_X . \end{aligned} \tag{8.8.24}$$

It follows (8.8.9) holds with $l_1 = 2c_3^{-1}c_4c_5$. ∎

Remark 8.8.1 *In general,*

$$c_0 \leq c_1 \ \ (i.e.,\ l_0 \leq l) \tag{8.8.25}$$

holds and $\frac{l}{l_0}$ *can be arbitrarily large (see Section 5.1), where* l *and* l_0 *are given by (8.8.22). If* $l = l_0$*, our Theorem 8.8.1 reduces to the corresponding Theorem 2.1 in [324, p. 126]. Otherwise our condition (5.1.56) is weaker than the corresponding one in [324] using the Newton-Kantorovich hypothesis (5.2.5). That is,*

$$h \leq 1 \implies h_1 \leq 1 \tag{8.8.26}$$

but not vice versa. As already shown in Section 5.1, finer estimates on the distances $\|u_d - \Pi_d(u_0)\|$ *and a more precise information on the location of the solution are provided here and under the same computational cost since in practive the evaluation of* c_1 *requires that of* c_0*. Note also that our parameter* d *will be smaller than the corresponding one in [324], which in turn implies that fewer computations and smaller dimension subspaces* S_d *are used to appproximate* u_d*. This observation is very important in computational mathematics [68].*

The above observations suggest that all results obtained in [324] can be improved if rewritten using weaker (5.1.56) instead of stronger (5.2.5). However we do not attempt this here (leaving this task to the motivated reader).

Instead we provide examples of nonlinear problems already reported in [324] where finite element methods apply along the lines of our Theorem above.

Example 8.8.1 *Find* $u \in H_0^1(\tau)$, $\tau = (b, c) \subseteq \mathbf{R}$ *such that*

$$\langle F(u), v\rangle = \int_\tau [g_0(x, u, u')v' + g(x, u, u')v]\, dx = 0, \ \ \textit{for all } v \in H_0^1(\tau), \tag{8.8.27}$$

where g_0, g_1 *are sufficiently smooth functions from* $\tau \times \mathbf{R} \times \mathbf{R}$ *to* $\mathbf{R}$.

Example 8.8.2 *For the* N*-dimensional case* ($N = 2, 3$)*: let* $D \subseteq \mathbf{R}^N$ *be a bounded domain with a Lipschitz boundary. Then consider the problem: find* $u \in H_0^1(d)$ *such that*

$$\langle F(u), v\rangle = \int_D [q_0(x, u, \nabla u)\nabla v + q(x, u, \nabla u)v]\, dx = 0, \ \ \textit{for all } v \in H_0^1(d), \tag{8.8.28}$$

where $q_0 : D \times \mathbf{R} \times \mathbf{R}^N \to \mathbf{R}^N$ *and* $q \in D \times \mathbf{R} \times \mathbf{R}^N \to \mathbf{R}$ *are sufficiently smooth functions. Since equations (8.8.27) and (8.8.28) are defined in divergence form, their finite element solutions are defined in a natural way.*

Results on iterative methods defined on generalized Banach spaces with a convergence structure [68], [231], [232] can be found in the exercises.

8.9 Exercises

8.1 Let g be an algorithm for finding a solution x^* of equation $F(x) = 0$, and x the approximation to x^* computed by g. Define the error for approximating x by

$$d(g, F) = \|x - x^*\|.$$

Consider the problem of approximating x^* when F satisfies some conditions. Algorithms based on these conditions cannot differentiate between operators in the class C of all operators satisfying these conditions we confort the class C instead of specific operators from C. Define

$$d_i = \inf_{g \in A} \sup_{F \in C} d(g, F)$$

where A is the class of all algorithms using i units of time. The time t needed to approximate x^* to with in error tolerance $\varepsilon > 0$ is the smallest i such that $d_i \leq \varepsilon$, and an algorithm is said to be optimal if

$$\sup_{F \in C} d(g, F) = d_t.$$

If for any algorithm using i units of time, there exist functions F_1, F_2 in C such that:
the minimum distance between any solution of F, and any solution of F_2 is greater or equal to 2ε then, show:

$$d_i \geq \varepsilon.$$

8.2 With the notation introduced in Exercise 8.1, assume:
(1) $F : [a, b] \to \mathbf{R}$ is continuous;
(2) $F(a) < 0$, $F(b) > 0$,
Then, show:

$$d_i = \tfrac{b-a}{2^{i+1}}.$$

8.3 With the notation introduced in Exercise 8.1, assume:
(1) $F : [a, b] \to \mathbf{R}$, $F'(x) \geq \alpha > 0$ for all $x \in [a, b]$;
(2) $F(a) < 0$, $F(b) > 0$.
Then, show:

$$d_i = \tfrac{b-n}{2^{i+1}}.$$

8.4 With the notation introduced in Exercises 8.1, assume:
(1) $F : [a, b] \to \mathbf{R}$, $F'(x) \leq b$ for all $x \in [a, b]$;
(2) $F(a) < 0$, $F(b) > 0$
Then, show:

$$d_i = \tfrac{b-a}{2^{i+1}}.$$

8.5 With the notation introduced in Exercis 8.1, assume:
(1) $F : [a, b] \to \mathbf{R}$, $b \geq F'(x) \geq \alpha > 0$ for all $x \in [a, b]$;
(2) $F(a) < 0$, $F(b) > 0$
Then, show:

$$d_i \geq (b-a)\left[\frac{\left(1-\frac{\alpha}{b}\right)^2}{2}\right]^{i+1}.$$

8.6 Assume:
(1) hypotheses of Exercise 8.5 hold;
(2) $\|F''(x)\| \leq \gamma$ for all $\gamma \in [a, b]$.
Then show that the problem of finding a solution x^* of equation $F(x) = 0$ can be solved superlinearly.

8.7. Let L, M, M_1 be operators such that $L \in C^1(V_1 \to V)$, $M_1 \in L_+(V)$, $M \in C(V_1 \to V)$, and x_n be points in D. It is convenient for us to define the sequences c_n, d_n, a_n, b_n $(n \geq 0)$ by

$$\begin{aligned} d_{n+1} &= (L + M + M_1)(d_n) + L'(|x_n|)\, c_n, \quad d_0 = 0, \\ c_n &= |x_{n+1} - x_n|, \\ a_n &= (L + M + M_1)^n (a) \quad \text{for some } a \in C, \\ b_n &= (L + M + M_1)^n (0), \end{aligned}$$

and the point b by

$$b = (L + M + M_1)^\infty (0).$$

Prove the result:

Let X be a Banach space with convergence structure (X, V, E) with $V = (V, C, \|\cdot\|_v)$, an operator $F \in C^1(D \to X)$ with $D \subseteq X$, an operator $Q \in C(D \to X)$, an operator $L \in C^1(V_1 \to V)$ with $V_1 \subseteq V$, an operator $M \in C(V_1 \to V)$, an operator $M_1 \in L_+(V)$, a point $a \in C$, and a null sequence $\{z_n\} \in D$ such that the following conditions are satisfied:

(C_6) the inclusions $U(a) \subseteq D$ and $[0, a] \subseteq V1$ are true;

(C_7) L is order-convex on $[0, a]$, and satisfies

$$L'(|x| + |y|) - L'(|x|) \in B(F'(x) - F'(x + y))$$

for all $x, y \in U(a)$ with $|x| + |y| \leq a$;

(C_8) M satisfies the conditions

$$0 \le (Q(x) - Q(x+y)),\ M(|x| + |y|) - M(|x|) \in E$$

for all $x, y \in U(a)$ with $|x| + |y| \le a$, and

$$M(w_1) - M(w_2) \le M(w_3) - M(w_4) \text{ and } M(w) \ge 0$$

for all w, w_1, w_2, w_3, $w_4 \in [0, a]$ with $w_1 \le w_3$, $w_2 \le w_4$, $w_2 \le w_1$, $w_4 \le w_3$;

(C_9) M_1, x_n, z_n satisfy the inequality

$$0 \le (F'(x_n)(z_n) - F'(x_{n-1})(z_{n-1}),\ M_1(d_n - d_{n-1})) \in E$$

for all $n \ge 1$;

(C_{10}) $L'(0) \in B(I - F'(0))$, and
$(-(F(0) + Q(0) + f'(0)(z_0)),\ L(0) + M(0) + M_1(0)) \in E$;

(C_{11}) $(L + M + M_1)(a) \le a$ with $0 \le L + M + M_1$; and

(C_{12}) $(M + M_1 + L'(a))^n a \to 0$ as $n \to \infty$.

Then,

(i) The sequence $(x_n, d_n) \in (X \times V)^N$ is well defined, remains in E^N, is monotone, and satisfies for all $n \ge 0$

$$d_n \le b$$

where b is the smallest fixed point of $L + M + M_1$ in $[0, a]$.

(ii) Moreover, the iteration $\{x_n\}$ $(n \ge 0)$ generated by

$$x_{n+1} = x_n + F'(x_n)^* [-(F(x_n) + Q(x_n))] - z_n,\ z_0 = 0$$

converges to a solution $x^* \in U(b)$ of the equation $F(x) + Q(x) = 0$, which is unique in $U(a)$.

(iii) Furthermore, the following estimates are true for all $n \ge 0$:

$$\begin{aligned} b_n &\le d_n \le b, \\ b_n &\le a_n, \\ |x_{n+1} - x_n| &\le d_{n+1} - d_n, \\ |x_n - x^*| &\le b - d_n, \end{aligned}$$

and

$$|x_n - x^*| \le a_n - b_n, \quad \text{for } M_1 = 0, \text{ and } z_n = 0\ (n \ge 0).$$

8.8. We will now introduce results on a posteriori estimates for the iteration introduced in exercise 8.7. It is convenient to define the operator

$$R_n(q) = (I - L'(|x_n|))^* S_n(q) + c_n$$

where,

$$S_n(q) = (L + M + M_1)(|x_n| + q) - (L + M + M_1)(|x_n|) - L'(|x_n|)(q),$$

and the interval

$$I_n = [0, a - |x_n|].$$

Show:

(a) operators S_n are monotone on I_n;

(b) operators $R_n : [0, a - d_n] \to [0, a - d_n]$ are well defined, and monotone. *Hint*: Verify the scheme:

$$\begin{aligned} d_n + c_n \leq d_{n+1} &\Longrightarrow R(a) - d_{n+1} \leq a - d_n - c_n \\ &\Longrightarrow S_n(a - d_n) + L'(|x_n|)(a - d_n - c_n) \leq a - d_n - c_n \quad (n \geq 0); \end{aligned}$$

(c) if $q \in I_n$ satisfy $R_n(q) \leq q$, then

$$c_n \leq R_n(q) = p \leq q,$$

and

$$R_{n+1}(p - c_n) \leq p - c_n \quad \text{for all } n \geq 0;$$

(d) under the hypotheses of Exercise 8.7, let $q_n \in I_n$ be a solution of $R_n(q) \leq q$, then

$$|x^* - x_m| \leq a_m \quad (m \geq n),$$

where

$$a_n = q_n \text{ and } a_{m+1} = R_m(a_m) - c_m;$$

and

(e) under the hypotheses of exercise 8.7, any solution $q \in I_n$ of $R_n(q) \leq q$ is such that

$$|x^* - x_n| s < R_n^{\infty}(0) \leq q.$$

8.9. Let $A \in L(X \to X)$ be a given operator. Define the operators $P, T(D \to X)$ by

$$P(x) = AT(x+u),$$
$$T(x) = G(x) + R(x), \quad P(x) = F(x) + Q(x),$$

and

$$F(x) = AG(x+u), \quad Q(x) = AR(x+u),$$

where $A \in L(X \to X)\, G$, R are as F, Q respectively. We deduce immediately that under the hypotheses of Exercise 8.7 the zero x^* of P is also a zero of AT, if $u = 0$.

We will now provide a monotonicity result to find a zero x^* of AT. the space X is assumed to be partially ordered and satisfies the conditions for V given in (C_1)–(C_5). Moreover, we set $X = V$, $D = C^2$ so that $|\cdot|$ turns out to be I.

Prove the result:

Let V be a partially ordered Banach space satisfying conditions (C_1)–(C_5), Y be a Banach space, $G \in C^1(D \to Y)$, $R \in C(D \to Y)$ with $D \subseteq V$, $A \in L(X \to V)$, $M \in C(D \to V)$, $M_1 \in L_+(V)$ and $u, v \in V$ such that:

(C_{13}) $[u, v] \subseteq D$;

(C_{14}) $I - AG'(u) + M + M_1 \in L_+(V)$;

(C_{15}) for all $w_1, w_2 \in [u, v] : w_1 \leq w_2 \Longrightarrow AG'(w_1) \geq AG'(w_2)$;

(C_{16}) $AT(u) + AG'(u)(z_0) \leq 0$, $AT(v) + AG'(v)(z_0) \geq 0$ and $AT(v) - M_1(v-u) \geq 0$;

(C_{17}) condition (C_8) is satisfied, and $M(v-u) \leq -Q(v-u)$;

(C_{18}) condition (C_9) is satisfied ;

(C_{19}) the following initial condition is satisfied

$$-(Q(0) + AG'(u)(z_0)) \leq (M + M_1)(0);$$

and

(C_{20}) $(I - AG'(v) + m + M_1)^n (v-u) \to 0$ as $n \to \infty$.

Then the Newton sequence

$$y_0 = u, \; y_{n+1} = y_n + (AG'(y_n))^* [-AT(y_n)] - z_n \quad (n \geq 0)$$

is well defined for all $n \geq 0$, monotone and converges to a unique zero x^* of AT in $[u, v]$.

8.10. Let X be a Banach space with convergence structure (X, V, E) with $V = (V, C, \|\cdot\|_v)$, an operator $F \in C^1(X_F \to X)$ with $X_F \subseteq X$, an operator $L \in C^1(V_L \to V)$ with $V_L \subseteq V$ and a point of C such that

(a) $U(a) \subseteq X_F$, $[0, a] \subseteq V_L$;

(b) L is order convex on $[0, a]$ and satisfies for $x, y \in U(a)$ with $|x| + |y| \leq a$;

$$L'(|x| + |y|) - L'(|x|) \in B(F'(x) - F'(x + y));$$

(c) $L'(0) \in B(I - F'(0))$, $(-F(0), L(0)) \in E$;

(d) $L(a) \leq a$;

(e) $L'(a)^n \to 0$ as $n \to \infty$.

Then show Newton's sequence $x_0 := 0$, $x_{n+1} = x_n + F'(x_n)^*(-F(x_n))$ is well defined, and converges to the unique zero z of F in $U(a)$.

8.11. Under the hypotheses of Exercise 8.10 consider the case of a Banach space with a real norm $\|\cdot\|$. Let $F'(0) = I$ and define a monotone operator

$$k : [0, a] \to \mathbb{R}\ \forall x \in U(a) : \|F''(x)\| \leq k(\|x\|)$$

and

$$L(t) = \|F(0)\| + \int_0^t ds \int_0^s d\theta k(\theta).$$

Show (d) above is equivalent to $\|F(0)\| + .5k(a)a^2 \leq a$. Under what conditions is this inequality true.

If conditions (a)–(c) or Exercise 7.4 are satisfied, and

$$\exists t \in (0, 1) : L(a) \leq ta,$$

then show there exists $a' \in [0, ta]$ satisfying (a)–(e). The zero $z \in U(a')$ is unique in $U(a)$.

8.12. Let $L \in L_+(V)$ and $a, e \in C$ be given such that: Let $a \leq e$ and $L^n e \to 0$ as $n \to \infty$. then show: operator

$$(I - L)^* : [0, a] \to [0, a],$$

is well defined and continuous.

8.13. Let $A \in L(X)$, $L \in B(A)$, $y \in D$, and $e \in C$ such that

$$Le + |y| \leq e \text{ and } L^n e \to 0 \text{ as } n \to \infty.$$

Then show $x := (I - A)^* y$ is well defined, $x \in D$, and $|x| \leq (I - L)^* |y| \leq e$.

8.14. Let V be a partially ordered Banach space, Y a Banach space, $G \in C^1(V_G \to Y)$, $A \in L(X \to Y)$ and $u, v \in V$ such that

(a) $[u, v] \subseteq V_G$;
(b) $I - AG'(u) \in L_+(V)$;
(c) $\forall w_1, w_2 \in [u, v] : w_1 \leq w_2 \Longrightarrow AG'(w_1) \geq AG'(w_2)$;
(d) $AG(u) \leq 0$ and $AG'(v) \geq 0$;
(e) $(I - AG'(v))^n (v - u) \to 0$ as $n \to \infty$.

Then show: the Newton sequence

$$u_0 = u,\ u_{n+1} := u_+ [AG'(u_n)]^* [-AG(u_n)] \quad (n \geq 0)$$

is well defined, monotone, and converges to the unique zero z of AG in $[u, v]$.

8.15. Consider the two-point boundary value problem

$$-x''(s) = 4\sin(x(s)) + f(s), \quad x(0) = x(1) = 0$$

as a possible application of Exercises 8.7–8.14 in the space $X = C[0,1]$ and $V = X$ with natural partial ordering; the operator $/\cdot/$ is defined by taking absolute values. Let $G \in L_+(C[0,1])$ be given by

$$Gx(s) = \tfrac{1}{2\sin(2)} \left\{ \int_0^s \sin(2t)\sin(2-2s)\,x(t)\,dt + \int_0^s \sin(2-2t)\sin(2s)\,x(t)\,dt \right\}$$

satisfying $Gx = y \Longleftrightarrow -y'' - 4y = x,\ y(0) = y(1) = 0$. Define the operator

$$F : C[0,1] \to C[0,1], \quad F(x) := x - G(4\sin(x) - 4x + f).$$

Let

$$L : C[0,1] \to C[0,1], \quad L(e) = 4G(e - \sin(e)) + |Gf|.$$

For $x, y, w \in C[0,1]$

$$|x|, |x| + |y| \leq .5\pi \Rightarrow$$
$$|[F'(x) - F'(x+y)]w| \leq [L'(|x| + |y|) - L'L'(|x|)]\,|w|.$$

Further we have $L(0) = |Gf|$ and $L'(0) = 0$. We have to determine $a \in C_+[0,1]$ with $|a| \leq .5\pi$ and $s \in (0,1)$ such that $L(a) = 4G(a - \sin(a)) + |Gf| \leq sa$. We seek a constant function as a solution. For $e_0(s) = 1$, we compute

$$p : \|Ge_0\|_\infty = .25\left(\tfrac{1}{\cos(e)} - 1\right).$$

Show that $a = te_0$ will be a suitable solution if

$$4p\,(t - \sin(t)) + \|Gf\|_\infty < t.$$

8.16. (a) Assume: given a Banach space X with a convergence structure (X, V, E) with $V = (V, C, \|\cdot\|_V)$, an operator $F \in C^1 (X_0 \subseteq X \to X)$, operators M, L_0, $L \in L^1 (V_0 \subseteq V \to V)$, and a point $p \in C$ such that the following conditions hold:

$$U(p) \subseteq X_0, \quad [0, p] \subseteq V_0;$$

M, L_0, L are order convex on $[0, p]$, and such that for $x, y, z \in U(p)$ with $|x| \le p$, $|y| + |x| \le p$

$$\begin{aligned}
L_0'(|x|) - L_0'(0) &\in B\left(F'(0) - F'(x)\right), \\
L'(|y| + |x|) - L'(|y|) &\in B\left(F'(y) - F'(y + x)\right), \\
M(p) &\le p \\
L_0(p_0) &\le M(p_0) \text{ for all } \; p_0 \in [0, p], \\
L_0'(p_0) &\le M'(p_0) \text{ for all } \; p_0 \in [0, p], \\
L_0' \in B(I - F'(0)), &\quad (-F(0), L_0(0)) \in E, \\
M'(p)^n(p) &\to 0 \text{ as } n \to \infty, \\
M'(d_n)(b - d_n) + L(d_n) &\le M(b) \text{ for all } n \ge 0,
\end{aligned}$$

where

$$d_0 = 0,\ d_{n+1} = L(d_n) + L_0'(|x_n|)(c_n), \quad c_n = |x_{n+1} - x_n| \quad (n \ge 0)$$

and

$$b = M^\infty(0).$$

Show: sequence $\{x_n\}$ $(n \ge 0)$ generated by Newton's method is well defined, remains in $E^{\mathbb{N}}$, is monotone, and converges to a unique zero x^* in $U(b)$, where b is the smallest fixed point of M in $[0, p]$. Moreover the following bounds hold for all $n \ge 0$

$$\begin{aligned}
d_n &\le b, \\
|x^* - x_n| &\le b - d_n, \\
|x^* - x_n| &\le M^n(p) - M^n(0), \\
c_n + d_n &\le d_{n+1},
\end{aligned}$$

and

$$c_n = |x_{n+1} - x_n|.$$

(b) If $r \in [0, p - |x_n|]$ satisfies $R_n(r) \leq r$ then show: the following holds for all $n \geq 0$:

$$c_n \leq R_n(r) = p \leq r, \tag{8.9.1}$$

and

$$R_{\{n+1\}}(p - c_n) \leq p - c_n$$

(c) Assume hypotheses of (a) hold and let $r_n \in [0, p - |x_n|]$ be a solution (8.9.1). Then show the following a posteriori estimates hold for all $n \geq 0$

$$|x^* - x_m| < q_m,$$

where

$$q_n = r_n, \quad q_{m+1} = R_m(q_m) - c_m \quad (m \geq n).$$

(d) Under hypotheses of (a) show any solution $r \in [0, p - |x_n|]$ of $R_n(r) \leq r$ yields the a posteriori estimate

$$|x^* - x_n| \leq R_n^{\infty}(0) \leq r \quad (n \geq 0).$$

Let X be partially ordered and set

$$X = V, \quad D = C^2 \text{ and } |\cdot| = I.$$

(e) Let X be partially ordered, Y a Banach space, $G \in C^1(X_0 \subseteq X \rightarrow Y)$, $A \in L(Y \rightarrow X)$ and $x, y \in X$ such that:

1. $[x, y] \in X_0$.
2. $I - AG'(x) \in L_+(X)$,
3. $z_1 \leq z_2 \Longrightarrow AG'(z_1) \leq AG'(z_2)$ for all $z_1, z_2 \in [x, y]$,
4. $AG(x) \leq 0, \quad AG(y) \geq 0$,
5. $(I - AG'(y))^n (y - x) \rightarrow 0$ as $n \rightarrow \infty$.

Show: sequence $\{y_n\}$ $(n \geq 0)$ generated by

$$y_0 = x, \quad y_{n+1} = y_n + AG'(y_n)^* [-AG(y_n)]$$

is well defined for all $n \geq 0$, monotone and converges to a unique zero y^* of AG in $[x, y]$.

8.17. Let there be given a Banach space X with convergence structure (X, V, E) where $V = (V, C, \|\cdot\|_V)$, and operator $F \in C''(X_0 \to X)$ with $X_0 \subseteq X$, and operator $M \in C'(V_0 \to V)$ with $V_0 \subseteq V$, and a point $p \in C$ satisfying:

$$U(p) \subseteq X_0, \quad [0, p] \subseteq V_0;$$

M is order-convex on $[0, p]$ and for all $x, y \in U(p)$ with $|x| + |y| \leq p$

$$M'(|x+y|) - M'(x) \in B([F''(x) - F''(x+y)](y))$$
$$M'(0) \in B(I - F'(0)), \quad (-F(0), M(0)) \in E;$$
$$M(p) \leq p;$$

and

$$M'(p)^n p \to 0 \quad \text{as} \quad n \to \infty.$$

Then, show sequence $(x_n, t_n) \in (X \times V)^{\mathbb{N}}$, where $\{x_n\}$ is generated by Newton's method and $\{t_n\}$ $(n \geq 0)$ is given by

$$t_0 = 0, \; t_{n+1} = M(t_n) + M'(|x_n|)(a_n), \quad a_n = |x_{n+1} - x_n|$$

is well defined for all $n \geq 0$, belongs in $E^{\mathbb{N}}$, and is monotone.

Moreover the following hold for all $n \geq 0$

$$t_n \geq b,$$

where,

$$b = M^{\infty}(0),$$

is the smallest fixed point of M in $[0, p]$.

b) Show corresponding results as in Exercises 8.16 (b)–(c).

8.18. Assume hypotheses of Exercise 8.16 hold for $M = L$.

Show: (a) conclusions (a)–(e) hold (under the revised hypothesis) and the last hypothesis in (a) can be dropped;

(b) error bounds $|x^* - x_n|$ $(n \geq 0)$ obtained in this setting are finer and the infromation on the location of the solution more precise than the corresponding ones in Exercise 8.10 provided that $L_0(p_0) < L(p_0)$ or $L_0'(p_0) < L'(p_0)$ for all $p \in [0, p]$.

As in Exercise 8.11 assume there exists a monotone operator $k_0 : [0, p] \to R$ such that

$$\|F'(x) - F'(0)\| \leq k_0(\|x\|)\|x\|, \quad \text{for all } x \in U(p),$$

and define operator L_0 by

$$L_0(t) = \|F(0)\| + \int_0^t ds \int_0^s d\theta k_0(\theta).$$

Sequence $\{d_n\}$ given by $d_0 = 0$, $d_{n+1} = L(d_n) + L_0'(|x_n|)\, c_n$ converges to some $p^* \in [0, p]$ provided that

$$\left(k_0(p) + \tfrac{k(p)}{2}\right) \|F(0)\| \leq 1.$$

Conclude that the above semilocal convergence condition is weaker than (d) in Exercise 8.10 or equivalently (for $p = a$)

$$2k(p)\, \|F(0)\| \leq 1.$$

Finally, conclude that in the setting of Exercise 8.7 and under the same computational cost we always obtain and under weaker conditions: finer error bounds on the distances $|x_n - x^*|$ $(n \geq 0)$ and a better information on the location of the solution x^* than in Exercise 8.7.

Chapter 9

The Newton-Kantorovich Theorem and Mathematical Programming

The Newton-Kantorovich theorem 4.2.4 with a few notable exceptions [293], [294], [282] has not been sufficiently utilized in the mathematical programming community. The purpose of this chapter is to provide a bridge between the two research communities by showing that the Newton-Kantorovich theorem can be used in analyzing LP and interior point methods, and can be used for obtaining optimal error bounds along the way.

In Sections 1 and 2 we show how to improve on the elegant works of Renegar, Shub in [294] and Potra in [282]. To avoid repetitions, we simply refer the reader to the above excellent works for the development of these methods. We simply start from the point where the hypotheses made can be replaced by ours which are weaker. The benefits of this approach have also been explained in the introduction and in Section 5.1.

We also note that the work in [294] is motivated by a Theorem of Smale in combination with the Newton-Kantorovich theorem, whereas the work in [282] differs from the above since it applies the latter theorem directly.

9.1 Case 1: Interior Point Methods

It has already been shown in [282] that the Newton-Kantorovich theorem can be used to construct and analyze optimal-complexity path following algorithms for linear complementary problems. Potra has chosen to apply this theorem to linear complementary problems because such problems provide a convenient framework for analyzing primal-dual interior point methods. Theoretical and experimental work conducted over the past decade has shown that primal-dual path following algorithms are among the best solution methods for linear programming (LP), quadratic programming (QP), and linear complementary problems (LCP) (see for example [285], [346]). Primal-dual path following algorithms are the basis of the best general-purpose practical methods, and they have important theoretical properties [304], [346], [354].

Potra, using the Newton-Kantorovich theorem 4.2.4, in particular showed how to construct path-following algorithms for LCP that have $O(\sqrt{n}L)$ iteration complexity [282].

Given a point x that approximates a point $x(\tau)$ on the central path of the LCP with complementarity gap τ, the algorithms compute a parameter $\theta \in (0,1)$ so that x satisfies the Newton-Kantorovich hypothesis (4.2.51) for the equation defining $x\left((1-\theta)\tau\right)$. It is proven that θ is bounded below by a multiple of $n^{-1/2}$. Since (4.2.51) is satisfied, the sequence generated by Newton's method (4.1.3) or by the modified Newton method (take $F'(x_n) = F'(x_0)$, $n \geq 0$) with starting x, will converge to $x\left((1-\theta)\tau\right)$. He showed that the number of steps required to obtain an acceptable approximation of $x\left((1-\theta)\tau\right)$ is bounded above by a number independent of n. Therefore, a point with complementarity less than ε can be obtained in at most $O\left(\sqrt{n}\log\left(\frac{\varepsilon}{\varepsilon_0}\right)\right)$ steps (for both methods), where ε_0 is the complementary gap of the starting point. For linear complementarity problems with rational input data of bit length L, this implies that an exact solution can be obtained in at most $O\left(\sqrt{n}L\right)$ iterations plus a rounding procedure including $O\left(n^3\right)$ arithmetic operations [346].

The differences between Potra's work and the earlier works by Renegar [293] and Renegar and Shub [294] have been stated in the introduction of this chapter and further analyzed in [282].

We also refer the reader to the excellent monograph of Nesterov and Nemirovskii [254] for an analysis of the construction of interior point methods for a larger class of problems than that considered in [282].

Below one can find our contribution.

Let $\|\cdot\|$ be a given norm on $\mathbf{R}^i$, i a natural integer, and x_0 be a point of D such that the closed ball of radius r centered at x_0,

$$\overline{U}(x_0, r) = \{x \in R^i \colon \|x - x_0\| \leq r\} \tag{9.1.1}$$

is included in $D \subseteq \mathbf{R}^i$, i.e.

$$\overline{U}(x_0, r) \subseteq D. \tag{9.1.2}$$

We assume that the Jacobian $F'(x_0)$ of $F : D \subseteq X \to X$ is nonsingular and that Lipschitz condition (4.2.50) is satisfied.

The Newton–Kantorovich Theorem 4.2.4 states that if

$$k = \ell\eta \leq \frac{1}{2}, \tag{9.1.3}$$

then there exists $x^* \in \overline{U}(x_0, r)$ with $F(x^*) = 0$. Moreover the sequences produced by Newton's method (4.1.3) and by the modified Newton method (4.1.4) are well defined and converge to x^*.

Define

$$k^0 = \bar{\ell}\eta \leq \frac{1}{2}, \tag{9.1.4}$$

where,

$$\bar{\ell} = \frac{\ell_0 + \ell}{2}, \tag{9.1.5}$$

where ℓ_0 is the constant appearing in the center Lipschitz condition.

Note that

$$k \leq \frac{1}{2} \Rightarrow k^0 \leq \frac{1}{2} \tag{9.1.6}$$

but not vice versa unless if $\ell_0 = \ell$.

Similarly by simply replacing ℓ with ℓ_0 (since the center-Lipschitz condition instead of (4.2.50) is actually needed in the proof) and condition (9.1.3) by the weaker

$$k^1 = \alpha\omega_0 \leq \frac{1}{2} \tag{9.1.7}$$

in the proof of Theorem 1 in [282] we show that method (**4.1.4**) also converges to x^* and the improved bounds

$$\|y_n - x^*\| \leq \frac{2\beta_0\lambda_0^2}{1-\lambda_0^2}\xi_0^{n-1} \quad (n \geq 1) \tag{9.1.8}$$

where

$$\beta_0 = \frac{\sqrt{1-2k^1}}{\omega_0}, \quad \lambda_0 = \frac{1-\sqrt{1-2k^1}-h^1}{k^1} \tag{9.1.9}$$

and

$$\xi_0 = 1 - \sqrt{1-2k^1}, \tag{9.1.10}$$

hold. In case $\ell_0 = \ell$ (9.1.9) reduces to (12) in [282]. Otherwise our error bounds are finer. Note also that

$$k \leq \frac{1}{2} \Rightarrow k^1 \leq \frac{1}{2} \tag{9.1.11}$$

but not vice versa unless if $\ell_0 = \ell$.

We can now describe the linear complementarity problem as follows: Given two matrices $Q, R \in \mathbf{R}^{n\times n}$ $(n \geq 2)$ and a vector $b \in \mathbf{R}^n$, the horizontal linear complementarity problem (HLCP) consists of approximating a pair of vectors (w, s) such that

$$\begin{aligned} ws &= 0 \\ Q(w) + R(s) &= b \\ w, s &\geq 0. \end{aligned} \tag{9.1.12}$$

The monotone linear complementarity problem (LCP) is obtained by taking $R = -I$ and Q positive semidefinite.

Moreover the linear programming problem (LP) and the quadratic programming problem (QP) can be formulated as HLCPs. That is, HLCP is a suitable way for studying interior point methods.

We assume HLCP (9.1.12) is monotone in the sense that:

$$Q(u) + R(v) = 0 \text{ implies } u^t v \geq 0, \text{ for all } u, v \in \mathbf{R}^n. \tag{9.1.13}$$

Condition (9.1.13) holds if the HLCP is a reformulation of a QP. If the HLCP is a reformuation of a LP then the following stronger condition holds:

$$Q(u) + R(v) = 0 \text{ implies } u^t v = 0, \text{ for all } u, v \in \mathbf{R}^n. \tag{9.1.14}$$

Then we say in this case that the HLCP is skew-symmetric.

If the HLCP has an interior point, i.e. there is $(w, s) \in \mathbf{R}^n_{++} \times \mathbf{R}^n_{++}$ satisfying $Q(w) + R(s) = b$, then for any parameter $\tau > 0$ the nonlinear system

$$\begin{aligned} ws &= \tau e \\ Q(w) + R(s) &= b \\ w, s &\geq 0 \end{aligned} \tag{9.1.15}$$

has a unique positive solution $x(\tau) = \left[w(\tau)^t, s(\tau)^t\right]^t$.

The set of all such solutions defines the central path C of the HLCP. It can be proved that $(w(\tau), s(\tau))$ converges to a solution of the HLCP as $\tau \to 0$. Such an approach for solving the HLCP is called the path following algorithm.

At a basic step of a path following algorithm, an approximation (w, s) of $(w(\tau), s(\tau))$ has already been computed for some $\tau > 0$. The algorithm determines the smaller value of the central path parameter $\tau_+ = (1 - \theta)\,\tau$, where the value $\theta \in (0, 1)$ is computed in some unspecified way. The approximation (w^t, s^t) of $(w(\tau_+), s(\tau_+))$ is computed. The procedure is then repeated with (w^+, s^+, τ^+) in place of $(w, s.\tau)$.

In order for us to relate the path following algorithm and the the Newton-Kantorovich theorem we introduce the notations

$$x = \begin{bmatrix} w \\ s \end{bmatrix}, \quad x(\tau) = \begin{bmatrix} w(\tau) \\ s(\tau) \end{bmatrix},$$

$$x^+ = \begin{bmatrix} w^+ \\ s^+ \end{bmatrix}, \quad x(\tau_+) = \begin{bmatrix} w(\tau_+) \\ s(\tau_+) \end{bmatrix}, \text{ etc.}$$

Then for any $\theta > 0$ we define the nonlinear operator

$$F_\sigma(x) = \left[\begin{array}{c} ws - \sigma e \\ Q(w) + R(s) - b \end{array} \right]. \tag{9.1.16}$$

Then system (9.1.15) defining $x(\tau)$ becomes

$$F_\sigma(x) = 0, \tag{9.1.17}$$

whereas the system defining $x(\tau_+)$ is given by

$$F_{(1-\theta)\tau}(x) = 0. \tag{9.1.18}$$

We assume that the initial guess x belongs in the interior of the feasible set of the HLCP

$$F^0 = \left\{ x = (w^t, s^t)^t \in \mathbf{R}^{2n}_{++} : Q(w) + R(s) = b \right\}. \tag{9.1.19}$$

In order to verify the Newton-Kantorovich hypothesis for equation (9.1.17) we introduce the quantity

$$\eta = \eta(x, \tau) = \left\| F'(x)^{-1} F_\tau(x) \right\|, \tag{9.1.20}$$

the measure of proximity

$$\begin{aligned} k &= k(x,\tau) = \eta\ell, \ \ell = \ell(x) \\ k^0 &= k^0(x,\tau) = \eta\bar{\ell}, \ \bar{\ell} = \bar{\ell}(x) \\ k^1 &= k^1(x,\tau) = \eta\ell_0, \ \ell_0 = \ell_0(x) \end{aligned} \tag{9.1.21}$$

and the normalized primal-dual gap

$$\mu = \mu(x) = \frac{w^t s}{\eta}. \tag{9.1.22}$$

If for a given interior point x and a given parameter τ we have $k^0(x,\tau) \le .5$ for the Newton-Kantorovich method or $k^1(x,\tau) \le .5$ for the modified Newton-Kantorovich method then corresponding sequences starting from x will converge to the point $x(\tau)$ on the central path. We can now describe our algorithm which is a weaker version of the one given in [282]:

Algorithm 9.1.1 *(using Newton-Kantorovich method).*
Given $0 < k_1^0 < k_2^0 < .5$, $\varepsilon > 0$, *and* $x_0 \in F^0$ *satisfying* $k^0(x_0, \mu(x_0)) \le k_1^0$;
Set $k^0 \leftarrow 0$ *and* $\tau_0 \leftarrow \mu(x_0)$;
repeat (outer iteration)
Set $(x, \tau) \leftarrow (x_k, \tau_k)$, $\bar{x} \leftarrow x_k$;
Determine the largest $\theta \in (0,1)$ *such that* $k^0(x, (1-\theta)\tau) \le k_2^0$;

Set $\tau \leftarrow (1-\theta)\tau$;
repeat (inner iteration)

$$\text{Set } x \leftarrow x - F'(x)^{-1}F_\tau(x) \tag{9.1.23}$$

until $k^0(x,\mu) \leq k_1^0$;
Set $(x_{k+1}, \tau_{k+1}) \leftarrow (x, \tau)$;
Set $k \leftarrow k+1$;
until $\left(w^k\right)^t s^k \leq \varepsilon$.

For the modified the Newton-Kantorovich algorithm k_1^0, k_2^0, k^0 should be replaced by k_1^1, k_2^1, k^1, and (9.1.23) by

$$\text{Set } x \leftarrow x - F'(\bar{x})^{-1}F_\tau(x)$$

respectively.

In order to obtain Algorithm 1 in [282] we need to replace k_1^0, k_2^0, k^0 by k_1, k_2, k respectively.

The above suggest that all results on interior methods obtained in [282] using (9.1.3) can now be rewritten using only the weaker (9.1.5) (or (9.1.8)).

We only state those results for which we will provide applications.

Let us introduce the notation

$$\Psi_i^a = \begin{cases} 1+\theta_i^a + \sqrt{2\theta_i^a + r_i^a}, & \text{if HLCP is monotone,} \\ 1+q_{ia} + \sqrt{2q_{ia}+q_{ia}^2}, & \text{if HLCP is skew-symmetric} \end{cases} \tag{9.1.24}$$

where $\sqrt{r_i^a} = \theta_i^a$, $\sqrt{t_i^a} = k_i^a$, $a = 0, 1$,

$$\theta_i^a = t_i\left[1 + \tfrac{t_i^a}{1-t_i^a}\right], \quad q_{ia} = \frac{t_i^a}{2}, \quad i = 1, 2. \tag{9.1.25}$$

Then by simply replacing k, k_1, k_2 by k^0, k_1^0, k_2^0 respectively in the corresponding results in [282] we obtain the following improvements:

Theorem 9.1.2 *The parameter θ determined at each outer iteration of algorithm 9.1.1 satisfies*

$$\theta \geq \frac{\chi^a}{\sqrt{n}} = \lambda^a$$

where

$$\chi^a = \begin{cases} \frac{\sqrt{2}(k_2^a - k_1^a)}{\sqrt{2+p^2 t_i}\sqrt{\psi_1^a}}, & \textit{if HLCP is skew-symmetric or if no simplified Newton-Kanorovich steps are performed,} \\ \frac{\sqrt{2}(k_2^a - k_1^a)}{(\sqrt{2}+pk_1^a)\sqrt{\psi_1^a}}, & \textit{otherwise} \end{cases} \tag{9.1.26}$$

where

$$p = \begin{cases} \sqrt{2}, & \text{if HLCP is monotone,} \\ 1, & \text{if HLCP is skew-symmetric.} \end{cases} \tag{9.1.27}$$

Clearly, the lower bound on λ^a on θ is an improvement over the corresponding one in [283, Corollary 4].

In the next result a bound on the number of steps of the inner iteration that depends only on k_1^0 and k_2^0 is provided.

Theorem 9.1.3 *If Newton-Kantorovich method is used in Algorithm 9.1.1 then each inner iteration terminates in at most $N^0\left(k_1^0, k_2^0\right)$ steps, where*

$$N^0\left(k_1^0, k_2^0\right) = \text{integer}\,part\left[\log_2\left(\frac{\log_2\left(x_{N^0}\right)}{\log_2\left[\left(1-\sqrt{1-2k_2^0}-k_2^0\right)/k_2^0\right]}\right)\right] \tag{9.1.28}$$

and

$$x_{N^0} = \frac{\left(1-pk_2^0/\sqrt{2}\right)\left[t_2^0-\left(1-\sqrt{1-2k_2^0}\right)^2\right]k_1^0}{2\sqrt{2}t_2^0\sqrt{1-2k_2^0}\left[\sqrt{\psi_2^0}+1-\sqrt{1-2k_2^0}\right]\left(1+k_1^0\right)}. \tag{9.1.29}$$

If the modified the Newton-Kantorovich method is used in Algorithm 9.1.1 then each iteration terminates in at most $S^0\left(k_1, k_2\right)$ steps, where

$$S^1\left(k_1^1, k_2^1\right) = \text{integer}\,part\left[\frac{\log_2\left(x_{S^1}\right)}{\log_2\left(1-\sqrt{1-\sqrt{1-2k_2^1}}\right)}\right]+1 \tag{9.1.30}$$

and

$$x_{S^1} = \frac{\left(1-pk_2^1/\sqrt{2}\right)\left[t_2^1-\left(1-\sqrt{1-2k_2^1}-k_2^1\right)^2\right]k_1^0}{2\sqrt{2}\sqrt{1-2k_2^1}\left(1-\sqrt{1-2k_2^1}-k_2^1\right)^2\left(\sqrt{\psi_2^1}+1-\sqrt{1-2k_2^1}\right)\left(1+k_1^1\right)}.$$

Clearly, if $k_1^1 = k_1^0 = k_1$, $k_2^1 = k_2^0 = k_2$, $k^1 = k^0 = k$, Theorem 9.1.2 reduces to the corresponding Theorem 2 in [282]. Otherwise the following improvement holds:

$$N^0\left(k_1^0, k_2^0\right) < N\left(k_1, k_2\right),\ N^0 < N,\ S^1\left(k_1^1, k_2^1\right) < S\left(k_1, k_2\right),\ \text{and } S^1 < S.$$

Since $\frac{k_1}{k_1^0}$, $\frac{k_2}{k_2^0}$, $\frac{k_1}{k_1^1}$, $\frac{k_2}{k_2^1}$ can be arbitrarily large for a given triplet η, ℓ and ℓ_0, the choices

$$k_1^0 = k_1^1 = .12,\quad k_2^0 = k_2^1 = .24 \quad \text{when} \quad k_1 = .21 \text{ and } k_2 = .42$$

and

$$k_1^0 = k_1^1 = .24, \quad k_2^0 = k_2^1 = .48 \quad \text{when} \quad k_1 = .245 \text{ and } k_2 = .49$$

are possible. Then using formulas (9.1.27), (9.1.28) and (9.1.30) for our results and (9)–(11) in [282], we obtain the following tables:

(a) If the HLCP is monotone and only Newton directions are performed, then:

Potra	Argyros
$\chi(.21, .42) > .17$	$\chi^0(.12, .24) > .1$
$\chi(.245, .49) > .199$	$\chi^0(.24, .48) > .196$

Potra	Argyros
$N(.21, .42) = 2$	$N^0(.12, .24) = 1$
$N(.245, .49) = 4$	$N^0(.24, .48) = 3$

(b) If the HLCP is monotone and Modified Newton directions are performed:

Potra	Argyros
$\chi(.21, .42) > .149$	$\chi^1(.12, .24) > .098$
$\chi(.245, .49) > .164$	$\chi^1(.24, .48) > .162$

Potra	Argyros
$S(.21, .42) = 5$	$S^1(.12, .24) = 1$
$S(.245, .49) = 18$	$S^1(.24, .48) = 12$

All the above improvements are obtained under weaker hypotheses and the same computational cost (in the case of Newton's method) or less computational cost (in the case of the modified Newton method) since in practice the computation of ℓ requires that of ℓ_0 and in general the computation of ℓ_0 is less expensive than that of ℓ.

9.2 Case 2: LP Methods

We are motivated by paper [294], where a unified complexity analysis for Newton LP methods was given. Here we show that it is possible under weaker hypotheses to provide a finer semilocal convergence analysis, which in turn can reduce the computational time for interior algorithms appearing in linear programming and convex quadratic programming.

Finally an example is provided to show that our results apply where the ones in [294] cannot.

Let us assume that $F\colon D \subseteq X \to Y$ is an analytic operator and $F'(x)^{-1}$ exists at $x = x_0 \in D$. Define

$$\eta = \eta(x_0) = \|F'(x_0)^{-1}F(x_0)\| \tag{9.2.1}$$

and

$$\gamma = \gamma(x_0) = \sup_{k\geq 2} \left\| \frac{1}{k!} F'(x_0)^{-1}F^{(k)}(x)\right\|^{\frac{1}{k-1}}. \tag{9.2.2}$$

Note that $\eta(x_0)$ is the step length at x_0 when method (4.1.3) is applied to approximate x^*.

Smale in [312] showed that if $D = X$, and

$$\gamma\eta \leq \frac{1}{8} = .125, \tag{9.2.3}$$

then Newton-Kantorovich method (4.1.3) converges to x^*, so that

$$\|x_{n+1} - x_n\| \leq 2\left(\frac{1}{2}\right)^n \|x_1 - x_0\| \quad (n \geq 0). \tag{9.2.4}$$

Rheinboldt in [296], using the Newton–Kantorovich theorem showed convergence assuming

$$D \subseteq X \quad \text{and} \quad \gamma\eta \leq .11909565. \tag{9.2.5}$$

We need the following semilocal convergence result for Newton-Kantorovich method (4.1.3) and twice Fréchet-differentiable operators:

Theorem 9.2.1 *Let $F\colon D \subseteq X \to Y$ be a twice Fréchet-differentiable operator. Assume there exist a point $x_0 \in D$ and parameters $\eta \geq 0$, $\ell_0 \geq 0$, $\ell \geq 0$, $\delta \in [0,1]$ such that:*

$$F'(x_0)^{-1} \in L(Y, X), \tag{9.2.6}$$

$$\|F'(x_0)^{-1}[F'(x) - F'(x_0)]\| \leq \ell_0\|x - x_0\|, \tag{9.2.7}$$

$$\|F'(x_0)^{-1}F''(x)\| \leq \ell \tag{9.2.8}$$

for all $x \in D$,

$$\overline{U}(x_0, t^*) \subseteq D, \tag{9.2.9}$$

and

$$h_\delta = (\ell + \delta \ell_0)\eta \leq \delta, \tag{9.2.10}$$

where,

$$t^* = \lim_{n\to\infty} t_n, \tag{9.2.11}$$

with

$$t_0 = 0,\ t_1 = \eta,\ t_{n+2} = t_{n+1} + \frac{\ell(t_{n+1} - t_n)^2}{2(1 - \ell_0 t_{n+1})} \quad (n \geq 0). \tag{9.2.12}$$

Then (a) sequence $\{x_n\}$ $(n \geq 0)$ generated by Newton-Kantorovich method (4.1.3) is well defined, remains in $\overline{U}(x_0, t^*)$ for all $n \geq 0$, and converges to a unique solution x^* of equation $F(x) = 0$ in $\overline{U}(x_0, t^*)$. Moreover the following estimates hold for all $n \geq 0$:

$$\|x_{n+2} - x_{n+1}\| \leq \frac{\ell \|x_{n+1} - x_n\|^2}{2(1 - \ell_0 \|x_{n+1} - x_0\|)} \leq t_{n+2} - t_{n+1}, \tag{9.2.13}$$

$$\|x_n - x^*\| \leq t^* - t_n \leq t_0^{**} - t_n \leq \alpha\eta, \tag{9.2.14}$$

where,

$$t_0^{**} = \left[1 + \frac{\delta^0}{2} + \frac{1}{1 - \frac{\delta}{2}} \frac{\delta}{2} \frac{\delta^0}{2}\right]\eta,\quad t^{**} = \alpha\eta,\ \alpha = \frac{2}{2 - \delta}, \tag{9.2.15}$$

and

$$\delta^0 = \frac{\ell\eta}{1 - \ell_0\eta}. \tag{9.2.16}$$

Furthermore the solution x^* is unique in $U(x_0, R)$, $R = \frac{2}{\ell_0} - t^*$ provided that

$$U(x_0, R) \subseteq D, \quad \textit{and} \quad \ell_0 \neq 0. \tag{9.2.17}$$

(b) If

$$H_\alpha = (\ell + 2\ell_0\alpha)\eta < 2, \tag{9.2.18}$$

then

$$0 \leq a = \frac{\ell\eta}{2(1 - \ell_0\alpha\eta)} < 1 \tag{9.2.19}$$

and

$$\|x_{n+1} - x_n\| \leq a_0 \|x_n - x_{n-1}\|^2 \leq a^{2^n - 1}\eta \ \textit{ for all } n \geq 0, \tag{9.2.20}$$

$$\|x_n - x^*\| \le t^* - t_n \le a^{2^n-1}\bar{\bar{b}}\eta \le a^{2^n-1}\bar{b}\eta \le a^{2^n-1}b\eta, \quad \textit{for all } n \ge 0 \quad (9.2.21)$$

where,

$$a_0 = \frac{\ell}{2(1-\ell_0\alpha\eta)}, \quad \bar{\bar{b}} = 1 + a^{2^n} + a^{2^{n+1}} + \cdots, \tag{9.2.22}$$

$$\bar{b} = 1 + \frac{a^{2^n}}{1-a^2} \quad \textit{and} \quad b = \frac{2-a^2}{a^2}. \tag{9.2.23}$$

Proof. Part (a) follows immediately as a special case of Theorem 5.1.6. For part (b) note that (9.2.18) implies (9.2.19), whereas (9.2.20) a consequence of Proposition 3 in [312, p. 193]. Finally (9.2.21) follows from (9.2.20). ■

Corollary 9.2.2 *If*

$$2a = \frac{\ell\eta}{1-\ell_0\alpha\eta} \le 1 \tag{9.2.24}$$

then under the rest of the hypotheses of Theorem 9.2.1

$$\|x_{n+1} - x_n\| \le 2\left(\frac{1}{2}\right)^{2^n} \eta \quad (n \ge 0). \tag{9.2.25}$$

That is x_0 is an approximate zero of F and sequence $\{x_n\}$ $(n \ge 0)$ converges to x^* at least R-quadratically with R_2 factor $\frac{1}{2}$ [1], [6].

Let $\gamma \ge 0$ and $\lambda \ge 0$. It is convenient for us to define parameters

$$\beta_0 = \gamma\eta, \quad \beta_1 = \gamma\lambda, \quad \beta = \gamma\alpha\eta \tag{9.2.26}$$

and functions f, f_1, g, h by

$$f(\delta, \beta_0, \beta_1, \beta) = \frac{(1-\beta_1)^4}{[2(1-\beta_1)^2-1]^2(1-\beta)^2}\left[\frac{2}{1-\beta} + \delta(2-\beta)\right]\left(\beta_0 + \frac{\beta_1}{1-\beta_1}\right), \tag{9.2.27}$$

$$f_1(\alpha, \beta_0, \beta_1) = \frac{\alpha(1-\beta_1)^2}{2(1-\beta_1)^2-1}\left(\beta_0 + \frac{\beta_1}{1-\beta_1}\right) + \beta_1 - \frac{[2(1-\beta_1)^2-1](1-\beta)^2}{(1-\beta_1)^2(2-\beta)}, \tag{9.2.28}$$

$$g(\alpha, \beta_0, \beta_1, \beta) = \frac{2(1-\beta_1)^4}{[2(1-\beta_1)^2-1]^2(1-\beta)^2}\left[\frac{1}{1-\beta} + \alpha(2-\beta)\right]\left(\beta_0 + \frac{\beta_1}{1-\beta_1}\right),$$

(9.2.29)

$$h(\alpha,\beta_0,\beta_1,\beta)=\frac{(1-\beta_1)^4}{[2(1-\beta_1)^2-1]^2(1-\beta)^2}\left[\frac{2}{1-\beta}+\alpha(2-\beta)\right]\left(\beta_0+\frac{\beta_1}{1-\beta_1}\right), \tag{9.2.30}$$

respectively, provided that $\beta\neq 1$, $\beta_1\neq 1$ and $\beta_1\neq\frac{2-\sqrt{2}}{2}$.

We can state and prove the main semilocal convergence theorem for method (4.1.3) involving analytic operators:

Theorem 9.2.3 *Let $F\colon D\subseteq X\to Y$ be an analytic operator. Assume there exist points $x_0,\overline{x}_0\in D$ and parameters $\eta\geq 0$, $\gamma\geq 0$, $\lambda\geq 0$, $\delta\in[0,1]$ such that*

$$\|x_0-\overline{x}_0\|\leq\lambda\leq\alpha\eta, \tag{9.2.31}$$

$$\alpha\gamma\eta<1, \tag{9.2.32}$$

$$2\lambda\gamma<2-\sqrt{2}, \tag{9.2.33}$$

$$f(\delta,\beta_0,\beta_1,\beta)\leq\delta, \tag{9.2.34}$$

$$f_1(\alpha,\beta_0,\beta_1)\leq 0, \tag{9.2.35}$$

$$\overline{U}(x_0,\alpha\eta)\subseteq D \tag{9.2.36}$$

and condition (9.2.6) hold. Then sequence $\{x_n\}$ $(n\geq 0)$ starting at $\overline{x}_0$ is well defined, remains in $\overline{U}(x_0,\alpha\eta)$ for all $n\geq 0$ and converges to a unique zero x^* of F in $\overline{U}(x_0,\alpha\eta)$, so that estimates (9.2.13) and (9.2.14) hold for η, ℓ, ℓ_0 replaced by

$$\overline{\eta}=\frac{(1-\beta_1)^2}{2(1-\beta_1)^2-1}\left[\eta+\frac{\lambda}{1-\beta_1}\right], \tag{9.2.37}$$

$$\overline{\ell}=\frac{2\gamma(1-\beta_1)^2}{[2(1-\beta_1)^2-1](1-\beta)^3}, \tag{9.2.38}$$

and

$$\overline{\ell}_0=\frac{(1-\beta_1)^2(2-\beta)\gamma}{[2(1-\beta_1)^2-1](1-\beta)^2}, \tag{9.2.39}$$

respectively. Moreover the solution x^* is unique in $U(x_0,\overline{R})$, $\overline{R}=\frac{2}{\ell_0}-t^*$ provided that

$$U(x_0,\overline{R})\subseteq D,\quad \textit{and}\quad \gamma\neq 0. \tag{9.2.40}$$

Furthermore if

$$g(\alpha,\beta_0,\beta_1,\beta)<2, \tag{9.2.41}$$

then estimates (9.2.20)–(9.2.21) hold (with η, ℓ, ℓ_0 replaced by $\overline{\eta}$, $\overline{\ell}$ and $\overline{\ell}_0$ respectively. Finally if

$$h(\alpha, \beta_0, \beta_1, \beta) \leq 1, \tag{9.2.42}$$

then $\overline{x}_0$ (and x_0) is an approximate zero of F and sequence $\{x_n\}$ $(n \geq 0)$ converges to x^* at least R-quadratically with R_2 factor $\frac{1}{2}$, so that estimates (9.2.24) and (9.2.25) hold.

Proof. We shall show that under the stated conditions, the hypotheses of Theorem 9.2.1 hold with $\overline{x}_0$, $\overline{\eta}$, $\overline{\ell}_0$ and ℓ replacing x_0, η, ℓ_0 and ℓ, respectively. We first obtain a bound on $\|x_0 - x^*\|$. For all $x \in U(x_0, \alpha\eta)$ we obtain in turn

$$\|F'(x_0)^{-1}F''(x)\| = \left\| F'(x_0)^{-1} \sum_{i=0}^{\infty} \frac{1}{i!} F^{(2+i)}(x_0)(x - x_0)^i \right\|$$

$$\leq \gamma \sum_{i=0}^{\infty} (i+2)(i+1)[\gamma\|x - x_0\|]^i$$

$$\leq \frac{2\gamma}{(1 - \gamma\|x - x_0\|)^3} \leq \frac{2\gamma}{(1 - \gamma\alpha\eta)^3} = \frac{2\gamma}{(1 - \beta)^3}. \tag{9.2.43}$$

According to Theorem 9.2.1 there exists a unique zero x^* of F in $U(x_0, t^*)$. Therefore Newton's sequence starting at $\overline{x}_0$ converges to the same zero x^*. We can have

$$\|F'(x_0)^{-1}[F'(\overline{x}_0) - F'(x_0)]\| = \left\| F'(x_0)^{-1} \sum_{i=1}^{\infty} \frac{1}{i!} F^{(i+1)}(x_0)(\overline{x}_0 - x_0)^i \right\|$$

$$\leq \sum_{i=1}^{\infty} (i+1)[\gamma\|\overline{x}_0 - x_0\|]^i \leq \frac{1}{(1 - \gamma\|\overline{x}_0 - x_0\|)^2} - 1$$

$$\leq \frac{1}{(1 - \gamma\lambda)^2} - 1 \leq \frac{1}{(1 - \beta_1)^2} - 1 < 1 \quad \text{(by (9.2.33))}. \tag{9.2.44}$$

It follows by the identity

$$B^{-1} = \sum_{i=0}^{\infty} (-1)^i [A^{-1}(B - A)]^i A^{-1}$$

and (9.2.44) that for any $x \in X$,

$$\|F'(\overline{x}_0)^{-1}x\| \leq \frac{(1 - \beta_1)^2}{2(1 - \beta_1)^2 - 1} \|F'(x_0)^{-1}x\|. \tag{9.2.45}$$

Hence, we can have

$$\begin{aligned}\|F'(\overline{x}_0)^{-1}F(\overline{x}_0)\| &\leq \frac{(1-\beta_1)^2}{2(1-\beta_1)^2-1}\|F'(x_0)^{-1}F(\overline{x}_0)\| \\ &\leq \frac{(1-\beta_1)^2}{2(1-\beta_1)^2-1}\left\|F'(x_0)^{-1}\sum_{i=0}^{\infty}\frac{1}{i!}F^{(i)}(x_0)(\overline{x}_0-x_0)^i\right\| \\ &\leq \frac{(1-\beta_1)^2}{2(1-\beta_1)^2-1}\left(\eta+\|\overline{x}_0-x_0\|\sum_{j=0}^{\infty}[\gamma\|\overline{x}_0-x_0\|]^j\right) \\ &\leq \frac{(1-\beta_1)^2}{2(1-\beta_1)^2-1}\left(\eta+\frac{\lambda}{1-\gamma\lambda}\right) \equiv \overline{\eta}.\end{aligned} \tag{9.2.46}$$

Moreover, for $x \in U(\overline{x}_0, \alpha\overline{\eta})$ we have $x \in U\big(x_0, \frac{1}{\overline{\ell}_0}\big)$, since

$$\|x-\overline{x}_0\| \leq \|x-x_0\|+\|x_0-\overline{x}_0\| \leq \alpha\overline{\eta}+\lambda \leq \frac{1}{\overline{\ell}_0} \quad \text{(by (9.2.35))}, \tag{9.2.47}$$

and $\overline{\ell}_0$ is given by (9.2.39) where we have used $\overline{x}_0$ for x in (9.2.44) and (9.2.45) (to obtain $\overline{\ell}_0$). Therefore, we get

$$\|F'(\overline{x}_0)^{-1}F''(x)\| \leq \frac{(1-\beta_1)^2}{2(1-\beta_1)^2-1}\|F'(x_0)^{-1}F''(x)\|$$

$$\leq \frac{(1-\beta_1)^2}{2(1-\beta_1)^2-1}\frac{2\gamma}{(1-\gamma\beta_1)^3} \equiv \overline{\ell}. \tag{9.2.48}$$

Theorem 9.2.1 and Corollary 9.2.2 can now apply with $\overline{x}_0$, $\overline{\eta}$, $\overline{\ell}_0$ and $\overline{\ell}$ since hypotheses (9.2.10), (9.2.18) and (9.2.24) become (9.2.34), (9.2.41) and (9.2.42) respectively. ■

Remark 9.2.1 *In general*

$$\ell_0 \leq \ell \tag{9.2.49}$$

holds and $\frac{\ell}{\ell_0}$ *can be arbitrarily large. Note that the famous Newton–Kantorovich hypothesis (4.2.51) is a special case of corresponding condition (9.2.10). Simply set* $\ell_0 = \ell$ *and* $\delta = 1$ *in (9.2.10) to obtain (4.2.51). The advantages of using condition (9.2.10) over (4.2.51) (or over (9.2.52) used in Theorem 3 in [294]) have been explained in Section 5.1. In particular here we note that even under the hypotheses of Proposition 2.2 which uses stronger hypotheses than Theorem 2.1, our results have the following advantages over the corresponding ones in Theorem 3 in [294, p. 12]: our conditions (9.2.10) and (9.2.24) for* $\alpha = \frac{3}{2}$ *become*

$$\left(\ell + \frac{2}{3}\ell_0\right)\eta \leq \frac{2}{3}, \tag{9.2.50}$$

and

$$\left(\ell + \frac{3}{2}\ell_0\right)\eta \leq 1, \tag{9.2.51}$$

respectively, which are weaker than condition

$$\ell\eta \leq \frac{4}{9} \tag{9.2.52}$$

given in [294, p. 12] (for sufficiently small ℓ_0*). Note that all the above benefits are obtained under the same computational cost since in practice the computation of* ℓ *requires that of* ℓ_0*. Consequently the same benefits carry over when we compare our Theorem 9.2.3 (based on Theorem 9.2.1 (or Corollary 9.2.2)) with Theorem 2 in [294, p. 2].*

Let us provide an example where conditions of Theorem 9.2.3 hold but corresponding crucial condition:

$$\eta \leq \frac{1}{2}\lambda \leq \frac{1}{40\gamma} \tag{9.2.53}$$

in Theorem 2 in [294, p. 2] fails.

Example 9.2.1 *Let*

$$\lambda = \frac{3}{2}\eta, \quad \delta = \frac{2}{3} \tag{9.2.54}$$

and

$$\beta_0 = \gamma\eta = \frac{1}{22} = .0\overline{45}. \tag{9.2.55}$$

Then using (9.2.15) and (9.2.26) we get

$$\alpha = \frac{3}{2} \quad \text{and} \quad \beta = \beta_1 = .06\overline{81}.$$

Conditions (9.2.34), (9.2.35), (9.2.41) and (9.2.42) are satisfied, since

$$\begin{aligned} f(\delta, \beta_0, \beta_1, \beta) &= .651987692 < \delta = \frac{2}{3}, \\ f_1(\alpha, \beta_0, \beta_1) &= -.10652731 \leq 0, \\ g(\alpha, \beta_0, \beta_1, \beta) &= 1.50775228 < 2, \end{aligned}$$

and

$$h(\alpha, \beta_0, \beta_1, \beta) = .943087149 < 1.$$

That is the conditions of our Theorem 9.2.3 hold but condition (9.2.53) required for the application of Theorem 2 in [294] does not hold. Hence the conclusions of our Theorem 9.2.3 hold but not the ones in Theorem 2 in [294] in this case.

These facts influence (widens) choices. Indeed by simply repeating the proofs given in Sections 2–6 in [294], with the above changes we can show:

Application 9.2.4 *"Optimal" sequences $\{t^{(i)}\}$ $(i \geq 0)$ given by*

$$t^{(i+1)} = \left(1 - \frac{1}{41\sqrt{m}}\right) t^{(i)} \quad \text{(LP Barrier Method, QP Barrier Method)},$$

$$t^{(i+1)} = \left(1 - \frac{1}{40\sqrt{m}}\right) t^{(i)} \quad \text{(Primal-dual LP Algorithm, Primal-dual QP Algorithm)}$$

can be replaced by wider

$$\bar{t}^{(i+1)} = \left(1 - \frac{1}{23\sqrt{m}}\right) \bar{t}^{(i)}$$

and

$$\bar{t}^{(i+1)} = \left(1 - \frac{1}{22\sqrt{m}}\right) \bar{t}^{(i)}$$

respectively.

The rest of the results in [294] can also be improved if rewritten using our Theorem 9.2.3 above. However we leave the details to the motivated reader.

The observations/improvements made here are important in computational mathematics and scientific computing.

9.3 Variational inequalities on finite dimensional spaces

In this section we provide a semilocal convergence analysis of Newton's method for solving nonlinear variational inequalities on finite dimensional spaces using the 2-norm. Earlier results in the literature [200], [298]–[301] with the exception of the work in [345] share the same drawback: before showing the convergence of the method, the existence and some properties of the solution to the original problem have to be assumed, which cannot be verified computationally (except in some special cases). Our convergence analysis overcomes this drawback and uses more precise majorizing sequences than in the works above (especially the work in [345]).

This way we provide under the same computational cost: weaker sufficient convergence conditions, finer error bounds on the distances involved and an at least as precise information on the location of the solution.

Let Ω be a nonempty subset of $\mathbf{R}^m$ and let F be an operator from $\mathbf{R}^m$ into itself. The variational inequality problem, denoted by $VI(\Omega, F)$, is to find a vector $x^* \in \Omega$ such that

$$(y - x^*)^T F(x^*) \geq 0 \quad \text{for all } y \in \Omega.$$

The most popular mehtod for generating a sequence $\{x_n\}$ $(n \geq 0)$ approximating x^* is undoubtedly Newton's method. This method computes a sequence of iterates $\{x_n\}$, such that x_{n+1} solves the nth linearized problem $VI(\Omega, F^n)$, where

$$F^n(x) = F(x_n) + F'(x_n)(x - x_n), \tag{9.3.1}$$

where $F'(x)$ denotes the Jacobian matrix of F, i.e. $F'(x) = \left(\frac{\partial F_i(x)}{\partial x_j}\right)$.

There is an extensive literature on results concerning the convergence of method (9.3.1): Robinson proposed Newton's method for variational inequalities and nonlinear complementarity problems [298] but did not give a convergence analysis. Later Josephy [200] analyzed the method for the variational inequalities with applications to PIES model and independently obtained convergence results in their technical reports respectively. A survey of the works mentioned above and the concepts introduced below as well as results by others and us can be found in [93], [263], [298]–[303], [345] and the references there.

In particular we are motivated by the results:

Theorem 9.3.1 *[263] Let Ω be a nonempty closed and convex subset of R^m. Let $F\colon R^m \to R^m$ be continuously differentiable, and x^* be a regular solution to $VI(\Omega, F)$. Then there exists a neighborhood of x^* such that if the starting point x_0 is chosen there, the Newton sequence $\{x_n\}$ is well defined and converges to x^*. Moreover if $F'(x)$ is Lipschitz continuous near x^*, then the convergence is quadratic.*

Corollary 9.3.2 *[263] Let Ω, F, F, x^* be as in Theorem 9.3.1 with $F'(x^*)$ being positive definite. Then there exists a neighborhood of x^* such that if the starting point is chosen there, the Newton sequence $\{x_n\}$ is well defined and converges to the solution x^*. Moreover, if $F'(x)$ is Lipschitz continuous near x^*, the convergence is quadratic.*

The above results share the same drawback:

Before showing the convergence of the method, the existence and some properties of the solution to the original problem has to be assumed, which cannot be verified computationally.

Recently, Wang in [345] using the Newton–Kantorovich theorem addressed this drawback by providing a semilocal convergence analysis for Newton method (9.3.1).

In our study we show that under weaker hypotheses and the same computational cost than in [345]: finer error bounds on the distances $\|x_n - x^*\|$, $\|x_{n+1} - x_n\|$ $(n \geq 0)$ and an at least as precise information on the location of the solution x^*.

We will need the following lemma. The matrix norm is subordinate to some vector norm.

Lemma 9.3.3 *[345]. Let B be an $n \times n$ real symmetric matrix. The following hold true:*

(a) B is positive definite if and only if there is a symmetric positive definite matrix A such that

$$\|A - B\| < \frac{1}{\|A^{-1}\|}; \tag{9.3.2}$$

(b) if there is a symmetric positive definite matrix A such that

$$\|A - B\| = \frac{1}{\|A^{-1}\|}, \tag{9.3.3}$$

then B is either semi-definite or positive definite.

We also need the three well known facts that follow which are stated here in order to make the paper as self-contained as possible:

Definition 9.3.4 *[298]. The operator $F\colon \mathbf{R}^m \to \mathbf{R}^m$ is said to be strongly monotone over $D \subseteq \mathbf{R}^m$ if there exists an $\ell > 0$ such that*

$$(F(x) - F(y))^T (x - y) \geq \ell \|x - y\|^2 \quad \textit{for all } x, y \in D, \tag{9.3.4}$$

where $\|\cdot\|$ denotes any vector norm in $\mathbf{R}^m$.

Theorem 9.3.5 *Let Ω be a nonempty, closed and convex subset of R^m and let F be a continuous operator from Ω to R^m. If F is strongly monotone with respect to Ω, then there exists a unique solution to the problem $VI(\Omega, F)$.*

Remark 9.3.1 *For affine operator $F(x) = M(x) + q$, where $M \in \mathbf{R}^{m\times m}$ and $q \in \mathbf{R}^m$, the strong monotonicity is equivalent to the positive definiteness of the matrix M. Therefore, if M is positive definite and $\Omega \subseteq \mathbf{R}^m$ is nonempty, closed and convex, the $VI(\Omega, F)$ has a unique solution.*

The proof of the following lemma can essentially be found in Application 5.1.4.

Lemma 9.3.6 *Assume:*

there exist parameters $\beta \geq 0$, $\ell \geq 0$, $\ell_0 \geq 0$, with $\ell_0 \leq \ell$, $\eta \geq 0$ and $\delta \in [0, 2)$ such that any of the following set of conditions holds true:

$$h_\delta = \beta(\ell + \delta\ell_0) \leq \delta, \quad \delta \in [0, 1] \tag{9.3.5}$$

or

$$h_\delta \leq \delta, \quad \delta \in [0, 2), \tag{9.3.6}$$

$$\frac{2\beta\ell_0\eta}{2-\delta} \leq 1 \tag{9.3.7}$$

and

$$\frac{\ell_0\delta^2}{2-\delta} \leq \ell \tag{9.3.8}$$

or

$$\beta\ell_0\eta \leq 1 - \frac{\delta}{2}, \quad \delta \in [\delta_0, 2), \tag{9.3.9}$$

where,

$$\delta_0 = \frac{-b + \sqrt{b^2 + 8b}}{2}, \quad b = \frac{\ell}{\ell_0}, \quad \ell_0 \neq 0. \tag{9.3.10}$$

Then scalar iteration $\{t_n\}$ $(n \geq 0)$ *given by*

$$t_0 = 0, \quad t_1 = \eta, \quad t_{n+2} = t_{n+1} + \frac{\beta\ell(t_{n+1} - t_n)^2}{2(1 - \beta\ell_0 t_{n+1})} \quad (n \geq 0), \tag{9.3.11}$$

is non-decreasing, bounded above by

$$t^{**} = \frac{2\eta}{2-\delta}, \tag{9.3.12}$$

and converges to some t^* *such that*

$$\eta \leq t^* \leq t^{**}. \tag{9.3.13}$$

Moreover the following estimates hold true for all $n \geq 0$

$$0 \leq t_{n+2} - t_{n+1} \leq \frac{\delta}{2}(t_{n+1} - t_n) \leq \left(\frac{\delta}{2}\right)^{n+1}\eta. \tag{9.3.14}$$

Let us recall $\tilde{A} = \frac{A+A^T}{2}$, where A is a real square matrix.

From now on we assume $\|\cdot\| = \|\cdot\|_2$. We can show the following semilocal convergence theorem of Newton's method (9.3.1) for variational inequalities.

Theorem 9.3.7 *Let* $D \subseteq R^m$ *be open, let* $\Omega \subseteq D$ *be nonempty, convex and closed. Assume:*

$F\colon D \to R^m$ *is continuously differentiable and*

$$\|F'(x) - F'(y)\| \leq \ell\|x - y\| \quad \textit{for all } x, y \in D_0, \tag{9.3.15}$$

where $D_0 \subseteq \Omega$ is convex;

there exists an initial guess $x_0 \in D_0$ such that $F'(x_0)$ is positive definite,

$$\|\tilde{F}'(x_0)^{-1}\| \leq \beta \tag{9.3.16}$$

and

$$\|F'(x) - F'(x_0)\| \leq \ell_0 \|x - x_0\| \quad \text{for all } x \in D_0. \tag{9.3.17}$$

Denote by x_1 the unique solution to $VI(\Omega, F^0)$ and

$$\|x_1 - x_0\| \leq \eta, \tag{9.3.18}$$

where,

$$F^0(x) = F(x_0) + F'(x_0)(x - x_0). \tag{9.3.19}$$

If hypotheses of Lemma 9.3.6 hold true and

$$\overline{U}(x_0, t^*) \subseteq D_0 \tag{9.3.20}$$

then sequence $\{x_n\}$ $(n \geq 0)$ generated by Newton's method (9.3.1) is well defined, remains in $\overline{U}(x_0, t^*)$ for all $n \geq 0$ and converges to a unique solution x^* in $\overline{U}(x_0, t^*)$ of the variational inequality $VI(\Omega, F)$.

Moreover the following estimates hold true for all $n \geq 0$:

$$\|x_{n+1} - x_n\| \leq t_{n+1} - t_n \tag{9.3.21}$$

and

$$\|x_n - x^*\| \leq t^* - t_n. \tag{9.3.22}$$

Proof. It is clear that $\beta \neq 0$. If $\ell = 0$, then $F'(x) = F'(x_0)$ for all $x \in D_0$, which implies $F(x) = F^0(x)$ by the mean value theorem (see Section 7.6). That is Newton's method (9.3.1) terminates at x_1 since $x_1 = x^*$. If $\eta = 0$, then $x_1 = x_0$, since x_1 solves $VI(\Omega, F^0)$ we obtain

$$(y - x_0)^T F(x_0) = (y - x_1)^T [F(x_0) + F'(x_0)](x_1 - x_0) \geq 0 \quad \text{for all } y \in \Omega,$$

which shows that x_0 is a solution of $VI(\Omega, F)$, and consequently the Newton method (1.1) terminates. Therefore we can assume without loss of generality $\ell \neq 0$ and $\eta \neq 0$. By (9.3.13) we have $\eta \leq t^*$ and since $\|x_1 - x_0\| \leq \eta$ we conclude $x_1 \in \overline{U}(x_0, t^*) \subseteq D_0$.

We shall show for all $k \geq 0$:

$$\|x_{k+1} - x_k\| \leq t_{k+1} - t_k \tag{9.3.23}$$

and

$$\overline{U}(x_{k+1}, t^* - t_{k+1}) \subseteq \overline{U}(x_k, t^* - t_k). \tag{9.3.24}$$

For every $z \in \overline{U}(x_1, t^* - t_1)$, estimate

$$\|z - x_0\| \leq \|z - x_1\| + \|x_1 - x_0\| \leq t^* - t_1 + t_1 = t^* - t_0,$$

implies $z \in \overline{U}(x_0, t^* - t_0)$.

We also have:

$$\|x_1 - x_0\| \leq \eta = t_1 - t_0.$$

That is (9.3.23) and (9.3.24) hold for $k = 0$. Given they hold for $n = 0, 1, \ldots, k$, then

$$\begin{aligned}\|x_{k+1} - x_0\| &\leq \sum_{i=1}^{k+1} \|x_i - x_{i-1}\| \leq \sum_{i=1}^{k+1} (t_i - t_{i-1}) \\ &= t_{k+1} - t_0 = t_{k+1} \leq t^*\end{aligned}$$

and

$$\|x_k + \theta(x_{k+1} - x_k) - x_0\| \leq t_k + \theta(t_{k+1} - t_k) \leq t^*, \quad \theta \in [0, 1].$$

We have

$$\begin{aligned}\|\tilde{F}'(x_k) - \tilde{F}'(x_0)\| &\leq \frac{1}{2}\left[\|F'(x_k) - F'(x_0)\| + \|(F'(x_k) - F'(x_0))^T\|\right] \\ &= \|F'(x_k) - F'(x_0)\| \leq \ell_0 \|x_k - x_0\| \\ &\leq \ell_0(t_k - t_0) = \ell_0 t_k < \frac{1}{\beta}.\end{aligned} \tag{9.3.25}$$

It follows by Lemma 9.3.3 that $\tilde{F}'(x_k)$ is positive definite, and consequently so is $F'(x_k)$. We know $VI(\Omega, F^k)$ is uniquely solvable. That is x_{k+1} is well defined. Using (9.3.2) and (9.3.25) we get

$$\|\tilde{F}'(x_k)^{-1}\| \leq \frac{\beta}{1 - \beta\ell_0\|x_k - x_0\|} \leq \frac{\beta}{1 - \beta\ell_0 t_k}. \tag{9.3.26}$$

Considering that $x_k, x_{k+1} \in \Omega$, and that x_k solves $VI(\Omega, F^{k-1})$ and x_{k+1} solves $VI(\Omega, F^k)$ we can write

$$\begin{aligned}&(x_{k+1} - x_k)^T F^{k-1}(x_k) \\ &\qquad = (x_{k+1} - x_k)^T \left[F(x_{k-1}) + F'(x_{k-1})(x_k - x_{k-1})\right] \geq 0,\end{aligned} \tag{9.3.27}$$

$$\begin{aligned}&(x_k - x_{k+1})^T F^k(x_{k+1}) \\ &\qquad = (x_k - x_{k+1})^T \left[F(x_k) + F'(x_k)(x_{k+1} - x_k)\right] \geq 0.\end{aligned} \tag{9.3.28}$$

Adding (9.3.27) to (9.3.28) we can obtain

$$\begin{aligned}(x_k - x_{k+1})^T F'(x_k)(x_k - x_{k+1}) \\ &\leq (x_k - x_{k+1})^T \left[F(x_k) - F(x_{k-1}) - F'(x_{k-1})(x_k - x_{k-1})\right] \\ &\leq \|x_k - x_{k+1}\| \, \|F(x_k) - F(x_{k-1}) - F'(x_{k-1})(x_k - x_{k-1})\|. \end{aligned} \tag{9.3.29}$$

It follows from (9.3.15) that

$$\|F(x_k) - F(x_{k-1}) - F'(x_{k-1})(x_k - x_{k-1})\| \leq \frac{\ell}{2}\|x_k - x_{k-1}\|^2. \tag{9.3.30}$$

Denote the eigenvalues of $\tilde{A}$ by $\lambda_1 \leq \lambda_2 \leq \cdots \leq \lambda_n$. We can get

$$x^T \tilde{A} x \geq \lambda_1 x^T x = \frac{\|x\|^2}{\rho(\tilde{A}^{-1})} \geq \frac{\|x\|^2}{\|\tilde{A}^{-1}\|}. \tag{9.3.31}$$

In view of $x^T A x = x^T \tilde{A} x$, we have

$$\begin{aligned}\frac{\|x_k - x_{k+1}\|^2}{\|\tilde{F}'(x_k)^{-1}\|} &\leq (x_k - x_{k+1})\tilde{F}'(x_k)(x_k - x_{k+1}) \\ &= (x_k - x_{k+1})^T F'(x_k)(x_k - x_{k+1}). \end{aligned} \tag{9.3.32}$$

Using (9.3.24), (9.3.30) and (9.3.32) we get

$$\frac{\|x_k - x_{k+1}\|^2}{\|\tilde{F}'(x_k)^{-1}\|} \leq \frac{\ell}{2}\|x_k - x_{k+1}\| \, \|x_k - x_{k-1}\|^2. \tag{9.3.33}$$

We assume $x_k \neq x_{k+1}$, otherwise we obtain the exact solution by setting $x_k = x^*$, and Newton's method is terminated. It follows from (9.3.33) and the induction hypothesis that

$$\|x_{k+1} - x_k\| \leq \frac{\ell}{2}\|\tilde{F}'(x_k)^{-1}\| \, \|x_k - x_{k-1}\|^2 \leq t_{k+1} - t_k, \tag{9.3.34}$$

which shows (9.3.23) for all $k \geq 0$. Moreover for every $w \in \overline{U}(x_{k+2}, t^* - t_{k+2})$ we have

$$\|w - x_k\| \leq \|w - x_{k+1}\| + \|x_{k+1} - x_k\| \leq t^* - t_{k+1} + t_{k+1} - t_k = t^* - t_{k+1},$$

which shows (9.3.24) for all $k \geq 0$.

By Lemma 9.3.6 $\{t_n\}$ $(n \geq 0)$ is a Cauchy sequence. It follows from (2.23) and (2.24) that $\{x_n\}$ $(n \geq 0)$ is a Cauchy sequence too in a Banach space X and as such it converges to some $x^* \in \overline{U}(x_0, t^*)$. Considering

$$(y - x_{k+1})^T \left[F(x_k) + F'(x_k)(x_{k+1} - x_k)\right] \geq 0 \text{ for all } y \in \Omega \tag{9.3.35}$$

and using the continuity of F and F' we deduce

$$(y - x^*)^T F(x^*) \geq 0 \quad \text{for all } y \in \Omega, \tag{9.3.36}$$

when $k \to \infty$. Hence, we conclude x^* solves $VI(\Omega, F)$.

Finally to show uniqueness let $y^* \in \overline{U}(x_0, t^*)$ be a solution of $VI(\Omega, F)$. In view of $x_k \in \Omega$ being a solution to $VI(\Omega, F^k)$ we have

$$(x_k - y^*)^T F(y^*) \geq 0, \tag{9.3.37}$$

$$(y^* - x_k)^T \left[F(x_{k-1}) + F'(x_{k-1})(x_k - x_{k-1})\right] \geq 0, \tag{9.3.38}$$

from which it follows

$$\|x_k - y^*\| \leq \frac{\ell\beta\|y^* - x_{k-1}\|^2}{2(1 - \ell_0\beta\|x_{k-1} - x_0\|)} \leq t^* - t_k \overset{k\to\infty}{\longrightarrow} 0. \tag{9.3.39}$$

That is $\lim_{k\to\infty} x_k = y^*$. But we showed $\lim_{k\to\infty} x_k = x^*$.

Hence, we conclude that

$$x^* = y^*.$$

■

Remark 9.3.2 *(a) In general*

$$\ell_0 \leq \ell \tag{9.3.40}$$

holds true and $\frac{\ell}{\ell_0}$ can be arbitrarily large. If $\ell_0 = \ell$ then our results reduce to the corresponding ones in [345]. Otherwise (i.e. if $\ell_0 < \ell$) they constitute an improvement. Indeed let scalar sequence $\{s_n\}$ be given by

$$s_0 = 0, \quad s_1 = \eta, \quad s_{n+2} = s_{n+1} + \frac{\ell\beta(s_{n+1} - s_n)^2}{2(1 - \ell\beta s_{n+1})} \quad (n \geq 0). \tag{9.3.41}$$

Sequence $\{s_n\}$ is the majorizing sequence essentially used in [345] instead of $\{t_n\}$. We showed in Section 5.1 in a more general setting:

$$t_n < s_n, \tag{9.3.42}$$

$$t_{n+1} - t_n < s_{n+1} - s_n, \tag{9.3.43}$$

$$t^* \leq s^* = \lim_{n\to\infty} s_n = \frac{1 - \sqrt{1-h}}{\ell} \tag{9.3.44}$$

where,

$$h = 2\beta\ell\eta. \tag{9.3.45}$$

Note that the crucial sufficient convergence condition for sequence $\{s_n\}$ is given by the famous Newton–Kantorovich-type hypothesis

$$h \leq 1. \tag{9.3.46}$$

Note that (9.3.46) implies corresponding hypotheses (see Lemma 9.3.6) but not vice versa unless $\ell_0 = \ell$.

Therefore under the same computational cost (since in practice the computation of ℓ requires the computation of ℓ_0) and weaker hypotheses than in [345] we provide finer error bounds on the distances $\|x_{n+1} - x_n\|$, $\|x_n - x^*\|$ and an at least as precise information on the location of the solution x^*.

(b) The uniqueness of solution x^* can be extended to $U(x_0, t_1^*)$ $t^* \leq t_1^*$ provided that

$$\beta(t^* + t_1^*) \leq 2 \quad \text{and} \quad \overline{U}(x_0, t_1^*) \subseteq D_0 \quad (\text{see.Section 5.1}) \tag{9.3.47}$$

(c) The results obtained here can be rewritten in affine invariant form if Theorem 9.3.7 is applied to operator $\tilde{F}'(x_0)^{-1}F(x)$ and $\beta\ell_0$, $\beta\ell$ are replaced by L_0 and L respectively. The advantages of this approach have been explained in [143].

We finish this study by stating two results without proofs. The proofs are identical to Theorem 2.14 and 2.15 in [345] with the exception that we are using (for simplicity (9.3.5) for $\delta = 1$):

$$h_1 = (\ell_0 + \ell)\eta \leq 1, \tag{9.3.48}$$

instead of (9.3.46) used there and the more precise estimate (9.3.26) instead of the one used in [345] (simply set $\ell_0 = \ell$ in (9.3.26) to obtain the one used in [345]).

Theorem 9.3.8 *Let $D \subseteq R^m$ be open, let $\Omega \subseteq D$ be nonempty, closed and convex. Assume $F\colon D \to R^m$ is continuously differentiable on a convex set $D_0 \subseteq \Omega$, (9.3.48) and (9.3.15) hold on D_0. Let $x^* \in D_0$ be a solution of $VI(\Omega, F)$, where $F'(x^*)$ is positive definite and*

$$\|\tilde{F}'(x^*)^{-1}\| \leq \beta^*. \tag{9.3.49}$$

If $\overline{U}(x^*, \frac{1}{\beta^*\ell_0}) \subseteq D_0$, then the conclusions of Theorem 9.3.7 hold for any starting point $x_0 \in \overline{U}(x^*, \frac{\theta}{\beta^*\ell_0})$, $0 < \theta < \frac{2-\sqrt{2}}{2}$.

Remark 9.3.3 *If $\ell_0 = \ell$ our result reduces to Theorem 2.14 in [345]. Otherwise it is an improvement since*

$$\overline{U}\left(x^*, \frac{\theta}{\beta^*\ell}\right) \subseteq \overline{U}\left(x^*, \frac{\theta}{\beta^*\ell_0}\right). \tag{9.3.50}$$

That is our result provides a wider range of initial choices x_0 than in [345].

Theorem 9.3.9 *If the conditions of Theorem 2.5 hold, then for any* $x \in U\left(x_0, \frac{1}{\beta \ell_0}\right)$, $F'(x)$ *is positive definite.*

Remark 9.3.4 *(a) If* $\ell_0 = \ell$ *our Theorem 9.3.7 reduces to Theorem 2.15 in [345]. Otherwise it is an improvement over it since*

$$U\left(x_0, \frac{1}{\beta \ell}\right) \subseteq U\left(x_0, \frac{1}{\beta \ell_0}\right). \tag{9.3.51}$$

(b) Note that Theorem 9.3.7 is more general than Corollary 9.3.2 (see also [345]).

9.4 Exercises

9.1. (a) [296] Use the Newton-Kantorovich theorem to show that if: $F : D \subseteq X \to Y$ is analytic on D, $x_0 \in D$ with $F'(x_0)^{-1} \in L(Y, X)$,

$$\gamma \eta \leq \frac{\alpha}{\sqrt{2}} = .11909565,$$

where α is the positive root of the cubic equation $\left(\sqrt{2}-1\right)(1-r)^3 - \sqrt{2}r = 0$ and $U(x_0, r_0) \subseteq D$, where $r_0 = \frac{\alpha}{\gamma}$, then $x_0 is$ an approximate zero of F.

(b) [296] If $D = X$ and $\gamma\eta \leq \alpha_1 = .15229240$, where $\alpha_1 is$ the positive zero of cubic equation $(1-r)^3 - 4r = 0$, then show x_0 is an approximate zero of F.

(c) [339] If $D = X$ and $\gamma\eta \leq 3 - 2\sqrt{2} = .171573$ show the quadratic convergence of Newton's method (4.1.3) but with ration not necessarily smaller or equal to .5.

9.2. [294] Let $F : D \subseteq X \to Y$ be a twice continuously Fréchet-differentiable operator. Assume there exist $x_0 \in D$, $\eta > 0, \ell > 0$ such that

$$\begin{aligned} F'(x_0)^{-1} &\in L(Y, X), \\ \left\|F'(x_0)^{-1}F(x_0)\right\| &\leq \eta, \\ \left\|F'(x_0)^{-1}F''(x)\right\| &\leq \ell, \text{ for all } x \in D, \\ \gamma\eta \leq \frac{4}{9} &\text{ and } U(x_0, \tfrac{3}{2}\eta) \subseteq D. \end{aligned}$$

Then show that the the Newton-Kantorovich method is well defined, remains in $U(x_0, \frac{3}{2}\eta)$ for all $n \geq 0$, converges to a unique zero x^* of F in $D \cap U(x_0, \frac{1}{\ell})$ and satisfies

$$\|x_n - x^*\| \leq 2\left(\frac{1}{2}\right)^{2^n} \eta, \quad (n \geq 1).$$

9.3. [294] Let $F : D \subseteq X \to Y$ be analytic. Assume there exist $x_0 \in D$, $\delta \geq 0$ such that

$$\begin{aligned} F'(x_0)^{-1} &\in L(Y, X), \\ \eta \leq \frac{1}{2}\delta &\leq \frac{1}{40\gamma}, \end{aligned}$$

and

$$\bar{U}(x_0, 4\delta) \subseteq D.$$

If $\|\bar{x} - x_0\| \leq \delta$ then show that the Newton-Kantorovich method (4.1.3) starting at $\bar{x}$ is well defined, and converges to a unique zero x^* of F in $\bar{U}(x_0, \frac{3}{2}\eta)$ so that

$$\|x_n - x^*\| \leq \frac{7}{2}\left(\frac{1}{2}\right)^{2^n} \delta.$$

9.4. Consider the LP barrier method [294]. Let $Int = \{x : A(x) > b\}$, A is a real matrix with a_i denoting the ith row of A, and let $h : Int \times \mathbf{R}_+ \to \mathbf{R}$ denote the map

$$h(x,t) = c^t x - t \sum \ln (a_i x - b_i).$$

For fixed t, the map $x \to h(x,t)$ is strictly convex having a unique minimum. The sequence of minimum as $t \to 0$ converges to the optimal solution of the LP. The algorithm simply computes a Newton-Kantorovich sequence $\{x_n\}$ $(n \geq 0)$, where

$$x_{n+1} = x_n - \nabla_x^2 h(x_n, t_{n+1})^{-1} \nabla_x h(x_n, t_{n+1})$$

and $t_n \to 0$.

Show:

(a) for suitable x_0 we can always choose $t_{n+1} = \left(1 - \frac{1}{41\sqrt{m}}\right) t_n$ and each x_n will be a "good" approximation to the minimum of the map $x \to h(x, t_n)$.

Hint: Let y be the minimum of $x \to h(x,t)$, assume $\|\bar{y} - y\| \leq \frac{1}{20}$ and show $\|\bar{y}' - y'\|' \leq \frac{1}{20}$, where $\|x\| = \left\|\Delta(y)^{-1} A(x)\right\|_2$ and for $t' > 0$, y' and $\|\cdot\|'$ are defined in an obvious way. Then use Exercise 9.3.

(b) An $O(\sqrt{m}L)$ iteration bound.

(c) the choice of sequence $\{t_n\}$ above can be improved if 41 is replaced by any $a \in \{22, 23, \ldots, 40\}$.

Hint: Use Theorem 9.2.3.

Note that this is an improvement over the choice in (a).

(d) An $O(\sqrt{m}L)$ iteration bound using choices of sequences $\{t_n\}$ given in (c).

(e) How to improve along the same lines as above the QP barrier method, the primal LP algorithm, the primal dual LP algorithm and the primal-dual QP algorithm as introduced in [294].

Chapter 10

Generalized equations

Recent results on generalized equations that compare favourably with earlier ones are presented in this chapter.

10.1 Strongly regular solutions

In this section we are concerned with the problem of approximating a strongly regular solution x^* of generalized equation

$$0 \in F(x) + N_C(x) \tag{10.1.1}$$

where F is a continuously Fréchet-differentiable operator defined on an open, nonempty convex subset D of a Banach space X with values in a Banach space Y, and $N_C(x)$ is a normal cone to C at x ($C \subseteq X$ closed and convex). Many problems in Mathematical Programming/Economics can be formulated as generalized equations [68], [99], [298], [302] (see also Chapter 7).

Here motivated by [298] we show that it is possible under the same hypotheses computational cost to provide a wider choice of initial guesses in the local case and under weaker conditions finer error bounds on the distances involved as well as a more precise information on the location of the solution in the semilocal case. We achieve all that by using more precise majorizing sequences, along the lines of our work in Chapter 5.

We first provide in a condensed form the preliminaries for generalized equations and strongly regular solutions, and further we provide a local convergence (see Theorem 10.1.6) and a semilocal convergence analysis (see Theorem 10.1.10) for method (4.1.3).

Finally we show that our results compare favorably with the earlier ones.

To make the section as self-contained as possible we need the definitions [298]. For a more detailed information on strongly regular equations see [298], [302] and the references there.

Definition 10.1.1 *Let $F\colon D \subseteq X \to Y$ be an operator with a continuous Fréchet-derivative F'. Let y be a fixed point in D. Then the linearization of F at y denoted by LF_y is defined by*

$$LF_y(x) = F(y) + F'(y)(x - y). \tag{10.1.2}$$

It follows from (10.1.2) that LFy is an affine operator passing through y having "slope" equal to that of F at y.

Consequently the linearization of a generalized equation at y can be obtained by replacing F by LF_y, formally:

Definition 10.1.2 *Let F, D, X, Y be as in the previous Definition. Then the linearization of the generalized equation*

$$0 \in F(x) + N_K(x) \tag{10.1.3}$$

at y is the generalized equation $0 \in LF_y(x) + N_K(x)$, where K is a non-empty, closed convex cone in D, and $N_K(x)$ be the normal cone to K at x.

Definition 10.1.3 *Let T be a set-valued operator. We define*

$$T^{-1}(y) = \{x \mid y \in T(x)\}. \tag{10.1.4}$$

It then follows $T^{-1}(0) = \{x \mid 0 \in T(x)\}$ is the solution set of the generalized equation $0 \in T(x)$. We pay special attention to the case T^{-1} when $T = LF_y + N_K$.

The following definition is due to Robinson [298] and describes the behavior of $(LF_y + N_K)^{-1}$ suitable for the convergence of Newton's method:

Definition 10.1.4 *Let $0 \in F(x) + N_K(x)$ have a solution x^*. Set $T = LF_{x^*} + N_K$. Then x^* is a strongly regular solution of $0 \in F(x) + N_K(x)$ if and only if there exists a neighborhood B_1 of the origin, and a neighborhood B_2 of x^*, such that the restriction $B_2 \cap T^{-1}$ to B_1 is a single-valued and Lipschitz continuous operator. Note that the operator $B_2 \cap T^{-1}$ is defined by*

$$(B_2 \cap T^{-1})(y) = B_2 \cap T^{-1}(y).$$

Finally we need a Corollary of Theorem 2.4 in [298].

Corollary 10.1.5 *Let $F\colon D \subseteq X \to Y$ have a Fréchet derivative F' on D. Suppose $x_0 \in D$ and the generalized equation*

$$0 \in LF_{x_0}(x) + N_C(x)$$

has a strongly regular solution with association Lipschitz constant d. Then there exist positive constants ρ, r and R such that for all $x \in \overline{U}(x_0, \rho)$ the following hold: $(LF_x + N_C)^{-1} \cap \overline{U}(x_1, r)$ restricted to $\overline{U}(0, R)$ is single-valued and Lipschitz continuous with modulus

$$\frac{d}{1 - d\|F'(x) - F'(x_0)\|}.$$

We can provide the following local convergence result for Newton's method involving generalized equations:

Theorem 10.1.6 *Let C be a nonempty, closed, convex subset of X, and let D be a nonempty, open, convex subset of X. Moreover let $F\colon D \to Y$ have a Lipschitz continuous Fréchet derivative F' with Lipschitz constant b. Suppose: generalized equation*

$$0 \in F(x) + N_C(x) \tag{10.1.5}$$

has a strongly regular solution x^* with associated Lipschitz constant d; operator F' has a center-Lipschitz constant (at $x = x^*$) b_0. Let ρ, r, and R be the positive constants introduced in Corollary 1. Let $x_0 \in D$ and define the notation:

$$\begin{aligned} r_0 &= \|x^* - x_0\| \\ S_x &= LF_x + N_C \\ a_0 &= d(1 - db_0r_0)^{-1} \\ h_0 &= \frac{1}{2}a_0br_0. \end{aligned}$$

Moreover suppose:

$$d(b + 2b_0)r_0 < 2, \tag{10.1.6}$$

$$\frac{1}{2}br_0^2 < R, \tag{10.1.7}$$

$$r_0 < \rho \tag{10.1.8}$$

$$\overline{U}(x^*, r_0) \subset D \cap \overline{U}(x^*, \rho). \tag{10.1.9}$$

Then Newton sequence $\{x_n\}$ $(n \geq 0)$ given by

$$x_{n+1} = (S_{x_n})^{-1}(0) \cap \overline{U}(x^*, r) \tag{10.1.10}$$

is well defined, remains in $\overline{U}(x^*, r_0)$ for all $n \geq 0$ and converges to x^*. Moreover the following estimates hold for all $n \geq 0$:

$$\|x_{n+1} - x^*\| \leq \frac{1}{2}a_0b\|x_n - x^*\|^2 \tag{10.1.11}$$

and

$$\|x_{n+1} - x^*\| \leq 2(a_0b)^{-1}h_0^{2^{n+1}}. \tag{10.1.12}$$

Note that $b_0 \leq b$ holds in general and that $\frac{b}{b_0}$ can be arbitrarily large.

Proof. We introduce

$$T_x = \overline{U}(x^*, r) \cap S_x^{-1}, \quad T_k = T_{x_k},$$
$$f(x) = d[1 - d\|F'(x) - F'(x^*)]^{-1}$$
$$J(x) = LF_x(x^*) - LF_{x^*}(x^*).$$

We will show:

$$\|T_x(u) - T_x(v)\| \leq a_0\|u - v\| \quad \text{for all } x \in \overline{U}(x^*, r_0) \tag{10.1.13}$$

and $u, v \in \overline{U}(0, R)$,

$$x^* \in S_x^{-1}(J(x)) \quad \text{for all } x \in D, \tag{10.1.14}$$

$$x^* = T_x(J(x)) \quad \text{for } x \in \overline{U}(x^*, r_0) \text{ and } J(x) \in \overline{U}(0, R) \tag{10.1.15}$$

and

$$\|J(x)\| \leq \frac{b}{2}\|x - x^*\|^2, \tag{10.1.16}$$

$$h < 1. \tag{10.1.17}$$

By Corollary 10.1.5 T_x is a single-valued and Lipschitz continuous with modulus $f(x)$ for $x \in \overline{U}(x^*, \rho)$. Since, for $x \in \overline{U}(x^*, r_0) \subset \overline{U}(x^*, \rho) \cap D$

$$\|F'(x) - F'(x^*)\| \leq b_0\|x - x_0\| \leq b_0 r_0$$

we get $f(x) \leq a_0$, which shows (10.1.13) because $g(x)$ is the Lipschitz constant for operator T_x. Let $x \in D$, then we get

$$\begin{aligned} 0 \in F(x^*) + N_C(x^*) &= LF_{x^*}(x^*) + N_C(x^*) \\ &= LF_x(x^*) - J(x) + N_C(x^*) \\ &= -J(x) + S_x(x^*) + N_C(x^*) \end{aligned}$$

so $J(x) \in S_x(x^*)$, which shows (10.1.14). Equation (10.1.15) follows from (10.1.14) and the fact that operator T_x is single-valued on $\overline{U}(x^*, r_0)$. Estimate (10.1.16) is standard under the stated hypotheses, whereas (10.1.17) follows immediately from hypothesis (10.1.6). By (10.1.8), (10.1.17), and (10.1.19) we conclude (10.1.13) holds for $x = x_n$. Using (10.1.15) and (10.1.20), we get $X^* = T_n(J(x_n))$. That is $x_{n+1} = T_n(0)$ is well defined, and by (10.1.13)

$$\begin{aligned} \|x_{n+1} - x^*\| &= \|T_0(0) - T_n(J(x_n))\| \leq a_0\|J(x_n)\| \\ &\leq \frac{1}{2}ba_0\|x_n - x^*\|^2 \quad \text{by (10.1.16), (10.1.19) and } \overline{U}(x^*, r_0) \subseteq D \\ &\leq \frac{1}{2}ba_0(h_0 r_0)^2 \quad \text{by (10.1.19)} \\ &= h_0^3 r_0 < h_0 r_0 < r_0. \end{aligned}$$

Furthermore by $x_{n+1} \in \overline{U}(x^*, r_0) \subseteq D$ and (10.1.16) we get

$$\begin{aligned} \|J(x_{n+1})\| &\leq \frac{1}{2} b \|x_{n+1} - x^*\|^2 \\ &\leq \frac{1}{2} b (h_0 r_0)^2 \leq \frac{1}{2} b r_0^2 < R, \end{aligned}$$

which completes the induction. Finally (10.1.11) follows from (10.1.19), and (10.1.12) from (10.1.11) using induction. ∎

Moreover using induction on $k \geq 1$ we show:

$$x_k = T_{k-1}(0) \text{ is well defined} \tag{10.1.18}$$

$$\|x_k - x^*\| \leq \frac{1}{2} a_0 b \|x_{k-1} - x^*\| \leq h r_0 \tag{10.1.19}$$

$$J(x_k) \in \overline{U}(0, R). \tag{10.1.20}$$

For $k = 1$: $T_0(0)$ is single-valued by Corollary 10.1.5, which shows (10.1.18). By (10.1.7) and (10.1.16)

$$\|J(x_0)\| \leq \frac{b}{2} \|x_0 - x^*\|^2 = \frac{b}{2} r_0^2 < R$$

and hence by (10.1.15), $x^* = T_0(J(x_0))$. Therefore we get

$$\begin{aligned} \|x_1 - x^*\| &= \|T_0(0) - T_0(J(x_0))\| \leq a_0 \|J(x_0)\| \quad \text{by (10.1.13)} \\ &\leq \frac{1}{2} b a_0 \|x_0 - x^*\|^2 = \frac{1}{2} a_0 b r_0^2 \quad \text{by (9) and (10.1.16)} \\ &= h_0 r_0 < r_0 \quad \text{by (10.1.17)}. \end{aligned}$$

It follows by (10.1.8) and (10.1.9) that $x_1 \in \overline{U}(x^*, r_0) \subseteq D$. By (10.1.16) we can obtain

$$\begin{aligned} \|J(x_1)\| &\leq \frac{b}{2} \|x_1 - x^*\|^2 \leq \frac{1}{2} b (h_0 r_0)^2 = \frac{1}{2} b r_0^2 h_0^2 \\ &< \frac{1}{2} b r_0^2 \quad \text{by (10.1.17)} \\ &< R \quad \text{by (10.1.7)}. \end{aligned}$$

Therefore (10.1.18)–(10.1.20) hold for $k = 1$. Let us assume (10.1.18)–(10.1.20) hold for all $k \leq n$. We must show (10.1.18)–(10.1.20) hold for $k = n + 1$.

The following two results can be found in Section 5.1.

Lemma 10.1.7 *Let $b_0 > 0$, $b > 0$, $d > 0$ and $\eta > 0$ with $b_0 \leq b$. Set $h = d\left(\frac{b_0 + b}{2}\right)\eta$ and suppose $h \leq \frac{1}{2}$. Define $t_0 = 0$, $t_{k+1} = G_0(t_k)$, $(k \geq 0)$, where*

$$G_0(t) = \frac{\frac{1}{2} d b t^2 - \eta}{d b_0 t - 1}.$$

Let $t^* = \lim_{n\to\infty} t_n$ *and* $t^{**} = 2\eta$. *Define also* $Q_0(s,t) = \frac{dbs^2}{2(1-db_0t)}$. *Then the following hold: (a)* $t_{k+1} = t_k + Q_0(t_k - t_{k-1}, t_k)$ $(k \geq 1)$; *(b)* G_0 *is monotone non-decreasing on* $[0, t^*]$; *(c)* t^* *is the smallest fixed point of* G_0 *on* $[0, t^{**}]$. *(d) sequence* $\{t_k\}$ *is monotonically increasing, and* $\lim_{n\to\infty} t_n$ *exists; (e)* $t_{k+1} - t_k \leq 2^{-k}\eta$ $(k \geq 0)$; *(f)* $t_{k+1} \leq 2\left[1 - \left(\frac{1}{2}\right)^{k+1}\right]\eta$ $(k \geq 0)$; *(g)* $|t^* - t_k| \leq 2a^{2^k} \cdot (bd2^k)^{-1}$ $(k \geq 0)$.

Lemma 10.1.8 *Let* $D \subseteq X$, *and let* $\{x_n\}$ $(n \geq 0)$ *be a sequence in* X. *With the notation using in Lemma 1 assume: (a)* $\|x_{k+2} - x_{k+1}\| \leq Q_0(\|x_{k+1} - x_k\|, \|x_{k+1} - x_0\|)$ *whenever* $x_k, x_{k+1} \in D$; *(b)* $x_0 \in D$; *(c)* $t_1 \geq \eta = \|x_1 - x_0\|$; *(d)* $x \in D$ *whenever* $\|x - x_0\| \leq t^*$; *(e) hypotheses of Lemma 1 hold. Then the following conclusions hold: (f)* $\|x_k - x_0\| \leq t^*$ $(k \geq 0)$; *(g) sequence* $\{x_n\}$ *converges to some* x^* *with* $\|x^* - x_0\| \leq t^*$; *(h)* $\|x^* - x_{k+1}\| \leq t^* - t_{k+1}$ $(k \geq 0)$ *(i)* $\|x_{k+1} - x_k\| \leq t_{k+1} - t_k$ $(k \geq 0)$.

Lemma 10.1.9 *Let* D *be an open, convex subset of* D. *Let* $F\colon D \to Y$ *be continuously differentiable. Suppose that for some* $\ell > 0$ $\|F'(u) - F'(v)\| \leq \ell\|u - v\|$ *whenever* $u, v \in D$. *Then*

$$\|F(u) - LF_v(u)\| \leq \frac{1}{2}\ell\|u - v\|^2 \quad \textit{for all } u, v \in D.$$

We can show the following semilocal convergence theorem for Newton's method:

Theorem 10.1.10 *Let* C, D, F, b, b_0 *and* S_x *be as in Theorem 10.1.6. Let* $x_0 \in D$ *and assume that the generalized equation* $0 \in S_{x_0}(x)$ *is strongly regular at* x_1. *Let* ρ, r, *and* R *be the positive constants associated with strong regularity, and* d *the associated Lipschitz constant, as given in Corollary 10.1.5. Let* $\eta := \|x_1 - x_0\|$ *and* $t^* := \lim_{n\to\infty} t_n$, *where* $a := d\left(\frac{b_0+b}{2}\right)\eta$. *Suppose that the following relations hold:* $h \leq \frac{1}{2}$, $\frac{1}{2}bm^2 \leq R$, *and* $\overline{U}(x_0, t^*) \subset D \cap \overline{U}(x_0, \rho)$. *Then the sequence* $\{x_n\}$ *defined by*

$$x_{n+1} := S_{x_n}^{-1}(0) \cap \overline{U}(x_0, r) \tag{10.1.21}$$

is well defined, converges to some $x^* \in \overline{U}(x_0, t^*)$ *which satisfies* $0 \in F(x^*) + N_C(x^*)$, *and* $\|x^* - x_n\| \leq t^* - t_n$ *for* $n = 1, 2, \ldots$. *Note that by* b_0 *here we mean the center Lipschitz constant of* F' *at* $x = x_0$.

Proof. We first define the following notation:

$$\begin{aligned}
g(x) :=& d(1 - db_0\|x - x_0\|)^{-1},\\
T_n :=& \overline{B}(x_1, r) \cap S_{x_n}^{-1},\\
LF_n :=& LF_{x_n},\\
J_n :=& F(x_n) - LF_{n-1}(x_n).
\end{aligned}$$

Before beginning the induction part of the proof, we prove the following relation holds:

$$x_{n+1} = T_{n+1}(J_{n+1}) \quad \text{where} \quad x_n, x_{n+1} \in \overline{U}(x_0, t^*), \tag{10.1.22}$$

$x_n := T_n(0)$, $J_{n+1} \in \overline{U}(0, R)$ and T_{n+1} restricted to $\overline{U}(0, R)$ is single-valued. By definition of T_n, $x_{n+1} \in \overline{U}(x_1, r)$ and $0 \in LF_n(x_{n+1}) + N_C(x_{n+1})$. Thus

$$\begin{aligned} 0 &\in LF_{n+1}(x_{n+1}) - (LF_{n+1}(x_{n+1})) - LF_n(x_{n+1}) + N_C(x_{n+1}) \\ &= LF_{n+1}(x_{n+1}) - J_{n+1} + N_C(x_{n+1}). \end{aligned}$$

Rearranging results in $J_{n+1} \in (LF_{n+1} + N_C)(x_{n+1})$. That is, $x_{n+1} \in (LF_{n+1} + N_C)^{-1}(J_{n+1})$. Since $J_{n+1} \in \overline{U}(0, R)$ and $x_{n+1} \in \overline{U}(x_1, r)$, we have $x_{n+1} = T_{n+1}(J_{n+1})$, which shows (10.1.22). We now proceed by induction to prove the following for all $n \geq 1$.

T_n restricted to $\overline{U}(0, R)$ is single-valued and Lipschitz continuous with Lipschitz constant $g(x_n)$. (10.1.23)

$$\|x_{n+1} - x_n\| \leq Q(\|x_n - x_{n-1}\|, \|x_n - x_0\|), \text{ where } x_{n+1} := T_n(0). \tag{10.1.24}$$

$$x_{n+1} \in \overline{U}(x_0, t^*). \tag{10.1.25}$$

$$\|J_{n+1}\| \leq R. \tag{10.1.26}$$

We first show that (10.1.23)–(10.1.26) hold for $n = 1$. By Lemma 10.1.7(d), $t_1 \leq t^*$. By definition, $t_1 = G_0(t_0) = m = \|x_1 - x_0\|$. Hence,

$$x_1 \in \overline{U}(x_0, t_1) \subset \overline{U}(x_0, t^*) \subset D \cap \overline{U}(x_0, \rho).$$

By Corollary 10.1.5, T_1 restricted to $\overline{U}(0, R)$ is a single-valued and Lipschitz continuous operator with modulus $d(1 - d\|F'(x_1) - F'(x_0)\|)^{-1}$. By hypothesis, $\|F'(x_1) - F'(x_0)\| \leq b_0\|x_1 - x_0\|$, so that $d(1 - d\|F'(x_1) - F'(x_0)\|)^{-1} \leq g(x_1)$. Hence, (10.1.23) holds for $n = 1$. Now define $x_2 = T_1(0)$. By Lemma 10.1.9, $\|J_1\| := \|F(x_1) - LF_0(x_1)\| \leq \frac{1}{2}b\|x_1 - x_0\|^2 = \frac{1}{2}b\eta^2 \leq R$. Thus, $J_1 \in \overline{U}(0, R)$ and by (10.1.23),

$$\begin{aligned} \|T_1(J_1) - T_1(0)\| &\leq g(x_1) \cdot \|J_1\| \leq \frac{1}{2} db\|x_1 - x_0\|^2(1 - db_0\|x_1 - x_0\|)^{-1} \\ &= Q_0(\|x_1 - x_0\|, \|x_1 - x_0\|). \end{aligned}$$

But $x_2 = T_1(0)$ and by (10.1.22), $x_1 = T_1(J_1)$, hence we have $\|x_2 - x_1\| = \|T_1(0) - T_1(J_1)\|$. This establishes (10.1.24) when $n = 1$. By Lemma 10.1.8(f), $x_2 \in \overline{U}(x_0, t^*)$. Thus, (10.1.25) holds for $n = 1$. Also, $x_2 \in \overline{U}(x_0, t^*) \subset D$, and thus by Lemmas 10.1.7(e), 10.1.8(i) and 1 we get:

$$\begin{aligned}\|J_2\| := \|F(x_2) - LF_1(x_2)\| &\leq \frac{1}{2}b\|x_2 - x_1\|^2 \\ &\leq \frac{1}{2}b(t_2 - t_1)^2 \\ &\leq \frac{1}{2}b\eta^2\, 2^{-1} \\ &\leq R.\end{aligned}$$

Thus, (10.1.26) holds for $n = 1$, and we have completed the induction when $n = 1$. Suppose that (10.1.23)–(10.1.26) hold for all $n \geq k$. We proceed to show that they also hold for $n = k + 1$. By (10.1.25), $x_{k+1} \in \overline{U}(x_0, t^*) \subset \overline{U}(x_0, \rho) \cap D$. By Corollary 10.1.5, T_{k+1} restricted to $\overline{U}(0, R)$ is single-valued and Lipschitz continuous with modulus $d(1 - d\|F'(x_{k+1}) - F'(x_0)\|)^{-1}$. By hypothesis, $\|F'(x_{k+1}) - F'(x_0)\| \leq b_0\|x_{k+1} - x_0\|$. Thus, $g(x_{k+1}) \geq d(1 - d\|F'(x_{k+1} - F'(x_0)\|)^{-1}$. Hence (10.1.23) holds for $n = k + 1$. Let $x_{k+2} := T_{k+1}(0)$. By (10.1.22)–(10.1.26), $x_{k+1} = T_{k+1}(J_{k+1})$. Thus,

$$\begin{aligned}\|x_{k+2} - x_{k+1}\| = \|T_{k+1}(0) - T_{k+1}(J_{k+1})\| &\leq g(x_{k+1})\|J_{k+1}\| \\ &\leq g(x_{k+1}) \cdot \frac{1}{2}b\|x_{k+1} - x_k\|^2 \\ &= Q_0(\|x_{k+1} - x_k\|, \|x_{k+1} - x_0\|).\end{aligned}$$

This establishes (10.1.24) for $n = k + 1$. It follows from Lemma 10.1.8(f) that $x_{k+2} \in \overline{U}(x_0, t^*) \subset D$, thus (10.1.25) holds for $n = k + 1$. Now, by Lemmas 10.1.7, 10.1.8(i) and (10.1.3) we get:

$$\begin{aligned}\|J_{k+2}\| &= \|F'(x_{k+2}) - LF_{k+1}(x_{k+2})\| \\ &\leq \frac{1}{2}b\|x_{k+2} - x_{k+1}\|^2 \\ &\leq \frac{1}{2}b(t_{k+2} - t_{k+1})^2 \\ &\leq \frac{1}{2}b\eta^2 \\ &\leq R.\end{aligned}$$

Thus, (10.1.26) holds for $n = k + 1$, which completes the induction. By Lemma 10.1.8, the sequence $\{x_n\}$ converges to $x^* \in \overline{U}(x_0, t^*)$ and $\|x^* - x_n\| \leq (2^n bd)^{-1}(2a)^{(2^n)}$. It remains to show that x^* is a solution to the generalized equation $0 \in F(x) + N_C(x)$. Let graph N_C denote the graph of N_C, that is, graph $N_C = \{(u, v) \mid v \in N_C(u)\}$. The assumptions on C imply that graph N_C is a closed subset of $X \times X$. By construction,

$$(x_{n+1}, -LF_n(x_{n+1})) \in \text{graph } N_C.$$

It follows from the continuity properties of F and F' on D that $(x_{n+1}, -LF_n(x_{n+1}))$ converges to $(x^*, -F(x^*))$. Since graph N_C is closed, $(x^*, -F(x^*))$ belongs to graph N_C, which implies $0 \in F(x^*) + N_C(x^*)$. ■

Remark 10.1.1 *Denote by $\{s_n\}$ the sequence $\{t_n\}$ when $b_0 = b$. Then our results reduce to the ones in [200]. Otherwise we have (see Chapter 5)*

$$\begin{aligned} t_n &< s_n \\ t_{n+1} - t_n &< s_{n+1} - s_n \\ t^* - t_n &\leq s^* - s_n \\ t^* &\leq s^* \\ h &< h_K = 2db\eta \\ r_J &= \frac{2}{3bd} < r_0, \end{aligned}$$

which justify the claims made in the introduction.

10.2 Quasi Methods

In this section we are concerned with the problem of approximating a strongly regular solution x^* of generalized equation

$$0 \in F(x) + N_C(x) \tag{10.2.1}$$

where F is a continuously Fréchet-differentiable operator defined on an open, convex, nonempty subset D of R^k ($k \geq 1$ an integer) with values in R^k, and

$$N_C(x) = \begin{cases} \{z \mid \langle z, q - x\rangle \leq 0 \text{ for all } q \in C\}, & \text{if } x \in C, \\ \emptyset, & \text{if } x \notin C. \end{cases}$$

We say point $x^* \in D$ satisfies the generalized equation (1) if and only if

$$x^* \in C$$

and

$$\langle F(x^*), q - x^*\rangle \geq 0 \quad \text{for all } q \in C.$$

Many problems in Mathematical Programming/Economics and engineering can be formulated as generalized equations of form (10.2.1). A survey on generalized equations and their solutions can be found in [68], [99], [298], [301], [302] and the references there.

Newton's method is undoubtedly the most popular method for approximating x^*.

However Newton's method involves the evaluation of F' at each step, which is an expensive task in general. That is why in such cases we replace $F'(x_n)$ with an affine operator $F(x_n) + B(\bullet - x_n)$ whose derivative B approximates $F'(x_n)$ and is easier to compute. The Secant approximations are such an example, and in general the popular so-called quasi-Newton methods [140], [201], [263], [298].

Newton's method for approximating x^* iteratively solves

$$0 = F(x_n) + F'(x_n)(x - x_n)$$

to find a solution x_{n+1}, whereas a quasi-Newton method also iteratively solves $0 = F(x_n) + B_n(x - x_n)$ where B_n approximates $F'(x_n)$ in some sense to be precised later. A survey on the convergence of quasi-Newton method can be found in [140], [201], [263], [298] and the references there.

Here motivated by the work in Chapter 5 we show that a larger convergence radius and finer error bounds on the distances involved can be found than before [201], [298], and under the same computational cost.

This observation is important in computational mathematics since it allows a wider choice of initial guesses x_0 and given in error tolerance $\varepsilon > 0$ finds the approximate solution using less computational steps.

We need a corollary to Theorem 2.4 of Robinson [298]:

Corollary 10.2.1 *Let F, F', D, C, R^k be as in the introduction. Assume equation (10.2.1) has a strongly regular solution $x^* \in D$ [201], [298], with associated Lipschitz constant d. Then there exist positive constants r, R, b and r_0 such that the following hold for an $k \times k$ matrix M and $\overline{x} \in R^k$.*

$$\overline{U}(x^*, r) \cap \left[F(\overline{x}) + A \cdot (\bullet - \overline{x}) + N_C\right]^{-1}$$

restricted to $\overline{U}(0, R)$ is single-valued and Lipschitz continuous with modulus $d(1 - d\|A - F'(x^*)\|)^{-1}$ provided $\overline{x} \in U(x^*, r_0)$ and $\|A - F'(x^*)\| < b$.

We can state and prove the following local convergence result for quasi-Newton methods:

Theorem 10.2.2 *Let D, C, F, F', N_C be as in the Introduction. Assume: $x^* \in D$ is a strongly regular solution of the generalized equation (1) with Lipschitz constant d; there exist positive constants ℓ_0 and ℓ such that*

$$\|F'(x) - F'(y)\| \leq \ell\|x - y\| \tag{10.2.2}$$

and

$$\|F'(x) - F'(x^*)\| \leq \ell_0\|x - x^*\| \tag{10.2.3}$$

for all $x, y \in D$. Let $N \in L(R^k, R^k)$ the space of linear operators from R^k into R^k. Let $\|\cdot\|_M$ denote a matrix norm and let $a > 0$ be such that $\|\cdot\| \leq a\|\cdot\|_M$, *where* $\|\cdot\|$ *is a matrix norm for a given vector norm on* R^k. *Moreover assume: for* $(x, B) \in D \times N$, $\overline{B} \in S(x, B)$, *where* S *is an update operator and* $\overline{x}$ *the vector closest to* x *in the set* $(F(x) + B(\bullet - x) + N_C)^{-1}(0)$ *exists, there exist positive constants* a_1 *and* a_2 *such that:*

$$\|\overline{B} - F'(x^*)\|_M \leq \left[1 + a_1 \max\{\|\overline{x} - x^*\|, \|x - x^*\|\}\right] \|B - F'(x^*)\|_M \\ + a_2 \max\{\|\overline{x} - x^*\|, \|x - x^*\|\},$$

hold. Let b, r_0, r *and* R *be as in Corollary 10.2.1. Furthermore assume: for a fixed* $c \in (0, 1)$ *and* r_0, b *reduced if necessary that the following hold:*

$$\left[\left(\frac{1}{2}\ell + \ell_0\right) r_0 + b\right] r_0 < R,$$

$$d(1 - db)^{-1} \left[\left(\frac{1}{2}\ell + \ell_0\right) r_0 + b\right] < c, \tag{10.2.4}$$

$$2a(a^{-1}ba_1 + a_2)r_0(1 - p)^{-1} < b, \tag{10.2.5}$$

$$\|B_0 - F'(x^*)\|_M < \frac{1}{2} ba^{-1},$$

$$\overline{U}(x^*, r_0) \subseteq D$$

and

$$x_0 \in U(x^*, r_0).$$

Then sequences of iterates $\{x_n\}$ *and matrices* $\{B_n\}$ *exist and satisfy for all* $n \geq 0$:

$$0 \in F'(x_n) + B_n(x_{n+1} - x_n) + N_C(x_n),$$
$$B_{n+1} \in S(x_n, B_n),$$
$$\|B_n - F'(x^*)\|_M < a^{-1}b,$$
$$\|x_{n+1} - x^*\| \leq c\|x_n - x^*\|,$$

and

$$\lim_{n \to \infty} x_n = x^*.$$

Proof. As in [201] we use the standard notation:

$$S_n = \overline{U}(x^*, r) \cap \left[F(x^*) + B_n(\bullet - x^*) + N_C\right]^{-1},$$
$$T_n = \overline{U}(x^*, r) \cap \left[F(x_n) + B_n(\bullet - x_n) + N_C\right]^{-1},$$
$$J_n = F(x^*) - F(x_n) - B_n(x^* - x_n)$$

and

$$d_0 = d(1 - db)^{-1}.$$

By hypothesis $\|B_0 - F'(x^*)\| \leq b$. It follows from Corollary 10.2.1, T_0 restricted to $\overline{U}(0, R)$ is single-valued and Lipschitz continuous such that the modulus

$$d[1 - d\|B_0 - F'(x^*)\|]^{-1} < d_0.$$

Similarly, S_0 restricted to $\overline{U}(0, R)$ is single-valued, Lipschitz continuous, with modulus d_0, and $S_0(0) = x^*$. Define $x_1 = T_0(0)$. Rewrite J_0 as

$$J_0 = F(x^*) - LF_{x_0}(x^*) + [F'(x^*) - B_0](x^* - x_0) + [F'(x_0) - F'(x^*)](x^* - x_0)$$

from which we can obtain in turn

$$\begin{aligned} \|J_0\| &\leq \|F(x^*) - LF_{x_0}(x^*)\| + \|F'(x^*) - B_0\| \, \|x^* - x_0\| \\ &\quad + \|F'(x^*) - F'(x_0)\| \, \|x^* - x_0\| \\ &\leq \frac{\ell}{2}\|x^* - x_0\|^2 + b\|x^* - x_0\| + \ell_0\|x^* - x_0\|^2 \\ &\leq \left[\left(\frac{1}{2}\ell + \ell_0\right) r_0 + b\right] \|x^* - x_0\| < R. \end{aligned}$$

It follows $J_0 \in F(x^*) + B_0(x_1 - x^*) + N_C(x_1)$, by the definition of T_0, $x_1 \in \overline{U}(x^*, r)$, and the estimate

$$0 \in F(x_0) + B_0(x_1 - x_0) + N_C(x_1) = F(x^*) + B_0(x_1 - x^*) - J_0 + N_C(x_1).$$

Hence, we deduce $x_1 \in S_0(J_0)$. It also follows from $\|J_0\| < R$ that $S_0(J_0)$ is a singleton set, and $x_1 = S_0(J_0)$. We can also obtain the estimate

$$\begin{aligned} \|x_1 - x^*\| &= \|S_0(J_0) - S_0(0)\| \leq d_0\|J_0\| \\ &\leq d_0 \left[\left(\frac{1}{2}\ell + \ell_0\right) r_0 + b\right] \|x^* - x_0\| < c\|x^* - x_0\|. \end{aligned}$$

Using induction on $i \geq 0$ we shall show:

$$\|B_i - F'(x^*)\|_M < a^{-1}b, \tag{10.2.6}$$

and

$$\|x_{i+1} - x^*\| \leq c\|x_i - x^*\| \leq c^i\|x_0 - x^*\|. \tag{10.2.7}$$

We showed above relations (10.2.6) and (10.2.7) hold for $i = 0$. Let us assume relations (10.2.6) and (10.2.7) hold for all $i \leq j - 1$ and $j \geq 1$. We shall show relations (10.2.6) and (10.2.7) hold for $i = j$. We have for all $i \leq j - 1$

$$h_i = \max\{\|x_{i+1} - x^*\|, \|x_i - x^*\|\} = \|x_i - x^*\| \quad (\text{since } c \in (0, 1))$$

and

$$\begin{aligned}\|B_{i+1} - F'(x^*)\|_M - \|B_i - F'(x^*)\|_M &\leq a_1 h_i \|B_i - F'(x^*)\|_M + a_2 h_i \\ &\leq (a_1 b a^{-1} + a_2) c^i \|x_0 - x^*\| \\ &\leq (a_1 b a^{-1} + a_2) c^i r_0.\end{aligned}$$

By taking a summation over i we get

$$\|B_j - F'(x^*)\|_M \leq \|B_0 - F'(x^*)\|_M + (a_1 b a^{-1} + a_2)(1-p)^{-1} r_0.$$

That is (10.2.6) holds. To show (10.2.7) we note

$$\|x_j - x^*\| \leq c^j \|x_0 - x^*\| \leq c^j r_0 < r_0.$$

It now follows from Corollary 1 that when T_j and S_j are single valued and Lipschitz continuous with modulus $d(1-d\|B_j - F'(x^*)\|)^{-1}$ provided they are both restricted on $\overline{U}(0,R)$. As with J_0 above we can rewrite J_j as

$$J_j = [F(x^*) - LF_{x_j}(x^*)] + [F'(x^*) - B_j](x^* - x_j) + [F'(x_j) - F'(x^*)](x^* - x_j),$$

and therefore we can get in turn

$$\begin{aligned}\|J_j\| &\leq \frac{\ell}{2}\|x_j - x^*\|^2 + b\|x_j - x^*\| + \ell_0 \|x^* - x_j\|^2 \\ &\leq \left[\left(\frac{\ell}{2} + \ell_0\right) r_0 + b\right] \|x_j - x^*\| < R.\end{aligned}$$

Moreover, $x_{j+1} \in S_j(J_j)$ and $\|J_j\| < R$, since

$$0 \in F(x_j) + B_j(x_{j+1} - x_j) + N_C(x_{j+1}) = F(x^*) + B_j(x_{j+1} - x^*) - J_j + N_C(x_{j+1}).$$

Hence, we deduce $x_{j+1} = S_j(J_j)$. Furthermore we can obtain

$$\begin{aligned}\|x_{j+1} - x^*\| &= \|S_j(J_j) - S_j(0)\| \\ &\leq d[1 - d\|B_j - F'(x^*)\|]^{-1} \|J_j\| \\ &\leq d_0 \left[\left(\frac{\ell}{2} + \ell_0\right) r_0 + b\right] \|x_j - x^*\| \\ &< c\|x_j - x^*\| \leq c^{j+1} \|x_0 - x^*\|,\end{aligned}$$

which completes the induction, and also shows $\lim_{j\to\infty} x_j = x^*$. ∎

We next show that the quasi-Newton sequence converges Q-superlinearly [140], [263]:

Theorem 10.2.3 *Let D, C, F, F', N_C, d, x^*, ℓ_0, ℓ be as in Theorem 10.2.2. Assume: $\{B_n\}$ is a bounded sequence of $k \times k$ matrices (in norm); for $x_0 \in D$ sequence $\{x_n\}$ converges to x^* and satisfies*

$$0 \in F(x_n) + B_n(x_{n+1} - x_n) + N_C(x_{n+1}) \quad \textit{for all } n \geq 0.$$

Define

$$E_n = B_n - F'(x^*), \; s_n = x_{n+1} - x_n, \; v_{n+1} = -F(x_n) - B_n s_n,$$
$$P_{n+1} = F(x_{n+1}) + v_{n+1} \quad \textit{and} \quad \Delta F = F(x_{n+1}) - F(x_n) - F'(x^*)(x_{n+1} - x_n).$$

Moreover assume

$$\lim_{n\to\infty} \frac{\|E_n s_n\|}{\|s_n\|} = 0.$$

Then sequence $\{x_n\}$ converges Q-superlinearly to x^*.

Proof. We can write

$$E_n s_n = F(x_{n+1}) - F(x_n) - F'(x^*)s_n - (v_{n+1} + F(x_{n+1})).$$

Moreover we have

$$\|\Delta F\| \leq \ell_0 \max\{\|x_{n+1} - x^*\|, \|x_n - x^*\|\} \cdot \|x_{n+1} - x_n\|.$$

Hence, we get

$$\begin{aligned}\lim_{n\to\infty} \frac{\|p_{n+1}\|}{\|s_n\|} &\leq \lim_{n\to\infty} \left(\frac{\|\Delta F\|}{\|s_n\|} + \frac{\|E_n s_n\|}{\|s_n\|} \right) \\ &\leq \ell_0 \lim_{n\to\infty} \max\{\|x_{n+1} - x^*\|, \|x_n - x^*\|\} + \lim_{n\to\infty} \frac{\|E_n s_n\|}{\|s_n\|} = 0.\end{aligned} \tag{10.2.8}$$

The rest of the proof is identical to Theorem 10.2.3 in [201] and is omitted. ∎

Remark 10.2.1 *In general*

$$\ell_0 \leq \ell \tag{10.2.9}$$

holds and $\frac{\ell}{\ell_0}$ can be arbitrarily large. If $\ell_0 = \ell$ our results reduce to the corresponding ones in [201]. Otherwise our convergence radius r_0 obtained by (10.2.4) and (10.2.5) is larger than the corresponding one $\overline{r}_0$ in [201] ($\overline{r}_0$ can also be obtained from (10.2.4) and (10.2.5) for $\ell_0 = \ell$). Moreover by (10.2.7)–(10.2.9) (for $\ell_0 < \ell$) our bounds on the distances $\|x_{n+1} - x^*\|$ $(n \geq 0)$ are finer.

All the above justify our approach whose usefullness (for regular equations) has also been explained in Section 5.1. The results extend immediately in a Banach space setting if R^k is replaced by a Banach space X along the lines of the works in Section 5.1.

10.3 Variational Inequalities under Weak Lipschitz Conditions

In this section we are concerned with the problem of approximating a locally unique solution x^* of the variational inequality of the form

$$F(x) + \partial\varphi(x) \ni 0, \tag{10.3.1}$$

where F is a Gateaux differentiable operator defined on closed set D of a Hilbert space H with values in H, $\varphi\colon H \to (-\infty, \infty]$ is a lower semicontinuous convex function with $S(\varphi) = \{\varphi \in H\colon \varphi(y) < \infty\} \neq \emptyset$, and $\partial\varphi(y)$ is the subgradient of φ at y, given by

$$\partial\varphi(y) = \{x \in H\colon \varphi(y) - \varphi(v) \leq \langle x, y - v\rangle \text{ for all } v \in S(\varphi)\}.$$

Many problems in computational sciences can be formulated as in (10.3.1) [148].

The modified Newton's method

$$F'(x_0)x_{n+1} + \partial\varphi(x_{n+1}) \ni F'(x_0)x_n - F(x_n) \tag{10.3.2}$$

as well as Newton's method

$$F'(x_n)x_{n+1} + \partial\varphi(x_{n+1}) \ni F'(x_n)x_n - F(x_n) \tag{10.3.3}$$

has already been used to approximate x^* [68], [148], [221], [235], [326].

If $\varphi = 0$ (10.3.2), and (10.3.3) reduce to the classical modified Newton and Newton method respectively.

The main hypothesis in all earlier results is a Lipschitz continuity condition on the Gateaux derivative F' of F, i.e. there exist a constant $\ell > 0$ such that for all $x, y \in D$

$$\|F'(x) - F'(y)\| \leq \ell\|x - y\|. \tag{10.3.4}$$

Here we replace (10.3.4) by the weaker center-Lipschitz condition

$$\|F'(x) - F'(x_0)\| \leq \ell_0\|x - x_0\| \quad \text{for all } x \in D. \tag{10.3.5}$$

Note that in general

$$\ell_0 \leq \ell \tag{10.3.6}$$

holds and $\frac{\ell}{\ell_0}$ can be arbitrarily large.

From the computational point of view it is easier to evaluate ℓ_0 than ℓ unless if $\ell_0 = \ell$. Therefore sufficient convergence conditions based on ℓ_0 rather than ℓ are weaker and can be used in cases not covered before.

This is our motivation for providing a semilocal convergence analysis for methods (10.3.2) and (10.3.3) based on (10.3.5) rather than (10.3.4).

Throughout this section we assume φ is a convex function.

We can now state and prove the semilocal convergence result for modified Newton's method (10.3.2):

Theorem 10.3.1 *Assume: there exist $x_0 \in H$, $a > 0$, $b > 0$, and $r \geq 0$ such that for all $y \in D$*

$$\partial\varphi(x_0) \ni 0, \tag{10.3.7}$$

$$\|F(x_0)\| \leq a, \tag{10.3.8}$$

$$\langle F'(x_0)y, y\rangle \geq b\|y\|^2, \tag{10.3.9}$$

$$2a\ell_0 < b^2, \tag{10.3.10}$$

$$\left| r - \frac{b}{\ell_0} \right| \leq r_0, \tag{10.3.11}$$

and

$$U(x_0, r) \subseteq D, \tag{10.3.12}$$

where,

$$r_0 = \frac{\sqrt{b(b - 2a\ell_0)}}{\ell_0}. \tag{10.3.13}$$

Then sequence $\{x_n\}$ $(n \geq 0)$ generated by modified Newton method (10.3.2) is well defined, remains in $U(x_0, r)$ for all $n \geq 0$ and converges to a unique solution x^* of inequality (10.3.1) in $U(x_0, r)$. Moreover the following estimates hold for all $n \geq 0$:

$$\|x_{n+2} - x_{n+1}\| \leq c\|x_{n+1} - x_n\|, \tag{10.3.14}$$

where $c \in [0, 1)$ is given by

$$c = 1 - \frac{\sqrt{b^2 - 2a\ell_0}}{b}. \tag{10.3.15}$$

Proof. Our proof is based on the contraction mapping theorem 3.1.4. Let $x \in U(x_0, r)$ and define operator $g(x)$ by the variational inequality

$$F'(x_0)g(x) + \partial\varphi(g(x)) \ni F'(x_0)x - F(x). \tag{10.3.16}$$

It follows by Theorem 3.5 in [221] and (10.3.9) that operator g is well defined on $U(x_0, r)$ (which is a closed set). The convexity of φ implies the monotonicity of $\partial\varphi$ in the sense that

$$h_1 \in \partial\varphi(y_1),\ h_2 \in \partial\varphi(y_2) \Rightarrow \langle h_1 - h_2, y_1 - y_2\rangle \geq 0. \tag{10.3.17}$$

Using (10.3.7), (10.3.16) and (10.3.17) we get

$$\langle F(x) - F'(x_0)(x - g(x)), g(x) - x_0\rangle \leq 0,$$

which can be rewritten as

$$\langle F'(x_0)(g(x)-x_0), g(x)-x_0\rangle \leq \langle F'(x_0)(x-x_0)-F(x), g(x)-x_0\rangle. \quad (10.3.18)$$

Moreover by the choice of r, (10.3.5), (10.3.8), (10.3.9) and (10.3.18) we obtain in turn

$$\begin{aligned}
\|g(x)-x_0\| &\leq \frac{1}{b}\|F(x)-F'(x_0)(x-x_0)\| \\
&= \frac{1}{b}\|F(x_0)+F(x)-F(x_0)-F'(x_0)(x-x_0)\| + \frac{1}{b}\|F(x_0)\| \\
&\quad + \frac{1}{b}\int_0^1 \|F'(x_0+t(x-x_0))-F'(x_0)\|\,\|x-x_0\| dt \\
&\leq \frac{a}{b}+\frac{\ell_0}{2b}\|x-x_0\|^2 \leq \frac{a}{b}+\frac{\ell_0}{2b}r^2 \leq r
\end{aligned} \quad (10.3.19)$$

(since $x_0+t(x-x_0)\in U(x_0,r)$ for all $t\in[0,1]$). It follows from (10.3.19) that operator g maps the closed set $U(x_0,r)$ into itself. Furthermore for any $x,y\in U(x_0,r)$, the monotonicity of $\partial\varphi$ and (10.3.2) we get

$$\langle F'(x_0)(g(x)-g(y))+F(x)-F(y)-F'(x_0)(x-y), g(x)-g(y)\rangle \leq 0,$$

which can be rewritten as

$$\begin{aligned}
&\langle F'(x_0)(g(x)-g(y)), g(x)-g(y)\rangle \\
&\leq \langle F'(x_0)(x-y)-F(x)+F(y), g(x)-g(y)\rangle.
\end{aligned} \quad (10.3.20)$$

Using (10.3.5), (10.3.8), (10.3.9), (10.3.11) and (10.3.20) we can get in turn

$$\begin{aligned}
\|g(x)-g(y)\| &\leq \frac{1}{b}\|F(x)-F(y)-F'(x_0)(x-y)\| \\
&\leq \frac{\ell_0}{b}\max\{\|x-x_0\|,\|y-x_0\|\}\|x-y\| \quad (10.3.21)\\
&\leq \frac{\ell_0 r}{b}\|x-y\| \leq c\|x-y\|,
\end{aligned}$$

which implies g is a contraction from the closed set $U(x_0,r)$ into itself. The result now follows from the contraction mapping principle. Note also that by setting $x_{n+1}=g(x_n)$ and using (10.3.21) we obtain (10.3.14). ■

We can now state and prove the semilocal convergence result for Newton's method (10.3.3):

Theorem 10.3.2 *Assume: there exist $x_0\in H$, $a>0$, $b>0$ such that for all $x,y\in D$ (10.3.7), (10.3.8), (10.3.12) holds, and*

$$\langle F'(x)y,y\rangle \geq b\|y\|^2; \quad (10.3.22)$$

Equation

$$G(p) = \left[\frac{\ell_0}{2(1-\frac{2\ell_0 p}{b})} + 1\right]\frac{a}{b} - p = 0, \tag{10.3.23}$$

has a unique positive zero $r \in \left(0, \frac{b}{2\ell_0}\right)$. *Then sequence* $\{x_n\}$ $(n \geq 0)$ *generated by Newton method (10.3.3) is well defined, remains in* $U(x_0, r)$ *for all* $n \geq 0$, *and converges to a unique solution* x^* *of inequality (10.3.1) in* $U(x_0, r)$. *Moreover the following estimates hold for all* $n \geq 0$:

$$\|x_{n+2} - x_{n+1}\| \leq q\|x_n - x_{n-1}\| \leq q^n\|x_1 - x_0\|, \tag{10.3.24}$$

and

$$\|x_n - x^*\| \leq \frac{q^n}{1-q}\frac{a}{b}, \tag{10.3.25}$$

where,

$$q = \frac{2\ell_0 r}{b}.$$

Proof. We first note that by the coercivity condition (10.3.22) and Theorem 3.5 in [221], the solutions x_{n+1} of (10.3.3) exist for all $n \geq 0$. By the monotonicity of $\partial\varphi$, (10.3.7) and (10.3.3) for $n = 1$, we get

$$\langle F(x_0) - F'(x_0)(x_0 - x_1), x_1 - x_0\rangle \leq 0,$$

or by rewriting

$$\langle F'(x_0)(x_1 - x_0), x_1 - x_0\rangle \leq \langle F(x_0), x_1 - x_0\rangle. \tag{10.3.26}$$

It follows from (10.3.8), (10.3.22) and (10.3.26) that we can obtain

$$\|x_1 - x_0\| \leq \frac{1}{b}\|F(x_0)\| \leq \frac{a}{b}. \tag{10.3.27}$$

By the monotonicity of $\partial\varphi$, (10.3.3), the corresponding equation for $n-1$ becomes

$$\begin{aligned} &\langle F'(x_n)(x_{n+1} - x_n), x_{n+1} - x_n\rangle \\ &\leq \langle F(x_n) - F(x_{n-1}) - F'(x_{n-1})(x_n - x_{n-1}, x_{n+1} - x_n\rangle. \end{aligned} \tag{10.3.28}$$

Using (10.3.5), (10.3.22) and (10.3.28) we can obtain in turn assuming $x_0, x_1, \ldots, x_n \in U(x_0, r)$

$$\|x_{n+1} - x_n\| \leq \frac{1}{b}\|F(x_n) - F(x_{n-1}) - F'(x_{n-1})(x_n - x_{n-1})\|$$

$$\leq \frac{1}{b}\left\| \int_0^1 [F'(x_{n-1}+t(x_n-x_{n-1}))-F'(x_0)](x_n-x_{n-1})dt \right\|$$

$$+\frac{1}{b}\|(F'(x_{n-1})-F'(x_0))(x_n-x_{n-1})\| \tag{10.3.29}$$

$$\leq \frac{\ell_0}{b}\left\{\int_0^1 [(1-t)\|x_{n-1}-x_0\|+t\|x_n-x_0\|]dt+\|x_{n-1}-x_0\|\right\}\|x_n-x_{n-1}\|$$

$$\leq \frac{\ell_0}{b}\left[\frac{1}{2}r+\frac{1}{2}r+r\right]\|x_n-x_{n-1}\| \leq \frac{2\ell_0 r}{b}\|x_n-x_{n-1}\| = q\|x_n-x_{n-1}\| \ (n\geq 2) \tag{10.3.30}$$

(or $\leq \frac{a\ell_0}{2b}\|x_1-x_0\|$ for $n=1$), which shows (10.3.24) for all $n\geq 0$. It follows by (10.3.30) that

$$\begin{aligned}
\|x_{n+1}-x_0\| &\leq \|x_{n+1}-x_n\|+\|x_n-x_{n-1}\|+\cdots+\|x_2-x_1\|+\|x_1-x_0\| \\
&\leq (q^{n-1}+q^{n-2}+\cdots+1)\|x_2-x_1\|+\|x_1-x_0\| \\
&\leq \frac{1-q^n}{1-q}\|x_2-x_1\|+\|x_1-x_0\| \leq \frac{\|x_2-x_1\|}{1-q}+\|x_1-x_0\| \\
&\leq \frac{a\ell_0}{2b(1-q)}+\frac{a}{b}=r
\end{aligned} \tag{10.3.31}$$

by the choice of r. That is $x_{n+1}\in U(x_0,r)$. Furthermore for all $m>0$, $n>0$ we have

$$\begin{aligned}
\|x_{n+m}-x_n\| &\leq \|x_{n+m}-x_{n+m-1}\| \\
&\qquad +\|x_{n+m-1}-x_{n+m-2}\|+\cdots+\|x_{n+1}-x_n\| \\
&\leq (q^{n+m-1}+\cdots+q^n)\|x_1-x_0\| \\
&\leq q^n(1+\cdots+q^{m-2}+q^{m-1})\|x_1-x_0\| \\
&\leq \frac{1-q^m}{1-q}q^n\frac{a}{b}.
\end{aligned} \tag{10.3.32}$$

It follows by (10.3.32) that sequence $\{x_n\}$ is Cauchy in a complete space H and as such it converges to some $x^*\in U(x_0,r)$. By letting $n\to\infty$ we obtain (10.3.25). We must show x^* solves (10.3.1). Let $y\in S(\varphi)$. Then by (10.3.3) and the lower semicontinuity of φ we get in turn

$$\varphi(x^*)\leq \lim_{n\to\infty}\inf\varphi(x_{n+1})\leq\varphi(y)+\lim_{n\to\infty}\inf\langle F'(x_n)(x_n-x_{n+1})-F(x_n),x_{n+1}-y\rangle$$

$$=\varphi(y)-\langle F(x^*),x^*-y\rangle.$$

By the definition of $\partial\varphi$ we get $-F(x^*) \in \partial\varphi(x^*)$, which implies x^* solves (10.3.1). Finally to show uniqueness let $y^* \in U(x_0, r)$ be a solution of (10.3.1). It follows by the monotonicity of $\partial\varphi$ that

$$\langle F(x^*) - F(y^*), x^* - y^* \rangle \leq 0,$$

and by the mean value theorem

$$\langle F'(tx^* + (1-t)y^*)(x^* - y^*), x^* - y^* \rangle \leq 0 \tag{10.3.33}$$

for all $t \in [0, 1]$. It then follows by (10.3.22) and (10.3.33) that

$$b\|x^* - y^*\|^2 \leq \langle F'(tx^* + (1-t)y^*)(x^* - y^*), x^* - y^* \rangle \leq 0,$$

which implies $x^* = y^*$. ∎

Remark 10.3.1 *(a) If equality holds in (10.3.6) then our Theorem 10.3.1 reduces to the corresponding Theorem 1 in [326, p. 432]. Otherwise our condition (10.3.10) is weaker than corresponding condition (10.3.13) in [326, p. 432], which simply reads*

$$2a\ell < b^2. \tag{10.3.34}$$

Note that corresponding error estimates are also smaller since $\ell_0 < \ell$. (b) Sufficient conditions for finding zeros of G can be found in [68]. For example one such set of conditions can be given by

$$r \geq (1 + \ell_0)\frac{a}{b} \tag{10.3.35}$$

provided that

$$a \leq \frac{b^2}{4\ell_0(1 + \ell_0)}. \tag{10.3.36}$$

(c) It follows by (10.3.25) and (10.3.27) that scalar $\{t_n\}$ $(n \geq 0)$ defined by

$$t_0 = 0, \ t_1 = \frac{a}{b}, \ t_{n+2} = t_{n+1} + \frac{\ell_0}{b}\left[2t_n + \frac{1}{2}(t_{n+1} - t_n)\right](t_{n+1} - t_n)$$

is a majorizing sequence for $\{x_n\}$ $(n \geq 0)$. Sufficient convergence conditions for sequences more general than $\{t_n\}$ have already been given in Section 5.1 and can replace condition (10.3.23) in Theorem 10.3.2 on such condition is: there exist $\theta \in [0, 1)$ such that for all $n \geq 1$

$$\frac{a}{b}\left\{\left[\frac{1-\theta^n}{1-\theta}\right] + \frac{1}{4}\theta^{n-1}\right\} \leq \frac{b\theta}{2\ell_0} \tag{10.3.37}$$

or the stronger but easier to verify

$$\frac{a}{b}\left[\frac{1}{1-\theta}+\frac{1}{4}\right] \le \frac{b\theta}{2\ell_0}. \tag{10.3.38}$$

In this case we have

$$\|x_{n+1}-x_n\| \le \theta(t_{n+1}-t_n), \tag{10.3.39}$$

and

$$\|x_n - x^*\| \le t^* - t_n, \quad t^* = \lim_{n\to\infty} t_n \tag{10.3.40}$$

replacing the less precise estimates (in general) (10.3.24) and (10.3.25) respectively. (d) Direct comparison between Theorem 10.3.2 and the corresponding Theorem 2 in [326, p. 434] cannot be made since we use only (10.3.5) instead of the stronger (10.3.4). Note however that condition in [326]

$$a < \frac{2b^2}{\ell} \tag{10.3.41}$$

corresponds to say (10.3.36), which implies that our condition is weaker provided that

$$\frac{\ell}{\ell_0(1+\ell_0)} > 8, \tag{10.3.42}$$

which can happen since as stated in the introduction $\frac{\ell}{\ell_0}$ *can be arbitrarily large. (e) It turns out that condition (10.3.5) can be weakened even further, and replaced as follows: Assume operator* $F'(x)$ *is simply continuous at* $x_0 \in D$. *It follows that for all* $\varepsilon > 0$, *there exists* $\delta > 0$ *such that if* $x \in U(x_0, \delta)$

$$\|F(x) - F'(x_0)\| < \varepsilon. \tag{10.3.43}$$

Then, we can have:

Case of Theorem 10.3.1. *Replace hypotheses (10.3.10) and (10.3.11) by*

$$r = \min\{\delta, r_1\}$$

where

$$r_1 = \frac{a}{b-\varepsilon}$$

provided that

$$b > \varepsilon.$$

Set $c = \frac{\varepsilon}{b}$. With the above changes the conclusions of Theorem 10.3.1 hold as it can easily be seen by the proof of this theorem (see in particular (10.3.19), where (10.3.43) is used instead of (10.3.5)). Similarly, in:

Case of Theorem 10.3.2. *Replace condition (10.3.23) by*

$$r = \min\{\delta, r_2\},$$

where,

$$r_2 = \frac{a}{b}\left[1 + \frac{\varepsilon}{1-q}\right], \quad q = \frac{2\varepsilon}{b}$$

provided that

$$b > 2\varepsilon.$$

Then again with the above changes the conclusions of Theorem 10.3.2 hold as it can easily be seen from the proof of this theorem (see in particular (10.3.29), where (10.3.43) is used instead of (10.3.5)). (f) The results obtained here can be extended from a Hilbert into a Banach space setting along the lines of our works in Section 5.1. However we leave the details of the minor modifications to the motivated reader.

10.4 Exercises

10.1. Consider the problem of approximating a locally unique solution of the variational inequality

$$F(x) + \partial\varphi(x) \ni 0, \tag{10.4.1}$$

where F is a Gâteaux differentiable operator defined on a Hilbert space H with values in H; $\varphi\colon H \to (-\infty, \infty]$ is a lower semicontinuous convex function.

We approximate solutions x^* of (10.4.1) using the generalized Newton method in the form

$$F'(x_n)(x_{n+1}) + \partial\varphi(x_{n+1}) \ni F'(x_n)(x_n) - F(x_n) \tag{10.4.2}$$

to generate a sequence $\{x_n\}$ $(n \geq 0)$ converging to x^*.
Define: the set

$$D(\varphi) = \{x \in H\colon \varphi(x) < \infty\} \quad \text{and assume } D(\varphi) \neq \phi;$$

the subgradient

$$\partial\varphi(x) = \{z \in H\colon \varphi(x) - \varphi(y) \leq \langle z, x-y\rangle, y \in D(\varphi)\};$$

and the set

$$D(\partial\varphi) = \{x \in D(\varphi)\colon \partial\varphi(x) \neq 0\}.$$

Function $\partial\varphi$ is multivalued and for any $\lambda > 0$, $(1+\lambda\partial\varphi)^{-1}$ exists (as a single valued function) and satisfies

$$\|(1+\lambda\partial\varphi)^{-1}(x) - (I+\lambda\partial\varphi)^{-1}(y)\| \le \|x-y\| \quad (x, y \in H).$$

Moreover $\partial\varphi$ is monotone:

$$f_1 \in \partial\varphi(x_1), f_2 \in \partial\varphi(x_2) \Rightarrow \langle f_1 - f_2, x_1 - x_2\rangle \ge 0.$$

Furthermore, we want $D(\bar{\varphi}) = D(\bar{\partial}\varphi)$, so that $D(\partial\varphi)$ is sufficient for our purposes.

We present the following local result for variational inequalities and twice Gâteaux differentiable operators:

(a) Let $F\colon H \to H$ be a twice Gâteaux differentiable function. Assume:

(1) variational inequality (10.4.1) has a solution x^*;

(2) there exist parameters $a \ge 0$, $b > 0$, $c > 0$ such that

$$\|F''(x) - F''(y)\| \le a\|x-y\|,$$
$$\|F''(x^*)\| \le b,$$

and

$$c\|y-z\|^2 \le \langle F'(x)(y-z), y-z\rangle$$

for all $x, y, z \in H$;

(3) $x_0 \in D(\varphi)$ and $x_0 \in U(x^*, r)$, where

$$r = 4c\left[b + \sqrt{b^2 + \tfrac{16ac}{3}}\right]^{-1}.$$

Then show: generalized Newton method (10.4.2) is well defined, remains in $U(x^*, r)$ and converges to x^* with

$$\|x_n - x^*\| \le p \cdot d^{2^n}, \quad (n \ge 0)$$

where,

$$p^{-1} = \tfrac{1}{c}\left[\tfrac{1}{3}a\|x^* - x_0\| + \tfrac{b}{2}\right] \quad \text{and} \quad d = p^{-1}\|x^* - x_0\|.$$

(b) We will approximate x^* using the generalized Newton method in the form

$$f''(x_n)(x_{n+1}) + \partial\varphi(x_{n+1}) \ni f''(x_n)(x_n) - \nabla f(x_n). \tag{10.4.3}$$

We present the following semilocal convergence result for variational inequalities involving twice Gâteaux differentiable operators. Let $f\colon H \to R$ be twice Gâteaux differentiable Assume:

(1) for $x_0 \in D(\varphi)$ there exist parameters $\alpha \geq 0$, $\beta > 0$, $c > 0$ such that

$$\langle (f'''(x) - f'''(x_0))(y, y), z \rangle \leq \alpha \|x - x_0\| \, \|y\|^2 \|z\|,$$

$$\|f'''(x_0)\| \leq \beta$$

and

$$c\|y - z\|^2 \leq \langle f''(x)(y - z), y - z \rangle$$

for all $x, y, z \in H$;

(2) the first two terms of (10.4.3) x_0 and x_1, are such that for

$$\eta \geq \|x_1 - x_0\|$$

$$\eta \leq \begin{cases} c[\beta + 2\sqrt{\alpha c}]^{-1}, & \beta^2 - 4\alpha c \neq 0 \\ c(2\beta)^{-1}, & \beta^2 - 4\alpha c = 0. \end{cases}$$

Then show: generalized Newton's method (10.4.3) is well defined, remains in $U(x_0, r_0)$ for all $\eta \geq 0$, where c_0 is the small zero of function δ,

$$\delta(r) = \alpha \eta r^2 - (c - \beta \eta) r + c\eta,$$

and converges to a unique solution x^* of inclusion $\nabla f(x) + \partial \varphi(x) \ni 0$. In particular $x^* \in \bar{U}(x_0, r_0)$. Moreover the following error bounds hold for all $n \geq 0$

$$\|x_n - x^*\| \leq \gamma d^{2^n},$$

where,

$$\gamma^{-1} = \tfrac{\alpha r_0 + \beta}{c} \quad \text{and} \quad d = \eta \gamma^{-1}.$$

10.2. Let M, $\langle \cdot, \cdot \rangle$, $\|\cdot\|$ denote the dual, inner product and norm of a Hilbert space H, respectively. Let C be a closed convex set in H. Consider an operator $a\colon H \times H \to [0, +\infty)$. If a is continuous bilinear and satisfies

$$a(x, y) \geq c_0 \|y\|^2, \quad y \in H, \tag{10.4.4}$$

and

$$a(x, y) \leq c_1 \|x\| \cdot \|y\|, \quad x, y \in H, \tag{10.4.5}$$

for some constants $c_0 > 0$, $c_1 > 0$ then a is called a coercive operator.

Given $z \in M$, there exists a unique solution $x \in C$ such that:

$$a(x, x - y) \geq \langle z, x - y \rangle, \quad y \in C. \tag{10.4.6}$$

Inequality (10.4.6) is called variational. It is well known that x^* can be obtained by the iterative procedure

$$x_{n+1} = P_C(x_n - \rho F(G(x_n) - z)), \tag{10.4.7}$$

where P_C is a projection of H into C, $\rho > 0$ is a constant, F is a canonical isomorphism from M onto H, defined by

$$\langle z, y\rangle = \langle F(z), y\rangle, \quad y \in H,\ z \in M, \tag{10.4.8}$$

and

$$a(x, y) = \langle G(x), y\rangle, \quad y \in H. \tag{10.4.9}$$

Given a point-to-set operator C from H into M we define the quasivariational inequality problem to be: find $x \in C(x)$ such that:

$$a(x, y - x) \geq \langle z, y - x\rangle \quad y \in C(x). \tag{10.4.10}$$

Here, we consider $C(x)$ to be of the form

$$C(x) = f(x) + C, \tag{10.4.11}$$

where f is a point-to-point operator satisfying

$$\|f(x^*) - f(y)\| \leq c_2\|x^* - y\|^\lambda \tag{10.4.12}$$

for some constants $c_2 \geq 0$, $\lambda \geq 1$, all $y \in H$ and x^* a solution of (10.4.10). We will compute the approximate solution to (10.4.10).

(a) Show: For fixed $z \in H$, $x \in C$ satisfies

$$\langle x - z, y - x\rangle \geq 0 \quad y \in C \tag{10.4.13}$$

$$\Leftrightarrow\ x = P_C(z), \tag{10.4.14}$$

where P_C is the projection of H into C.

(b) P_C given by (10.4.14) is non-expansive, that is

$$\|P_C(x) - P_C(y)\| \leq \|x - y\|, \quad x, y \in H. \tag{10.4.15}$$

(c) For C given by (10.4.11), $x \in C(x)$ satisfies (10.4.10) $\Leftrightarrow$

$$x = f(x) + P_C(x - \rho F(G(x) - z)). \tag{10.4.16}$$

Result (c) suggests the iterative procedure

$$x_{n+1} = f(x_n) + P_C\big(x_n - \rho F(G_n) - z) - f(x_n)\big) \tag{10.4.17}$$

for approximating solutions of (10.4.10).

Let us define the expression

$$\theta = \theta(\lambda, \rho) = 2c_2\|x_0 - x\|^{\lambda - 1} + \sqrt{1 + \rho^2 c_1^2 - 2c_0\rho}\,.$$

It is simple algebra to show that $\theta \in [0, 1)$ in the following cases:

(1) $\lambda = 1$, $c_2 \le \frac{1}{2}$, $c_0 \ge 2c_1\sqrt{c_2(1-c_2)}$, $0 < \rho < \frac{c_0+\sqrt{c_0^2-4c_1^2c_2(1-c_2)}}{c_1^2}$;

(2) $\lambda > 1$, $c_0 \le c_1$, $\|x_0 - x\| < \left[\frac{1}{2c_2}\left(1+\sqrt{1-\left(\frac{c_0}{c_1}\right)^2}\right)\right]^{1/\lambda-1} = c_3$,

$$0 < \rho < \frac{c_0+\sqrt{c_0^2+qc_1^2}}{c_1^2}, \quad q = \left(1 - 2c_2\|x_0 - x\|^{\lambda-1}\right)^2 - 1;$$

(3) $\lambda > 1$, $c_0 > c_1$, $\|x_0 - x\| \le \left(\frac{1}{2c_2}\right)^{1/\lambda-1} = c_4$, $0 < \rho < \frac{c_0+\sqrt{c_0^2+qc_1^2}}{c_1^2}$.

Denote by H_0, H_1 the sets

$$H_0 = \{y \in H \mid \|y - x^*\| \le c_3\} \quad \text{and} \quad H_1 = \{y \in H \mid \|y - x^*\| \le c_4\}.$$

(d) Let operator f satisfy (10.4.12) and C be a non-empty closed convex subset of H. If $a(x, y)$ is a coercive, continuous bilinear operator on H, x^* and x_{n+1} are solutions of (10.4.10) and (10.4.17) respectively, then x_{n+1} converges to x^* strongly in H if (1) or (2) or (3) above hold.

It follows from (d) that a solution x^* of (10.4.10) can be approximated by the iterative procedure

(1) $x^* \in C(x^*)$ is given,

(2) $x_{n+1} = f(x_n) + P_C\big(x_n - \rho F(G(x_n) - z) - f(x_n)\big)$,
where ρ, x_0 are as in (1) or (2) or (3).

If $\lambda = 1$ our result (d) reduces to Theorem 3.2 in [337] (provided that (10.4.12) is replaced by $\|f(x) - f(y)\| \le c_2^1\|x - y\|$ for all $x, y \in H$). Note also that since $c_2^1 \ge c_2$ in general our error bounds on the distances $\|x_n - x^*\|$ $(n \ge 0)$ are smaller. Moreover, if $C(x)$ is independent of x, then $f = 0$ and $c_2 = 0$, in which case (c) and (d) reduce to the ones in [336].

10.3. Let $x_0 \in D$ and $R > 0$ be such that $D \equiv U(x_0, R)$. Suppose that f is m-times Fréchet-differentiable on D, and its mth derivative $f^{(m)}$ is in a certain sense uniformly continuous:

$$\|f^{(m)}(x) - f^{(m)}(x_0)\| \le w(\|x - x_0\|), \quad \text{for all } x \in D, \tag{10.4.18}$$

for some monotonically increasing positive function w satisfying

$$\lim_{t\to\infty} w(r) = 0, \tag{10.4.19}$$

or, even more generally, that

$$\|f^{(m)}(x) - f^{(m)}(x_0)\| \le w(r, \|x - x_0\|), \quad \text{for all } x \in \bar{D},\ r \in (0, R), \tag{10.4.20}$$

for some monotonically increasing in both variables positive function w satisfying

$$\lim_{t\to 0} w(r,t) = 0, \quad r \in [0,R]. \tag{10.4.21}$$

Let us define function θ on $[0,R]$ by

$$\theta(r) = \tfrac{1}{c+\alpha}\Big[\eta + \tfrac{\alpha_m}{m!}r^m + \cdots + \tfrac{\alpha_2}{2!}r^2 + \int_0^{v_{m-2}} \cdots \int_0^{v_1} w(v_{m-1})(r - v_1)dv_1 \cdots dv_{m-1}\Big] - r \tag{10.4.22}$$

for some constants α, c, η, α_i, $i = 2, \ldots, m$; the equation,

$$\theta(r) = 0; \tag{10.4.23}$$

and the scalar iteration $\{r_n\}$ $(n \geq 0)$ by

$$r_0 = 0, \quad r_{n+1} = r_n - \tfrac{\theta(r_n)}{\theta'(r_n)}. \tag{10.4.24}$$

Let g be a maximal monotone operator satisfying $L(z)+g(z) \ni y$, and suppose: (10.4.18) holds, there exist α_i $(i = 2, \ldots, m)$ such that

$$\|F^{(i)}(x_0)\| \leq \alpha_i, \tag{10.4.25}$$

and equation (10.4.23) has a unique $r^* \in [0,R]$ and $\theta(R) \leq 0$.

Then show: the generalized Newton's method $\{x_n\}$ $(n \geq 0)$ generated by

$$f'(x_n)x_{n+1} + g(x_{n+1}) \ni f'(x_n)(x_n) - f(x_n) \quad (n \geq 0), \ (x_0 \in D)$$

is well defined, remains in $V(x_0, r^*)$ for all $n \geq 0$, and converges to a solution x^* of

$$f(x) + g(x) \ni x.$$

Moreover, the following estimates hold for all $n \geq 0$:

$$\|x_{n+1} - x_n\| \leq r_{n+1} - r_n, \tag{10.4.26}$$

and

$$\|x_n - x^*\| \leq r^* - r_n, \quad r^* = \lim_{n\to\infty} r_n. \tag{10.4.27}$$

10.4. Let $x_0 \in D$ and $R > 0$ be such that $D \equiv U(x_0, R)$. Suppose that f is Fréchet-differentiable on D, and its derivative f' is in a certain sense uniformly continuous as an operator from D into $L(H,H)$; the space of linear operators from H into H. In particular we assume:

$$\|f'(x) - f'(y)\| \leq w(\|x-y\|), \quad x, y \in D, \tag{10.4.28}$$

for some monotonically increasing positive function w satisfying

$$\lim_{t\to\infty} w(r) = 0,$$

or, even more generally, that

$$\|f'(x) - f'(y)\| \leq w(r, \|x-y\|), \quad x, y \in \bar{D},\ r \in (0,R), \tag{10.4.29}$$

for some monotonically increasing in both variables positive function w satisfying

$$\lim_{t\to 0} w(r,t) = 0, \quad r \in [0,R].$$

Conditions of this type have been studied in the special cases $w(t) = dt^{\lambda}$, $w(r,t) = d(r)t^{\lambda}$, $\lambda \in [0,1]$ for regular equations; and for $w(t) = dt$ for generalized equations of the form $f(x)+g(x) \ni x$. The advantages of using (10.4.28) or (10.4.29) have been explained in great detail in the excellent paper [6]. It is useful to pass from the function w to

$$\bar{w}(r) = \sup\{w(t) + w(s) : t + s = r\}.$$

The function may be calculated explicitly in some cases. For example if $w(r) = dr^{\lambda}$ $(0 < \lambda \leq 1)$, then $\bar{w}(r) = 2^{1-\lambda}dr^{\lambda}$. More generally, if w is a concave function on $[0,R]$, then $\bar{w}(r) = 2w\left(\frac{r}{2}\right)$, and $\bar{w}$ is increasing, convex and $\bar{w}(r) \geq w(r)$, $r \in [0,R]$.

Let us define the functions $\theta, \bar{\theta}$ on $[0,R]$ by

$$\theta(r) = \tfrac{1}{c+\alpha}\left[\eta + \int_0^r w(t)dt\right] - r, \quad \text{for some } \alpha > 0, \eta > 0, c > 0$$

$$\bar{\theta}(r) = \tfrac{1}{c+\alpha}\left[\eta + \int_0^r \bar{w}(t)dt\right] - r$$

and the equations

$$\theta(r) = 0,$$
$$\bar{\theta}(r) = 0.$$

Let g be a maximal monotone operator satisfying

there exists $c > -\alpha$ such that $\langle f'(z)(x), x\rangle \geq c\|x\|^2$, for all $x \in H, z \in D$, and suppose: (10.4.28) holds and equation $\bar{\theta}(r) = 0$ has a unique solution $r^* \in [0, R]$.

Then, show: the generalized Newton method $\{x_n\}$ $(n \geq 0)$ generated by

$$f'(x_n)x_{n+1} + g(x_{n+1}) \ni f'(x_n)(x_n) - f(x_n) \quad (n \geq 0), \ (x_0 \in D)$$

is well defined, remains in $U(x_0, r^*)$ for all $n \geq 0$, and converges to a solution x^* of $f(x) + g(x) \ni x$. Moreover, the following estimates hold for all $n \geq 0$:

$$\|x_{n+1} - x_n\| \leq r_{n+1} - r_n$$
$$\|x_n - x^*\| \leq r^* - r_n,$$

where,

$$r_0 = 0, \ r_{n+1} = r_n - \frac{\bar{\theta}(r_n)}{\theta'(r_n)},$$

and

$$\lim_{n \to \infty} r_n = r^*.$$

Chapter 11

Monotone convergence of point to set-mapping

In this chapter we discuss the monotonicity of the iteration sequence generated by the algorithmic model

$$x_{k+1} \in f_k(x_0, x_1, ..., x_k) \qquad (k \geq l-1). \tag{11.0.1}$$

Here, we assume again that for $k \geq l-1$, $f_k : X^{k+1} \to 2^X$, where X is a subset of a partially ordered space L, and let $\leq$ denote the order in L, (see also Chapter 2, [97]–[101])

11.1 General Theorems

Our main results will be based on a certain type of isotony of the algorithmic mappings.

Definition 11.1.1 *The sequence of point-to-set mapping $g_k : X^{k+1} \to 2^L$ is called increasingly isotone from the left on X, if for any sequence $\{x^{(k)}\}$ such that $x^{(k)} \in X$ $(k \geq 0)$ and $x^{(0)} \leq x^{(1)} \leq x^{(2)} \leq ...$, $y_1 \leq y_2$ for all $y_1 \in g_k(x^{(0)}, x^{(1)}, ..., x^{(k)})$, $y_2 \in g_{k+1}(x^{(0)}, x^{(1)}, ..., x^{(k+1)})$ and $k \geq l-1$.*

Consider the special case of single variable mappings g_k. The above properties are now reduced to the following. Assume that $x^{(0)}, x^{(1)} \in X$, $x^{(0)} \leq x^{(1)}$, $y_1 \in g_k(x^{(0)})$, and $y_2 \in g_{k+1}(x^{(1)})$, then $y_1 \leq y_2$.

Similarly, the sequence of point-to-set mappings $g_k : X^{k+1} \to 2^L$ is called increasingly isotone from the right on X, if for any sequence $\{x^{(k)}\}$ such that $x^{(k)} \in X$ $(k \geq 0)$ and $x^{(0)} \leq x^{(1)} \leq x^{(2)} \leq ...$, $y_1 \geq y_2$ for all $y_1 \in g_k(x^{(0)}, x^{(1)}, ..., x^{(k)})$, $y_2 \in g_{k+1}(x^{(0)}, x^{(1)}, ..., x^{(k+1)})$ and $k \geq l-1$.

Our main result is the following:

Theorem 11.1.2 *Assume that the sequence of mappings f_k is increasingly isotone from the left; futhermore, $x_0 \le x_1 \le ... \le x_{l-1} \le x_l$ ($x_k \in X$, $0 \le k \le l-1$). Then for all $k \ge 0$, $x_{k+1} \ge x_k$.*

Proof. By induction, assume that for all indices i ($i < k$), $x_{i+1} \ge x_i$. The relations

$$x_k \in f_{k-1}(x_0 \le x_1 \le ... \le x_{k-1}) \qquad \text{and} \qquad x_{k+1} \in f_k(x_0 \le x_1 \le ... \le x_k)$$

and Definition 11.1.1 imply that $x_{k+1} \ge x_k$. Because this inequality holds for $k \le l-1$, the proof is complete. ■

Consider next the modified algorithmic model

$$y_{k+1} \in f_k(y_k, y_{k-1}, ..., y_1, y_0), \tag{11.1.1}$$

which differs from process (11.0.1) only in the order of the variables in the right-hand side. Using finite induction again similarly to Theorem 11.1.2, one may easily prove the following theorem.

Theorem 11.1.3 *Assume that the sequence of mappings f_k is increasingly isotone from the right; furthermore, $y_0 \ge y_1 \ge ... \ge y_{l-1} \ge y_l$ ($y_k \in X$, $0 \le k \le l-1$). The for all $k \ge 0$, $y_{k+1} \le y_k$.*

Corollary 11.1.4 *Assume that $X \subseteq \mathbb{R}^n$; furthermore, $x_k \to s^*$ and $y_k \to s^*$ as $k \to \infty$. It is also assumed that " $\le$ " is the usual partial order of vectors; i.e., $a = (a^{(i)}) \le b = (b^{(i)})$ if and only if for all i, $a^{(i)} \le b^{(i)}$. Under the conditions of Theorems 11.1.2 and 11.1.3,*

$$x_k \le s^* \le y_k.$$

This relation is very useful in constructing an error estimator and a stopping rule for methods (11.0.1) and (11.1.1), since for all coordinates of vectors x_k, y_k and s^*,

$$0 \le s^{*(i)} - x_k^{(i)} \le y_k^{(i)} - x_k^{(i)}$$

and

$$0 \le y_k^{(i)} - s^{*(i)} \le y_k^{(i)} - x_k^{(i)}.$$

Hence, if each element of $y_k - x_k$ is less than an error tolerance e, then all elements of x_k and y_k are closer than e to the corresponding elements of the limit vector.

11.2 Convergence in linear spaces

Assume now that X is a subset of a partially ordered linear space L. In addition, the partial order satisfies the following conditions:

(F1). $x \le y$ $(x, y \in L)$ implies that $-x \ge y$;

(F2). If $x^{(1)}, x^{(2)}, y^{(1)}, y^{(2)} \in L$, $x^{(1)} \le y^{(1)}$ and $x^{(2)} \le y^{(2)}$, then $x^{(1)} + x^{(2)} \le y^{(1)} + y^{(2)}$.

Assume that for $k \ge l-1$ and $p = 1, 2$, mapping $K_k^{(p)}$, $H_k^{(p)}$ are point-to-set mapping defined on X^{k+1}, and for all $(x^{(1)}, x^{(2)}, ..., x^{(k+1)}) \in X^{k+1}$, $K_k^{(p)}(x^{(1)}, x^{(2)}, ..., x^{(k+1)})$, and $H_k^{(p)}(x^{(1)}, x^{(2)}, ..., x^{(k+1)})$ are nonempty subsets of L; i.e., $K_k^{(p)}$, and $H_k^{(p)} : X^{k+1} \to 2^L$. Consider next the algorithmic model:

$$x_{k+1} = t_{k+1}^{(1)} - s_{k+1}^{(1)} \qquad \text{and} \qquad y_{k+1} = t_{k+1}^{(2)} - s_{k+1}^{(2)}, \tag{11.2.1}$$

where

$$t_{k+1}^{(1)} \in K_k^{(1)}(x_0, x_1, ..., x_k), \;\; s_{k+1}^{(1)} \in H_k^{(1)}(y_k, y_{k-1}, ..., y_0)$$

and

$$t_{k+1}^{(2)} \in K_k^{(2)}(y_k, y_{k-1}, ..., y_0), \;\; s_{k+1}^{(1)} \in H_k^{(2)}(x_0, x_1, ..., x_k).$$

In the formulation of our next theorem we will need what follows.

Definition 11.2.1 *A point-to-set mapping $g : X^{k+1} \to 2^L$ is called increasingly isotone on X, if for all $x^{(1)} \le x^{(2)} \le ... \le x^{(k+2)}$ $(x^{(i)} \in X,\ i = 1, 2, ..., k+2)$, relations $y_1 \in g(x^{(1)}, x^{(2)}, ..., x^{(k+1)})$ and $y_2 \in g(x^{(2)}, x^{(3)}, ..., x^{(k+2)}$ imply that $y_1 \le y_2$.*

Increasingly isotone functions have the following property.

Lemma 11.2.2 *Assume that mapping $g : X^{k+1} \to 2^L$ is increasingly isotone. Then, for all $x^{(i)}$ and $y^{(i)} \in X$ $(i = 1, 2, ..., k+1)$ such that*

$$x^{(1)} \le x^{(2)} \le ..., x^{(k+1)} \le y^{(1)} \le y^{(2)} \le ... \le y^{(k+1)}$$

and for any

$$x \in g(x^{(1)}, x^{(2)}, ..., x^{(k+1)}) \;\; \text{and} \;\; y \in g(y^{(1)}, y^{(2)}, ..., y^{(k+1)}),\ x \le y.$$

Proof. Let $y_i \in g(x^{(i+1)}, ..., x^{(k+1)}, y^{(1)}, ..., y^{(i)})$ be arbitrary for $i = 1, 2, ..., k$. Then

$$x \le y_1 \le y_2 \le ... \le y_k \le y,$$

which completes the proof. ∎

Remark 11.2.1 *In the literature a point-to-set mapping $g : X^{k+1} \to 2^L$ is called isotone on X, if for all $x^{(i)}$ and $y^{(i)}$ such that $x^{(i)} \le y^{(i)}$ $(i = 1, 2, ..., k+1)$ and for all $x \in g(x^{(1)}, ..., x^{(k+1)})$ and $y \in g(y^{(1)}, ..., y^{(k+1)})$, $x \le y$. It is obvious that an isotone mapping is increasingly isotone, but the reverse is not necessarily true as the following example illustrates.*

Example 11.2.1 *Define $L = \mathbb{R}^1$, $X = [0,1]$, $k = 1$, $\leq$ to be the usual order of real numbers, and*

$$g(x^{(1)}, x^{(2)}) = \begin{cases} x^{(2)}, & \textit{if } x^{(1)} \geq 2x^{(2)} - 1 \\ x^{(1)} - x^{(2)} + 1, & \textit{if } x^{(1)} < 2x^{(2)} - 1. \end{cases}$$

We will now verify that g is increasingly isotone, but not isotone on X. Select arbitrary $x^{(1)} \leq x^{(2)} \leq x^{(3)}$ from the unit interval. Note first that $g(x^{(1)}, x^{(2)}) \leq x^{(2)}$, since if $x^{(1)} \geq 2x^{(2)} - 1$, then $g(x^{(1)}, x^{(2)}) = x^{(2)}$, and if $x^{(1)} < 2x^{(2)} - 1$, then $g(x^{(1)}, x^{(2)}) = x^{(1)} - x^{(2)} + 1 < 2x^{(2)} - 1 - x^{(2)} + 1 = x^{(2)}$. Note next that $g(x^{(2)}, x^{(3)}) \geq x^{(2)}$, since if $x^{(2)} \geq 2x^{(3)} - 1$, then $g(x^{(2)}, x^{(3)}) = x^{(3)} \geq x^{(2)}$, and if $x^{(2)} < 2x^{(3)} - 1$, then $g(x^{(2)}, x^{(3)}) = x^{(2)} - x^{(3)} + 1 \geq x^{(2)}$. Consequently, $g(x^{(1)}, x^{(2)}) \leq x^{(2)} \leq g(x^{(2)}, x^{(3)})$. Hence, g is increasingly isotone; however, it is not isotone on X, since for any points $(t, 1)$ and $(t, 1-e)$ (where $t, e > 0$ and $t + 2e < 1$),

$$g(t,1) = t - 1 + 1 = t < g(t, 1-e) = t - (1-e) + 1 = t + e,$$

but $1 > 1 - e$.

Consider now the following assumptions:

(G_1) Sequences of mapping $K_k^{(1)}$ and $H_k^{(2)}$ are increasingly isotone from the left. Futhermore, $K_k^{(1)}$ and $H_k^{(2)}$ are increasingly isotone from the right.

(G_2) For all k and $t^{(0)} \leq t^{(1)} \leq ... \leq t^{(k)} \leq \bar{t}^{(k)} \leq ... \leq \bar{t}^{(1)} \leq \bar{t}^{(0)}$ ($t^{(i)}$ and $\bar{t}^{(i)} \in X$, $i = 0, 1, ..., k$),

$$\leqq K_k^{(2)}(\bar{t}^{(k)}, ..., \bar{t}^{(0)})$$

and

$$H_k^{(1)}(\bar{t}^{(k)}, ..., \bar{t}^{(0)}) \geqq H_k^{(2)}(t^{(0)}, t^{(1)}, ..., t^{(k)}).$$

(G_3) The initial points are selected so that

$$x_0 \leq x_1 \leq ... \leq x_{l-1} \leq x_l \leq y_l \leq y_{l-1} \leq ... \leq y_1 \leq y_0;$$

(G_4) $\langle x_0, y_0 \rangle = \{x \mid x \in L,\, x_0 \leq x \leq y_0\} \subseteq X$.

Theorem 11.2.3 *Under assumtions (F_1), (F_2) and (G_1) through (G_4), x_k and y_k are in X for all $k \geq 0$. Furthermore,*

$$x_k \leq x_{k+1} \leq y_{k+1} \leq y_k.$$

Proof. By induction, assume that $y_0 \geqq x_{i+1} \geq x_i$ and $x_0 \leqq y_{i+1} \leq y_i$ for all $i < k$. Therefore, points x_i and y_i $(i = 0, 1, ..., k)$ are in X. Note next that assumption (G_1) implies that

$$t_{k+1}^{(1)} \geq t_k^{(1)},\ t_{k+1}^{(2)} \leq t_k^{(2)},\ s_{k+1}^{(1)} \leq s_k^{(1)} \text{ and } s_{k+1}^{(2)} \geq s_k^{(2)},$$

and hence,

$$x_{k+1} = t_{k+1}^{(1)} - s_{k+1}^{(1)} \geq t_k^{(1)} - s_k^{(1)} = x_k$$

and

$$y_{k+1} = t_{k+1}^{(2)} - s_{k+1}^{(2)} \leq t_k^{(2)} - s_k^{(2)} = y_k.$$

Assume next that $x_i \leq y_i$ $(i \leq k)$. Then, assumption (G_2) implies that

$$t_{k+1}^{(1)} \leq t_{k+1}^{(2)} \text{ and } s_{k+1}^{(1)} \geq s_{k+1}^{(2)}.$$

Therefore,

$$x_{k+1} = t_{k+1}^{(1)} - s_{k+1}^{(1)} \leq t_{k+1}^{(2)} - s_{k+1}^{(2)} = y_{k+1}.$$

Thus, the proof is complete. ■

Introduce next the point-to-set mappings

$$f_k^{(p)}(x^{(1)}, ..., x^{(k+1)}) = \{t - s \mid t \in K_k^{(p)}(x^{(1)}, ..., x^{(k+1)}),\ s \in H_k^{(p)}(x^{(1)}, ..., x^{(k+1)})\}.$$

Assume that z is a common fixed point of mappings $f_k^{(1)}$ and $f_k^{(2)}$; i.e., for $p = 1, 2$, $z \in f_k^{(p)}(z, z, ..., z)$, furthermore (G_5). Mappings $K_k^{(1)} K_k^{(2)} H_k^{(1)} H_k^{(2)}$ are increasingly isotone.

Theorem 11.2.4 *Assume that $x_1 \leq z \leq y_1$. Then, under assumptions (F_1), (F_2) and (G_1) through (G_5),*

$$x_k \leq z \leq y_k \textit{ for all } k \geq 0.$$

Proof. By induction, assume that $x_i \leq z \leq y_i$, for $i \leq k$. Then, for $p = 1, 2$ and for all $t^{(p)} \in K_k^{(p)}(z, z, ..., z)$ and $s^{(p)} \in H_k^{(p)}(z, z, ..., z)$,

$$t_{k+1}^{(1)} \leq t^{(1)},\ t^{(2)} \leq t_{k+1}^{(2)} \quad \text{and} \quad s_{k+1}^{(1)} \geq s^{(1)},\ s^{(2)} \geq s_{k+1}^{(2)}.$$

Therefore,

$$x_{k+1} = t_{k+1}^{(1)} - s_{k+1}^{(1)} \leq t^{(p)} - s^{(p)} \leq t_{k+1}^{(2)} - s_{k+1}^{(2)} = y_{k+1}, \quad (p = 1, 2)$$

and since $z = t - s$ with some t and s, the proof is completed. ■

Remark 11.2.2 *Two important special cases of algorithm (11.2.1) can be given:*

1 *If* $K_k^{(1)} = K_k^{(2)} = K_k$ *and* $H_k^{(1)} = H_k^{(2)} = H_k$*, then* $f_k^{(1)} = f_k^{(2)} = f_k$*, and so* z *is a common fixed point of the mapping* f_k*.*

2 *Assume next that* K_k *and* H_k *are* $l-$*variable functions. Then one may select*

$$K_k^{(1)}(t^{(0)}, ..., t^{(k)}) = K_k(t^{(k-l+1)}, ..., t^{(k)}),$$
$$K_k^{(2)}(t^{(0)}, ..., t^{(k)}) = K_k(t^{(0)}, ..., t^{(l-1)}),$$
$$H_k^{(1)}(t^{(0)}, ..., t^{(k)}) = H_k(t^{(0)}, ..., t^{(l-1)})$$

and

$$H_k^{(2)}(t^{(0)}, ..., t^{(k)}) = H_k(t^{(k-l+1)}, ..., t^{(k)}).$$

Note that the resulting process for $l = 1$ was discussed in Ortega and Rheinboldt ([263], Section 13.2).

Corollary 11.2.5 *Assume that in addition* L *is a Hausdorff topological space which satisfies the first axiom of countabi;ity, and*

(F_3) *If* $x_k \in L$ $(k \geq 0)$ *such that* $x_k \leq x$ *with some* $x \in L$ *and* $x_k \leq x_{k+1}$ *for all* $k \geq 0$*, then the sequence* $\{x_k\}$ *converges to a limit point* $x^* \in L$*; furthermore,* $x^* \leq x$*.*

(F_4) *If* $x_k \in L$ $(k \geqq 0)$ *and* $x_k \to x$*, then* $-x_k \to -x$*.*

Note first that assumptions (F_1) through (F_4) imply that any sequence $\{y_k\}$ with the properties $y_k \in L$, $y_k \geq y_{k+1}$, and $y_k \geq y$ $(k \geq 0)$ is convergent to a limit point $y^* \in L$; furthermore, $y^* \geq y$. Since under the conditions of the theorem $\{x_k\}$ is increasing with an upper bound y_0 and $\{y_k\}$ decreases with a lower bound x_0, both sequences are convergent. If x^* and y^* denote the limit points and z is a common fixed point of mappings f_k such that $x_l \leq z \leq y_l$ then $x^* \leq z \leq y^*$.

Consider the special $l-$step algorithm given as special case 2 above. Assume that for all $k \geq l - 1$, mappings K_k and H_k are closed. Then, x^* is the smallest fixed point and y^* is the largest fixed point in $\langle x_l, y_l \rangle$.

As a further special case we mention the following lemma.

Lemma 11.2.6 *Let* B *be a partially ordered topological space and let* x. y *be two points of* B *such that* $x \leq y$*. If* $f : \langle x, y \rangle \to B$ *is a continuous isotone mappings having the property that* $x \leq f(x)$ *and* $y \geq f(y)$*, then a point* $z \in \langle x, y \rangle$ *exists such that* $z = f(z)$*.*

Proof. Select $X = \langle x, y \rangle$, $l = 1$, $H_k = 0$ and $K_k = f$, and consider the iteration sequences $\{x_k\}$ and $\{y_k\}$ starting from $x_0 = x$ and $y_0 = y$. Then, both sequences are convergent, and the limits are fixed points. ■

Remark 11.2.3 *This result is known as Theorem 2.2.2, (see, Chapter 2).*

11.3 Applications

Consider the simple iteration method

$$x_{k+!} = f(x_k), \tag{11.3.1}$$

where $f : [a, b] \to \mathbb{R}$ with the additional properties:

$$f(a) > a,\ f(b) < b,\ 0 \leq f'(x) \leq q < 1 \quad \text{for all} \times \in [a, b],$$

where q is a given constant. Consider the iteration sequences

$$x_{k+1} = f(x_k), \qquad x_0 = a$$

and

$$y_{k+1} = f(y_k), \qquad y_0 = b \tag{11.3.2}$$

Since f is increasingly, Theorems 11.1.2 and 11.1.3 apply to both sequences $\{x_k\}$ and $\{y_k\}$ that converge to the unique fixed point s^*; furthermore, sequence $\{x_k\}$ is increasing and $\{y_k\}$ decreases. This situation is illustrated in Figure 11.3.1.

Next we drop the assumption $f'(x) \leqq q < 1$, but retain all other conditions. In other words, f is continuous, increasing in $[a, b]$,

$$f(a) > a \qquad f(b) < b.$$

Notice that Theorems 11.1.2 and 11.1.3 still can be applied for the iteration sequence (11.3.2). However, the uniqueness of the fixed point is no longer true. This situation is shown in Figure 11.3.2.

Because$f$ is continuous, there is at least one fixed point in the interval (a, b). Alternatively, Theorems 11.2.3 and 11.2.4 may also be applied with $K_k^{(1)} = K_k^{(2)} \equiv f$ and $H_k^{(1)} = H_k^{(2)} = 0$, which has the following additional consequence. Let s^* and s^{**} denote the smallest and the largest fixed points between a and b. Because $a < s^*$ and $b > s^{**}$, $f(a) \leqq f(s^*) = s^*$ and $f(b) \geqq f(s^{**}) = s^{**}$, thus, s^* and s^{**} are between x_1 and y_1. Therefore, for all $k \geqq 1$, $x_k \leqq s^* \leqq s^{**} \leqq y_k$; futhermore, $x_k \to s^*$ and $y_k \to s^{**}$. If the fixed point is unique, the uniqueness of the fixed point can be estabilished by showing that sequences $\{x_k\}$ and $\{y_k\}$ have the same limit.

Finally, we mention that the above facts can be easily extended to the more general case when $f : \mathbb{R}^n \to \mathbb{R}^n$.

Example 11.3.1 *We now illustrate procedure (11.3.2) in the case of the single variable nonlinear equation*

$$x = \frac{1}{10}e^x + 1.$$

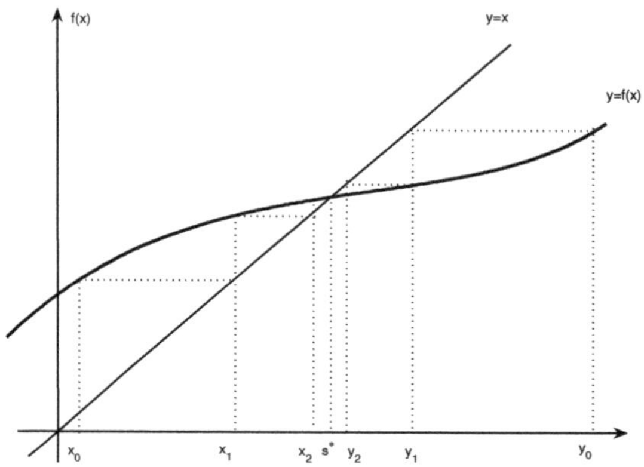

Figure 11.1: Monotone fixed point iterations.

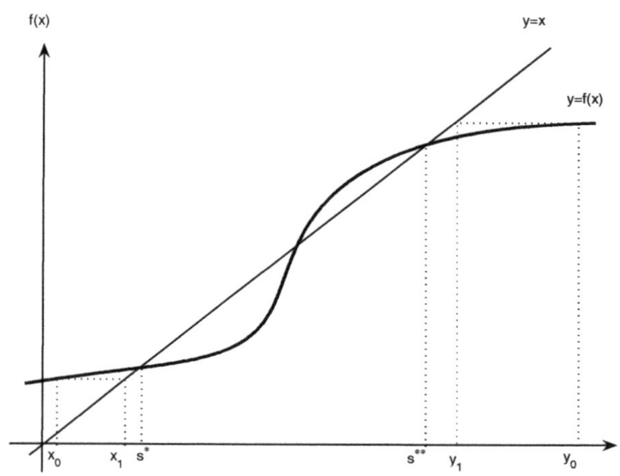

Figure 11.2: The case of multiple fixed points

It is easy to see that

$$\frac{1}{10}e^1 + 1 > 1 \qquad \textit{and} \qquad \frac{1}{10}e^2 + 1 < 2.$$

Further,

$$\frac{d}{dx}\left(\frac{1}{10}e^x + 1\right) = \frac{1}{10}e^x \in (0, 0.75).$$

Therefore, all conditions are satisfied. Select $x_0 = 1$ *and* $y_0 = 2$. *Then,*

$$\begin{aligned}
x_1 &= \frac{1}{10}e^1 + 1 \approx 1.27183; & y_1 &= \frac{1}{10}e^2 + 1 \approx 1.73891;\\
x_2 &= \frac{1}{10}e^{1.27183} + 1 \approx 1.35674; & y_2 &= \frac{1}{10}e^{1.73891} + 1 \approx 1.56911;\\
x_3 &= \frac{1}{10}e^{1.35674} + 1 \approx 1.38835; & y_3 &= \frac{1}{10}e^{1.56911} + 1 \approx 1.48024;\\
x_4 &= \frac{1}{10}e^{1.38835} + 1 \approx 1.40082; & y_4 &= \frac{1}{10}e^{1.48024} + 1 \approx 1.43940;\\
x_3 &= \frac{1}{10}e^{1.40082} + 1 \approx 1.40585; & y_5 &= \frac{1}{10}e^{1.43940} + 1 \approx 1.42182,
\end{aligned}$$

and so on. The monotonicity of both iteration sequences, as expected, is clearly visible.

11.4 Exercises

1. Consider the problem of approximating a locally unique zero s^* of the equation

   $$g(x) = 0,$$

   where g is a given real function defined on a closed interval $[a, b]$. The Newton iterates $\{x_n\}$, $\{y_n\}$, $n \geq 0$ are defined as

   $$g(x_n) + g'(x_n)(x_{n+1} - x_n) = 0 \qquad x_0 = a$$

   and

   $$g(y_n) + g'(y_n)(y_{n+1} - y_n) = 0 \qquad y_0 = b.$$

   Use Theorems 11.1.2 and 11.1.3 to find sufficient conditions for the monotone convergence of the above sequences to $s^* \in [a, b]$.

2. Examine the above problem, but using the secant iterations $\{x_n\}$, $\{y_n\}$, $n \geq -1$ given by

   $$g(x_n) + \frac{g(x_n) - g(x_{n-1})}{x_n - x_{n-1}}(x_{n+1} - x_n) = 0 \qquad x_1 = a,\ x_0 = \text{given}$$

   $$g(y_n) + \frac{g(y_n) - g(y_{n-1})}{y_n - y_{n-1}}(y_{n+1} - y_n) = 0 \qquad y_1 = b,\ y_0 = \text{given}.$$

3. Examine the above problem, but using the Newton-like iteration, where $\{x_n\}$, $\{y_n\}$, $n \geq 0$ are given by

   $$g(x_n) + h(x_n)(x_{n+1} - x_n) = 0 \qquad x_0 = a$$

and

$$g(y_n) + h(y_n)(y_{n+1} - y_n) = 0 \qquad y_0 = b$$

with a suitably chosen function h.

4. Examine the above problem, but using the Newton-like iteration, where $\{x_n\}$, $\{y_n\}$, $n \geq 0$ are given by

$$g(x_n) + c_n(x_{n+1} - x_n) = 0 \qquad x_0 = a$$

and

$$g(y_n) + d_n(y_{n+1} - y_n) = 0 \qquad y_0 = b$$

with a suitably chosen real sequences $\{c_n\}$, $\{d_n\}$.

5. Generalize Problems 1 to 4 to $\mathbb{R}^n$.

6. Generalize Problems 1 to 4 to $\mathbb{R}^n$ by choosing g to be defined on a given subset of a partially ordered topological space. (For a definition, see Chapter 2).

7. Compare the answers given in Problem 1 to 6 with the results of Chapter 2.

8. To find a zero s^* for $g(x) = 0$, where g is a real function defined on $[a, b]$, rewrite the equation as

$$x = x + cg(x) = f(x)$$

for some constant $c \neq 0$. Consider the sequences

$$x_{n+1} = f(x_n) \qquad x_0 = a$$

and

$$y_{n+1} = f(y_n) \qquad y_0 = b.$$

Use Theorems 11.1.2 and 11.1.3 to find sufficient conditions for the monotone convergence of the above sequences to a locally unique fixed point $s^* \in [a, b]$ ot the equation $x = f(x)$.

Chapter 12

Fixed points of point-to-set mappings

In this chapter special monotonicity concepts are used to compare iteration sequences and common fixed points of point-to set maps. The results of this chapter are very useful in comparing the solutions of linear and nonlinear algebraic equations as well as the solutions of differential and integral equations (see also Chapter 3).

12.1 General theorems

Assume again that X is a subset of a partially ordered space L. Let P denote a partially ordered parameter set, and for all $p \in P$ and $k \geq l-1$, let the point-to-set mappings $f_k^{(p)} : X^{k+1} \to 2^X$. For each $p \in P$, define the iteration algorithm

$$x_{k+1}^{(p)} \in f_k^{(p)}(x_0^{(p)}, x_1^{(p)}, ..., x_k^{(p)}). \tag{12.1.1}$$

Introduce the following conditions:

(H_1) If p_1, $p_2 \in P$ such that $p_1 \leq p_2$, then for all $x^{(i)} \in X$ $(i \geq 0)$ and $y_j \in f_k^{(p)}(x_0^{(p)}, x_1^{(p)}, ..., x_k^{(k)})$ $(j = 1, 2)$, $y_1 \leq y_2$;

(H_2) If $p \in P$ and $t^{(i)}$, $s^{(i)} \in X$ such that $t^{(i)} \leq s^{(i)}$ $(i \geq 0)$ then $y_t \leq y_s$ for all $k \geq l-1$, where $y_t \in f_k^{(p)}(t^{(0)}, t^{(1)}, ..., t^{(k)})$ and $y_s \in f_k^{(p)}(s^{(0)}, t^{(1)}, ..., s^{(k)})$ are arbitrary;

(H_3) For $k = 0, 1, ..., l-1$ and all $p_1, p_2 \in P$ such that $p_1 \leq p_2$, $x_k^{(p_1)} \leq x_k^{(p_2)}$.

Assumption (H_1) shows that for fixed k, mappings $f_k^{(p)}$ are increasing in p, and assumption (H_2) means that all mappings $f_k^{(p)}$ are isotone. Condition (H_3) requires that the initial points are selected so that they are increasing in p.

Our first result shows that under the above assumptions the entire iteration sequence (12.1.1) is increasing in p.

Theorem 12.1.1 *Assume that conditions (H_1) through (H_3) hold. Then $p_1, p_2 \in P$ and $p_1 \le p_2$ imply that for all $k \ge 0$, $x_k^{(p_1)} \le x_k^{(p_2)}$.*

Proof. By induction, assume that $x_i^{(p_1)} \le x_i^{(p_2)}$, $(i < k)$. Then condition (H_2) implies that $x_k^{(p_1)} \le y$ for all $y \in f_{k-1}^{(p_1)}(x_0^{(p_2)}, ..., x_{k-1}^{(p_2)})$, and from assumption (H_1) we conclude that $y \le x_k^{(p_2)}$. Thus, $x_k^{(p_1)} \le x_k^{(p_2)}$, which completes the proof. ∎

Remark 12.1.1 *The proof of the theorem remains valid in the slightly more general case, when it is assumed only that condition (H_2) holds either for p_1 or for p_2.*

Assume next that L is a metric space. Furthermore,

(H_4) If $t_k, s_k \in X$ such that $t_k \le s_k$ $(k \ge 0)$ and $t_k \to t^*$, $s_k \to s^*$ as $k \to \infty$ with some $t^*, s^* \in L$, then $t^* \le s^*$;

(H_5) For all $p \in P$, iteration sequence $x_k^{(p)}$ converges to a limit $x^{(p)} \in L$.

Our next result compares the limit points $x^{(p)^*}$.

Theorem 12.1.2 *Under assumptions (H_1) through (H_5), $x^{(p_1)^*} \le x^{(p_2)^*}$ if $p_1 \le p_2$.*

Proof. From Theorem 12.1.1 we know that $x_k^{(p_1)} \le x_k^{(p_2)}$ for all $k \ge 0$, and therefore condition (H_4) implies the assertion. ∎

Corollary 12.1.3 *Assume that for all $p \in P$, algorithm (12.1.1) is an $l-$step process, mappings $f_k^{(p)}$ are also closed. Then, $x^{(p)^*}$ is a common fixed point of mappings $f_k^{(p)}$ $(k \ge l-1)$, and it is increasing in p.*

As the conclusion of this section a different class of comparison theorems is introduced. Consider the iteration sequences $x_k^{(1)}$, $x_k^{(2)} \in X$ $(k \ge 0)$, and assume that

(I_1) $x_i^{(1)} \le x_i^{(2)}$ for $i = 0, 1, ..., l-1$;

(I_2) For all $k \ge l$ there exist mappings $\phi_k : X^{k+1} \to 2^L$ such that

- $\phi_k(x^{(0)}, x^{(1)}, ..., x^{(k)})$ is decreasing in variables $x^{(i)}$ $(0 \le i < k)$, and is strictly increasing in $x^{(k)}$;
- $y^{(1)} \le y^{(2)}$ $\forall k \ge l$, $y^{(1)} \in \phi_k(x_0^{(1)}, x_1^{(1)}, ..., x_k^{(1)})$ and $y^{(2)} \in \phi_k(x_0^{(2)}, x_1^{(2)}, ..., x_k^{(2)})$.

Theorem 12.1.4 *Under conditions* (I_1) *and* (I_2), $x_k^{(1)} \le x_k^{(2)}$ *for all* $k \ge 0$.

Proof. By induction, assume that $x_i^{(1)} \le x_i^{(2)}$ for $i < k$, but $x_k^{(1)} > x_k^{(2)}$. Let $y_i \in \phi_k(x_0^{(2)}, ..., x_{i-1}^{(2)}, x_i^{(1)}, ..., x_k^{(1)})$ be arbitrary for $i = 0, 1, ..., k$, and select any $y_{k+1} \in \phi_k(x_0^{(2)}, x_1^{(2)}, ..., x_k^{(2)})$. Then the monotonicity of ϕ_k implies

$$y_0 \ge y_1 \ge ... \ge y_k \ge y_{k+1},$$

which contradicts the second bulleted section of assumption (I_2). Hence, $x_k^{(1)} \le x_k^{(2)}$, and since $x_i^{(1)} \le x_i^{(2)}$ for $i = 0, 1, ..., l-1$, the proof is complete. ■

Remark 12.1.2 *Note that this theorem generalizes the main result of Redheffer and Walter [291].*

12.2 Fixed points

Consider once again the general $l-$step algorithm

$$x_{k+1} \in f_k(x_{k+1-l}, x_{k+2-l}, ..., x_k), \tag{12.2.1}$$

and assume that for $k \ge l-1$, $f_k : X^l \to 2^X$, where X is a subset of a partially ordered metric space L. Assume that

(J_1) The algorithm converges to an $s^* \in L$ with arbitrary initial points $x_0, x_1, ..., x_{l-1} \in X$;

(J_2) For all $k \ge l-1$, mappings f_k is isotone;

(J_3) If $x_k, x \in X$, $x \le x_k$ $(k \ge 0)$ such that $x_k \to x^*$ $(x^* \in L)$ as $k \to \infty$, then $x \le x^*$.

The main result of this section can be formulated as the following theorem.

Theorem 12.2.1 *Assume that conditions* (J_1) *through* (J_3) *hold. Furthermore, for an* $x \in X$, $x \le y$ *with all* $k \ge l-1$ *and* $y \in f_k(x, x, ..., x)$. *Then,* $x \le s^*$.

Proof. Consider the iteration sequence starting with initial vectors $x_i = x$ $(i = 0, 1, ..., l-1)$. First we show that $x \le x_k$ for all $k \ge 0$. By induction, assume that $x \le x_i$ $(i < k)$. Since $x_k \in f_{k-1}(x_{k-l}, ..., x_{k-2}, x_{k-1})$, assumption (J_2) implies that for all $y \in f_{k-1}(x, x, ..., x)$, $y \le x_k$. Then, from the assumption of the theorem we conclude that $x \le x_k$. Because $x_k \to s^*$, condition (J_3) implies the assertion. ■

Remark 12.2.1 *The assertion of the theorem remains true if* (J_2) *is weakened as follows. If* x, $x^{(i)} \in X$ *such that* $x \le x^{(i)}$ $(i = 1, 2, ..., l)$, *then for all* $k \ge l$ *and* $y \in f_k(x, x, ..., x)$ *and* $y^* \in f_k(x^{(1)}, x^{(2)}, ..., x^{(l)})$, $y \le y^*$.

The theorem can be easiy extended to the general algorithm (12.1.1). The details are omitted.

12.3 Applications

Case Study 1 Consider the first order difference equations

$$\mathbf{x}_{k+1} = \mathbf{g}(x_k) \qquad (12.3.1)$$
$$\mathbf{z}_{k+1} = \mathbf{h}(z_k),$$

where $\mathbf{g}$ and $\mathbf{h} : D \to D$, with D being a subset of $\mathbf{R}^n$. Assume that for all $\mathbf{x} \in D$, $\mathbf{g}(\mathbf{x}) \leqq \mathbf{h}(\mathbf{x})$, and $\mathbf{x}$, $\bar{\mathbf{x}} \in D$, $\mathbf{x} \leqq \bar{\mathbf{x}}$ imply that $\mathbf{g}(\mathbf{x}) \leqq \mathbf{g}(\bar{\mathbf{x}})$ and $\mathbf{h}(\mathbf{x}) \leqq \mathbf{h}(\bar{\mathbf{x}})$. Then Theorem 12.1.1 implies that if the initial terms $\mathbf{x}_0$ and $\mathbf{z}_0$ are selected in such a way that $\mathbf{x}_0 \leqq \mathbf{z}_0$, then for the entire iteration sequence, $\mathbf{x}_k \leqq \mathbf{z}_k$ $(k \geq 0)$. If $\mathbf{x}_k \to \mathbf{x}^*$ and $\mathbf{z}_k \to \mathbf{z}^*$ as $k \to \infty$, then $\mathbf{x}^* \leq \mathbf{z}^*$. This inequality is illustrated in Figure 12.1.

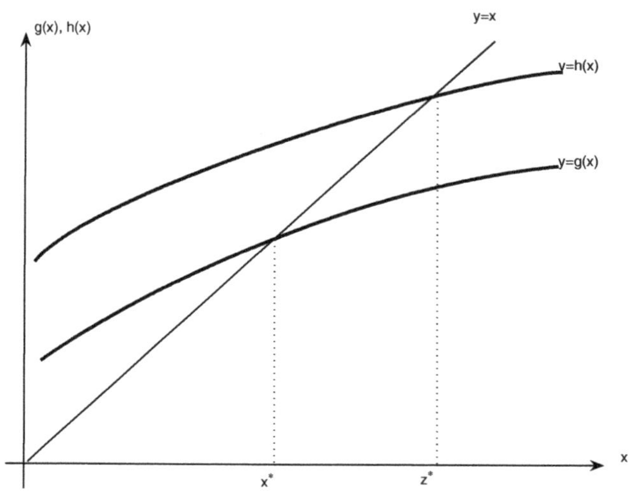

Figure 12.1: Comparison of fixed points on the real line.

Case Study 2 Assume that the above equations are linear, i.e., they have the special form

$$\mathbf{x}_{k+1} = \mathbf{A}x_k + \mathbf{b} \qquad (12.3.2)$$
$$\mathbf{z}_{k+1} = \mathbf{B}\mathbf{z}_k + \mathbf{d}.$$

If we assume that matrices $\mathbf{A}$, $\mathbf{B}$ and vectors $\mathbf{b}$ and $\mathbf{d}$ are non-negative, and that $\mathbf{A} \leqq \mathbf{B}$ and $\mathbf{b} \leqq \mathbf{d}$, then $\mathbf{x}_0 \leqq \mathbf{z}_0$ implies that $\mathbf{x}_k \leqq \mathbf{z}_k$ for all $k \geq 0$. This assertion is a simple consequence of Theorem 12.1.1 and the previous case study.

Case Study 3 Consider the linear input-output system

$$\mathbf{y} = \mathbf{x} - \mathbf{A}\mathbf{x}, \qquad (12.3.3)$$

where $\mathbf{A} \geq \mathbf{0}$ is an $n \times n$ constant matrix such that its spectral radius $\rho(\mathbf{A})$ is less than one. Here, $\mathbf{x}$ and $\mathbf{y}$ denote the gross and net production vectors, respectively. It is known that for all $\mathbf{y} \in \mathbf{R}^n$, equation (12.3.3) has a unique solution $\mathbf{x}(\mathbf{y})$, which can be obtained as the limit of the iteration process

$$\mathbf{x}_{k+1} = \mathbf{A}\mathbf{x}_k + \mathbf{y} \tag{12.3.4}$$

with arbitrary initial point $\mathbf{x}_0 \in \mathbf{R}^n$.

By selecting $X = \mathbf{R}^n$, $l = 1$, $\mathbf{f}(\mathbf{x}) = \mathbf{A}\mathbf{x} + \mathbf{y}$, we obtain the following result. If for any vector $\mathbf{u}$, $\mathbf{u} \leq \mathbf{A}\mathbf{u} + \mathbf{y}$, then $\mathbf{u} \leq \mathbf{x}(\mathbf{y})$; and if for a vector $\mathbf{v}$, $\mathbf{v} \geq \mathbf{A}\mathbf{v} + \mathbf{y}$, then $\mathbf{v} \geq \mathbf{x}(\mathbf{y})$. Here, " $\leq$ " the partial order of vectors, or $\mathbf{u} = (u^i) \leq \mathbf{v} = (v^i)$ if for all i, $u^i \leq v^i$.

Case Study 4 Consider again the linear input-output system (12.3.3) with the same assumptions as before. As we have seen, the iteration process (12.3.4) converges to the unique solution $\mathbf{x}(\mathbf{y})$. Since $\mathbf{A} \geq \mathbf{0}$, function $\mathbf{A}\mathbf{x} + \mathbf{y}$ is increasing in $\mathbf{x}$ and $\mathbf{y}$. Therefore Theorem 12.1.1 with the selection $p = \mathbf{y}$ implies that $\mathbf{x}(\mathbf{y}_1) \leq \mathbf{x}(\mathbf{y}_2)$, if $\mathbf{y}_1 \leq \mathbf{y}_2$. That is, higher net production requires higher gross production.

Case Study 5 As a simple consequence of Theorem 12.2.1, this case proves a useful result which can be applied in practical cases to determine wether a matrix has a nonnegative inverse. The assertion is the following.

Theorem 12.3.1 *Let* $\mathbf{A}$ *be an* $n \times n$ *real constant matrix.* $\mathbf{A}$ *is nonsingular and* $\mathbf{A}^{-1} \geq \mathbf{0}$ *if and only if an* $n \times n$ *real constant matrix* $\mathbf{D}$ *exists such that*

(i) $\mathbf{D} \geq \mathbf{0}$;

(ii) $\mathbf{I} - \mathbf{D}\mathbf{A} \geq \mathbf{0}$;

(iii) $\rho(\mathbf{I} - \mathbf{D}\mathbf{A}) < 1$.

Proof. 1. If $\mathbf{A}^{-1}$ exists and is nonnegative, then $\mathbf{D} = \mathbf{A}^{-1}$ satisfies the above conditions. 2. From (iii) we know that $\mathbf{D}\mathbf{A}$ is nonsingular; therefore, $\mathbf{A}^{-1}$ exists. In order to show that $\mathbf{A}^{-1} \geq \mathbf{0}$, we verify that equation

$$\mathbf{A}\mathbf{x} = \mathbf{b}$$

has a non-negative solution $\mathbf{x}$ for all $\mathbf{b} \geq \mathbf{0}$. Note first that this equation is equivalent to the following

$$\mathbf{x} = (\mathbf{I} - \mathbf{D}\mathbf{A})\mathbf{x} + \mathbf{D}\mathbf{b}. \tag{12.3.5}$$

Condition (iii) implies that the iteration sequence generated by the mapping $\mathbf{f}(\mathbf{x}) = (\mathbf{I} - \mathbf{D}\mathbf{A})\mathbf{x}$ converges to $\mathbf{s}^* = \mathbf{0}$. Starting from arbitrary initial vector $\mathbf{x}_0$. If the partial order is selected as " $\geq$ " (instead of " $\leq$ "), then for the solution $\mathbf{x}^*$ of equation (12.3.5) $\mathbf{x}^* \geq \mathbf{f}(\mathbf{x}^*)$, and therefore Theorem 12.2.1 implies that $\mathbf{x}^* \geq \mathbf{s}^* = \mathbf{0}$. Hence, the solution $\mathbf{x}^*$ of equation (12.3.5) is non-negative, which completes the proof. ∎

In the matrix theoretical literature, a special class of matrices satisfying the conditions of this theorem, is often discussed.

Definition 12.3.2 *An $n \times n$ real matrix $\mathbf{A} = (a_{ij})$ is called an M-matrix, if $\mathbf{D} = \mathrm{diag}(a_{11}^{-1}, a_{22}^{-1}, \ldots, a_{nn}^{-1})$ satisfies conditions (i), (ii), (iii).*

Notice that assumptions (i) and (ii) imply that for all i, $a_{ii} > 0$ and for all $i \neq j$, $a_{ij} \leq 0$. For the sake of simplicity, let **NP** denote the set of all $n \times n$ real matrice whose off-diagonal elements are all non-positive, and let **MM** denote the set of all $n \times n$ M-matrices. Obviously, $\mathbf{MM} \subset \mathbf{NP}$. A comprehensive summary of the mostly known characterizations of M-matrices is given as follows.

Theorem 12.3.3 *Let $\mathbf{A} \in \mathbf{NP}$. Then the following conditions are equivalent to each others:*

1. *there exists a vector $\mathbf{x} \geq \mathbf{0}$ such that $\mathbf{Ax} > \mathbf{0}$;*
2. *there exists a vector $\mathbf{x} > \mathbf{0}$ such that $\mathbf{Ax} > \mathbf{0}$;*
3. *there exists a diagonal matrix $\mathbf{D}$ with positive diagonal such that $\mathbf{AD1} > \mathbf{0}$;*
4. *there exists a diagonal matrix $\mathbf{D}$ with positive diagonal such that $\mathbf{AD}$ is strictly diagonally dominant;*
5. *for each diagonal matrix $\mathbf{R}$ such that $\mathbf{R} \geq \mathbf{A}$, $\mathbf{R}^{-1}$ exists and $\rho(\mathbf{R}^{-1}(\mathbf{A}_d - \mathbf{A})) < 1$, where $\mathbf{A}_d$ is the diagonal of $\mathbf{A}$;*
6. *if $\mathbf{B} \geq \mathbf{A}$ and $\mathbf{B} \in \mathbf{NP}$, then $\mathbf{B}^{-1}$ exists;*
7. *each real eigenvalue of $\mathbf{A}$ is positive;*
8. *each principal minor of $\mathbf{A}$ is positive;*
9. *there exists a strictly increasing sequence $\emptyset \neq M_1 \subseteq M_2 \subseteq \ldots \subseteq M_r = \{1, 2, \ldots, n\}$ such that the principal minors $\mathbf{A}(M_i)$ are positive, where $\mathbf{A}(M_i)$ is obtained from $\mathbf{A}$ by eliminating all rows and columns with indices $i \notin M_i$;*
10. *there exists a permutation matrix $\mathbf{P}$ such that $\mathbf{PAP}^{-1}$ can be decomposed as $\mathbf{LU}$, where $\mathbf{L} \in \mathbf{NP}$ is a lower triangular matrix with positive diagonal, and $\mathbf{U} \in \mathbf{NP}$ is an upper triangular matrix with positive diagonal;*
11. *inverse $\mathbf{A}^{-1}$ exists and is non-negative.*

Proof. $1 \Rightarrow 2$: The continuity of $\mathbf{Ax}$ implies that for sufficiently small $\varepsilon > 0$, $\mathbf{A}(\mathbf{x} + \varepsilon\mathbf{1}) > \mathbf{0}$. Here $\mathbf{x} + \varepsilon\mathbf{1} > 0$. $2 \Rightarrow 3$: Let $\mathbf{Ax} > \mathbf{0}$, and define $\mathbf{D} = \mathbf{diag}(\mathbf{x}^{(1)}, \ldots, x^{(n)})$, where $x^{(i)}$ denotes the ith element of $\mathbf{x}$. Then, $\mathbf{AD1} = \mathbf{Ax} > \mathbf{0}$. $3 \Rightarrow 4$: Let $\mathbf{D}$ be a diagonal matrix with positive diagonal such that $\mathbf{AD1} > \mathbf{0}$. Then, with th enotation

$$\mathbf{AD} = \mathbf{W} = (w_{ij}), \quad \sum_{j=1}^{n} w_{ij} > 0, \text{ i.e., } w_{ij} > \sum_{j \neq i} - w_{ij} = \sum_{j \neq i} |w_{ij}|,$$

since for $i \neq j$, $w_{ij} \leq 0$. 4 $\Rightarrow$ 5: Let $\mathbf{W}_D$ denote the diagonal of $\mathbf{W}$. Then the row norm of $\mathbf{I} - \mathbf{W}_D^{-1}\mathbf{W}$ is less than one, therefore $\rho(\mathbf{I} - \mathbf{W}_D^{-1}\mathbf{W}) < 1$, and

$$\begin{aligned}\rho(\mathbf{I} - \mathbf{A}_D^{-1}\mathbf{A}) = \rho(\mathbf{D}^{-1}(\mathbf{I} - \mathbf{A}_D^{-1}\mathbf{A})\mathbf{D}) &= \rho(\mathbf{I} - \mathbf{D}^{-1}\mathbf{A}_D^{-1}\mathbf{A}\mathbf{D}) \\ &= \rho(\mathbf{I} - \mathbf{W}_D^{-1}\mathbf{W}) < 1,\end{aligned}$$

since $\mathbf{W}_D = \mathbf{A}_D\mathbf{D}$. Let $\mathbf{R} \geq A$ be a diagonal matrix. Then, $\mathbf{R} \geq \mathbf{A}_D$, and therefore $\mathbf{R}^{-1} \leq \mathbf{A}_D^{-1}$. The Perron-Frobenius theorem (see, for example, Luenberger, 1979) implies that

$$\rho(\mathbf{R}^{-1}(\mathbf{A}_D - \mathbf{A})) \leq \rho(\mathbf{A}_D^{-1}(\mathbf{A}_D - \mathbf{A})) = \rho(\mathbf{I} - \mathbf{A}_D^{-1}\mathbf{A}) < 1.$$

5 $\Rightarrow$ 6: Let $\mathbf{B} \geq \mathbf{A}$ such that $\mathbf{B} \in \mathbf{NP}$. Let $\mathbf{B}_D$ denote the diagonal of $\mathbf{B}$. Notice that $\mathbf{A}_D > 0$, since otherwise a singular $\mathbf{R}$ could be selected with at least one zero diagonal element. Inequality $\mathbf{B}_D \geq \mathbf{A}_D$ implies that $\mathbf{B}_D^{-1}$ exists and is positive. Therefore

$$\mathbf{0} \leq \mathbf{B}_D^{-1}(\mathbf{B}_D - \mathbf{B}) \leq \mathbf{B}_D^{-1}(\mathbf{A}_D - \mathbf{A}),$$

which implies that

$$\rho(\mathbf{I} - \mathbf{B}_D^{-1}\mathbf{B}_D) = \rho(\mathbf{B}_D^{-1}(\mathbf{B_D} - \mathbf{B})) \leq \rho(\mathbf{B}_D^{-1}(\mathbf{A_D} - \mathbf{A})) < 1$$

since the selection $\mathbf{R} = \mathbf{B}$ satisfies condition 5. Hence, $\mathbf{B}_D^{-1}\mathbf{B_D}$ is invertible, which implies that $\mathbf{B}^{-1}$ exists. 6 $\Rightarrow$ 7: Assume that 6 holds, and let $w \leq 0$. Then, matrix $\mathbf{B} = \mathbf{A} - w\mathbf{I}$ satisfies inequality $\mathbf{B} \geq \mathbf{A}$. Since $\mathbf{B}^{-1}$ exists, w is not an eigenvalue of $\mathbf{A}$. 7 $\Rightarrow$ 8: Let $M \subset \{1, 2, ..., n\}$, and define matrix $\mathbf{B}$ as

$$b_{ij} = \begin{cases} a_{ij} & i, j \in M \\ a_{ii} & i \notin M \\ 0 & \text{otherwise.} \end{cases}$$

Clearly, $\mathbf{B} \geq \mathbf{A}$ and $\mathbf{B} \in \mathbf{NP}$. It is easy to prove by finite onduction that $\det(\mathbf{B}) > 0$. Notice that $\det(\mathbf{B})$ is the product of $\det \mathbf{A}(M)$ and the a_{ii}'s with $i \notin M$, and it is also the product of all eigenvalues of $\mathbf{B}$. We will show that all real eigenvalues of $\mathbf{B}$ are positive. Let β denote an eigenvalue of $\mathbf{B}$, and assume that $\beta \leq 0$. Then $\bar{\mathbf{B}} = \mathbf{B} - \beta\mathbf{I} \geq \mathbf{A}$, and we will next prove that $\mathbf{B} - \beta\mathbf{I}$ is nonsingular, which implies that β *is* not an eigenvalue of $\mathbf{B}$. There exists a $\sigma > 0$ such that $\mathbf{I} - \sigma\bar{\mathbf{B}} \geq \mathbf{0}$, and

$$\mathbf{V} = \mathbf{I} - \sigma\mathbf{A} \geq \mathbf{I} - \sigma\bar{\mathbf{B}} = \mathbf{U} \geq \mathbf{0}.$$

The Perron-Frobenius theory and assumption 7 imply that $0 \leq \rho(\mathbf{U}) \leq \rho(\mathbf{V}) < 1$, therefore, $(\mathbf{I} - \mathbf{U})^{-1}$ exists. Since $\mathbf{I} - \mathbf{U} = \sigma\bar{\mathbf{B}}$, $\bar{\mathbf{B}}$ is nonsingular. 8 $\Rightarrow$ 9: This is trivial. 9 $\Rightarrow$ 10: This assertion follows from the well-known conditions of the existence of **LU** decomposition (see, for example, Szidarovsky and Yakowitz, [314]. 10 $\Rightarrow$ 11: Assume that $\mathbf{A}$ may be written as $\mathbf{A} = \mathbf{LU}$ where $\mathbf{L}$ and $\mathbf{U}$ satisfy the given properties. Then both $\mathbf{L}^{-1}$ and $\mathbf{U}^{-1}$ exist and it is easy to show that $\mathbf{L}^{-1}$ and $\mathbf{U}^{-1}$ are nonnegative. Therefore, $\mathbf{A}^{-1} = (\mathbf{LU})^{-1} = \mathbf{U}^{-1}\mathbf{L}^{-1} \geq \mathbf{0}$. 11 $\Rightarrow$ 1: Select $\mathbf{x} = \mathbf{A}^{-1}1 \geq \mathbf{0}$, then $\mathbf{Ax} = \mathbf{1} > \mathbf{0}$. Thus, the proof is complete. ■

Corollary 12.3.4 *An $n \times n$ real matrix is an $M-$matrix if and only if it satisfies any one of the conditions of the theorem.*

Proof. If $\mathbf{A}$ is an $M-$matrix, then it satisfies 11, and if a matrix satisfies any one of the assumptions, then it satisfies 5. Select $\mathbf{R}$ as the diagonal of $\mathbf{A}$, the condition 5 is equivalent to the conditions of Theorem 12.3.1. ∎

Case Study 6 A nice application of the theory of M-matrices is contained here. Consider equations $\mathbf{g}(\mathbf{x}) = \mathbf{b}^{(1)}$ and $\mathbf{g}(\mathbf{x}) = \mathbf{b}^{(2)}$, where $\mathbf{g} : D \to \mathbf{R}^n$ with $D \subseteq \mathbf{R}^n$ being convex, $\mathbf{b}^{(1)}$ and $\mathbf{b}^{(2)}$ are in $R(\mathbf{g})$, furthermore $\mathbf{b}^{(1)} \leqq \mathbf{b}^{(2)}$. Assume that $\mathbf{g}$ is differentiable, and for all $\mathbf{x} \in D$ and i and j,

$$\frac{\partial g_i}{\partial x_i}(\mathbf{x}) > 0, \quad \frac{\partial g_i}{\partial x_j}(\mathbf{x}) \le 0 \quad (j \ne i),$$

and

$$\frac{\partial g_i}{\partial x_i}(\mathbf{x}) > \sum_{j \ne i} \left| \frac{\partial g_i}{\partial x_j}(\mathbf{x}) \right| .$$

12.4 Exercises

12.1. Assume that conditions (H_1), (H_3) hold. Moreover, (H_2) holds either for p_1 or for p_2. Then show that $p_1, p_2 \in P$ and $p_1 \le p_2$ imply that for all $k \ge 0$ $x_k^{(p_1)} \le x_k^{(p_2)}$.

12.2. Let $u_n : N \to \mathbf{R}$ and $u : \mathbf{R}^+ \to \mathbf{R}$ be given by

$$\delta u_n = u_{n+1} - u_n, \quad \delta u(t) = u(t+1) - u(t).$$

If $f : \mathbf{R} \times \mathbf{R} \to \mathbf{R}$ has the property that $p \ge q \Longrightarrow f(t,p) \ge f(t,q)$ it is said that $f(t,\uparrow)$ is increasing, and if $-f$ has this property it is said that $f(t,\downarrow)$ is decreasing. In either case, the condition is required for all $t \in \mathbf{R}^+$. Suppose that

(i) $f(t,\uparrow)$ is increasing and

$$\delta u_n \le f(n, u_n), \quad \delta v_n \ge f(n, v_n), \quad n \in \mathbf{N}$$

or

(ii) $f(t,\downarrow)$ is decreasing and

$$\delta u_n \le f(n, u_{n+1}), \quad \delta v_n \ge f(n, v_{n+1}), \quad n \in \mathbf{N}.$$

Then show that $u_0 \le v_0 \Longrightarrow u_n \le v_n$ for all $n \in \mathbf{N}$.

12.3. Let u and v be functions $\mathbf{R}^+ \to \mathbf{R}$. Suppose

(i) $f(t, \uparrow)$ is increasing and

$$\delta u(t) \leq f(t, u(t)), \quad \delta v(t) \geq f(t, v(t)), \quad t \in \mathbf{R}^+$$

or

(ii) $f(t, \downarrow)$ is decreasing and

$$\delta u(t) \leq f(t, u(t+1)), \quad \delta v(t) \geq f(t, v(t+1)), \quad t \in \mathbf{R}^+.$$

Then show that $u(t) \leq v(t)$ for all $0 \leq t < 1 \Longrightarrow u(t) \leq v(t)$ for $t \in \mathbf{R}^+$.

12.4 Let $u_-(t) = \inf_{t \leq q \leq t+1} u(q)$, $u_+(t) = \sup_{t \leq q \leq t+1} u(q)$. Suppose that f and u satisfy hypothesis (i) or (ii) of the previous problem and suppose further that $f(t, M) \leq 0$ for $t \in \mathbf{R}^+$, where M is a constant. Then show that $u_+(0) \leq M \Rightarrow u(t) \leq M, t \in \mathbf{R}^+$.

12.5 Let u be a locally bounded solution of

$$h(t)\delta u(t) \leq g(t) - u(t+1), \quad t \in \mathbf{R}^+$$

where $g > 0$ and $h > 0$ for large t. Suppose v is positive for large t and satisfies the conditions

$$(h(t) + 1)\left(v'(t) + \frac{1}{2}v''_-(t)\right) + v(t) \geq g(t), \quad t \ll 1.$$

Then show that

$$u(t) \leq O(v(t)), \quad as\ t \to \infty.$$

12.6 It is said that $f : \mathbf{R}^+ \to \mathbf{R}^+$ is admissible if constants a, T exist such that

$$f(t) > 0 \text{ and } \frac{f(s)}{f(t)} \leq a \text{ for } s, t \geq T, |s - t| \leq 1.$$

Let f and h be admissible functions such that

$$\lim_{t \to \infty} \sup h(t) f'(t) < \infty$$

and let u be a locally bounded, nonnegative solution of

$$f(t)\delta u(t) \leq g(t) - u(t+1)^{1+a}$$

where $a > 0$. Then show that

$$g(t) = O(f(t)^{-1-1/a}) \Rightarrow u(t) = O(f(t)^{-1/a}), \quad \text{as } t \to \infty.$$

12.7 Let $\varphi(t, p, q)$ be decreasing in p and strictly increasing in q. Then show that if

$$u_0 \le v_0$$

and

$$\varphi(n, u_n, u_{n+1}) \le \varphi(n, v_n, v_{n+1}), \quad \text{for all } n \ge 0$$

then

$$u_n \le v_n, \quad \text{for all } n \ge 0.$$

12.8 Consider the inequality

$$c(1+t)^r \delta u(t) \le g(t) - u(t+1)^{1+a}, \quad t \in \mathbf{R}^+,$$

where g is a positive function $\mathbf{R}^+ \to \mathbf{R}$ and c, a, r are real constants with $c > 0$ and $a \ge 0$. Assume u is locally bounded and $u \ge 0$. Let $a = 0$. If $r < 1$ (c, m arbitrary) or if $r = 1$ and $m \ge -\frac{1}{c'}$ then show that

$$g(t) = O(t^m) \Rightarrow u(t) = O(t^m).$$

12.9 Extend Theorem 12.2.1 for the general algorithm (11.0.1).

12.10 Find results similar to those introduced in case study 8 for systems of ordinary differential equations.

Chapter 13

Special topics

We exploit further the conclusions of Section 5.1.

13.1 Newton-Mysovskikh conditions

The Newton–Mysovskikh theorem [248], [263] provides sufficient conditions for the semilocal convergence of Newton's method to a locally unique solution of an equation in a Banach space setting. In this section we show that under weaker hypotheses and a more precise error analysis than before [248], [263] weaker sufficient conditions can be obtained for the local as well as the semilocal convergence of Newton's method. Error bounds on the distances involved as well as a larger radius of convergence are also obtained. Some numerical examples are also provided.

In particular we are motivated by the affine convariant Newton–Mysovskikh theorem (ACNMT) (see also Theorem 13.1.1 that follows) which basically states that if the initial guess x_0 is close enough to the solution x^* then Newton's method (4.1.3) converges quadratically to x^*. The basic condition is given by

$$\|F'(z)^{-1}[F'(y) - F'(x)](y - x)\| \leq w\|y - x\|^2 \text{ for all } x, y, z \in D. \tag{13.1.1}$$

Here we have noticed (see Theorem 13.3.1) that condition (13.1.1) is not used at full strength to prove convergence of method (4.3.1). Indeed a much weaker hypothesis (see (13.1.4)) can replace (13.1.1) in the proof of the theorem. This observation has the following advantages (semilocal case):

(a) the applicability of (ACNMT) is extended;

(b) finer error bounds on the distances $\|x_{n+1} - x_n\|$, $\|x_n - x^*\|$ $(n \geq 0)$;

(c) a more precise information on the location of the solution x^*;

(d) the Lipschitz constant is at least as small and easier to compute.

Local case:

(a) a larger radius of convergence;

(b) finer error bounds on the distances $\|x_n - x^*\|$ $(n \geq 0)$.

The above advantages are obtained under not only weaker hypotheses but under less or the same computational cost.

Some numerical examples are also provided for both the semilocal and local case where our results apply where other ones in [248], [263] fail.

We can show the following weaker semilocal convergence version of the Affine Covariant Newton–Mysovskikh Theorem (ACNMT):

Theorem 13.1.1 *Let $F\colon D \subseteq X \to Y$ be a Fréchet continuously differentiable operator, and let $x_0 \in D$ be an initial guess.*

Assume further that the following conditions hold true:

$$F'(x)^{-1} \in L(Y, X) \quad (x \in D), \tag{13.1.2}$$

$$\|F'(x_0)^{-1}F(x_0)\| \leq \eta, \tag{13.1.3}$$

$$\|F'(y)^{-1}[F'(x + t(y - x)) - F'(x)](y - x)\| \leq w_1\|y - x\|^2 t \tag{13.1.4}$$

for all $x, y \in D$, $t \in [0, 1]$

$$2h_1 = w_1\|F'(x_0)^{-1}F(x_0)\| \leq w_1\eta < 2, \tag{13.1.5}$$

$$\overline{U}(x_0, r_1) \subseteq D, \tag{13.1.6}$$

where

$$r_1 = \frac{\|F'(x_0)^{-1}F(x_0)\|}{1 - \frac{h_1}{2}}. \tag{13.1.7}$$

Then sequence $\{x_n\}$ $(n \geq 0)$ generated by Newton's method (4.1.3) is well defined, remains in $\overline{U}(x_0, r_1)$ for all $n \geq 0$, and there exists a unique $x^* \in \overline{U}(x_0, r_1)$ such that $F(x^*) = 0$ and $x_n \to x^*$ as $n \to \infty$ with

$$\|x_{n+1} - x_n\| \leq \frac{1}{2}w_1\|x_n - x_{n-1}\|^2, \tag{13.1.8}$$

$$\|x^* - x_{n+1}\| \leq \varepsilon_n\|x^* - x_{n-1}\|^2 \quad \text{for all } n \geq 1, \tag{13.1.9}$$

where,

$$\varepsilon_n = \frac{1}{2}w_1\left[1 + \sum_{j=1}^{\infty}\left(h_1^{2^n}\right)^{2^j}\right] \leq \frac{1}{2}\frac{w_1}{1 - h_1^{2^n}}. \tag{13.1.10}$$

Proof. Using (4.1.3) and (13.1.4) we obtain in turn

$$\begin{aligned}
x_{k+1} - x_k = {} & F'(x_k)^{-1}\big[F(x_k) - F(x_{k-1}) - F'(x_{k-1})(x_k - x_{k-1})\big] \\
& - F'(x_k)^{-1}\int_0^1 \big[F'(x_{k-1} + t(x_k - x_{k-1})) - \\
& - F'(x_{k-1})\big](x_k - x_{k-1})dt
\end{aligned} \tag{13.1.11}$$

and

$$\begin{aligned}\|x_{k+1} - x_k\| &\leq \left\| F'(x_k)^{-1} \int_0^1 \big[F'(x_{k-1} + t(x_k - x_{k-1}))- \right. \\ &\quad \left. - \; F'(x_{k-1})\big](x_k - x_{k-1})dt \right\| \\ &\leq \frac{1}{2} w_1 \|x_k - x_{k-1}\|^2, \end{aligned} \tag{13.1.12}$$

which shows (13.1.8).

We now prove $\{x_n\}$ is a Cauchy sequence in $\overline{U}(x_0, r_1)$. By induction on k we will show:

$$\|x_{k+1} - x_k\| \leq \frac{2}{w_1} h_1^{2^k}, \quad k \in N_0. \tag{13.1.13}$$

For $k = 0$, we have by (4.1.3), (13.1.3) and (13.1.5)

$$\|x_1 - x_0\| = \|F'(x_0)^{-1} F(x_0)\| \leq \eta = \frac{2}{w_1} h_1.$$

Assuming (13.1.13) to be true for some $k \in N$, we get

$$\begin{aligned}\|x_{k+2} - x_{k+1}\| &\leq \frac{1}{2} w_1 \|x_{k+1} - x_k\|^2 \leq \frac{1}{2} w_1 \left(\frac{2}{w_1} h_1^{2^k} \right)^2 \\ &= \frac{2}{w_1} h_1^{2^{k+1}}., \end{aligned} \tag{13.1.14}$$

and

$$\begin{aligned}\|x_{k+1} - x_0\| &\leq \|x_{k+1} - x_k\| + \cdots + \|x_1 - x_0\| \\ &\leq \frac{2}{w_1} \big(h_1^{2^k} + \cdots + h_1\big) = \frac{2}{w_1} h_1 \sum_{j=0}^{\infty} h_1^{2^j - 1} = r_1. \end{aligned} \tag{13.1.15}$$

it follows from (13.1.15) that $x_{k+1} \in \overline{U}(x_0, r_1)$:

Similarly it can be shown that

$$\|x_{m+k} - x_m\| \to 0 \quad \text{as } m \to \infty. \tag{13.1.16}$$

That is $\{x_n\}$ is a Cauchy sequence in $\overline{U}(x_0, r_1)$ and as such it converges to some $x^* \in \overline{U}(x_0, r_1)$.

Hence, we get

$$x_{k+1} - x_k = -F'(x_k)^{-1} F(x_k) \overset{k \to \infty}{\longrightarrow} -F'(x^*)^{-1} F(x^*) = 0$$

and thus $F(x^*) = 0$. To show (13.1.9), set $h_1^k = \frac{1}{2}w_1\|x_k - x_{k-1}\|$. Then we get in turn:

$$\begin{aligned}\|x_k - x^*\| &= \lim_{k<m\to\infty} \|x_k - x_m\| \\ &\le \lim_{k<m\to\infty} \left[\|x_m - x_{m-1}\| + \cdots + \|x_{k+1} - x_k\|\right] \\ &\le \frac{2}{w_1} \lim_{k<m\to\infty} \left[h^{m-1} + \cdots + h^k\right] \\ &= \frac{2h^k}{w_1} \lim_{k<m\to\infty} \left[1 + \frac{h^{k+1}}{h^k} + \cdots + \frac{h^{m-1}}{h^k}\right]. \end{aligned} \tag{13.1.17}$$

We can also have

$$h_1^k \le \left(\frac{w_1}{2}\right)^2 \|x_{k-1} - x_{k-2}\|^2 = \left(h_1^{k-1}\right)^2, \tag{13.1.18}$$

whence

$$h_1^{k+m} \le \left(h_1^k\right)^{2^m}, \quad k \in N_0. \tag{13.1.19}$$

That is

$$\begin{aligned}\|x_k - x^*\| &\le \frac{1}{2}w_1\|x_{k-1} - x_{k-2}\|^2\left[1 + (h^k)^2 + \cdots\right] \\ &\le \frac{1}{2}w_1\left[1 + \sum_{j=1}^{\infty}(h_1^k)^{2^j}\right]\|x_{k-1} - x_{k-2}\|^2, \end{aligned} \tag{13.1.20}$$

which shows (13.1.9).

Finally to show uniqueness, let $y^* \in \overline{U}(x_0, r_1)$ be a solution of equation $F(x) = 0$. Then, we have for all $t \in [0,1]$:

$$x^* + t(y^* - x^*) - x_0 = (1-t)(x^* - x_0) + t(y^* - x_0)$$

and

$$\begin{aligned}\|x^* + t(y^* - x^*) - x_0\| &\le (1-t)\|x^* - x_0\| + t\|y^* - x_0\| \\ &\le (1-t)r_1 + tr_1 = r_1. \end{aligned}$$

That is

$$x^* + t(y^* - x) \in \overline{U}(x_0, r_1), \quad (t \in [0,1]). \tag{13.1.21}$$

Using the identity

$$\begin{aligned} 0 = F(y^*) - F(x^*) &= \int_0^1 F'(x^* + t(y^* - x^*))(y^* - x^*)dt \\ &= M(y^* - x^*), \end{aligned} \tag{13.1.22}$$

where,

$$M = \int_0^1 F'(x^* + t(y^* - x^*))dt$$

and hypothesis (13.1.2) we deduce that for $x_t = x^* + t(y^* - x^*)$ linear operator M is invertible. Hence, we conclude

$$x^* = y^*.$$

■

Remark 13.1.1 *Clearly condition (13.1.1) implies (13.1.4) but not vice versa. That is (13.1.4) is a weaker condition than (13.1.1). Note that*

$$w_1 \leq w \tag{13.1.23}$$

holds in general and $\frac{w}{w_1}$ can be arbitrarily large.

The classical Affine Covariant Newton–Mysovskikh Theorem has been shown using (13.1.1), h, r instead of (13.1.4), h_1 and r_1 respectively, where

$$2h = w\|F'(x_0)^{-1}F(x_0)\| \leq w\eta < 2 \tag{13.1.24}$$

and

$$r = \frac{\|F'(x_0)^{-1}F(x_0)\|}{1 - \frac{h}{2}}. \tag{13.1.25}$$

Note that

$$2h < 2 \Rightarrow 2h_1 < 2 \tag{13.1.26}$$

but not vice versa unless if $w_1 = w$. If strict inequality holds in (13.1.23) then

$$r < r \tag{13.1.27}$$

and our error estimates (13.1.8)–(13.1.10) are finer than the corresponding ones in [248], [263] obtained by replacing w_1 by w in (13.1.8)–(13.1.10). Note that all the above advantages are obtained under less computational cost since in practice the computation of w requires that of w_1.

Remark 13.1.2 *Although the computation of constant w_1 in (13.1.4) is in practice easier than the computation of w in (13.1.1) one may want to replace (13.1.4) by the pair of conditions*

$$\|F'(x_0)^{-1}[F'(x + t(y - x)) - F'(x)](y - x)\| \leq w_2\|y - x\|^2 t \tag{13.1.28}$$

and

$$\|F'(x_0)^{-1}[F'(x) - F'(x_0)]\| \le w_0\|x - x_0\| \tag{13.1.29}$$

for all $x, y \in D$, $t \in [0,1]$.

Using (13.1.28) and (13.1.29) we showed in Section 5.1 that Newton's sequence converges to a unique solution x^* *of equation (4.1.3) in* $\overline{U}(x_0, \frac{2\eta}{2-\delta})$ *provided that*

$$h_2 = (\delta w_0 + w_2)\eta \le 1, \quad \delta \in [0,1]. \tag{13.1.30}$$

Clearly these results will be weaker provided

$$\frac{w_2}{1 - w_0\|z - x_0\|} < w, \quad z \in \overline{U}\left(x_0, \frac{2\eta}{2-\delta}\right), \tag{13.1.31}$$

since,

$$\begin{aligned}\|F'(z)^{-1}[F'(y) - F'(x)](y-x)\| &\le \|F'(z)^{-1}F'(x_0)\| \\ &\quad \cdot \|F'(x_0)^{-1}[F'(y) - F'(x)](y-x)\| \\ &\le \frac{w_2\|y-x\|^2}{1 - w_0\|z - x_0\|}\end{aligned} \tag{13.1.32}$$

by the Banach Lemma on invertible operators and the estimate

$$\|F'(z)^{-1}F'(x_0)\| \le \frac{1}{1 - w_0\|z - x_0\|}. \tag{13.1.33}$$

Condition (13.1.31) certainly holds if

$$\|x - x_0\| < \frac{1}{w_0}\left[1 - \frac{w_2}{w}\right]$$

or

$$\frac{2\eta}{2-\delta} < \frac{1}{w_0}\left[1 - \frac{w_2}{w}\right] \tag{13.1.34}$$

which can happen for sufficiently small η *and since* $\frac{w_2}{w}$ *can be arbitrarily small.*

We can provide an example to show that (13.1.5) and (13.1.30) hold but (13.1.24) fails.

Example 13.1.1 *Let* $X = Y = \mathbf{R}$, $D = [a, 2-a]$, $a \in [0, \frac{1}{2})$, $x_0 = 1$ *and define function* F *on* D *by*

$$F(x) = x^3 - a. \tag{13.1.35}$$

Using (13.1.1), (13.1.3), (13.1.4), (13.1.28) and (13.1.29) we obtain

$$w = \frac{2(2-a)}{a^2}, \ \eta = \frac{1}{3}(1-a), \ w_0 = 3-a, \ w_1 = \frac{2}{a^2} \text{ and } w_2 = 2(2-a). \tag{13.1.36}$$

Condition (13.1.24) does not hold since

$$2h = \frac{2}{3a^2}(1-a)(2-a) > 2 \quad \textit{for all } a \in \left[0, \frac{1}{2}\right). \tag{13.1.37}$$

That is there is no guarantee that Newton's method starting at x_0 converges to x^*. However condition (13.1.5) holds, since

$$2h_1 = \frac{2}{3a}(1-a) < 2 \quad \textit{for all } a \in \left(\frac{\sqrt{13}-1}{6}, \frac{1}{2}\right). \tag{13.1.38}$$

Finally, say for $\delta = 1$, condition (13.1.30) holds since

$$h_2 = \frac{1}{3}(1-a)[2(2-a)+3-a] \leq 1 \quad \textit{for all } a \in \left[\frac{5-\sqrt{13}}{3}, \frac{1}{2}\right). \tag{13.1.39}$$

In order for us to cover the local convergence of Newton's method under Newton–Mysovskikh-type conditions, let x^* be a simple solution of equation $F(x) = 0$. Assume:

$$\|F'(x)^{-1}[F'(x^* + t(x-x^*)) - F'(x)](x-x^*)\| \leq v(1-t)\|x-x^*\|^2 \tag{13.1.40}$$

for all $x \in \overline{U}(x^*, r^*)$, $t \in [0,1]$ where,

$$r^* = \frac{2}{v} \quad (v \neq 0). \tag{13.1.41}$$

Set also

$$r_1^* = \frac{2}{w} \quad (w \neq 0). \tag{13.1.42}$$

We can show the following local convergence theorem for Newton's method (4.1.3):

Theorem 13.1.2 *Let $F\colon D \subseteq X \to Y$ be a Fréchet-differentiable operator. Assume*

$$F'(x)^{-1} \in L(Y, X) \quad (x \in D);$$

there exists $x^* \in D$ such that $F(x^*) = 0$;
if
(a) condition (13.1.40) holds and

$$\overline{U}(x^*, r^*) \subseteq D \tag{13.1.43}$$

then, sequence $\{x_n\}$ generated by Newton's method (4.1.3) is well defined, remains in $U(x^*, r^*)$ for all $n \geq 0$ and converges to x^* provided that $x_0 \in U(x^*, r^*)$ with

$$\|x_{n+1} - x^*\| \leq \frac{v}{2}\|x_n - x^*\|^2 \quad (n \geq 0); \tag{13.1.44}$$

(b) condition (13.1.1) holds and

$$\overline{U}(x^*, r_1^*) \subseteq D, \tag{13.1.45}$$

then, sequence $\{x_n\}$ generated by Newton's method (4.1.3) is well defined, remains in $U(x^*, r_1^*)$ for all $n \geq 0$ and converges to x^* provided that $x_0 \in U(x^*, r_1^*)$ with

$$\|x_{n+1} - x^*\| \leq \frac{w}{2}\|x_n - x^*\|^2 \quad (n \geq 0). \tag{13.1.46}$$

Proof. (a) Using (13.1.42), (13.1.43) and induction on the integer k we get

$$\|x_{k+1} - x^*\| \leq \frac{v}{2}\|x_k - x^*\|^2 < \|x_k - x^*\| < r^* \tag{13.1.47}$$

which shows $x_k \in U(x^*, r^*)$ for all k and $\lim_{k\to\infty} x_k = x^*$.

(b) Similarly using (3), (45) and induction on the integer k we get

$$\|x_{k+1} - x^*\| \leq \frac{w}{2}\|x_k - x^*\|^2 < \|x_k - x^*\| < r^*. \tag{13.1.48}$$

Let us further introduce the conditions:

$$\|F'(x^*)^{-1}[F'(x^* + t(x - x^*)) - F'(x)](x - x^*)\| \leq v_1(1-t)\|x - x^*\|^2 \tag{13.1.49}$$

and

$$\|F'(x^*)^{-1}[F'(x) - F'(x^*)]\| \leq v_0\|x - x^*\|. \tag{13.1.50}$$

Then since

$$\|F'(x)^{-1}z\| \leq \|[F'(x)^{-1}F'(x^*)]\| \, \|[F'(x^*)^{-1}z]\| \tag{13.1.51}$$

if we set

$$r_2^* = \frac{2}{2v_0 + v_1} \tag{13.1.52}$$

then we showed (see Section 5.1) that Newton's method converges to x^* provided that $x_0 \in U(x^*, r_2^*)$ and

$$U(x^*, r_2^*) \subseteq D. \tag{13.1.53}$$

In general

$$v_1 \leq v \leq w \tag{13.1.54}$$

holds and $\frac{v}{v_1}$, $\frac{w}{v_1}$, $\frac{w}{v}$ can be arbitrarily large. Therefore by (13.1.41) and (13.1.42)

$$r_1^* \leq r^*.$$

■

We can now provide an example where we compare radii r^*, r_1^* and r_2^*.

Example 13.1.2 *Let $X = Y = \mathbf{R}$, $D = \overline{U}(0,1)$, $x^* = 0$, and define function F on D by*

$$F(x) = e^x - 1. \tag{13.1.55}$$

Using (13.1.1), (13.1.40), (13.1.49), (13.1.50) we obtain

$$w = e, \quad v = e - 1, \quad v_0 = e - 1 \text{ and } v_1 = e. \tag{13.1.56}$$

By (13.1.41), (13.1.42), (13.1.52), (13.1.56) we get

$$r_1^* = .735758882$$
$$r^* = 1.163953414$$

and

$$r_2^* = .324947231.$$

We need to reset $r^* = 1$ so that (13.1.53) holds.

It follows that since $r_1^* < r^*$, our approach provides a wider choice of initial guesses x_0 than in the local convergence results using the standard condition (13.1.1).

13.2 A conjugate gradient solver for the normal equations

In this section a weaker version of the affine covariant Lipschitz condition appearing in connection with Newton–Mysovskikh-type theorems for the semilocal convergence of Newton methods is introduced. It turns out under weaker hypotheses and less computational cost than before [142], [194], [263] a semilocal convergence analysis can be provided for Newton methods. This approach in particular on one hand extends the usage of the conjugate gradient solver for normal equations and on the other hand provides finer error estimates on the distances involved. A simple numerical example justifies our approach.

In this study we are concerned with the problem of approximating a locally unique solution x^* of equation

$$F(x) = 0, \tag{13.2.1}$$

where operator F is defined on an open convex subset D of the n-dimensional Euclidean space $\mathbf{R}^n$ with values in itself.

The most popular method for approximating x^* is undoubtedly Newton's method.

Newton's method computes iterates successively as the solution of linear algebraic systems

$$\begin{aligned} F'(x^k)\Delta x^k &= -F(x^k), \quad k \in \mathbb{N}_0, \\ x^{k+1} &= x^k + \Delta x^k. \end{aligned} \tag{13.2.2}$$

The classical convergence theorems of Newton–Kantorovich and Newton–Mysovsikikh and its affine covariant, affine contravariant, and affine conjugate versions assume the exact solution of (13.2.2).

In practice however, in particular if the dimension is large, (13.2.2) will be solved by an iterative method. In this case, we end up with an outer/inner iteration, where the outer iterations are the Newton steps and the inner iterations result from the application of an iterative scheme to (13.2.2). It is important to tune the outer and inner iterations and to keep track of the iteration errors.

With regard to affine covariance, affine contravariance, and affine conjugacy the iterative scheme for the inner iterations has to be chosen in such a way, that it easily provides information about the

- error norm in case of affine covariance,
- residual norm in case of affine contravariance, and
- energy norm in case of affine conjugacy.

Except for convex optimization, we cannot expect $F'(x)$, $x \in D$, to be symmetric positive definite. Hence, for affine covariance and affine contravariance we have to pick iterative solvers that are designed for nonsymmetric matrices. Appropriate candidates are

- CTGNE (<u>C</u>onjugate <u>G</u>radient for the <u>N</u>ormal <u>E</u>quations) in case of affine covariance,
- GMRES (<u>G</u>eneralized <u>M</u>inimum <u>RES</u>idual) in case of affine contravariance, and
- PCG (<u>P</u>reconditioned <u>C</u>onjuate <u>G</u>radient) in case of affine conjugacy.

Although our results can be extended to all solvers mentioned above we only report on CGN.

In this study we are motivated by the elegant work of R. Hoppe in [194] (see also related work by E. Cătinaş [118], [119]). Hoppe's work is based on the following assumption: Suppose that $F\colon D \subseteq \mathbf{R}^n \to \mathbf{R}^n$ is continuously differentiable on D with invertible Fréchet derivative $F'(x)$, $x \in \mathbb{R}^n$. Assume further that the following affine covariant Lipschitz condition is satisfied:

$$\|F'(z)^{-1}[F'(y) - F'(x)]v\| \leq w\|y - x\| \, \|v\| \text{ for all } x, y, z \in D, \ v \in \mathbf{R}^n. \tag{13.2.3}$$

Condition (13.2.3) together with the sufficient convergence Newton–Mysovskikh hypothesis

$$H = w\|\Delta x^0\| < 2 \tag{13.2.4}$$

form the basic hypotheses of the main results (Theorems 3.1 and 3.2 in [194]). However there are even simple scalar examples where hypothesis (13.2.4) does not hold and consequently there is no guarantee that method (13.2.2) converges to x^*:

Example 13.2.1 *Let* $n = 1$, $x^0 = 1$, $D = [a, 2-a]$, $a \in \left[0, \frac{1}{2}\right)$ *and define function* F *on* D *by*

$$F(x) = x^3 - a. \tag{13.2.5}$$

It follows by (13.2.2), (13.2.3) and (13.2.5) that

$$\|\Delta x^0\| = \frac{1}{3}(1-a), \quad w = \frac{2(2-a)}{a^2}. \tag{13.2.6}$$

By (13.2.4) and (13.2.6) we obtain

$$H = \frac{2}{3a^2}(1-a)(2-a). \tag{13.2.7}$$

However (13.2.4) does not hold since $H > 2$ *for all* $a \in \left[0, \frac{1}{2}\right)$. *That is the results obtained in [194] cannot be applied in this case.*

We noticed that all the results obtained in [194] hold in a weaker setting if (13.2.3) and (13.2.4) are replaced by the weaker (13.2.10) and (13.2.25) respectively. Moreover error bounds $\|x^{k+1} - x^k\|$, $\|x^k - x^*\|$ *and the relative error are finer. This information is important in computational mathematics [99], [142], [194], [263].*

To make the section as self-contained as possible we reintroduce some of the terminology and concepts in [263].

Affine Covariant Inexact Newton Methods

CGNE (<u>C</u>onjugate <u>G</u>radient for the <u>N</u>ormal <u>E</u>quations)

We assume $A \in R^{n \times n}$ to be a regular, nonsymmetric matrix and $b \in R^n$ to be given and look for $y^* \in R^n$ as the unique solution of the linear algebraic system

$$Ay = b.$$

As the name already suggests, CGNE is the conjugate gradient method applied to the normal equations:

It solves the system

$$AA^T z = b,$$

for z and then computes y according to

$$y = A^T z.$$

The implementation of CGNE is as follows:

CGNE Initialization:

Given an initial guess $y_0 \in R^n$, compute the residual $r_0 = b - Ay_0$ and set

$$p_0 = r_0, \qquad p_0 = 0,$$
$$\beta_0 = 0, \qquad \sigma_0 = \|r_0\|^2.$$

CGNE Iteration Loop: For $1 \le i \le i_{\max}$ compute

$$p_i = A^T r_{i-1} + \beta_{i-1} p_{i-1}, \qquad \alpha_i = \frac{\sigma_{i-1}}{\|p_i\|^2},$$

$$y_i = y_{i-1} - \alpha_i p_i, \qquad \gamma_{i-1}^2 = \alpha_i \sigma_{i-1},$$
$$r_i = r_{i-1} - \alpha_i A p_i, \qquad \sigma_i = \|r_i\|^2,$$
$$\beta_i = \frac{\sigma_i}{\sigma_{i-1}}.$$

CGNE has the error minimizing property

$$\|y^* - y_i\| = \min_{v \in \mathcal{K}_i(A^T r_0, A^T A)} \|y^* - v\|,$$

where $K_i(A^T r_0, A^T A)$ stands for the Krylov subspace

$$\mathcal{K}_i(A^T r_0, A^T A) := \operatorname{span}\{A^T r_0, (A^T A)A^T r_0, \ldots, (A^T A)^{i-1} A^T r_0\}.$$

Lemma 13.2.1 *[194]. Representation of the iteration error. Let $\varepsilon_i := \|y^* - y_i\|^2$ be the square of the CGNE iteration error with respect to the ith iterate. Then, there holds*

$$\varepsilon_i = \sum_{j=i}^{n-1} \gamma_j^2.$$

Proof. CGNE has the Galerkin orthogonality

$$(y_i - y_0, y_{i+m} - y_i) = 0, \quad m \in \mathbb{N}.$$

Setting $m = 1$, this implies the orthogonal decomposition

$$\|y_{i+1} - y_0\|^2 = \|y_{i+1} - y_i\|^2 + \|y_i - y_0\|^2,$$

which readily gives

$$\|y_i - y_0\|^2 = \sum_{j=0}^{i-1} \|y_{j+1} - y_j\|^2 = \sum_{j=0}^{i-1} \gamma_j^2.$$

On the other hand, observing $y_n = y^*$, for $m = n - i$ the Galerkin orthogonality yields

$$\underbrace{\|y^* - y_0\|^2}_{= \sum_{j=0}^{n-1} \gamma_j^2} = \underbrace{\|y^* - y_i\|^2}_{= \varepsilon_i^2} + \underbrace{\|y_i - y_0\|^2}_{= \sum_{j=0}^{i-1} \gamma_j^2}.$$

■

Computable lower bound for the iteration error

It follows readily from Lemma 13.2.1 that the computable quantity

$$[\varepsilon_i] := \sum_{j=i}^{i+m} \gamma_j^2, \quad m \in \mathbb{N},$$

provides a lower bound for the iteration error.

In practice, we will test the relative error norm according to

$$\delta_i := \frac{\|y^* - y_i\|}{\|y_i\|} \approx \frac{\sqrt{[\varepsilon_i]}}{\|y_i\|} \leq \overline{\delta},$$

where $\overline{\delta}$ is a user specified accuracy.

Convergence of Affine Covariant Inexact Newton Methods

We denote by $\delta x^k \in \mathbb{R}^n$ the result of an inner iteration, e.g., CGNE, for the solution of (13.2.2). Then, it is easy to see that the iteration error $\delta x^k - \Delta x^k$ satisfies the error equation

$$F'(x^k)(\delta x^k - \Delta x^k) = F(x^k) + F'(x^k)\delta x^k =: r^k. \tag{13.2.8}$$

We will measure the impact of the inexact solution of (13.2.2) by the **relative error**

$$\delta_k := \frac{\|\delta x^k - \Delta x^k\|}{\|\delta x^k\|}. \tag{13.2.9}$$

Theorem 13.2.2 *Affine covariant convergence theorem for the inexact Newton method. Part I: Linear Converence*

Suppose that $F\colon D \subset R^n \to R^n$ is continuously differentiable on D with invertible Fréchet derivatives $F'(x)$, $x \in R^n$. Assume further that the following affine covariant Lipschitz condition is satisfied

$$\|F'(y)^{-1}\big(F'(x) + t(y-x) - F'(x)\big)v\| \leq w_1\|y-x\|t\|v\|, \tag{13.2.10}$$

where $x, y, z \in D$, $v \in R^n$, $t \in [0,1]$.

Assume that $x^0 \in D$ is an initial guess for the outer Newton iteration and that $\delta x^0 = 0$ is chosen as the start iterate for the inner iteration. Consider the Kantorovich quantities

$$h_k := \omega_1\|\Delta x^k\|, \quad h_k^\delta := \omega_1\|\delta x^k\| = \frac{h_k}{\sqrt{1+\delta_k^2}} \tag{13.2.11}$$

associated with the outer and inner iteration.

Assume that

$$h_0 < 2\overline{\Theta}, \quad 0 \le \overline{\Theta} < 1, \tag{13.2.12}$$

and control the inner iterations according to

$$\vartheta(h_k, \delta_k) := \frac{\frac{1}{2}h_k^\delta + \delta_k(1 + h_k^\delta)}{\sqrt{1+\delta_k^2}} \le \overline{\Theta} < 1, \tag{13.2.13}$$

which implies linear convergence.

Note that a necessary condition for $\vartheta(h_k, \delta_k) \le \overline{\Theta}$ is that it holds true for $\delta_k = 0$, which is satisfied due to assumption (13.2.2).

Then, there holds:

Theorem 13.2.3 *The Newton CGNE iterates x^k, $k \in N_0$ stay in*

$$\overline{B}(x^0, \rho), \quad \rho := \frac{\|\delta x^0\|}{1 - \overline{\Theta}} \tag{13.2.14}$$

and converges linearly to some $x^* \in \overline{B}(x^0, \rho)$ with $f(x^*) = 0$.

The exact Newton increments decrease monotonically according to

$$\frac{\|\Delta x^{k+1}\|}{\|\Delta x^k\|} \le \overline{\Theta}, \tag{13.2.15}$$

whereas for the inexact Newton increments we have

$$\frac{\|\delta x^{k+1}\|}{\|\delta x^k\|} \le \frac{\sqrt{1+\delta_k^2}}{\sqrt{1+\delta_{k+1}^2}}\overline{\Theta} \le \overline{\Theta}. \tag{13.2.16}$$

Proof. Using (13.2.2) we can obtain in turn:

$$\begin{aligned}
\|\Delta x^{k+1}\| &= \|F'(x^{k+1})^{-1}F(x^{k+1})\| \\
&= \|F'(x^{k+1})^{-1}\big[F(x^{k+1}) - F(x^k)\big] + F'(x^{k+1})^{-1} \underbrace{F(x^k)}_{= r^k - F'(x^k)\delta x^k} \| \\
&= \|F'(x^{k+1})^{-1}\big[(x^{k+1}) - F(x^k) - F'(x^k)\delta x^k\big]\| \\
&\quad + \|F'(x^{k+1})^{-1} \underbrace{r^k}_{= F'(x^k)(\delta x^k - \Delta x^k)} \| \\
&\le \underbrace{\int_0^1 \|F'(x^{k+1})^{-1}\big[F'(x^k + t\delta x^k) - F'(x^k)\big]\delta x^k\|dt}_{=:\mathbf{I}} \\
&\quad + \underbrace{\|F'(x^{k+1})^{-1}F'(x^k)(\delta x^k - \Delta x^k)\|}_{=:\mathbf{II}}.
\end{aligned}$$

(13.2.17)

Using the affine covariant Lipschitz condition (13.2.10), the first term on the right-hand side in (13.2.17) can be estimated according to

$$\mathbf{I} \le \omega_1 \|\delta x^k\|^2 \int_0^1 t\,dt = \frac{1}{2}\omega_1 \|\delta x^k\|^2. \tag{13.2.18}$$

For the second term we obtain by the same argument

$$\begin{aligned} \mathbf{II} &= \|F'(x^{k+1})^{-1}\big[F'(x^k)(\delta x^k - \Delta x^k) \pm F'(x^{k+1})(\delta x^k - \Delta x^k)\big]\| \\ &\le \|F'(x^{k+1})^{-1}(F'(x^{k+1}) - F'(x^k))(\delta x^k - \Delta x^k)\| \\ &\quad + \|F'(x^{k+1})^{-1}F'(x^{k+1})(\delta x^k - \Delta x^k)\| \\ &\le \frac{1}{2}\omega_1 \|\delta x^k\|\,\|\delta x^k - \Delta x^k\| + \|\delta x^k - \Delta x^k\|^2. \end{aligned} \tag{13.2.19}$$

Combining (13.2.18) and (13.2.19) yields

$$\begin{aligned} \frac{\|\delta x^{k+1}\|}{\|\delta x^k\|} &\le \frac{1}{2}\underbrace{\omega_1\|\delta x^k\|}_{=h_k^\delta} + \underbrace{\frac{1}{2}\omega_1\|\delta x^k\|\frac{\|\delta x^k - \Delta x^k\|}{\|\delta x^k\|}}_{=\delta_k\ h_k^\delta} \\ &\quad + \underbrace{\frac{\|\delta x^k - \Delta x^k\|}{\|\delta x^k\|}}_{=\delta_k} \le h_k^\delta + \delta_k(1 + h_k^\delta). \end{aligned}$$

Observing (13.2.11), we finally get

$$\frac{\|\Delta x^{k+1}\|}{\|\Delta x^k\|} \le \vartheta(h_k, \delta_k) = \frac{\frac{1}{2}h_k^\delta + \delta_k(1 + h_k^\delta)}{\sqrt{1+\delta_k^2}} \le \overline{\Theta} < 1, \tag{13.2.20}$$

which implies linear convergence.

Note that a necessary condition for $\vartheta(h_k, \delta_k) \le \overline{\Theta}$ is that it holds true for $\delta_k = 0$, which is satisfied due to assumption (13.2.12).

For the contraction of the inexact Newton increments we get

$$\frac{\|\delta x^{k+1}\|}{\|\delta x^k\|} = \sqrt{\frac{1+\delta_k^2}{1+\delta_{k+1}^2}}\frac{\|\Delta x^{k+1}\|}{\|\Delta x^k\|} \le \sqrt{\frac{1+\delta_k^2}{1+\delta_{k+1}^2}}\overline{\Theta} \le \overline{\Theta}. \tag{13.2.21}$$

It can easily be shown that $\{x^k\}_{\mathbb{N}_0}$ is a Cauchy sequence in $\overline{B}(x^0, \rho)$. Consequently, there exists $x^* \in \overline{B}(x^0, \rho)$ such that $x^k \to x^*$ $(k \to \infty)$. Since

$$\underbrace{F'(x^k)\delta x^k}_{\to 0} = \underbrace{F(x^k) + r^k}_{\to F(x^*)},$$

we conclude $F(x^*) = 0$. ■

Remark 13.2.1 *In general*

$$w_1 \leq w \tag{13.2.22}$$

holds and $\frac{w}{w_1}$ *can be arbitrarily large. If* $w_1 = w$ *and (13.2.10) is replaced by stronger (13.2.3) then our Theorem 13.2.2 reduces to Theorem 3.1 in [194, p. 20]. Otherwise our Theorem 13.2.2 improves Theorem 3.1 in [194]; under weaker hypotheses and less computational cost since in practice the computation of* w_1 *is less expensive than* w. *Note also that*

$$H < 2 \Rightarrow h_0 < 2 \tag{13.2.23}$$

but not vice versa unless if $w_1 = w$.

Returning back to Example 13.2.1 and using (13.2.10) we obtain

$$w_1 = \frac{2}{a} < \frac{2(2-a)}{a^2} = w \quad \textit{for all } a \in \left(0, \frac{1}{2}\right). \tag{13.2.24}$$

Crucial hypothesis

$$h_0 < 2 \tag{13.2.25}$$

becomes

$$\frac{2(1-a)}{3a} < 2 \tag{13.2.26}$$

which holds for all $a \in \left(\frac{1}{4}, \frac{1}{2}\right)$. *That is the conclusion of our Theorem 13.2.2 hold whereas the corresponding ones of Theorem 3.1 in [194] do not.*

In the next result we examine the uniqueness of the solution x^* in $\overline{U}(x^0, \rho)$ not studied in [194]:

Proposition 13.2.4 *Under the hypotheses of Theorem 13.2.2 the solution* x^* *of equation* $F(x) = 0$ *is unique in* $\overline{U}(x^0, \rho)$.

Proof. Let y^* be a solution of equation $F(x) = 0$ in $\overline{U}(x^0, \rho)$. Then

$$x^* + t(y^* - x^*) - x_0 = (1-t)(x^* - x_0) + t(y^* - x_0)$$

and

$$\begin{aligned} \|x^* + t(y^* - x^*) - x_0\| &\leq (1-t)\|x^* - x_0\| + t\|y^* - x_0\| \\ &\leq (1-t)\rho + t\rho = \rho. \end{aligned} \tag{13.2.27}$$

That is

$$x^* + t(y^* - x^*) \in \overline{U}(x^0, \rho). \tag{13.2.28}$$

Using the identity

$$0 = F(y^*) - F(x^*) = \int_0^1 F'(x^* + t(y^* - x^*))(y^* - x^*)\,dt = M(y^* - x^*), \quad (13.2.29)$$

and the hypothesis $F'(z)^{-1}$ is invertible for all $z \in \overline{U}(x^0, \rho)$ we conclude that for $z = x^* + t(y^* - x^*)$ linear operator M is invertible.

It follows from (13.2.29) that

$$x^* = y^*.$$

■

Theorem 13.2.5 *Affine covariant convergence theorem for the inexact Newton method. Part II: Quadratic convergence.*

Under the same assumptions on $F\colon D \subset \mathbf{R}^n \to \mathbf{R}^n$ as in Theorem 13.2.2 suppose that the initial guess $x^0 \in D$ satisfies

$$h_0 < \frac{2}{1+\rho} \quad (13.2.30)$$

for some appropriate $\rho > 0$ and control the inner iterations such that

$$\delta_k \le \frac{\rho}{2}\frac{h_k^\delta}{1+h_k^\delta}. \quad (13.2.31)$$

Then, there holds:

Theorem 13.2.6 *The Newton CGNE iterates x^k, $k \in N_0$ stay in*

$$\overline{B}(x^0, \overline{\rho}), \quad \overline{\rho} := \frac{\|\delta x^0\|}{1 - \frac{1+\rho}{2}h_0} \quad (13.2.32)$$

and converge quadratically to some $x^* \in \overline{B}(x^0, \overline{\rho})$ with $F(x^*) = 0$.

The exact Newton increments and the inexact Newton increments decrease quadratically according to

$$\|\Delta x^{k+1}\| \le \frac{1+\rho}{2}\omega_1 \|\Delta x^k\|^2, \quad (13.2.33)$$

$$\|\delta x^{k+1}\| \le \frac{1+\rho}{2}\omega_1 \|\delta x^k\|^2. \quad (13.2.34)$$

Proof. We proceed as in the proof of Theorem 13.2.2 to obtain

$$\frac{\|\Delta x^{k+1}\|}{\|\Delta x^k\|} \le \vartheta(h_k, \delta_k) = \frac{\frac{1}{2}h_k^\delta + \delta_k(1+h_k^\delta)}{\sqrt{1+\delta_k^2}}$$

and

$$\frac{\|\delta x^{k+1}\|}{\|\delta x^k\|} = \sqrt{\frac{1+\delta_k^2}{1+\delta_{k+1}^2}}\,\frac{\|\Delta x^{k+1}\|}{\|\Delta x^k\|}.$$

In view of (13.2.31) we get the further estimates

$$\frac{\|\Delta x^{k+1}\|}{\|\Delta x^k\|} \le \frac{1+\rho}{2}\,\frac{h_k}{1+\delta_k^2} \le \frac{1+\rho}{2}h_k$$

and

$$\frac{\|\delta x^{k+1}\|}{\|\delta x^k\|} \le \frac{1+\rho}{2}\,\frac{h_k^\delta}{\sqrt{1+\delta_{k+1}^2}} \le \frac{1+\rho}{2}h_k^\delta,$$

from which (13.2.33) and (13.2.34) follow by the definition of the Kantorovich quantities.

In order to deduce quadratic convergence we have to make sure that the initial increments ($k=0$) are small enough, i.e.,

$$\frac{1+\rho}{2}h_0^\delta \le \frac{1+\rho}{2}h_0 < 1. \tag{13.2.35}$$

Furthermore, (13.2.34) and (13.2.35) allow us to show that the iterates x^k, $k \in \mathbb{N}$ stay in $\overline{B}(x^0, \overline{\rho})$. Indeed, (13.2.34) implies

$$\|\delta x^j\| \le \frac{1+\rho}{2}h_{j-1}\|\delta x^{j-1}\| \le \frac{1+\rho}{2}h_0\|\delta x^{j-1}\|, \quad j \in \mathbb{N},$$

and hence

$$\|x^k - x^*\| \le \sum_{j=0}^{k}\|\delta x^j\| \le \|\delta x^0\|\sum_{j=0}^{k}\Big(\frac{1+\rho}{2}h_0\Big)^j \le \frac{\|\delta x^0\|}{1-\frac{1+\rho}{2}h_0}.$$

■

Algorithmic Aspects of Affine Covariant Inexact Newton Methods

(i) Convergence monitor

Let us assume that the quantity $\overline{\Theta} < 1$ in both the linear convergence mode and the quadratic convergence mode has been specified and let us further assume that we use CGNE with $\delta x_0^k = 0$ in the inner iteration.

Then, (13.2.16) suggests the monotonicity test

$$\tilde{\Theta}_k := \sqrt{\frac{1+\overline{\delta}_{k+1}^2}{1+\overline{\delta}_k^2}\,\frac{\|\delta x^{k+1}\|}{\|\delta x^k\|}} \le \overline{\Theta}, \tag{13.2.36}$$

where $\overline{\delta}_k^2$ and $\overline{\delta}_{k+1}^2$ are computationally available estimates of δ_k^2 and δ_{k+1}^2.

(ii) Termination criterion

We recall that the termination criterion for the exact Newton iteration with respect to a user specified accuracy $XTOL$ is given by

$$\frac{\|\Delta x^k\|}{1-\Theta_{k-1}^2} \leq \text{XTOL}\,.$$

According to (13.2.11) we have

$$\|\Delta x^k\| = \sqrt{1+\delta_k^2}\,\|\delta x^k\|.$$

Consequently, replacing Θ_{k-1} and δ_k by the computable quantities $\tilde{\Theta}_{k-1}$ and $\overline{\delta}_k$, we arrive at the termination criterion

$$\frac{\sqrt{1+\overline{\delta}_k^2}}{1-\tilde{\Theta}_{k-1}^2} \leq \text{XTOL}\,. \tag{13.2.37}$$

(iii) Balancing outer and inner iterations

According to (13.2.13) of Theorem 13.2.2, in the linear convergence mode the adaptive termination criterion for the inner iteration is

$$\vartheta(h_k,\delta_k) := \frac{\frac{1}{2}h_k^\delta + \delta_k(1+h_k^\delta)}{\sqrt{1+\delta_k^2}} \leq \overline{\Theta} < 1.$$

On the other hand, in view of (13.2.31) of Theorem 13.2.5, in the quadratic convergence mode the termination criterion is

$$\delta_k \leq \frac{\rho}{2}\,\frac{h_k^\delta}{1+h_k^\delta}\,.$$

Since the theoretical Kantorovich quantities (cf. (13.2.11))

$$h_k^\delta = \omega_1\|\delta x^k\| = \frac{h_k}{\sqrt{1+\delta_k^2}}$$

are not directly accessible, we have to replace them by computationally available estimates $[h_k^\delta]$.

We recall that for h_k we have the a priori estimate

$$[h_k] = 2\Theta_{k-1}^2 \leq h_k.$$

Consequently, replacing δ_k by $\overline{\delta}_k$, h_k by $[h_k]$, and Θ_{k-1} by $\tilde{\Theta}_{k-1}$ (cf. (13.2.36)), we get the a priori estimates

$$[h_k^\delta] = \frac{[h_k]}{\sqrt{1+\overline{\delta}_k^2}}\,, \quad [h_k] = 2\tilde{\Theta}_{k-1}^2, \quad k\in\mathbb{N}. \tag{13.2.38}$$

For $k = 0$, we choose $\overline{\delta}_0 = \delta_0 = \frac{1}{4}$.

In practice, for $k \geq 1$ we begin with the quadratic convergence mode and switch to the linear convergence mode as soon as the approximate contraction factor $\tilde{\Theta}_k$ is below some prespecified threshold value $\overline{\Theta} \leq \frac{1}{2}$.

(iii)$_1$ Quadratic convergence mode

The computationally realizable termination criterion for the inner iteration in the quadratic convergence mode is

$$\overline{\delta}_k \leq \frac{\rho}{2} \frac{[h_k^\delta]}{1 + [h_k^\delta]} \,. \tag{13.2.39}$$

Inserting (13.2.38) into (13.2.39), we obtain a simple nonlinear equation in $\overline{\delta}_k$.

Remark 13.2.2 *Validity of the approximate termination criterion. Observing that the right-hand side in (13.2.39) is a monotonically increasing function of $[h_k^\delta]$, and taking $[h_k^\delta] \leq h_k^\delta$ into account, it follows that for $\delta_k \leq \overline{\delta}_k$ the approximate termination criterion (13.2.39) implies the exact termination criterion (13.2.31).*

Remark 13.2.3 *Computational work in the quadratic convergence mode. Since $\overline{\delta}_k \to 0$ $(k \to \infty)$ is enforced, it follows that:*

The more the iterates x^k approach the solution x^*, the more computational work is required for the inner iterations to guarantee quadratic convergence of the outer iteration.

(iii)$_2$ Linear convergence mode

We switch to the linear convergence mode, once the criterion

$$\tilde{\Theta}_k < \overline{\Theta} \tag{13.2.40}$$

is met.

The computationally realizable termination criterion for the inner iteration in the linear convergence mode is

$$[\vartheta(h_k, \delta_k)] := \vartheta([h_k], \overline{\delta}_k) = \frac{\frac{1}{2}[h_k^\delta] + \overline{\delta}_k(1 + [h_k^\delta])}{\sqrt{1 + \overline{\delta}_k^2}} \leq \overline{\Theta} \,. \tag{13.2.41}$$

Remark 13.2.4 *Validity of the approximate termination criterion. Since the right-hand side in (13.2.41) is a monotonically increasing function in $[h_k^\delta]$ and $[h_k^\delta] \leq h_k^\delta$, the estimate provided by (13.2.41) may be too small and thus result in an overestimation of δ_k. However, since the exact quantities and their a priori estimates both tend to zero as k approaches infinity, asymptotically we may rely on (13.2.41).*

In practice, we require the monotonicity test (13.2.36) in CGNE and run the inner iterations until $\overline{\delta}_k$ satisfies (13.2.41) or divergence occurs, i.e.,

$$\tilde{\Theta}_k > 2\overline{\Theta} \,.$$

Remark 13.2.5 *Computational work in the linear convergence mode. As opposed to the quadratic convergence mode, we observe*

The more the iterates x^k approach the solution x^*, the less computational work is required for the inner iterations to guarantee linear convergence of the outer iteration.

Remark 13.2.6 *The results obtained here extend immediately in a Banach space setting provided $F\colon D \subseteq X \to Y$, where X and Y are Banach spaces (see Section 5.1).*

13.3 A finer mesh independence principle for Newton's method

This study is motivated by optimization considerations and presents a finer than before [145] affine invariant theory on the asymptotic mesh independence of Newton's method for discretized nonlinear operator equations. Using a weaker version of the Newton–Mysovskikh theorem (NMT) (see also Section 13.1), for finding solutions of nonlinear operator equations in a Banach space setting we provide a finer mesh independence of Newton's method with the following advantages over earlier works: (a) The usage of mesh independence is extended under weaker hypotheses; (b) error bounds are finer; (c) the information on the location of the solution is more precise; (d) Lipschitz constants are smaller; (e) discretization algorithms are faster and terminate earlier.

We are concerned with the problem of approximating a locally unique solution x^* of equation

$$F(x) = 0, \tag{13.3.1}$$

where F is a continuously Fréchet-differentiable operator defined on a convex subset D of a Banach space X with values in a Banach space Y.

The most popular iteration process for approximating x^* is undoubtedly Newton's method

$$F'(x^k)\Delta x^k = -F(x^k), \quad x^{k+1} = x^k + \Delta x^k, \quad k \geq 0, \ x_0 \in D. \tag{13.3.2}$$

The mesh independence principle [2] states that finite dimensional Newton methods, when applied to solve nonlinear equations on successively finer discretizations with comparable initial guesses, show almost the same convergence behavior on all sufficiently fine discretizations. The mesh independence principle in various forms has been used in mesh design by several authors [2], [22], [35], [44], [48], [51], [144], [145], [228]. Those papers however, lacked certain important features in the theoretical characterization that prohibited their application to discretized partial differential equations. This drawback was first avoided in [144], where the "asymptotic mesh independence" term was used.

Recently in the elegant paper [145] which is the motivation for our study a different approach is suggested, which is affine invariant but much simpler and also more natural from the algorithmic implementation point of view. The foundation of this paper is the Newton–Mysovskikh semilocal convergence theorem (see Section 13.1) which is based on the following: affine covariant Lipschitz condition

$$\|F'(z)^{-1}[F'(y) - F'(x)]v\| \leq w\|y - x\| \, \|v\| \text{ for all } x, y, z \in D, \ v \in X; \tag{13.3.3}$$

Newton–Mysovskikh conditions

$$H = w^1 \|\Delta x^0\| < 2 \tag{13.3.4}$$

and

$$\overline{U}(x^0, \rho^1) \subseteq D, \tag{13.3.5}$$

where,

$$\rho^1 = \frac{\|\Delta x^0\|}{1 - \frac{H}{2}}. \tag{13.3.6}$$

Under these conditions Newton's method (13.3.2) converges quadratically to x^* (see also Theorem 13.1.1).

However there are even simple scalar examples where hypothesis (13.3.4) does not hold and consequently there is no guarantee that method (13.3.2) starting at x^0 converges to x^* (see Example 13.1.1).

We noticed that all the results obtained in [145] hold in a weaker setting if (13.3.3)–(13.3.5) are replaced by the weaker (13.3.7)–(13.3.9) respectively. Indeed condition (13.3.5) is never used at full extend since weaker (13.3.7) is really needed in the proof of the theorem.

We provide the main semilocal convergence result for Newton's method (13.3.2):

Theorem 13.3.1 *Let F, D, X, Y be as in the introduction. Let $\|\cdot\|$ denote the norm in the domain space X. Suppose that $F'(x)$ is invertible for each $x \in D$. Assume that the following affine covariant Lipschitz condition is satisfied:*

$$\|F'(y)^{-1}[F'(x + t(y - x)) - F'(x)]v\| \leq w\|y - x\| \, \|v\| t \quad \textit{for all } x, y \in D, \quad v \in X. \tag{13.3.7}$$

For the initial guess $x^0 \in D$ assume that

$$h_0 = w\|\Delta x^0\| < 2 \tag{13.3.8}$$

and

$$\overline{U}(x^0, \rho) \subseteq D \tag{13.3.9}$$

for

$$\rho = \frac{\|\Delta x^0\|}{1 - \frac{h_0}{2}}. \tag{13.3.10}$$

Then sequence $\{x^k\}$ generated by Newton's method (13.3.2) is well defined, remains in $\overline{U}(x^0, \rho)$ for all $k \geq 0$ and converges to a unique solution $x^* \in \overline{U}(x^0, \rho)$. Moreover the following estimates hold for all $k \geq 1$:

$$\|x^{k+1} - x^k\| \leq \frac{1}{2} w \|x^k - x^{k-1}\|^2. \tag{13.3.11}$$

Proof. Simply repeat the proof of the Newton–Mysovskikh theorem (see Section 13.1) step by step and notice that weaker (13.3.7) can replace stronger condition (13.3.3) throughout the proof of the theorem.. ■

Remark 13.3.1 *In general*

$$w \leq w^1 \tag{13.3.12}$$

holds and $\frac{w^1}{w}$ can be arbitrarily large. If $w^1 = w$ and (13.3.7) is replaced by stronger condition (13.3.3) then our Theorem 13.3.1 reduces to the Newton–Mysovskikh theorem. Otherwise our Theorem 13.3.1 improves the (NMT) under weaker hypotheses and less computational cost since in practice the computation of w is less expensive than w^1. Note also that

$$H < 2 \Rightarrow h_0 < 2 \tag{13.3.13}$$

but not vice versa unless if $w^1 = w$.

Returning back to Example 13.1.1

Example 13.3.1 *Using (13.3.7) we get*

$$w = \frac{2}{a} < w^1 \quad \text{for all } a \in \left(0, \frac{1}{2}\right). \tag{13.3.14}$$

Crucial hypothesis (13.3.8) becomes

$$\frac{2(1-a)}{3a} < 2, \tag{13.3.15}$$

which holds for all $a \in \left(\frac{1}{4}, \frac{1}{2}\right)$. That is the conclusions of our Theorem 13.3.1 hold whereas the corresponding ones in the (NMT) in [145] do not.

The mesh independence of Newton's method (9.3.2)

Except in some special cases in practice, we can only solve discretized nonlinear equations of finite dimension, at best on a sequence of successively finer mesh levels, say

$$F_j(x_j) = 0, \quad j = 0, 1, \dots . \tag{13.3.16}$$

where $F_j\colon D_j \to Y_j$ denotes a nonlinear mapping defined on a convex domain $D_j \subset X_j$ of a finite-dimensional subspace $X_j \subset X$ with values in a finite dimensional space Y_j. The corresponding finite dimensional ordinary Newton method reads

$$F_j'(x_j^k)\Delta x_j^k = -F_j(x_j^k), \quad x_j^{k+1} = x_j^k + \Delta x_j^k, \quad k = 0, 1, \dots .$$

In each Newton step, a system of linear equations must be solved. This system can equally well be interpreted either as a discretization of the linearized operator equation (13.3.1) or as a linearization of the discrete nonlinear system (13.3.2). Again we assume that Theorem 13.3.1 holds, this time for the finite dimensional mapping F_j. Let ω_j denote the corresponding affine invariant Lipschitz constant. Then the quadratic convergence of this Newton method is governed by the relation

$$\|x_j^{k+1} - x_j^k\| \le \frac{1}{2}\omega_j \|x_j^k - x_j^{k-1}\|^2.$$

Under the assumptions of Theorem 13.3.1 there exist unique discrete solutions x_j^* on each level j. Of course, we want to choose appropriate discretization schemes such that

$$\lim_{j\to\infty} x_j^* = x^*. \tag{13.3.17}$$

We introduce a family of linear projections

$$\pi_j\colon X \to X_j, \quad j = 0, 1, \dots$$

assumed to satisfy the *stability condition*

$$q_j = \sup_{x\in W^*,\ x\neq 0} \frac{\|\pi_j x\|}{\|x\|} \le \overline{q} < \infty, \quad j = 0, 1, \dots . \tag{13.3.18}$$

The projection property $\pi_j^2 = \pi_j$ immediately gives rise to the lower bound

$$q_j \ge 1. \tag{13.3.19}$$

As a measure of the *approximation quality* we define

$$\delta_j = \sup_{x\in W^*,\ x\neq 0} \frac{\|x - \pi_j x\|}{\|x\|}, \quad j = 0, 1, \dots . \tag{13.3.20}$$

The rather natural idea that a refinement of the discretization improves the approximation quality is expressed by the asymptotic assumption

$$\lim_{j\to\infty} \delta_j = 0. \tag{13.3.21}$$

We now study the comparative behavior of discrete versus continuous Newton sequences.

We will consider the phenomenon of mesh independence of Newton's method in two steps. First, we will show that the discrete Newton sequence tracks the continuous Newton sequence closely, with a maximal distance bounded in terms of the mesh size; both of the Newton sequences behave nearly identical until, eventually, a small neighborhood of the solution is reached. Second, we prove the existence of affine invariant Lipschitz constants ω_j for the discretized problems, which approach the Lipschitz constant ω of the continuous problem in the limit $j \to \infty$; again, the distance can be bounded in terms of the mesh size. Upon combining these two lines, we finally establish the existence of locally unique discrete solutions x_j^* in a vicinity of the continuous solution x^*.

To begin with, we prove the following nonlinear perturbation lemma.

Lemma 13.3.2 *Consider two Newton sequences $\{x^k\}, \{y^k\}$ starting at initial guesses x^0, y^0 and continuing as*

$$x^{k+1} = x^k + \Delta x^k, \quad y^{k+1} = y^k + \Delta y^k,$$

where Δx^k, Δy^k are the corresponding ordinary Newton corrections. Assume the affine invariant Lipschitz condition (13.3.1) is satisfied. Then the following contraction result holds:

$$\|x^{k+1} - y^{k+1}\| \le \omega\left(\frac{1}{2}\|x^k - y^k\| + \|\Delta x^k\|\right)\|x^k - y^k\|. \tag{13.3.22}$$

Proof. Dropping the iteration index k we start with

$$\begin{aligned}
&x + \Delta x - y - \Delta y \\
&\quad = x - F'(x)^{-1}F(x) - y + F'(y)^{-1}F(y) \\
&\quad = x - F'(x)^{-1}F(x) + F'(x)^{-1}F(y) - F'(x)^{-1}F(y) - y + F'(y)^{-1}F(y) \\
&\quad = x - y - F'(x)^{-1}(F(x) - F(y)) + F'(x)^{-1}(F'(y) - F'(x))F'(y)^{-1}F(y) \\
&\quad = F'(x)^{-1}\left(F'(x)(x-y) - \int_{t=0}^{1} F'(y + t(x-y))(x-y)dt\right) \\
&\qquad + F'(x)^{-1}(F'(y) - F'(x))\Delta y.
\end{aligned}$$

Upon using assumption (13.3.7) we conclude that

$$\begin{aligned}
&\|x^{k+1} - y^{k+1}\| \\
&\quad \le \int_{t=0}^{1} \|F'(x^k)^{-1}\big(F'(x^k) - F'(y^k + t(x^k - y^k))\big)(x^k - y^k)\|\, dt \\
&\qquad + \|F'(x^k)^{-1}(F'(y^k) - F'(x^k))\Delta y^k\| \\
&\quad \le \frac{\omega}{2}\|x^k - y^k\|^2 + \omega\|x^k - y^k\|\, \|\Delta y^k\|,
\end{aligned}$$

which confirms (13.3.22).

With the above auxiliary result, we are now ready to study the relative behavior of discrete versus continuous Newton sequences. ■

The proofs of the rest of the results as identical to the corresponding ones in [145] are omitted. We simply replace the (NMT) with our Theorem 13.1.1.

Theorem 13.3.3 *In addition to the notation as already introduced, let* $x^0 = x_j^0 \in X_j$ *denote a given starting value such that the assumptions of Theorem 13.1.1 hold for the continuous Newton iteration, including*

$$h_0 = \omega\|\Delta x^0\| < 2.$$

For the discrete mapping F_j *and all arguments* $x_j \in D_j = S\big(x^0, \rho + 2/\omega\big) \cap X_j$ *define*

$$F_j'(x_j)\Delta x_j = -F_j(x_j), \quad F'(x_j)\Delta x = -F(x_j). \tag{13.3.23}$$

Assume that the discretization is fine enough such that

$$\|\Delta x_j - \Delta x\| \le \delta_j \le \frac{1}{2\omega} \tag{13.3.24}$$

uniformly for $x_j \in D_j$. *Then the following cases occur:*

1. *If* $h_0 \le 1 - \sqrt{1 - 2\omega\delta_j}$, *then*

$$\|x_j^k - x^k\| < 2\delta_j \le \frac{1}{\omega}, \quad k = 0, 1, \ldots .$$

2. *If* $1 - \sqrt{1 - 2\omega\delta_j} < h_0 < 1 + \sqrt{1 - 2\omega\delta_j}$, *then*

$$\|x_j^k - x^k\| \le \frac{1}{\omega}\big(1 + \sqrt{1 - 2\omega\delta_j}\big) < \frac{2}{\omega}, \quad k = 0, 1, \ldots .$$

In both cases I and II, the asymptotic result

$$\limsup_{k\to\infty} \|x_j^k - x^k\| \le \frac{1}{\omega}\big(1 - \sqrt{1 - 2\omega\delta_j}\big) < 2\delta_j \le \frac{1}{\omega}$$

holds.

We are interested in the question whether a discrete solution point x_j^* exists. The above tracking theorem, however, only supplies the following result.

Corollary 13.3.4 *Under the assumptions of Theorem 13.3.3, there exists at least one accumulation point*

$$\hat{x}_j \in \overline{S}(x^*, 2\delta_j) \cap X_j \subset S\left(x^*, \frac{1}{\omega}\right) \cap X_j,$$

which need not be a solution point of the discrete equations $F_j(x_j) = 0$.

In order to prove more, Theorem 13.3.1 directs us to study whether a Lipschitz condition of the kind (13.3.7) additionally holds.

Lemma 13.3.5 *Let Theorem 13.3.1 hold for the mapping* $F\colon X \to Y$. *For collinear* x_j, y_j, $y_j + v_j \in X_j$, *define quantities* $w_j \in X_j$ *and* $w \in X$ *according to*

$$F'(x_j)w = \big(F'(y_j + v_j) - F'(y_j)\big)v_j, \tag{13.3.25}$$

$$F_j'(x_j)w_j = \big(F_j'(y_j + v_j) - F_j'(y_j)\big)v_j. \tag{13.3.26}$$

Assume that the discretization method satisfies

$$\|w - w_j\| \leq \sigma_j \|v_j\|^2. \tag{13.3.27}$$

Then there exists constants

$$\omega_j \leq \omega + \sigma_j, \tag{13.3.28}$$

such that the affine invariant Lipschitz condition

$$\|w_j\| \leq \omega_j \|v_j\|^2$$

holds for the discrete Newton process.

Proof. The proof is a simple application of the triangle inequality

$$\|w_j\| \leq \|w\| + \|w_j - w\| \leq \omega\|v_j\|^2 + \sigma_j\|v_j\|^2 = (\omega + \sigma_j)\|v_j\|^2.$$

Finally, the existence of a unique solution x_j^* is a direct consequence. ■

Corollary 13.3.6 *Under the assumptions of Theorem 13.3.3 and Lemma 13.3.5 the discrete Newton sequence* $\{x_j^k\}$, $k = 0, 1, \ldots$ *converges q-quadratically to a unique discrete solution point*

$$x_j^* \in \overline{S}(x^*, 2\delta_j) \cap X_j \subset S\left(x^*, \frac{1}{\omega}\right) \cap X_j.$$

Proof. We just need to apply Theorem 13.3.1 to the finite dimensional mapping F_j with the starting value $x_j^0 = x^0$, and the affine invariant Lipschitz constant ω_j from (13.3.8). ■

Summarizing, we come to the following conclusion, at least in terms of the analyzed upper bounds: If the asymptotic properties

$$\lim_{j\to\infty} \delta_j = 0, \quad \lim_{j\to\infty} \sigma_j = 0,$$

can be shown to hold, then the convergence speed of the discrete ordinary Newton method is asymptotically just the one for the continuous ordinary Newton method. Moreover, if related initial guesses x^0 and x_j^0 and a common termination criterion are chosen, then even the number of iterations will be nearly the same.

13.4 Exercises

13.1. Show the classical affine covariant Newton-Mysovskikh theorem using condition (13.1.1) instead of (13.1.4).

13.2. Show Theorem 13.2.2 using condition (13.2.3) instead of (13.2.10).

13.3. Provide a numerical example to verify comments made in Remarks 13.2.4, 13.2.5.

13.4. Show how to go about extending the results of Section 13.2 to hold in a Banach space setting.

13.5. Show that under conditions (13.3.3)–(13.3.6) Newton Kantorovich method (4.1.3) converges quadratically to x^*.

13.6. Complete the details in the proof of Theorems 13.3.1, 13.3.3 and Corollary 13.3.4.

13.7. Provide a numerical example to justify the results of Section 13.3.

Bibliography

[1] Alefeld, G., Potra, F.A and Shen, Z., On the existence theorems of Kantorovich, Moore and Miranda, Computing Supplementum **15**, (2001) 21–28.

[2] Allgower, E. L., Böhmer, K., Potra, F. A., Rheinboldt, W. C., A mesh-independence principle for operator equations and their discretizations, SIAM J. Numer. Anal., **23,** (1986), no. 1, 160–169.

[3] Amat, S., Busquier, S., and Candela, V., A class of quasi-Newton generalized Steffensen methods on Banach spaces, *J. Comput. Appl. Math.* **149** (2002), 397–408.

[4] Amer, S.M., On the solution of nonlinear singular integral equations with shift in generalized Hölder space, Chaos. Soliton. Fractals, **12**, no. 7, (2001), 1323–1334.

[5] Anselone, P.M. and Moore, R.H., An extension of the Newton-Kantorovich method for solving nonlinear equations with applications to elasticity, J. Math. Anal. Appl. 13 (1996), 476–501

[6] Appell, J., DePascale, E., Lysenko, J.V. and Zabrejko, P.P., New results on Newton–Kantorovich approximations with applications to nonlinear integral equations, Numer. Funct. Anal. and Optimiz. 18, (1 & 2) (1997), 1–17.

[7] Argyros, I.K., Quadratic equations and applications to Chandrasekhar's and related equations, Bull. Austral. Math. Soc. **32** (1985), 275–297

[8] Argyros, I.K., On the cardinality of solutions of multilinear differential equations and applications, Intern. J. Math. and MAth. Sci. **9**, 4 (1986), 757–766

[9] Argyros, I.K., On the approximation of some nonlinear equations, Aequationes Mathematicae, **32** (1987), 87–95.

[10] Argyros, I.K., On polynomial equations in Banach space, perturbation techniques and applications, Intern. J. MAth. and Math. Sci. **10**, 1 (1987), 69–78.

[11] Argyros, I.K., Newton-like methods under mild differentiability conditions with error analysis, Bull. Austral. Math. Soc. **37** (1987), 131–147.

[12] Argyros, I.K., On Newton's method and nondiscrete mathematical induction, *Bull. Austral. Math. Soc.* **38** (1988), 131–140.

[13] Argyros, I.K., On the Secant method and fixed points of nonlinear equations, Monatschfte für Mathematik, **106** (1988), 85–94.

[14] Argyros, I.K., On a class of nonlinear integral equations arising in neutron transport, Aequationes Mathematicae, **36** (1988), 99–111

[15] Argyros, I.K., On the number of solutions of some integral equations arising in radiative transfer, Intern. J. Math. and Math. Sci. **12**, 2 (1989), 297–304

[16] Argyros, I.K., Improved error bounds for a certain class of Newton-like methods, J. Approx. Th. and Its Applic. **61** (1990), 80–98

[17] Argyros, I.K., Error founds for the modified secant method, BIT, **20** (1990), 92–200.

[18] Argyros, I.K., On the solution of equations with nondifferentiable operators and the Ptak error estimates, BIT, **30** (1990), 752–754.

[19] Argyros, I.K., On some projection methods for the approximation of implicit functions,Appl. Math. Letters, **32** (1990), 5–7.

[20] Argyros, I.K., The Newton–Kantorovich method under mild differentiability conditions and the Ptâk error estimates, *Monatschefte für Mathematik* **109**, 3 (1990), 110–128.

[21] Argyros, I.K., The secant method in generalized Banach spaces, Appl. Math. and Comp. **39** (1990), 111–121.

[22] Argyros, I, K. A mesh independence principle for operator equations and their discretizations under mild differentiability conditions, *Computing* **45**, (1990), 265–268.

[23] Argyros, I.K., On the convergence of some projection methods with perturbation, J. Comp. and Appl. Math., **36** (1991), 255–258.

[24] Argyros, I.K., On an application of the Zincenko method to the approximation of implicit functions, Public. Math. Debrecen, **39** (3–4) (1991), 1–7.

[25] Argyros, I.K., On an iterative algorithm for solving nonlinear equations, Beitrage zür Numerischen Math., **10**, 1 (1991), 83–92.

[26] Argyros, I.K., On a class of quadratic equations with perturbation, Funct. et Approx. Comm. Math. **XX** (1992), 51–63.

[27] Argyros, I.K., Improved error bounds for the modified secant method, Intern. J. Computer Math., **43**, (1&2) (1992), 99–109.

[28] Argyros, I.K., Some generalized projection methods for solving operator equations, J. Comp. and Appl. Math. **39**, 1 (1992), 1–6.

[29] Argyros, I.K., On the convergence of generalized Newton-methods and implicit functions, J. Comp. and Appl. Math. **43** (1992), 335–342.

[30] Argyros, I.K., On the convergence of inexact Newton-like methods, Publ. Math. Debrecen **42**, (1&2) (1992), 1–7.

[31] Argyros, I.K., On the convergence of a Chebysheff-Halley-type method under Newton-Kantorovich hypothesis, Appl. Math. Letters **5**, 5 (1993), 71–74.

[32] Argyros, I.K., Newton-like methods in partially ordered linear spaces, J. Approx. Th. and its Applic. **9**, 1 (1993), 1–10.

[33] Argyros, I.K., On the solution of undetermined systems of nonlinear equations in Euclidean spaces, Pure Math. Appl. **4**, 3 (1993), 199–209.

[34] Argyros, I.K., A convergence theorem for Newton-like methods under generalized Chen–Yamamato-type assumptions, *Appl. Math. Comp.* **61**, 1 (1994), 25–37.

[35] Argyros, I.K., On the discretization of Newton-like methods, *Internat. J. Computer. Math.* **52** (1994), 161–170.

[36] Argyros, I.K., A unified approach for constructing fast two-step Newton-like methods, Mh. Math. **119** (1995), 1–22.

[37] Argyros, I.K., Results on controlling the residuals of perturbed Newton-like methods on Banach spaces with aconvergence structure, Southwest J. Pure Appl. Math. **1** (1995), 32–38.

[38] Argyros, I.K., On the method of tangent hyperbolas, J. Appr. Th. Appl. **12**, 1 (1996), 78–96.

[39] Argyros, I.K., On an extension of the mesh-independence principle for operator equations in Banach space, Appl. Math. Lett. **9**, 3 (1996), 1–7.

[40] Argyros, I.K., A generalization of Edelstein's theorem on fixed points and applications, Southwest J. Pure Appl. Math. **2** (1996), 60–64.

[41] Argyros, I.K., Chebysheff-Halley-like methods in Banach spaces, Korean J. Comp. Appl. Math. **4**, 1 (1997), 83–107.

[42] Argyros, I.K., Concerning the convergence of inexact Newton methods, J. Comp. Appl. Math. **79** (1997), 235–247.

[43] Argyros, I.K., General ways of constructing accelerating Newton-like iterations on partially ordered topological spaces, Southwest J. Pure Appl. Math. **2** (1997), 1–12.

[44] Argyros, I.K., On a new Newton–Mysovskii-type theorem with applications to inexact Newton-like methods and their discretizations, *IMA J. Num. Anal.* **18** (1997), 43–47.

[45] Argyros, I.K., On the convergence of two-step methods generated by point-to-point operators, Appl. Math. Comput. **82**, 1 (1997), 85–96.

[46] Argyros, I.K., Improved error bounds for Newton-like iterations under Chen–Yamamoto assumptions, *Appl. Math. Letters*, **10**, 4 (1997), 97–100.

[47] Argyros, I.K., Inexact Newton methods and nondifferentiable operator equations on Banach spaces with a convergence structure, *Approx. Th. Applic.* **13**, 3 (1997), 91–104.

[48] Argyros, I.K., A mesh independence principle for inexact Newton-like methods and their discretizations under generalized Lipschitz conditions, Appl. Math. Comp. **87** (1997), 15–48.

[49] Argyros, I.K., Concerning the convergence of inexact Newton methods, J. Comp. Appl. Math. **79** (1997), 235–247.

[50] Argyros, I.K., Smoothness and perturbed Newton-like methods, Pure Math. Appl. **8**,1 (1997), 13–28.

[51] Argyros, I.K., The asymptotic mesh independence principle for inexact Newton-Galerkin-like methods, Pure Math. Applic. **8**, (2 & 3) (1997), 169–194.

[52] Argyros, I.K., On the convergence of a certain class of iterative procedures under relaxed conditions with applications, J. Comp. Appl. Math. **94** (1998), 13–21.

[53] Argyros, I.K., Sufficient conditions for constructing methods faster than Newton's, Appl. Math. Comp. **93** (1998), 169–181.

[54] Argyros, I.K., The Theory and Application of Abstract Polynomial Equations, St. Lucie/CRC/Lewis Publ. Mathematics Series, Boca Raton, Florida, 1998.

[55] Argyros, I.K., A new convergence theorem for the Jarratt method in Banach spaces, Computers and Mathematics with Applications, **36**, 8 (1998), 13–18.

[56] Argyros, I.K., Improving the order and rates of convergence for the Super-Halley method in Banach spaces, Comp. Appl. Math. **5**, 2 (1998), 465–474.

[57] Argyros, I.K., Improved error bounds for a Chebysheff-Halley-type method, Acta Math. Hungarica, **84**, 3 (1999), 211–221.

[58] Argyros, I.K., Relations between forcing sequences and inexact Newton iterates in Banach space, *Computing* **63** (1999), 131–144.

[59] Argyros, I.K., Convergence domains for some iterative processes in Banach spaces using outer and generalized inverses, *Comput. Anal. Applic.* **1**, 1 (1999), 87–104.

[60] Argyros, I.K., Concerning the convergence of a modified Newton-like method, Journal for Analysis and its Applications (ZAA), **18**, 3 (1999), 1–8.

[61] Argyros, I.K., Convergence domains for some iterative processes in Banach spaces using outer and generalized inverses, *J. Comput. Anal. and Applic.* **1**, 1 (1999), 87–104.

[62] Argyros, I.K., Concerning the radius of convergence of Newton's method and applications, Korean J. Comp. Appl. Math. **6**, 3 (1999), 451–462.

[63] Argyros, I.K., Convergence rates for inexact Newton-like methods of singular points and applications, Appl. Math. Comp. **102** (1999), 185–201.

[64] Argyros, I.K., Choosing the forcing sequences for inexact Newton methods in Banach space, Comput. Appl. Math. **19**, 1 (2000), 79–89.

[65] Argyros, I.K., A survey of efficient numerical methods for solving equations and applications, Kyung Moon Publ., Seoul, Korea, 2000.

[66] Argyros, I.K., Local convergence of inexact Newton-like iterative methods and applications, *Computers and Mathematics with Application* **39** (2000), 69–75.

[67] Argyros, I.K., A mesh independence principle for perturbed Newton-like methods and their discretizations,Korean J. Comp. Appl. Math. **7**, 1 (2000), 139–159.

[68] Argyros, I.K., Advances in the efficiency of computational methods and applications, World Scientific Publ. Co., River Edge, NJ, 2000.

[69] Argyros, I.K., Newton methods on Banach spaces with a convergence structure and applications, *Computers and Math. with Appl. Intern. J. Pergamon Press* **40**, 1 (2000), 37–48.

[70] Argyros, I.K., Forcing sequences and inexact Newton iterates in Banach space, Appl. Math. Letters. **13** (2000), 77–80.

[71] Argyros, I.K., Choosing the forcing sequences for inexact Newton methods in Banach space, Comput. Appl. Math., **19**, 1 (2000), 79–89.

[72] Argyros, I.K., Local convergence of inexact Newton-like iterative methods and applications, Computers and Mathematics with Application **39** (2000), 69–75.

[73] Argyros, I.K., Semilocal convergence theorems for a certain class of iterative procedures using outer or generalized inverses, Korean J. Comp. Appl. Math. **7**, 1 (2000), 29–40.

[74] Argyros, I.K., The effect of rounding errors on a certain class of iterative methods, Applicationes Mathematicae, **27**, 3 (2000), 369–375.

[75] Argyros, I.K., Local convergence of Newton's method for nonlinear equations using outer or generalized inverses, Chechoslovak Math. J., **50**, (125), (2000), 603–614.

[76] Argyros, I.K., On a class of nonlinear implicit quasivariational inequalities, Pan American Math. J. **10**, 4 (2000), 101–109.

[77] Argyros, I.K., A new semilocal convergence theorem for Newton's method in Banach space using hypotheses on the second Fréchet-derivative, *J. Comput. Appl. Math.* **139**, (2001), 369–373.

[78] Argyros, I.K., On the radius of convergence of Newton's method, *Intern. J. Comput. Math.* **77**, 3 (2001), 389–400.

[79] Argyros, I.K., A Newton-Kantorovich theorem for equations involving m-Frechet differentiable operators and applications in radiative transfer, Journ. Comp. Appl. Math. **131**, 1–2 (2001), 149–159.

[80] Argyros, I.K., Semilocal convergence theorems for Newton's method using outer inverses and hypotheses on the second Fréchet-derivative, Monatshefte fur Mathematik, **132** (2001), 183–195.

[81] Argyros, I.K., On general auxiliary problem principle and nonlinear mixed variational inequalities, Nonlinear Functional Analysis and Applications **6**, 2 (2001), 247–256.

[82] Argyros, I.K., On an iterative procedure for approximating solutions of quasi variational inequalities, Advances in Nonlinear Variational Inequalities, **4**, 2 (2001), 39–42.

[83] Argyros, I.K., On generalized variational inequalities, *Advances in Nonlinear Variational Inequalities*, **4**, 2 (2001), 75–78.

[84] Argyros, I.K., On a semilocal convergence theorem for a class of quasi variational inequalities, *Advances in Nonlinear Inequalities*, **4**, 2 (2001), 43–46.

[85] Argyros, I.K., On the convergence of a Newton-like method based on m-Fréchet-differentiable operators and applications in radiative transfer, *J. Comput. Anal. Applic.*, **4**, 2 (2002), 141–154.

[86] Argyros, I.K., A unifying semilocal convergence theorem for Newton-like methods based on center Lipschitz conditions, Comput. Appl. Math. **21**, 3 (2002), 789–796

[87] Argyros, I.K., A unifying semilocal convergence theorem for Newton-like methods based on center Lipschitz conditions, *Comput. Appl. Math.*, **21**, 3 (2002), 789–796.

[88] Argyros, I.K., A semilocal convergence analysis for the method of tangent hyperbolas, *Journal of Concrete and Applicable Analysis*, **1**, 2 (2002), 135–144.

[89] Argyros, I.K., New and generalized convergence conditions for the Newton-Kantorovich method, *J. Appl. Anal.*, **9**, 2 (2003), 287–299.

[90] Argyros, I.K., On the convergence and application of Newton's method under weak Hölder continuity assumptions, *International Journal of Computer Mathematics*, **80**, 5 (2003), 767–780.

[91] Argyros, I.K., On a theorem of L.V. Kantorovich concerning Newton's method, *Journ. Comp. Appl. Math.*, **155**, (2003), 223–230.

[92] Argyros, I.K., An improved error analysis for Newton–like methods under generalized conditions, *J. Comput. Appl. Math.*, ***157***, 1 (2003), 169–185.

[93] Argyros, I.K., A unifying local-semilocal convergence analysis and applications for two-point Newton-like methods in Banach spaces, J. Math. Anal. Appl., **298**, 2, (2004), 374–397.

[94] Argyros, I.K., New sufficient conditions for the secant method, Chechoslovak Math. J., **55**, 130 (2005), 175–187.

[95] Argyros, I.K., and Chen, D., A fourth order iterative method in Banach spaces, *Appl. Math. Lett.*, **6**, 4 (1993), 97–98.

[96] Argyros, I.K., and Chen, D., A note on the Halley method in Banach spaces, *Appl. Math. Comp.*, **58**, (1993), 215–224.

[97] Argyros, I.K. and Szidarovszky, F., On the monotone of general Newton-like methods, *Bull. Austral. Math. Soc.*, **45**, (1992), 489–502.

[98] Argyros, I.K. and Szidarovszky, F., Convergence of general iteration schemes, *J. Math. Anal. and Applic.* **168**, (1992), 42–62.

[99] Argyros, I.K. and Szidarovszky, F., The Theory and Application of Iteration Methods, C.R.C. Press Inc., Boca Raton, Florida, 1993.

[100] Argyros, I.K. and Szidarovszky, F., On the convergence of modified contractions, *J. Cmput. Appl. Math.*, **55**, 2 (1994), 97–108.

[101] Atkinson, K.E., The numerical evaluation of fixed points for completely continuous operators, *SIAM J. Num. Anal.*, **10**, (1973), 799–807.

[102] Atkinson, K.E., A Survey of Numerical Methods for the Solution of Fredholm Integral Equations of the Second Kind, SIAM, Philadelphia, 1976.

[103] Avila, J.H., The feasibility of continuation methods for nonlinear equations, SIAM J. Numer. Anal., **11**, 1, (1974), 102–122.

[104] Begehr, H. and Efendiev, M.A., On the asymptotical meromorphic solutions for nonlinear Riemann-Hilbert problems, Math. Proc. Cambridge Philos. Soc., **127**, 1, (1999), 159–172.

[105] Belluce, L.P. and Kirk, W.A., Fixed point theorems for a certain class of nonexpansive mappings, Proc. Amer. Math. Soc., **50**, (1969), 144–146.

[106] Ben-Israel, A. , A Newton-Raphson method for the solution of systems of operators, J. Math. Anal. Appl., **15,** (1966), 243–252.

[107] Ben-Israel, A. and Greville, T.N.E., Generalized Inverses: Theory and Applications, John Wiley and Sons, 1974.

[108] Berinde, V., Remarks on the convergence of the Newton-Raphson method, Revue d'analyse Numer. Th. Appr. **24**, (1 & 2) (1995), 15–21.

[109] Brent, R.P., Algorithms for Minimization Without Derivatives, Prentice Hall, Englewood Cliffs, New Jersey, 1973.

[110] Browder, F.E., and Petryshyn, W.V., The solution by iteration of linear functional equations in Banach spaces, *Bull. Amer. Math. Soc.* **72** (1996), 566–570.

[111] Browder, F.E., Nonexpansive nonlinear operators in Banach spaces, Proc. Nat. Acad. Sci. USA, **54**, (1965), 1011–1014

[112] Brown, P. N., A local convergence theory for combined inexact-Newton/finite-difference projection methods, *SIAM J. Numer. Anal.* **24**, (1987), 407–434.

[113] Brown, P. N. and Saad, Y., Convergence theory of nonlinear Newton-Krylov algorithms, *SIAM J. Optimiz.* **4**, 2 (1994), 297–230.

[114] Broyden, C.G., Dennis, J.E. and Moré, J., On the local and superlinear convergence of quasi-Newton methods, J. Inst. Math. Appl., **12**, (1973), 223–246.

[115] Candela, V. and Marquina, A., Recurrence relations for rational cubic methods I: The Halley method, *Computing*, **44**, (1990), 169–184.

[116] Caponetti, D, De Pascale, E, Zabreuiko, P.P., On the Newton-Kantorovich method in K-normed spaces, Rend. Circ. Mat. Palermo, Ser. II, **49** (2000), no. 3, 545–560.

[117] Cătinaş, E., On some iterative methods for solving nonlinear equations, *Rev. Anal. Numér. Théor. Approx.*, **23**, 1 (1994), 47–53.

[118] Cătinaş, E., Inexact perturbed Newton methods, and applications to a class of Krylov solvers, J. Optim. Theory Appl., **108** (2001) no. 3, 543–570

[119] Cătinaş, E., On the superlinear convergence of the successive approximations method, J. Optim. Theory Appl., **113** (2002) no. 3, 473–485.

[120] Chen, D., Kantorovich-Ostrowski convergence theorems and optimal error bounds for Jarratt's iterative method, *Intern. J. Computer. Math.* **31,** (3 & 4), (1990), 221–235.

[121] Chen, X. and Yamamoto, T., Convergence domains of certain iterative methods for solving nonlinear equations, *Numer. Funct. Anal. and Optimiz.*, **10** (1 & 2) (1989), 37–48.

[122] Chen, X. and Nashed, M.Z., Convergence of Newton-like methods for singular operator equations using outer inverses, *Numer. Math.*, **66** (1993), 235–257.

[123] Chen, X., Nashed, Z and Qi, L., Convergence of Newton's method for singular smooth and nonsmooth equations using adaptive outer inverses, SIAM J. Optim. **7** (1997), no. 2, 445–462.

[124] Chui, C.K. and Quak, F., Wavelets on a bounded interval. In: Numerical Methods of Approximation Theory, Vol. 9 (eds: D. Braess and Larry L. Schumaker, Intern. Ser. Num. Math., Vol. 105) Basel: Birkhäuser Verlag, pp. 53–75.

[125] Chow, S.N. and Hale, J.K., Methods of Bifurcation Theory, springer-Verlag, New-York, 1962.

[126] Cianciaruso, F, De Pascale, E., Zabrejko, P. P., Some remarks on the Newton-Kantorovich approximations, Atti. Sem. Mat. Fis. Univ. Modena, **48** (2000), 207–215.

[127] Cohen, G., Auxiliary problem principle and decomposition of optimization problems, J. Optim. Theory Appl., **32** (1980), no. 3, 277–305.

[128] Cohen, G., Auxiliary problem principle extended to variational inequalities, J. Optim. Theory Appl., **59**, (1988), no. 2, 325–333.

[129] Collatz, L., Functional Analysis and Numerisch Mathematik, Springer-Verlag, New York, 1964.

[130] Danes, J., Fixed point theorems, Nemyckii and Uryson operators, and continuity of nonlinear mappings, Comment. Math. Univ. Carolinae, **11**, (1970), 481–500.

[131] Danfu, H and Xinghua, W., The error estimates of Halley's method (submitted).

[132] Darbo, G., Punti uniti in trasformationa codominio non compatto, Rend. Sem. Mat. Univ. Padova, **24**, (1955), 84–92.

[133] Daubechies, I., Ten Lectures in Wavelets, (Conf. Board Math. Sci. (CBMS) Vol. 61), Society for Industrial and Applied Mathematics (SIAM), Philadelphia, PA, 1992.

[134] Davis, H. T., Introduction to nonlinear differential and integral equations, Dover Publications, Inc., New York, 1962.

[135] Decker, D. W., Keller, H. B., Kelley, C. T., Convergence rates of Newton's method at singular points, SIAM J. Numer. Anal., **20**, (1983), no. 2, 296–314.

[136] Dedieu, J.P., Estimations for the separation number of a polynomial system, J. Symbolic Comput., **24**, 6, (1997), 683–693.

[137] Dembo, R. S., Eisenstat, S. C., Steihaug, T., Inexact Newton methods, SIAM J. Numer. Anal., **19**, (1982), no. 2, 400–408.

[138] Dennis, J. E., Toward a unified convergence theory for Newton-like methods, In: Nonlinear Functional Anal. and Appl. (L.B. Rall, Ed.) Academic Press, New York, 1971.

[139] Dennis, J.E. and Moré, J., Quasi-Newton methods: Motivation and theory, SIAM Review, **19**, (1997), 46–89.

[140] Dennis, J. E., and Moré, J., Quasi-Newton methods: Motivation and theory, SIAM Review, **19**, (1997), 46–89.

[141] De Pascale, E. and Zabrejko, P.P., New convergence criteria for the Newton-Kantorovich method and some applications to nonlinear integral equations, Rend. Sem. Mat. Univ. Padova, **100,** (1998), 211–230.

[142] Deuflhard, P., Newton Methods for Nonlinear Problems, Springer-Verlag, New-York, 2004.

[143] Deuflhard, P., Heindl, G., Affine invariant convergence theorems for Newton's method, and extensions to related methods, SIAM J. Numer. Anal., **16**, (1979), no. 1, 1–10.

[144] Deuflhard, P., Potra, F. A., Asymptotic mesh independence of Newton-Galerkin methods and a refined Mysovskii theorem. SIAM J. Numer. Anal., **29**, (1992), no. 5, 1395–1412.

[145] Deuflhard, P., Schiela, A., and Weiser, M., Asymptotic mesh independence of Newton's method revisited, SIAM J. Numer. Anal. **42**, 5, (2005), 1830–1845.

[146] Döring, B., Iterative lösung gewisser randwertprobleme und integralgleichungen, Apl. Mat., **24**, (1976), 1–31.

[147] Dunford, N., Schwartz, J. T., Linear operators. Part I, Int. Publ. Leyden, (1963).

[148] Duvuat, G. and Lions, J.L., Inequalities in Physics and Mechanics, Springer-Verlag, Berlin, 1976.

[149] Edelstein, M., On fixed and periodic points under contractive mappings., J. London Math. Soc., **37**, (1962), 74–79.

[150] Edelstein, M., A remark on a theorem of M.A. Krasnoselskii, Amer. Math. Monthly, **73**, (1966), 509–510.

[151] Eisenstat, S. C., Walker, H. F., Globally convergent of inexact Newton methods, SIAM J. Optim., **4**, (1994), no. 2, 393–422.

[152] Eisenstat, S. C., Walker, H. F., Choosing the forcing terms in an inexact Newton method, SIAM J. Sci. Comput., **17**, (1996), no. 1, 16–32.

[153] Ezquerro, J. A., Hernández, M. A., Avoiding the computation of the second Fréchet-derivative in the convex acceleration of Newton's method., J. Comput. Appl. Math., **96**, (1998), 1–12.

[154] Ezquerro, J.A. and Hernandez, M.A., An efficient study of convergence for a fourth order two-point iteration in Banach space (submitted).

[155] Ezquerro, J. A., Hernández, M. A., On a convex acceleration of Newton's method, J. Optim. Theory Appl., **100**, (1999), no. 2, 311–326.

[156] Ezquerro, J. A., Hernández, M. A., On the application of a fourth order two-point method to Chandrasekhar's integral equation, Aequationes Math., **62**, (2001), no. 1–2, 39–47.

[157] Ezquerro, J. A., Hernández, M. A., Salanova, M. A., A discretization scheme for some conservative problems, Proceedings of the 8th International Congress on Computational and Applied Mathematics, ICCAM-98 (Leuven), J. Comput. Appl. Math., **115**, (2000), no. 1–2, 181–192.

[158] Ezquerro, J. A., Hernández, M. A., Salanova, M. A., Recurrence relations for the midpoint method, Tamkang J. Math., **31**, (2000), no. 1, 33–41.

[159] Ezquerro, J. A., Gutiérrez, J. M., Hernández, M. A., Salanova, M. A., A bi-parametric family of inverse free multipoint iterations, Comput. Appl. Math., **19**, (2000), no. 1, 109–124.

[160] Ezquerro, J. A., Hernández, M. A., A special type of Hammerstein integral equations, Int. Math. J., **1**, 6, (2002), 557–566.

[161] Ezquerro, J. A., Gutiérrez, J. M., Hernández, M. A. and Salanova, M. A., On the approximation of an inverse free Jarratt type approximation to nonlinear equations of Hammerstein type, (submitted).

[162] Ferreira, O. P. and Svaiter, B.F., Kantorovich's theorem on Newton's method in Riemannian manifolds, J. Complexity, **18**, (2002), no. 1, 304–329.

[163] Foerster, H., Frommer, A. and Mayer, G., Inexact Newton methods on a vector supercomputer, J. Comp. Appl. Math, **58**, (1995), 237–253.

[164] Fujimoto, T., Global asymptotic stability of nonlinear difference equations I, Econ. Letters, **22**, (1987), 247–250.

[165] Fujimoto, T., Global asymptotic stability of nonlinear difference equations II, Econ. Letters, **23**, (1987), 275–277.

[166] Galperin, A. and Waksman, Z., Regular smoothness and Newton's method, Numer Funct. Anal. Optimiz., **15**, (7 & 8), (1994), 813–858.

[167] Gander, W., On Halley's iteration method, Amer. Math. Monthly, **92**, (1985), 131–134.

[168] Gill, P.E., Murray, W. and Wright, M.A., Practical Optimization, Academic Press, 1991, London.

[169] Glowinski, R, Lions, J. L. and Trémolières, R., Numerical analysis of variational inequalities, North-Holland, Amsterdam, 1982.

[170] Gragg, W. B.; Tapia, R. A., Optimal error bounds for the Newton-Kantorovich theorem, SIAM J. Numer. Anal. **11**, (1974), 10–13.

[171] Graves, L. M., Riemann integration and Taylor's theorem in general analysis, Trans. Amer. Math. Soc. **29**, (1927), no. 1, 163–177.

[172] Grigat, E., Sachs, G., Predictor-corrector continuation method for optimal control problems, In Variational Calculus, optimal control and applications, (Trassenheide, 1996), vol. **124** of Internat. Ser. Numer. Math., 223–232, Birkhäuser, Verlag, 1998.

[173] Gutierez, J.M., A new semilocal convergence theorem for Newton's method, *J. Comput. Appl. Math.* **79** (1997), 131–145.

[174] Gutiérrez, J.M., Hernández, M.A., and Salanova, M.A., Accessibility of solutions by Newton's method, *Internat. J. Comput. Math.* **57** (1995), 239–247.

[175] Gutiérrez, J. M., Hernández, M. A., Salanova, M. A., Resolution of quadratic equations in Banach spaces, Numer. Funct. Anal. Optim., **17**, (1996), (1 & 2), 113–121.

[176] Gwinner, J., Generalized Stirling-Newton methods, In W. Oettli, K. Ritter (eds), Optimization and Operations Research, Oberwolfach, 1975, Lecture Notes Economics and Mathematical Systems, **11**, (1976), 99–135.

[177] Hadeller, K.P., Shadowing orbits and Kantorovich's theorem, Numer. Math. **73**, (1996), 65–73.

[178] Han, S., Pang, J. and Rangaraj, N., Globally convergent Newton methods for nonsmooth equations, Math. Oper. Res., **17**, 3, (1992), 586–607.

[179] Han, D.F. and Wang, X., Convergence of a deformed Newton method, Appl. Math. Comput., **94**, (1998), 65–72.

[180] Hartman, P., Ordinary differential equations, John Wiley & Sons, Inc., New York-London-Sydney, 1964. xiv+612 pp.

[181] Häubler, W.M. A Kantorovich-type convergence analysis for the Gauss-Newton method, *Numer. Math.* **48** (1986), 119–125.

[182] Helgason, S., Differential Geometry, Lie Groups and Symmetric Spaces, Pergamon Press, Oxford, 1982.

[183] Hellinger, E.; Toeplitz, O. Integralgleichungen und Gleichungen mit unendlichvielen Unbekannten, (German), Chelsea Publishing Company, New York, (1953), 1335–1616.

[184] Hernández, M. A., A note on Halley's method, Num. Math. **59**, 3 (1991), 273–276.

[185] Hernández, M. A., Newton's Raphson's method and convexity, Zb. Rad. Prirod.-Mat. Fax. Ser. Mat. 22, 1, (1992), 159–166.

[186] Hernández, M. A., Salanova, M. A., A family of Chebyshev-Halley type methods, Intern. J. Comp. Math., **47**, (1993), 59–63.

[187] Hernández, M. A., Relaxing convergence conditions for Newton's method, J. Math. Anal. Appl, **249**, (2000), no. 2, 463–475.

[188] Hernández, M. A., Chebyshev's approximation algorithms and applications, Comput. Math. Appl., **41** (2001), no. 3–4, 433–445.

[189] Hernández, M. A., Rubio, M. J. and Ezquerro, J. A., Secant-like methods for solving nonlinear integral equations of the Hammerstein type, J. Comput. Appl. Math., **115,** (2000), 245–254.

[190] Hernández, M. A., Rubio, M. J.,Semilocal convergence of the secant method under mild convergence conditions of differentiability, Comput. Math. Appl., **44**, (2002), no. 3–4, 277–285.

[191] Hernández, M. A. and Salanova, M. A., Sufficient conditions for semilocal convergence of a fourth order multipoint iterative method for solving equations in Banach spaces, Southwest J. Pure Appl. Math., (1999), 29–40.

[192] Hille, E. and Philips, R.S., Functional Analysis and Smigroups, Amer. Math. Soc. Coll. Publ., New York, 1957.

[193] Higle, J. L. and Sen, S., On the convergence of algorithms with applications to stochastic and nondifferentiable optimization, SIE Working Paper #89-027, UNiversity of Arizona (1989).

[194] Hoppe, R.H.W., Numerical methods for large-scale nonlinear systems, Handouts published on-line on the web, (2005), 1–108.

[195] Huang, Z., Newton method under weak Lipschitz continuous derivative in Banach spaces, Appl. Math. Comput., **140**, (2003), 115–126.

[196] Hu, Z. and Ding, J., A new semilocal convergence theorem for Newton's method involving twice differentiability only at a point, J. Comput. Appl. Math., (2005).

[197] Ip, Chi-M. and Kyparisis, J., Local convergence for quasi-Newton methods for B-differentiable operators, Math. Prog., **56**, (1992), 71–89.

[198] Ishikawa, S., Fixed points by a new iteration method, Proc. Amer. Math. Soc., **44**, (1974), 147–150.

[199] Jarrat, P., Some efficient fourth order multipoint methods for solving equations, BIT, **9**, (1969), 119–124.

[200] Josephy, N.H., Newton's method for generalized equations, Technical Summary Report No. 1965, Mathematics Research Center, University of Wisconsin-Madison, June 1979, available from National Technical Information Service, Springfield, VA 22161, under Accession No. A077 096.

[201] Josephy, N.H., Quasi-Newton method for generalized equations, Technical Summary Report No. 1966, Mathematics Research Center, University of Wisconsin-Madison, June 1979, available from National Technical Information Service, Springfield, VA 22161, under Accession No. A077 097.

[202] Josephy, N.H., A Newton method for the PIES energy model, Technical Summary Report No. 1971, Mathematics Research Center, University of Wisconsin-Madison, June 1979, available from National Technical Information Service, Springfield, VA 22161, under Accession No. A077 102.

[203] Josephy, N.H., Hogan's PIES example and Lemke's algorithm, Technical Summary Report No. 1972, Mathematics Research Center, University of Wisconsin-Madison, June 1979, available from National Technical Information Service, Springfield, VA 22161, under Accession No. A077 103.

[204] Kanno, S., Convergence theorems for the method of tangent hyperbolas, Math. Japon., **37**, (1992), no. 4, 711–722.

[205] Kantorovich, L.V., The method of succesive approximation for functional equations, Acta Math., **71**, (1939), 63–97.

[206] Kantorovich, L. V., On Newton's method for functional equations (Russian), Dokl. Akad. Nauk. SSSR **59**, (1948), 1237–1240.

[207] Kantorovich, L. V., Akilov, G. P., Functional analysis in normed spaces, Pergamon Press, New York (1964).

[208] King, R.F., Tangent methods for nonlinear equations, Numer. Math., **18**, (1972), 298–304.

[209] Kojima, M. and Shindo, S., Extensions of Newton and quasi-Newton methods to systems of PC^1 equations, J. Oper. Res. Soc. Japan, **29**, (1986), 352–374.

[210] Krasnosel'skii, M. A., Positive solutions of operator equations, Goz. Isdat. Fiz. Mat. Moscow 1962; Transl. by R. Flaherty and L. Boron, P. Noordhoff, Groningen 1964.

[211] Krasnosel'skii, M. A., Topological Methods in the Theory of Nonlinear Integral Equations, Pergamon Press, London, 1966.

[212] Krasnosel'skii, M. A. and Rutitskii, Ya. B., On the theory of equations with concave operators, SMZh, **10**, 3, (1969).

[213] Krasnosel'skii, M.A., Vainikko, G.M., Zabreiko, P.P., and Rutiskii, Ya.B., and Stetsenko, V.Ya., *Approximate Solution of Operator Equations*, Wolters-Noordhoff Publishing, Groningen, 1972.

[214] Krasnosel'skii, M. A. and Zabrejko, P.P., Geometrical Methods of Nonlinear Analysis, Springer-Verlag, New York, 1984.

[215] Kummer, B., Newton's method for non-differentiable functions. In J. Guddat et al (Eds.) Advances in Mathematical Optimization, Academie-Verlag, Berlin, **45**, (1988), 114–125.

[216] Kung, H. T., The complexity of obtaining starting points for solving operator equations by Newton's method, Technical report, nr.044-422, Carnegie-Mellon Univ., Pittsburgh, Pa., october, 1975, Article in Traub, J.F., Analytic computational complexity.

[217] Kuratowski, C., Sur les espaces complets, Fund. Math., **15**, (1930), 301–309.

[218] Kwon, U. K. and Redheffer, R. M., Remarks on linear equations in Banach space, Arch. Rational Mech. Anal., **32**, (1969), 247–254.

[219] Lancaster, P., Error analysis for the Newton-Raphson method, Numer. Math., **9**, (1968), 55–68.

[220] Laumen, M., A Kantorovich theorem for the structured PSB update in a Hilbert space, J. Optim. Theory Appl., **105**, 2, (2000), 391–415.

[221] Lions, J.L. and Stampachia, G., Variational Inequalities, Comm. Pure Appl. Math., **20**, (1967), 493–519.

[222] Liskovetz, O.A., Variational methods for solving nonstable problems, Nauka Technika, Minsk, (1981) (in Russian)

[223] Liusternik, L. A. and Sobolev, V. J., Elements of functional analysis, Ungar Publ., 1961.

[224] Mann, W.R., Mean value methods in iteration, Proc. Amer. Math. Soc., **4**, (1953), 506–510.

[225] Marcotte, P., Wu, J. H., On the convergence of projection methods, J. Optim. Theory Appl., **85,** (1995), no. 2, 347–362.

[226] Matveev, A.F., Yunganns, P., On the construction of an approximate solution of a nonlinear integral equation of permeable profile, Differ. Uravn., **33**, 9, (1997), 1242–1252, 1295.

[227] Mayer, J., A generalized theorem of Miranda and the theorem of Newton-Kantorovich, Numer. Funct. Anal. Optim., **23,** (2002), no. 3–4, 333–357.

[228] McCormick, S. F., A revised mesh refinement strategy for Newton's method applied to two-point boundary value problems, Lecture Notes in Mathemaics 674, Springer-Verlag, Berlin, (1978), 15–23.

[229] Meyer, P. W., Die Anwendung Verallgemeinerter Normen zer Fehlerabschätzung Bei Iteration Sverfahren, Dissertation, Düsseldort, 1980.

[230] Meyer, P. W., Das modifizierte Newton-Verfahren in verallgemeinerten Banach-Räumen, Numer. Math., **43**, (1984), no. 1, 91–104.

[231] Meyer, P. W., Newton's method in generalized Banach spaces. Numer. Funct. Anal. Optim., **9**, (1987), (3 & 4), 244–259.

[232] Meyer, P. W., A unifying theorem on Newton's method, Numer. Funct. Anal. Optim., **13**, (1992), no. 5-6, (1992), 463–473.

[233] Miel, G. J., Majorizing sequences and error bounds for iterative methods, Math. Comp., **34**, (1980), no. 149, 185–202.

[234] Migovich, F.M., On the convergence of projection-iterative methods for solving nonlinear operator equations, Dopov. Akad. Nauk. Ukr. RSR, Ser. A, **1**, (1970), 20–23.

[235] Minty, G.J., On the monotonicity of the gradient of a convex function, Pacific J. Math., **14**, (1964), 243–247.

[236] Miranda, C., Un osservatione su un teorema d, Brouwer, Ball. Unione Mat. Ital., Serr. 11, **3**, (1940), 5–7.

[237] Mirsky, L., An Introduction to linear algebra, Clarendon Press, Oxford, England, 1955.

[238] Moore, R. H., Approximations to nonlinear operator equations and Newton's method, Numer. Math., **12**, (1968), 23–34.

[239] Moore, R. E., A test for existence of solutions to nonlinear systems, SIAM J. Numer. Anal., **14**, (1977), no. 4, 611–615.

[240] Moore, R. E., Methods and applications of interval analysis, SIAM Publications, Philadelphia, Pa., 1979.

[241] Moret, I., A note on Newton-type iterative methods, Computing **33,** (1984), 65–73.

[242] Moret, I., On the behaviour of approximate Newton methods, Computing, **37**, (1986), no. 3, 185–193.

[243] Moret, I., On a general iterative scheme for Newton-type methods, Numer. Funct. Anal. Optim., **9**, (1987-88), (10–12), 1115–1137.

[244] Moret, I., A Kantorovich-type theorem for inexact Newton methods, Numer. Funct. Anal. Optim. 10 (1989), (3 & 4), 351–365.

[245] Mukaidani, Hl, Shimomura, T., Mizukanmi, K., Asymtotic expansions and a new numerical algorithm of the algebraic Riccati equation for multiparameter singularly perturbed systems, J. Math. Anal. Appl., **267**, 1, (2002), 209–234.

[246] Muroya, Y., Practical monotonous iterations for nonlinear equations, Mem. Fac. Sci. Kyushu Univ., Ser. A, **22**, (1968), 56–73.

[247] Muroya, Y., Left subinverses of matrices and monotonous iterations for nonlinear equations, Memoirs of the Faculty of Science and Engineering, Waseda University, **34**, (1970), 157–171.

[248] Mysovskii, I., On the convergence of Newton's method, Trudy Mat. Inst. Steklov, **28**, (1949), 145–147 (in Russian).

[249] Nagatou, K., Yamamoto, N., Nakao, M.T., An approach to the numerical verification of solutions for nonlinear elliptic problems with local uniqueness, Numer. Funct. Anal. Optim., **20**, 5–6, (1999), 543–565.

[250] Natanson, I.P., The theory of functions of a Real Variable (Russian), Gostehizdat, Moscow, (1957).

[251] Nayakkankuppam, M.V., Overton, V.I., Conditioning of semidefinite programs, Math. Program., **85**,3, Ser. A., (1999), 525–540.

[252] Necepurenko, M.T., On Chebysheff's method for functional equations (Russian), Usephi, Mat. Nauk, **9**, (1954), 163–170.

[253] Nerekenov, T.K., Necessary and sufficient conditions for uryson and nemytskii operators to satisfy a Lipschitz condition (Russian), VINITI 1459, **81**, Alma-Ata, (1981).

[254] Nesterov, Y., Nemirovsky, A., Interior Point Polynomial Methods in Convex Programming, SIAM, Philadelphia, 1994.

[255] Neumaier, A. and Shen, Z., The Krawczyk operator and Kantorovich's theorem, J. Math. Anal. Appl., **149,** (1990), no. 2, 437–443.

[256] Nguen, D.F. and Zabrejko, P.P., The majorant method in the theory of the Newton-Kantorovich approximations and the Ptâk error estimates, Numer. Funct. Anal. and Optimiz., **9**, (5 & 6), (1987), 671–686.

[257] Noble, B., The numerical solution of nonlinear integral equations and related topics, University Press, Madison, WI, (1964).

[258] Noor, K. I. and Noor, M. A., Iterative methods for a class of variational inequalities, in Numerical Analysis of singular perturbation problems, Hemker and Miller, eds., Academic Press, New-York, (1985), 441–448.

[259] Noor, M. A., An iterative scheme for a class of quasivariational inequalities, J. Math. Anal. and Appl., **110**, (1985), no. 2, 463–468.

[260] Noor, M.A., Generalized variational inequalities, Appl. Math. Letters, **1**, (1988), 119–122.

[261] Ojnarov, R. and Otel'baev, M., A criterion for a Uryson operator to be a contraction (Russian), Dokl. Akad. Nauk. SSSR, **255**, (1980), 1316–1318.

[262] Okuguchi, K., Expectations and stability in oligopoly models, Springer-Verlag, New York, (1976).

[263] Ortega, J.M. and Rheinboldt, W.C., Iterative solution of nonlinear equations in several variables, Academic Press, New York, (1970).

[264] Ostrowski, A.M., Solution of equations in euclidean and Banach spaces, Academic Press, New York, (1973).

[265] Owren, B. and Welfert, B., The Newton iteration on Lie groups, BIT, **40**, 1, (2000), 121–145.

[266] Pandian, M.C., A convergence test and componentwise error estimates for Newton-type methods, SIAM J. NUm. anal., **22**, (1985), 779–791.

[267] Pang, Jong-Shi, Newton's method for B-differentiable equations, Math. Oper. Research, **15**, 2, (1990), 311–341.

[268] Păvăloiu, I., Sur la méthode de Steffensen pour la résolution des équations opérationnelles non linéaires, Rev. Roumaine Math. Pures Appl., **13**, (1968), 6, 857–861.

[269] Păvăloiu, I., Rezolvarea equaţiilor prin interpolare. Dacia Publ. cluj-Napoca, Romania, (1981).

[270] Păvăloiu, I., Sur une généralisation de la méthode de Steffensen, Rev. Anal. Numér. Théor. Approx., **21**, (1992), no. 1, 59–65.

[271] Păvăloiu, I., A converging theorem concerning the chord method, Rev. Anal. Numér. Théor. Approx., **22**, 1 (1993), 83–85.

[272] Păvăloiu, I., Bilateral approximations for the solutions of scalar equations., Rev. Anal. Numér. Théor. Approx., **23**, (1994), no. 1, 95–100.

[273] Petryshin, W.V. and Williamson, T.E., Strong and weak convergence of the sequence of successive approximation for quasi-nonexpansive mappings, J. Math. Anal. Appl., **43**, (1973), 459–497.

[274] Potra, F.A., On the convergence of a class of Newton-like methods, Iterative solution of nonlinear systems of equations, Lecture Notes in Math., 953, Springer-Verlag, New York, 1982.

[275] Potra, F.A., An error analysis for the secant method, Numer. Math., **38**, (1982), 427–445.

[276] Potra, F.A., On the aposteriori error estimates for Netwon's method, Beitrage Numer. Math., **12**, (1984), 125–138.

[277] Potra, F. A. On an iterative algorithm of order 1.839... for solving nonlinear operator equations, Numer. Funct. Anal. Optim., **7**, (1984–1985), no. 1, 75–106.

[278] Potra, F.A., Sharp error bounds for a class of Newton-like methods, Libertas Mathematica, **5**, (1985), 71–84.

[279] Potra, F. A., Newton-like methods with monotone convergence for solving nonlinear operator equations, Nonlinear Anal., Theory Methods and Applications, **11**, (1987), no. 6, 697–717.

[280] Potra, F. A., On Q-order and R-order of convergence, SIAM J. Optim. Theory Appl., **63**, (1989), no. 3, 415–431.

[281] Potra, F.A., A path-following method for linear complementarity problems based on the affine invariant Kantorovich theorem, ZIB-Report 00-30, Konrad-Zuse-Zentrum, Berlin, August 2000.

[282] Potra, F.A., The Kantorovich theorem and interior point methods, Math. Progr. Ser. A, **102**, (2005), 47–50.

[283] Potra, F.A. and Pták, V., Sharp error bounds for Newton's method, Numer. Math., **34**, (1980), no. 1, 63–72.

[284] Potra, F. A. and Pták, V., Nondiscrete induction and iterative processes, Pitman Publ., London, 1984.

[285] Potra, F.A., Wright, S.J., Interior-point methods, J. Comp. Appl. Math., **124**, (2000), 281–302.

[286] Pták, V., The rate of convergence of Newton's process., Numer. Math., **25**, (1976), no. 3, 279–285.

[287] Qi, L. and Sun, J., A nonsmooth version of Newton's method, Math. Progr., **58**, (1993), 353–367.

[288] Rall, L. B., Computational solution of nonlinear operator equations, Wiley, New York, (1968).

[289] Rall, L. B., Nonlinear functional analysis and applications, Academic Press, New York, (1971).

[290] Rall, L. B., A comparison of the existence theorems of Kantorovich and Moore, SIAM J. Numer. Anal., **17**, (1980), no. 1, 148–161.

[291] Ralph, D., On branching numbers of normal manifolds, Technical Report 92-1283, Department of Computer Science, Cornell University, Ithaca, NY, May 1992.

[292] Redheffer, R. and Walter, W., A comparison theorem for difference inequalities, J. Diff. Eq., **44**, (1982), 111–117.

[293] Renegar, J., A polynomial-time algorithm, based on Newton's method, for linear programming, Math. Progr., **40**, 1, Ser. A, (1988), 59–93.

[294] Renegar, J., Shub, M., Unified complexity analysis for Newton LP methods, Math. Progr., **53**, 1, Ser. A, (1992), 1–16.

[295] Rheinboldt, W. C., An adaptive continuation process for solving systems of nonlinear equations, Publish Academy of Sciences, Banach Ctr. Publ., **3**, (1977), pp.129–142.

[296] Rheinboldt, W. C., On a theorem of S. Smale about Newton's method for analytic mappings, Appl. Math. Lett., **1**, (1988), no. 1, 69–72.

[297] Robinson, S.M., Generalized equations and their solutions, part I: Basic theory, Math. Progr. Study, **10**, (1979), 128–141.

[298] Robinson, S.M., Strongly regular generalized equations, Math. Oper. Res., **5**, (1980), 43–62.

[299] Robinson, S. M., Generalized equations. In: (A. Bachem, M. Grötschel and B. Korte, eds.) Mathematical programming: the state of the art, Springer, Berlin, 1982, pp. 346–367.

[300] Robinson, S.M., Implicit B-differentiability in generalized equations, Technical Summary Report 2854, Mathematics Research Center, University of Winsconsin-Madison, August 1985.

[301] Robinson, S.M., Normal maps induced by linear transformations, Math. Oper. Res., **17**, (1992), 691–714.

[302] Robinson, S.M., Newton's method for a class of nonsmooth functions, Set-Valued Analysis, **2**, (1994), 291–305.

[303] Rockne, J., Newton's method under mild differentiability conditions with error analysis, Numer. Math., **18**, (1972), 401–412.

[304] Roos, C., Vial, J.-Ph., Terlaky, T., Theory and algorithms for linear optimization: an interior point approach, Wiley Interscience Series in Discrete Mathematics and Optimization, John Wiley and Sons, 1997.

[305] Safiev, R. A., The method of tangent hyperbolas, sov. Math. Dokl., **4**, (1963), 482–485.

[306] Schmidt, J. W., Monotone einschliessung mit Regula-Falsi bei konvexen functionen, ZAMM, **50**, (1970), 640–643.

[307] Schmidt, J. W. and Leonhardt, H., Eingrenzung von lösungen mit hilfe der Regula-Falsi, Computing, **6**, (1970), 318–329.

[308] Schomber, H., Monotonically convergent iterative methods for nonlinear systems of equations, Numer. Math., **32**, (1979), 97–104.

[309] Shapiro, A., On concepts of directional differentiability, Research Report, Department of Mathematics, Applied Mathematics and Astronomy, University of South Africa, Pretoria, South Africa, July 1988, J. Optim. Theory Appl., **66**, 3, (1990), 477–487.

[310] Slugin, S. N., Approximate solution of operator equations on the basis of Caplygin method, (Russian), Dokl. Nauk SSSR, **103**, (1955), 565–568.

[311] Slugin, S. N., Monotonic processes of bilateral approximation in a partially ordered convergence group, Soviet. Math., **3**, (1962), 1547–1551.

[312] Smale, S., Newton's method estimates from data at one point. In the merging of disciplines in pure, applied, and computational mathematics, Springer Verlag, New York, 1986, pp.185–196,.

[313] Stirling, J., Methodus differentialis: sive tractatus de summatione et interpolations serierum infiniterum, W. Boyer, London, 1730.

[314] Stoffer, D., Palmer, K.J., Rigorous verification of chaotic behaviour of maps using validated shadowing, Nonlinearity, **12**, 6, (1999), 1683–1698.

[315] Szidarovszky, F. and Yakowitz, S., Principles and procedures in numerical analysis, Plenum, New York, 1978.

[316] Szidarovszky, F. and Bahill, T., Linear systems theory, CRC Press, Boca Raton, FL, (1992).

[317] Tan, K.K., Xu, H.K., Approximating fixed points of nonexpansive mappings by the Ishikawa iteration process, J. Math. Anal. Appl., **178**, (1993), 301–308.

[318] Taylor, A.E., Introduction to functional analysis, Wiley, New York, (1957).

[319] Tishyadhigama, S., Polak, E., Klessig, R., A comparative study of several convergence conditions for algorithms modeled by point-to-set maps, Math. Programming Stud., **10**, (1979), 172–190.

[320] Törnig, W., Monoton konvergente Iterationsverfahren zür Lösung michtlinearer differenzen–randwertprobleme, Beiträge zür Numer. Math., **4**, (1975), 245–257.

[321] Traub, J. F., Iterative methods for the solution of equations, Prentice-Hall Series in Automatic Computation Prentice-Hall, Inc., Englewood Cliffs, N.J. 1964, xviii+310 pp.

[322] Traub, J. F., Analytic computational complexity, Academic Press, New York-London, 1975.

[323] Triconi, F.G., Integral Equations, Interscience Publ., 1957.

[324] Tsuchiya, T., An application of the Kantorovich theorem to nonlinear finite element analysis, Numer. Math., **84**, 1, (1999), 121–141.

[325] Uko, L.U., Remarks on the generalized Newton method., Math. Programming, **59,** (1993), 405–412.

[326] Uko, L.U., The generalized Newton's method and variational inequalities, Bolletino U.M.I., **7**, 9-A, (1995), 429–437.

[327] Uko, L. U., Generalized equations and the generalized Newton method., Mathematical Programming, **73**, (1996), 251–268.

[328] Ulm, S., Iteration methods with divided differences of the second order, (Russian), Dokl. Akad. Nauk SSSR, **158**, (1964), 55–58.

[329] Urabe, M., Convergence of numerical iteration in solution of equations, J. Sci. Hiroshima Univ., Ser. A, **19**, (1976), 479–489.

[330] Uzawa, H., The stability of dynamic processes, Econometrica, **29**, (1961), 617–631.

[331] Vainberg, M.M., Variational method and method of monotone operators, Nauka, M. (1972) (in Russian).

[332] Vandergraft, J. S., Newton's method for convex operators in partially ordered spaces, SIAM J. Numer. Anal., **4**, (1967), 406–432.

[333] Varga, R. S., Matrix iterative analysis, Prentice-Hall, Englewood Cliffs, NJ, 1962.

[334] Verma, R. U., Nonlinear variational and constrained hemivariational inequalities involving relaxed operators., Z. Angew. Math. Mech., **77**, (1997), no. 5, 387–391.

[335] Verma, R. U., Approximation-solvability of nonlinear variational inequalities involving partially relaxed monotone (PRM) mappings., Adv. Nonlinear Var. Inequal., **2**, (1999), no. 2, 137–148.

[336] Verma, R. U., A class of projection-contraction methods applied to monotone variational inequalities., Appl. Math. Lett., **13**, (2000), no. 8, 55–62.

[337] Verma, R. U., Generalized multivalued implicit variational inequalities involving the Verma class of mappings., Math. Sci. Res. Hot-Line, **5,** (2001), no. 2, 57–64.

[338] Walker, H.F. and Watson, L.T., Large change secant update methods for undetermined systems, SIAM J. Numer. Anal., **27**, (1990), 1227–1262.

[339] Wang, D. and Zhao, F., The theory os Smale's point estimation and its applications, J. Comput. Appl. Math. **60**, (1995), 253–269.

[340] Wang, X., Convergence of Newton's method and inverse function theorem in Banach space, Math. Comp., **68**, (1999), 169–186.

[341] Wang, X., Convergence of Newton's method and uniqueness of the solution of equations in Banach space, IMA J. Numer. Anal., **20**, (2000), 123–134.

[342] Wang, X., Chong, Li and Ming-Jun Lai, A unified convergence theory for Newton type methods for zeros of nonlinear operators in Banach spaces, BIT, **42**, 1, (2002), 206–213.

[343] Wang, X. and Chong, Li, Convergence of Newton's method and uniqueness of the solution of equations in Banach space II, Acta Mathematica Sinica, English Series, **19**, 2, (2003), 405–412.

[344] Wang, Z., Semilocal convergence of Newton's method for finite-dimensional variational inequalities and nonlinear complementarity problems, Ph.D. Dissertation, Universität Karlsruhe, Germany, 2005.

[345] Werner, W., Über ein Verfahren der Ordnung $1 + \sqrt{2}$ zur Nullstellenbestimmunng, Numer. Math., **32**, (1979), 333–342.

[346] Wright, S.J., Primal-Dual Interior-Point Methods, SIAM, Philadelphia, 1997.

[347] Wu, J. W and Brown, D. P., Global asymptotic stability in discrete systems, J. Math. Anal. Appl., **140**, (1989), no. 1, 224–227.

[348] Xue, Z.Q., Zhou, H.Y. and Cho, Y.J., Iterative solution of nonlinear equations for m-accretive operators in Banach spaces, J. Non. Con. Anal., **1**, 3, (2000), 313–320.

[349] Yamamoto, T., Error bounds for computed eigenvalues and eigenvectors, Numer. Math., **39**, (1980), 189–199.

[350] Yamamoto, T., A method for finding sharp error bounds for Newton's method under the Kantorovich assumptions, Numer. Math., **44**, (1986), 203–220.

[351] Yamamoto, T., A convergence theorem for Newton-like methods in Banach spaces, Numer. Math., **51**, (1987), 545–557.

[352] Yamamoto, T., On the method of tangent hyperbolas in Banach spaces, J. Comput. Appl. Math., **21**, (1988), 75–86.

[353] Yamamoto, T. and Chen, Z., Convergence domains of certain iterative methods for solving nonlinear equations, Numer. Funct. Anal. Optim., **10**, (1989), 34–48.

[354] Ye, Y., Interior Point Algorithms: Theory and Analysis, Wiley-Interscience Series in Discrete Mathematics and Optimization, John Wiley and Sons, 1997.

[355] Ypma, T.J., Numerical solution of systems of nonlinear algebraic equations, Ph. D. thesis, Oxford, 1982.

[356] Ypma, T.J., Affine invariant convergence results for Newton's methods, BIT, **22**, (1982), 108–118.

[357] Ypma, T.J., The effect of rounding error on Newton-like methods, IMA J. NUmer. Anal., **3**, (1983), 109–118.

[358] Ypma, T. J., Convergence of Newton-like iterative methods, Numer. Math., **45**, (1984), 241–251.

[359] Ypma, T. J., Local convergence of inexact Newton methods, SIAM J. Numer. Anal., **21**, (1984), 583–590.

[360] Yau, L., Ben-Israel, A., The Newton and Halley methods for complex roots, Amer. Math. Monthly, **105,** (1998), no. 9, 806–818.

[361] Zaanen, A. C., Linear analysis, North-Holland Publ., Amsterdam, 1953.

[362] Zabrejko, P.P. and Majorova, N.L., On the solvability of nonlinear Uryson integral equations (Russian), Kach. Pribl. Metody Issled. Oper. Uravn., **3**, (1978), 61–73.

[363] Zabrejko, P. P.; Nguen, D. F., The majorant method in the theory of Newton-Kantorovich approximations and the Pták error estimates, Numer. Funct. Anal. Optim., **9**, (1987), no. 5–6, 671–684.

[364] Zabrejko, P. P. and Zlepko, P.P., On majorants of Uryson integral operators (Russian), Kach. Pribl. Metody Issled. Oper. Uravn., **8** (1983), 67–76.

[365] Zabrejko, P. P., K-metric and K-normed linear spaces: A survey, Collect. Math., **48,** (1997), no. 4–6, 825–859.

[366] Zeidler, E., Nonlinear Functional Analysis and Its Applications, I: Fixed Point Theorems, Springer-Verlag, New York, 1986.

[367] Zincenko, A.I., A class of approximate methods for solving operation equations with nondifferentiable operators, Dopovidi Akad. Nauk Ukrain. RSR (1963), 156–161.

[368] Zlepko, P.P. and Migovich, F.M., an application of a modification of the Newton-Kantorovich method to the approximate construction of implicit functions (Ukrainian), Ukrainskii Mathematischeskii Zhürnal, **30**, 2 (1978), 222–226.

[369] Zuhe, Shen, Wolfe, M. A., A note on the comparison of the Kantorovich and Moore theorems, Nonlinear Anal., **15,** (1990), no. 3, 229–232.

Appendix A

Glossary of Symbols

$\mathbb{R}^n$	real n-dimensional space
$\mathbb{C}^n$	complex n-dimensional space
$X \times Y, X \times X = X^2$	Cartesian product space of X and Y
$e^1, \ldots, e^n$	the coordinate vectors of $\mathbb{R}^n$
$x = (x_1, \ldots, x_n)^T$	column vector with component x_i
x^T	the transpose of x
$\{x_n\}_{n \geq 0}$	sequence of points from X
$\|\cdot\|$	norm on X
$\|\cdot\|_p$	L_p norm
$\lvert \cdot \rvert$	absolute value symbol
$//$	norm symbol of a generalized Banach space X
$\langle x, y \rangle$	set $\{z \in X \lvert z = tx + (1-t)\, y,\ t \in [0,1]\}$
$U(x_0, R)$	open ball $\{z \in X \lvert\ \|x_0 - z\| < R\}$
$\bar{U}(x_0, R)$	closed ball $\{z \in X \lvert\ \|x_0 - z\| \leq R\}$
$U(R) = U(0, R)$	ball centered at the zero element in X and of radius R
$U, \bar{U}$	open, closed balls, respectively no particular reference to X, x_0 or R
$M = \{m_{ij}\}$	matrix $1 \leq i, j \leq n$
M^{-1}	inverse of M
M^+	generalized inverse of M
$\det M$ or $\lvert M \rvert$	determinant of M
M^k	the kth power of M
$\operatorname{rank} M$	rank of M
I	identity matrix (operator)
L	linear operator
L^{-1}	inverse
null L	null set of L
rad L	radical set of L
$F : D \subseteq X \to Y$	an operator with domain D included in X, and values in Y
$F'(x)$, $F''(x)$	first, second Fréchet-derivatives of F evaluated at x

Index

www.ingramcontent.com/pod-product-compliance
Ingram Content Group UK Ltd.
Pitfield, Milton Keynes, MK11 3LW, UK
UKHW020257051026
15470UKWH00020B/391

9 780444 531629